Boundary Value PROBLEMS of Applied Mathematics

Boundary Value PROBLEMS of Applied Mathematics

John L. Troutman
Syracuse University

Maurino P. Bautista
Rochester Institute of Technology

SECOND EDITION

Dover Publications, Inc.
Mineola, New York

Bibliographical Note

This Dover edition, first published in 2017, is a corrected and enlarged republication of the work originally published by the PWS Publishing Company, Boston, Massachusetts, in 1994. The authors have updated an Appendix and supplied a new Preface to the Dover edition.

International Standard Book Number

ISBN-13: 978-0-486-81222-9
ISBN-10: 0-486-81222-7

Manufactured in the United States by LSC Communications
81222701 2017
www.doverpublications.com

Contents

7 Series Methods 269

8 Integral Transform Methods 322

9 Sturm–Liouville Problems 371

Appendices

Applications

Preface to the Dover Edition

We accepted with great pleasure Dover's offer to republish this book on boundary value problems, which has been out of print for some time. We still believe in its original purpose, that of showing how classical mathematical methods can be used to attack problems of interest in many contemporary fields. We also remain convinced that study of the interaction between mathematical and physical concepts benefits the understanding of both. That this is true historically is undeniable, and the list on the facing page indicates its enduring validity.

The original book had only one printing, and subsequently disappeared in successive takeovers by several ever larger publishing houses (a fate shared by many of its contemporaries). It could not be improved gradually in successive editions, but fortunately, John Grafton and his colleagues at Dover have allowed modest alterations to repair misprints, notational discrepancies, etc. Moreover, they have consented to our updating the original supplement on numerical methods, which now encourages the reader to use the latest version of Maple to obtain graphs and other illustrations similar to those that illuminate the text.

We also appreciate the encouraging comments from reviewers, students, and colleagues, many of whom have used our book in class, and in particular, those from Karl Barth and Phillip Church at Syracuse University and Douglas Meadows and Tamas Wiandt at Rochester Institute of Technology for many illuminating conversations. Finally, we wish to thank our wives, Patricia Brookes and Kim Bautista, who have endured our revisiting this previous acquaintance.

John L. Troutman
Maurino P. Bautista
March 1, 2017

Preface

This book introduces boundary value problems to those who have some knowledge of multidimensional calculus and ordinary differential equations. Boundary value problems arise when we seek mathematical descriptions for physical processes that relate occurrences within a space-time region to conditions on the boundary of the region. The study of these problems (now in its third century) has supplied mathematical foundations for all areas of applied science where continuum assumptions prevail, including elasticity, fluid mechanics, gas dynamics, electromagnetics, and general field theory. In addition, this study has produced important mathematical tools that admit insight derived from physical associations.

A text on boundary value problems should address both physical and mathematical aspects of its subject and present these in a manner that can appeal to engineers, mathematicians, and physicists, among others. Ideally, its user should learn how to model relevant physical processes mathematically, and how to develop methods that solve problems from contemporary technology. Moreover, he or she should learn to ask about the extent that mathematical solutions mirror physical reality and discover the rich scientific heritage that surrounds these questions. The need for such a text has long been recognized and the present work emerged from several decades of combined experience in mathematics, applied mathematics, and engineering, both in the classroom and in the laboratory.

In this book, we let mathematics help shape the questions put to the physical world and watch the resulting interaction. Using this approach, we discover that physical phenomena from various fields can be classified in terms of the mathematical description of an underlying process. We learn to appreciate the mathematical advantages of operating within a linear framework and develop superposition methods to handle related problems. For these problems we find solutions in the form of series or integrals, consider how to approximate them numerically, and use computer graphics to illustrate the results. In a modern spirit, we show how the fast Fourier transform speeds approximation, and why linear results might help design transistors, heart-pumps, earthquake-proof buildings, and ultralight aircraft. We also discuss some aspects of nonlinear wave behavior.

Part I provides material for a standard introductory course that includes the usual topics — Fourier series, separation methods, derivation of the heat, potential, and wave equations together with series solutions to representative problems, and d'Alembert's solution of the wave equation. Among other novel features in this part are an efficient pattern recognition approach to separation of variables in simple cases and the prediction of solution behavior through compatibility considerations. Those who already have a reasonable knowledge of Fourier series and separation methods might begin with Section 2.2, then take up Chapter 3 and continue into Part II. There we develop more sophisticated methods, including Sturm–Liouville theory and integral transform techniques, and apply these to problems from various fields.

The main text is supported by foundational material from Chapter 0, and by material from the appendices that includes a review of linear ordinary differential equations. There is also a supplement in the back of the book that explains how to treat standard problems using current computer software.

In order to present its subject effectively, the book employs a sliding scale-of-difficulty that encourages the user to become gradually more sophisticated both conceptually and technically. Moreover, with the possible exception of Chapter 5, material toward the end of a chapter is of less importance to subsequent exposition than that at the beginning. Chapter 0 is outside this general scheme as are occasional starred (*) sections, proofs, examples, and problems, which can be examined lightly or omitted entirely on first exposure.

These features make the book a convenient text at various levels, as indicated in the following table:

	Junior–Senior introduction	If Fourier series and separation methods known
One Quarter	Ch 1–4	Ch 3–6
One Semester	Ch 1–5	Ch 3–8
Two Quarters	Ch 1–7	Ch 3–9
Two Semesters	Ch 1–9	

Obviously, in each case a course that concentrates on initial sections of chapters permits broader coverage.

The text is essentially self-contained and its results are established without appeal to either the Lebesgue integral, the Dirac delta "function," or the contour integral methods of complex analysis. It incorporates the simplified approach to minimizing integral functions that was explored in a previous book [Tr], and contains some other results that are not easily available elsewhere. It should provide both an introduction to its subject and a useful reference for subsequent investigations.

The bibliography lists alphabetically by author(s) a substantial number of the reference books that were consulted during preparation of this text. Each work is identified there and elsewhere by initial letters of its author(s) name(s) in square brackets; for example, in the previous paragraph, [Tr] refers to Troutman.

Acknowledgments

This work has benefited greatly by the contributions of Maurino Bautista from the Rochester Institute of Technology. He supplied not only the excellent computer graphic illustrations and the computer supplement but also several interesting examples and many hours of helpful conversation. Without his continuing interest, the book might not have been completed.

For aid in transforming the manuscript into publishable form, thanks are due to its principal typist Ruth Dewey, our departmental staff, Audrey Burian, Sue Light, and Anne Wildman, a most encouraging editor, Alex Kugushev, and the editors and staff of PWS, especially Pamela Rockwell.

The author also received assistance and encouragement in many forms from students, colleagues, and (at times anonymous) reviewers. Gratitude is expressed to:

Dick Anderson
University of Vermont

Uday Banerjee
Syracuse University

Sidney Birnbaum
California State Polytechnic University

Jay Bourland
Colorado State University

Howard Card
Syracuse University

Phillip Crooke
Vanderbilt University

Ruth Favro
Lawrence Technological University

Wendell Fleming
Brown University

Gerald Folland
University of Washington

Larry Lardy
Syracuse University

Jimmy Lawson
Louisiana State University

Alan Levy
Syracuse University

Jacques Lewalle
Syracuse University

Francis Narcowich
Texas A & M University

Dan Waterman
Syracuse University

Finally, the author wishes to express appreciation to his wife Patricia Brookes for her support and understanding.

John L. Troutman
Syracuse, N.Y.

0

Foundations

Before we enter a building, we like to know that it rests on a firm foundation. The principal purpose of this brief chapter is to give assurance that this book rests on a foundation that can withstand inspection. Think of this chapter as the basement of the book and visit it only as needed.

In Section 0.1, we provide notation and terminology for a useful working vocabulary, along with a few standard results. In the remaining sections we expose the twin pillars that support our mathematical investigation of boundary value problems: the divergence theorem and uniform convergence. The divergence theorem supplies one means of relating a function on a set to its values on the boundary of the set. A simple version of the theorem is stated without proof in Section 0.2. By contrast, uniform convergence involves a test whose outcome has important consequences. When the test is passed, quite complicated limiting operations may be interchanged. If the test is not passed, such interchange is suspect, and, for example, term-by-term differentiation of a convergent series of functions, or exchange of orders of integration, might not be possible. Uniform convergence of a series can often be established through the *M*-test, which is presented in Section 0.3.

0.1

Terminology and Related Properties

(a) Geometry in $\mathbb{R}^d$: In this text, $\mathbb{R}$ represents the set of real numbers and $\mathbb{R}^d$ denotes d-dimensional euclidean space with points (or vectors) $\mathbf{x} = (x_1, x_2, \ldots, x_d)$, which obey the usual laws of componentwise addition and (real) scalar multiplication. By this we mean that

$$\mathbf{x} + \mathbf{y} \stackrel{\text{def}}{=} (x_1 + y_1, x_2 + y_2, \ldots, x_d + y_d)$$

$$a\mathbf{x} \stackrel{\text{def}}{=} (ax_1, ax_2, \ldots, ax_d)$$

for any real numbers a, $x_1, \ldots, x_d$, $y_1, \ldots, y_d$. We set $\mathbf{0} = (0, 0, \ldots, 0)$ and

$(-1)\mathbf{x} = -\mathbf{x}$; we also identify $\mathbb{R}^1$ with $\mathbb{R}$. Each $\mathbf{x} \in \mathbb{R}^d$ has euclidean length $|\mathbf{x}|$, where

$$|\mathbf{x}|^2 = \sum_{i=1}^{d} (x_i)^2 = \mathbf{x} \cdot \mathbf{x}.$$

The **dot product**

$$\mathbf{x} \cdot \mathbf{y} \stackrel{\text{def}}{=} \sum_{i=1}^{d} x_i y_i = x_1 y_1 + x_2 y_2 + \cdots + x_d y_d,$$

satisfies

$$|\mathbf{x} \cdot \mathbf{y}| \leq |\mathbf{x}|\,|\mathbf{y}|, \qquad \textit{(Cauchy inequality)}$$

with equality iff (if and only if) one of the vectors $\mathbf{x}$ or $\mathbf{y}$ is a scalar multiple of the other. This inequality follows on observation that

$$0 \leq \sum_{i,j=1}^{d} (x_i y_j - x_j y_i)^2 = 2[|\mathbf{x}|^2 |\mathbf{y}|^2 - (\mathbf{x} \cdot \mathbf{y})^2]$$

and implies that

$$|\mathbf{x} + \mathbf{y}| \leq |\mathbf{x}| + |\mathbf{y}|. \qquad \textit{(Triangle inequality)}$$

Cauchy's inequality also has the important consequence that for any real numbers $x_1, y_1, x_2, y_2, \ldots,$

$$\sum_{i=1}^{\infty} |x_i y_i| \leq \left(\sum_{i=1}^{\infty} x_i^2\right)^{1/2} \left(\sum_{j=1}^{\infty} y_j^2\right)^{1/2}$$

(Extended Cauchy inequality)

Because of the triangle inequality, $|\mathbf{x} - \mathbf{y}|$ may be thought of as the **distance** between $\mathbf{x}$ and $\mathbf{y}$ in $\mathbb{R}^d$.

Of course, $\mathbb{R}^2$ may be visualized as a plane and $\mathbb{R}^3$ considered as ordinary euclidean space. Certain subsets of $\mathbb{R}^d$ occur frequently enough to be granted names appropriated from $\mathbb{R}^3$. In particular, $B_r(\mathbf{c}) = \{\mathbf{x} : |\mathbf{x} - \mathbf{c}| < r\}$ is called the d-dimensional **open ball** of **finite** radius r and center $\mathbf{c}$. All balls and their subsets are said to be **bounded**; all other sets are said to be **unbounded**.

Most of this is probably familiar, but as you know, calculus involves differentiating and integrating functions defined on sets—or at least trying to perform those operations. However, unless both functions and sets are suitably "nice," we will not succeed, and, roughly speaking, functions and sets that are too large or too complicated present problems. In the following paragraphs we describe some sets and functions that are just nice enough to let us make progress while keeping us out of trouble.

Special (d-1)-dimensional subsets of $\mathbb{R}^d$ are the **surface** of the ball $B_r(\mathbf{c})$ given by $S_r(\mathbf{c}) = \{\mathbf{x} : |\mathbf{x} - \mathbf{c}| = r\}$, and, when $d > 1$, the **cylinder** $\{\mathbf{x} : \Sigma_{i=2}^{d}(x_i - c_i)^2 = r^2\}$ of radius r with center at $\mathbf{c}$ and axis parallel to the x_1 axis. (Cylinders with other axes through $\mathbf{c}$ can be envisioned.) Also, for each nonzero $\mathbf{c}$, the sets $\{\mathbf{x} : \mathbf{c} \cdot \mathbf{x} = a\}$ are parallel **hyperplanes** distinguished by the value of $a \in \mathbb{R}$; that for $a = 0$ passes through the origin. When $d > 1$, each of these sets has two "sides" relative to $\mathbb{R}^d$ and it separates $\mathbb{R}^d$ into two complementary regions, one on each "side." A cylinder or the surface of a ball separates regions that we identify as interior and exterior, and for each $\alpha \in \mathbb{R}$, the hyperplane $\{\mathbf{x} : \mathbf{c} \cdot \mathbf{x} = \alpha\}$ separates the region "above," where $\mathbf{c} \cdot \mathbf{x} > \alpha$ from that "below," where $\mathbf{c} \cdot \mathbf{x} < \alpha$. In each case, we can join points in the *same* complementary region by a path that does not cross the $(d\text{-}1)$-dimensional set.

(b) Domains and simple domains: To construct more interesting objects, we combine these sets. Observe that the union U of a *finite* number of these basic $(d\text{-}1)$-dimensional sets subdivides $\mathbb{R}^d$ into sets of geometrical character. The complement $\mathbb{R}^d \sim U$ consists of a *finite* number of "pieces," each distinguished by the fact that we can join any two points within it by a path that does not cross U. Each *bounded* piece of $\mathbb{R}^d \sim U$ will be called a (geometrically) **simple domain** of $\mathbb{R}^d$. Figure 0.1 shows some examples in $\mathbb{R}^2$ and in $\mathbb{R}^3$ (the most important cases for this book), but in higher dimensions you must use your imagination.

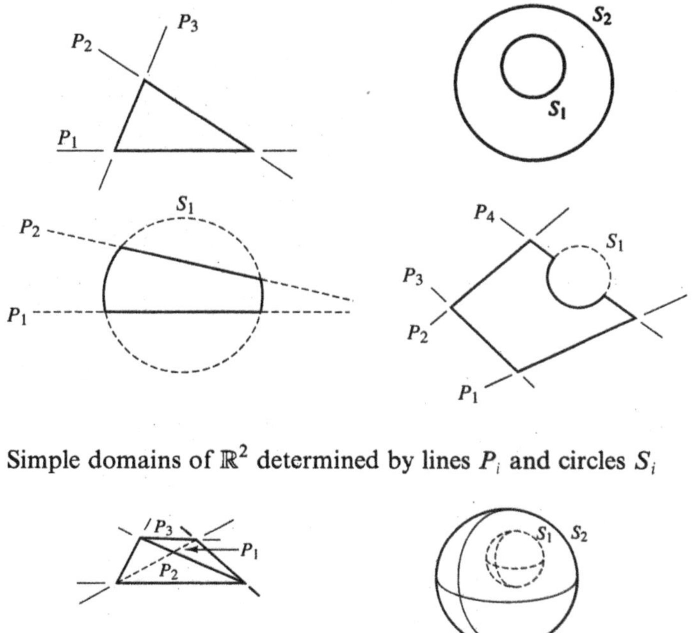

Simple domains of $\mathbb{R}^2$ determined by lines P_i and circles S_i

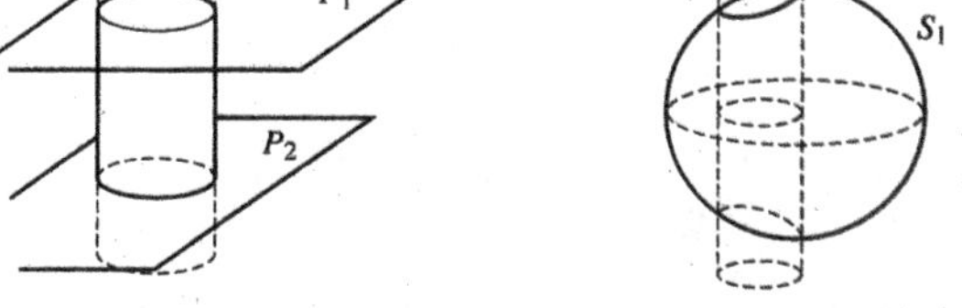

Simple domains of $\mathbb{R}^3$ determined by planes P_i, spherical surfaces S_i, and cylindrical surfaces C_i

———***Figure 0.1***———

As we see from Figure 0.1, a simple domain V can have corners or edges, but in any case its boundary is composed of parts of familiar geometrical objects. Moreover, V contains the boundaries of other simple domains whose boundary pieces are parallel to those of V and as close to them as desired.

To describe these sets and their relationship, we need more precise language.

(0.1) Definition: *A domain V of $\mathbb{R}^d$ is a subset of $\mathbb{R}^d$ such that each pair of its points belongs to a finite chain of successively overlapping open balls contained within V.*

Its **boundary** S (denoted ∂V) *consists of those points $\mathbf{s} \in \mathbb{R}^d \sim V$ such that each open ball with center $\mathbf{s}$ contains points in V; $\bar{V} \stackrel{\text{def}}{=} V \cup S$ is the* **closure** *of V.* ■

In a domain V of $\mathbb{R}^d$ we can approach a point from any direction and consider differentiating a function defined on V in that direction (the chain requirement lets us use the derivatives to relate function values at different points in V). Domains can be unbounded and their boundaries can be extremely complicated, but more tractable sets are obtained in a geometric subdivision of $\mathbb{R}^d$ by U as described above. Then each "piece" of $\mathbb{R}^d \sim U$ is a domain V whose boundary S is a geometrically describable subset of U, and when V is also bounded, V is a simple domain.

In $\mathbb{R}^1$ each simple domain is just a *bounded* open interval of the form $(a, b) = \{x : a < x < b\}$, with boundary $\{a, b\}$ and closure $[a, b] = \{x : a \leq x \leq b\}$. Such intervals provide the standard sets for integration. Similarly, in higher dimensions our simple domains, their boundaries, and their closures provide nice sets for integration.

(c) Function classes: Now suppose that V is a fixed domain of $\mathbb{R}^d$ with boundary S and closure $\bar{V} = V \cup S$, and that u is a real-valued function on V. To decide whether u is a nice function, we consider its graph, which would show that smoothness is determined by the degree to which u is continuously differentiable and the extent to which this extends to $\bar{V}$. Specifically, for $k = 1, 2, \ldots$ we say that $u \in C^k(V)$ when u and all its partial derivatives of degree $\leq k$ are continuous on V† and that $u \in C^k(\bar{V})$ when u and each of these derivatives has an extension that is continuous on $\bar{V}$. In unambiguous situations, we will use the same notation for a function on V and its continuous extension to $\bar{V}$. When the domain V is understood, we may use abbreviations such as $u \in C^k$ or "u is C^k" to imply that $u \in C^k(V)$.

These sets are related by the inclusions

$$C^k(\bar{V}) \subseteq C^k(V) \subseteq C^{k-1}(V) \cdots \subseteq C^1(V) \subseteq C(V),$$

and each is a **subspace** of the vector space of all real-valued functions on V. By this last statement, we mean, for example, that if u and v are functions in $C^k(V)$,

† We say that a real-valued function f is **continuous on a set** $T \subseteq \mathbb{R}^d$ (denoted $f \in C(T)$) when for each $\mathbf{t}_0 \in T$, and $\epsilon > 0$, there is a $\delta > 0$ such that $\mathbf{t} \in T$ and $|\mathbf{t} - \mathbf{t}_0| < \delta \Rightarrow |f(\mathbf{t}) - f(\mathbf{t}_0)| < \epsilon$. (If δ does not depend on $\mathbf{t}_0$, we say that f is **uniformly continuous** on T.)

then their linear combination $au + bv$ is also in $C^k(V)$, for any real constants a and b.

(d) Integration; orthogonality: In this text we use only Riemann integration. Functions that are Riemann integrable over a bounded set must be bounded and have finite integrals over that set. Then we can show that their sums and products — in particular, their *squares* — are also integrable over the set. If V is a simple domain, then each $f \in C(\bar{V})$ is Riemann integrable over $\bar{V}$ and over its boundary S. Remember that S consists of pieces of familiar geometrical objects, such as line segments, rectangles, circular arcs, and spherical segments.

We will also encounter *improper* Riemann integrals. The function $f(x) = 1/\sqrt{x}$ blows up on the interval $(0, 1]$, but it has an improper integral on this interval defined by

$$\int_0^1 f(x)\,dx = \lim_{\epsilon \to 0^+} \int_\epsilon^1 f(x)\,dx = \lim_{\epsilon \to 0^+} 2\sqrt{x}\Big|_\epsilon^1 = 2;$$

however, in the same sense $\int_0^1 f^2(x)\,dx = +\infty$.

Similarly, when a function is not Riemann integrable over a set, we may be able to use limits of its proper integrals over subsets to define its improper integral. Integrals over domains and over unbounded sets will always be considered as improper integrals.

If f and g are integrable over an interval $[a, b]$, then they satisfy the Cauchy-like inequality

$$\int_a^b |fg|\,dx \le \left(\int_a^b f^2\,dx\right)^{1/2}\left(\int_a^b g^2\,dy\right)^{1/2}.$$

(*Schwarz inequality*)

If $f \ge 0$ and $g \ge 0$ (the only relevant case), this inequality follows from the fact that

$$0 \le \int_a^b \int_a^b [f(x)g(y) - f(y)g(x)]^2\,dx\,dy$$
$$= 2\left(\int_a^b f^2\,dx\right)\left(\int_a^b g^2\,dy\right) - 2\left(\int_a^b fg\,dx\right)^2,$$

when the dummy variables are used appropriately. The Schwarz inequality also holds when its integrals are improper as well as for integrals over more general sets.

In case $\int_a^b f(x)g(x)\,dx = 0$, the functions f and g are said to be **orthogonal** on the interval $[a, b]$. To better appreciate this terminology, think of f and g as vectors with xth components $f(x)$ and $g(x)$, respectively, for each $x \in [a, b]$. Then, in analogy to $\mathbb{R}^d$, we can regard

$$[f, g] \overset{\text{def}}{=} \int_a^b f(x)g(x)\,dx$$

as a dot product between f and g (on $[a, b]$) so that f and g are orthogonal on $[a, b]$ precisely when $[f, g] = 0$. Using this product, we can geometrize the *function-space* $C[a,b]$ (or, more generally, the space of Riemann integrable functions on $[a,b]$). For example, we can define the **length** of f on $[a,b]$ by

$$\|f\| = [f,f]^{1/2} = \left(\int_a^b f^2(x)\,dx\right)^{1/2}$$

and restate the Schwarz inequality in the suggestive form

$$|[f,g]| \le \|f\|\,\|g\|.$$

(e) Mean-square convergence: In this text we will often need to represent a given function f by a *series* of the form

$$f(x) = \sum_{n=1}^{\infty} c_n y_n(x), \qquad a \le x \le b \tag{1}$$

for some constants $c_1, c_2, \ldots$, where the y_n are **pairwise orthogonal** functions on (a,b), in that $[y_n, y_m] = 0$, $n \neq m$. When each y_n is of unit length, we can geometrize this representation if for each n we take

$$c_n = [f, y_n] = \int_a^b f(x) y_n(x)\,dx \qquad \textit{(as in Section 0.1d)}$$

and think of $c_n y_n$ as the "component of f in the direction of the unit vector" y_n in a space of *infinite* dimension. Note that this component has length $|c_n|$.

Then, as Figure 0.2 suggests, for $N = 1, 2, \ldots$, we might set

$$F_N(x) = \sum_{n=1}^{N} c_n y_n(x), \qquad a \le x \le b,$$

not be surprised that $\|F_N\|^2 = \Sigma_{n=1}^N c_n^2$ (since $y_1, y_2, \ldots, y_N$ are mutually "perpendicular"), and try to establish the Pythagorean identity

$$\|f - F_N\|^2 = \|f\|^2 - \|F_N\|^2 = \|f\|^2 - \sum_{n=1}^{N} c_n^2,$$

assuming that $\|f\|$ is finite. The real questions concern what happens as $N \to \infty$. Does $\|f - F_N\|^2 = \int_a^b (f - F_N)^2(x)\,dx \to 0$? Or, equivalently, does

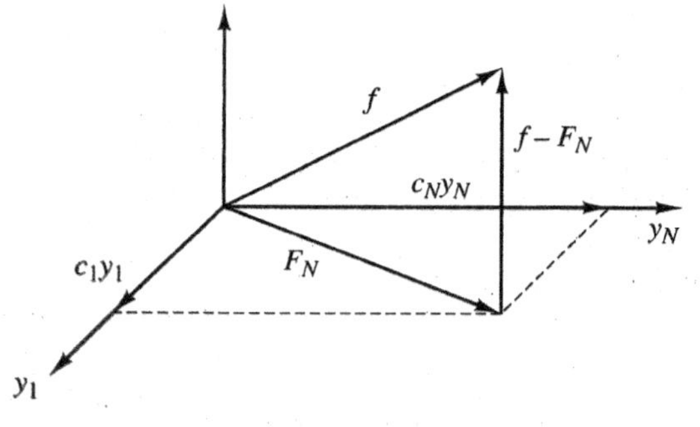

Figure 0.2

$\Sigma_{n=1}^{\infty} c_n^2 = \| f \|^2$? If so, we say that series (1) converges to f in the *mean-square* sense, and if this holds whenever $\| f \|$ is finite, we say that $\{y_n\}_{n=1}^{\infty}$ is a **complete** sequence of functions.

Even if series (1) converges in the mean-square sense, it might not converge at a single value of x, or if it does, its sum need not be $f(x)$. But for almost a century, this geometrical approach has guided investigations of convergence. In particular, you will see various elements of this approach reflected in this text. However, since it can be notationally awkward to force the analysis into the required framework, you may be invited to think geometrically about analytical expressions that suggest those above but are not identical to them.

(f) The gradient and the Laplacian: If $u \in C^1(V)$ where at $\mathbf{x} = (x_1, x_2, \ldots, x_d) \in V$, u has the value $u(\mathbf{x})$, we write $u = u(x_1, \ldots, x_d)$ and denote the first partial derivatives of u by

$$u_{x_j} = \frac{\partial u}{\partial x_j}, \; j = 1, 2, \ldots, d.$$

In Cartesian coordinates $\mathbf{x} = (x_1, x_2, \ldots, x_d)$, u_{x_j} provides the jth component of ∇u, the **gradient** of u, which is a d-dimensional vector-valued function on V with components in $C(V)$. The directional derivative of u at $\mathbf{x} \in V$, in the direction of the unit vector $\mathbf{n}$, is given by $\partial u/\partial n = \nabla u(\mathbf{x}) \cdot \mathbf{n}$, and from the Cauchy inequality (Section 0.1a), we see that $|\partial u/\partial n| \le |\nabla u(\mathbf{x})|$, with equality when $\nabla u(\mathbf{x}) \ne 0$, iff $\mathbf{n}$ and $\nabla u(\mathbf{x})$ are vectors in the same or opposite directions.

Similarly, if $u \in C^2(V)$, then u has continuous partial derivatives

$$u_{x_i x_j} = \frac{\partial}{\partial x_j}\left(\frac{\partial u}{\partial x_i}\right), \qquad i, j = 1, 2, \ldots, d,$$

and it is true (but *not* evident) that $u_{x_i x_j} = u_{x_j x_i}$, even when $i \ne j$. The combination

$$\nabla^2 u = u_{x_1 x_1} + u_{x_2 x_2} + \cdots + u_{x_d x_d}$$

is called the **Laplacian** of u.†

Corresponding notation and the possibility of interchange apply to $u \in C^k(V)$ if $k > 2$. For example, then $u_{x_1 x_2 x_3} = u_{x_3 x_1 x_2} = u_{x_2 x_3 x_1}$.

(g) Composite functions and the chain rule: Under the polar coordinate transformation

$$x = r\cos\theta, \qquad y = r\sin\theta$$

a function $u = u(x, y)$ of the rectangular coordinates $(x, y) \in \mathbb{R}^2$ becomes the *new* (composite) function

$$\tilde{u}(r, \theta) \stackrel{\text{def}}{=} u(r\cos\theta, r\sin\theta)$$

of the coordinates $(r, \theta) \in \mathbb{R}^2$. (See Figure 0.3.)

---

† ∇ itself may be regarded as a vector operator of differentiation; however, it is only in Cartesian coordinate systems for $\mathbf{x}$ that ∇ takes the simple form $\left(\frac{\partial}{\partial x_1}, \frac{\partial}{\partial x_2}, \ldots, \frac{\partial}{\partial x_d}\right)$ and ∇^2 takes the simple form $\frac{\partial^2}{\partial x_1^2} + \frac{\partial^2}{\partial x_2^2} + \cdots + \frac{\partial^2}{\partial x_d^2}$.

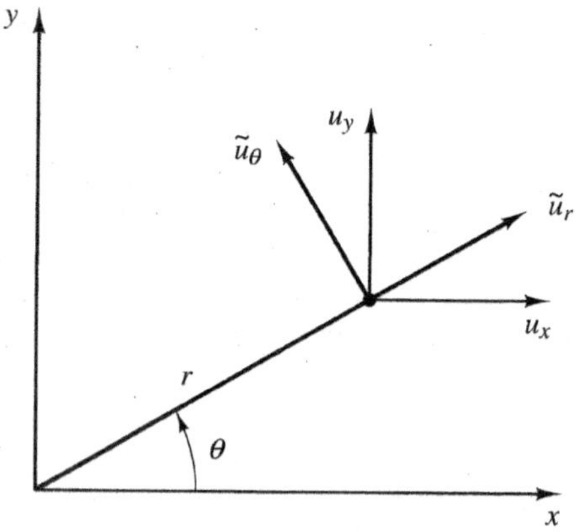

Figure 0.3

When u is C^1 with partial derivatives $u_x = u_x(x, y)$ and $u_y = u_y(x, y)$, then $\tilde{u}$ is also C^1 and has partial derivatives $\tilde{u}_r = \tilde{u}_r(r, \theta)$ and $\tilde{u}_\theta = \tilde{u}_\theta(r, \theta)$, which are related to those of u by means of the **chain rule** [Ed], as follows:

$$\tilde{u}_r = u_x \frac{\partial x}{\partial r} + u_y \frac{\partial y}{\partial r} = u_x \cos\theta + u_y \sin\theta$$

$$\tilde{u}_\theta = u_x \frac{\partial x}{\partial \theta} + u_y \frac{\partial y}{\partial \theta} = u_x(-r\sin\theta) + u_y(r\cos\theta).$$

Here, the partial derivatives of u are evaluated at $(r\cos\theta, r\sin\theta)$, so that the right side of each equation is indeed a function of r and θ. If u is C^2, then $\tilde{u}$ is also C^2, and its second partial derivatives can be related to derivatives of u by differentiating these last expressions, where the chain rule is used on the composite functions $u_x(r\cos\theta, r\sin\theta)$ and $u_y(r\cos\theta, r\sin\theta)$. (We can also equate the mixed partials u_{xy} and u_{yx}.)

In this manner we can verify that in polar coordinates the Laplacian takes the form

$$\nabla^2 = \frac{\partial^2}{\partial r^2} + \frac{1}{r}\frac{\partial}{\partial r} + \frac{1}{r^2}\frac{\partial^2}{\partial \theta^2}, \qquad r > 0$$

since $$\nabla^2 u = u_{xx} + u_{yy} = \tilde{u}_{rr} + \frac{1}{r}\tilde{u}_r + \frac{1}{r^2}\tilde{u}_{\theta\theta}, \qquad r > 0.$$

Once these principles are grasped — and the notation is understood — we can extend these results to other transformations and to higher dimensions. In particular, we can establish the important fact that the Laplacian takes the *same* form in each Cartesian coordinate system for $\mathbb{R}^d$.

0.2

The Divergence Theorem

If (a, b) is a bounded interval of $\mathbb{R}$ and $f \in D^1[a, b] \overset{\text{def}}{=} C[a, b] \cap C^1(a, b)$, then by the Fundamental Theorem of Calculus,

$$\int_a^b f'(x)\,dx = f(x)\Big|_a^b = f(b) - f(a),$$

where the Riemann integral is possibly improper. We wish to generalize this result to higher dimensions. The simplest version to state is that where (a, b) is replaced by a simple domain V of $\mathbb{R}^d$, and f is replaced by a vector field $\mathbf{f}$ with d components $f_1, f_2, \ldots, f_d$. Then we have the following result which is proved in [Ed].

(0.2) Divergence Theorem: *If V is a simple domain of $\mathbb{R}^d$ $(d > 1)$ with boundary S and $\mathbf{f}$ is a d-dimensional vector field with each component in $D^1(\bar{V}) \stackrel{\text{def}}{=} C(\bar{V}) \cap C^1(V)$, then*

$$\int_V \nabla \cdot \mathbf{f}\, dV = \int_S \mathbf{f} \cdot \mathbf{n}\, dS,$$

where the integral over V may be improper. ■

Remarks: $\mathbf{n}$ is the unit normal vector on S directed outward from V. It is well defined and continuous except on an edge-set negligible with respect to $(d\text{-}1)$-dimensional surface integration on S.† dV is the d-dimensional element of volume in $\mathbb{R}^d$, and $\nabla \cdot \mathbf{f}$ is the **divergence** of the field $\mathbf{f}$. In Cartesian coordinates $(x_1, x_2, \ldots, x_d)$ where $dV = dx_1\, dx_2 \ldots dx_d$, the divergence of $\mathbf{f}$ is given by

$$\nabla \cdot \mathbf{f} \stackrel{\text{def}}{=} \frac{\partial f_1}{\partial x_1} + \frac{\partial f_2}{\partial x_2} + \cdots + \frac{\partial f_d}{\partial x_d}.$$

The surface integral is an ordinary integral, but the volume integral is defined as the limit of proper integrals over subdomains with "parallel" boundary pieces in V.

For example, in $\mathbb{R}^2$ a rectangular domain V has boundary S composed of four line segments (see Figure 0.4). $\mathbf{n}$ is well defined and locally constant (except at the

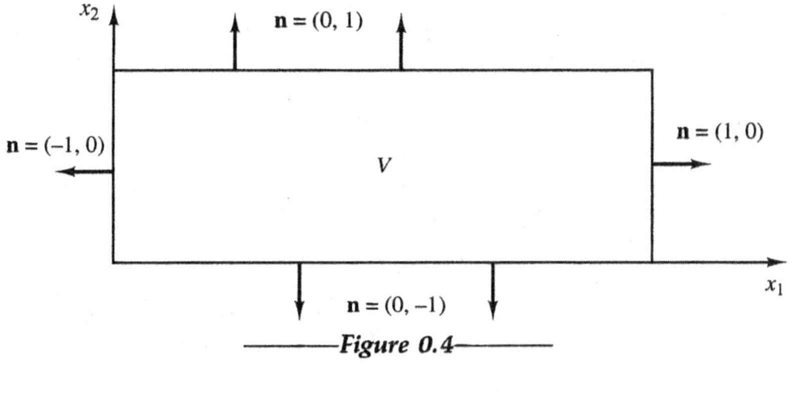

Figure 0.4

† To appreciate what this means, look back at the examples of simple domains in Figure 0.1. In fact, simple domains were invented to meet these conditions and to give meaning to the integral terms of the divergence theorem.

vertices). Any pair of functions f_1 and f_2 in $D^1(\bar{V})$ supplies the components of a vector field $\mathbf{f} = (f_1, f_2)$.

It is easy to verify that the theorem holds for a d-dimensional box oriented along the coordinate axes, and therefore it holds for domains consisting of finite unions of nonoverlapping coordinate boxes. The difficulty comes with using such configurations of boxes to approximate other simple domains. The theorem also holds for certain nonsimple bounded domains, but there are domains with boundaries so jagged that $\mathbf{n}$ cannot be assigned anywhere.

On comparing the divergence theorem with its predecessor, we see that $\nabla \cdot \mathbf{f}$ is a kind of field derivative. When V is a domain of $\mathbb{R}^d$ and $u \in C^2(V)$, then $\nabla \cdot \nabla u = \nabla^2 u$ is the Laplacian of u. Moreover, if $v \in C^1(V)$, then u and v satisfy the important identity

$$\nabla \cdot (v\nabla u) = v\nabla^2 u + (\nabla v) \cdot (\nabla u),$$

as is easily verified in Cartesian coordinates.

0.3

Limit Interchange: Uniform Convergence

Processes such as differentiating under an integral sign, reversing order of integrations, and differentiating or integrating a series involve the interchange of operations defined by limits. Interchange need *not* be possible: $F(x, n) = nx/(nx + 1)$ is defined for $x > 0$ and $n = 1, 2, \ldots$; however, $\lim_{x\to 0} \lim_{n\to\infty} nx/(nx + 1) = 1$, while $\lim_{n\to\infty} \lim_{x\to 0} nx/(nx + 1) = 0$.

More generally, suppose that we have a pair of sets U, V, and a real-valued function

$$F(u, v) \qquad \text{defined when} \qquad u \in U, v \in V,$$

for which the finite (pointwise) limit

$$F_1(v) = \lim_{u\to u_0} F(u, v) \qquad \text{exists for each} \qquad v \in V.$$

Here, u_0 need not be in U, but we suppose that U contains points u as "near u_0" as we wish. For example, a large integer N is considered to be "near ∞," and in Riemann integration, a suitably refined partition u is considered to be "near" a hypothetical idealized version u_0 of vanishingly small mesh size.

Uniform Convergence

(0.3) Definition: *We say that as $u \to u_0$, $F(u, v)$ converges to its limit $F_1(v)$* **uniformly** *on V when for each $\epsilon > 0$, by taking u sufficiently "near u_0," we can make $|F(u, v) - F_1(v)| < \epsilon$ for all $v \in V$,* **simultaneously**. ∎

In the case that $u_0 \in U \subseteq \mathbb{R}^d$, uniform convergence is assured if for each $\epsilon > 0$, there is a $\delta > 0$ such that

$$|u - u_0| < \delta \Rightarrow |F(u, v) - F_1(v)| < \epsilon, \qquad \forall v \in V.$$

To discuss series, however, we must take $u_0 = \infty$ and think of $u = n$ as approaching infinity. Fortunately, it may be easy to establish the *uniform convergence of a series* of functions through the following result:

(0.4) M-Test (Weierstrass): *If f_n is a real-valued function on a set V with $|f_n(v)| \leq M_n$, $v \in V$, $n = 1, 2, \ldots$, where $\Sigma_{n=1}^{\infty} M_n < \infty$, then the series*

$$F(v) = \sum_{n=1}^{\infty} f_n(v) = \lim_{N\to\infty} \sum_{n=1}^{N} f_n(v)$$

converges absolutely and uniformly on V.

Proof: Since $|f_n(v)| \leq M_n$, we know by comparison that the series in question converges (absolutely) for each $v \in V$. Then with $F_N(v) \stackrel{\text{def}}{=} \Sigma_{n=1}^{N} f_n(v)$, we have the *uniform* estimate

$$\begin{aligned} |F(v) - F_N(v)| &= \left|\sum_{n=1}^{\infty} f_n(v) - \sum_{n=1}^{N} f_n(v)\right| \\ &= \left|\sum_{n=N+1}^{\infty} f_n(v)\right| \\ &\leq \sum_{n=N}^{\infty} |f_n(v)| \leq \sum_{n=N}^{\infty} M_n < \epsilon, \end{aligned}$$

if $N \geq N_\epsilon$, by the hypothesized convergence of the series of "majorants" M_n. ■

Uniform convergence provides a basis for important statements about series.

(0.5) Theorem: *For each $n = 1, 2, \ldots$, let f_n be a real-valued function on a set V, and suppose that the series $F(v) = \Sigma_{n=1}^{\infty} f_n(v)$ converges* **uniformly** *on V.*

(A) If each f_n is continuous on V, then F is continuous on V.

(B) If each f_n is integrable over V, a **bounded** *set in $\mathbb{R}^d$, then F is integrable over V and $\int_V F(v)\,dV = \Sigma_{n=1}^{\infty} \int_V f_n(v)\,dV$.*

(C) If each f_n is continuously differentiable in $V = (a, b) \subseteq \mathbb{R}$ and the **series of derivatives** $\Sigma_{n=1}^{\infty} f_n'(v)$ **converges uniformly** *on V, then $F \in C^1(V)$ and $F'(v) = \Sigma_{n=1}^{\infty} f_n'(v)$, $v \in V$.*

Proof: Conclusions A and B are direct consequences of the limit interchange principle established below. Conclusion C follows when we use B to integrate the differentiated series from x_0 to $x \in (a, b)$ and then apply the Fundamental Theorem of Calculus. ■

There are similar estimates of uniformity and corresponding conclusions when the series of terms $f_n(v)$ is replaced by integrals of $f(u, v)$ over sets $U \subseteq \mathbb{R}^d$. (See [Ap].)

Example 1:

The trigonometric series $F(t) = \Sigma_{n=1}^{\infty} \dfrac{\sin nt}{n^3}$, $t \in \mathbb{R}$, converges absolutely and uniformly on $\mathbb{R}$ by the M-test since

$$|f_n(t)| = \left|\frac{\sin nt}{n^3}\right| \leq \frac{1}{n^3} = M_n, \quad n = 1, 2, \ldots, \quad \text{and} \quad \sum_{n=1}^{\infty} \frac{1}{n^3} < +\infty.$$

Consequently, we are assured that its sum F is continuous on $\mathbb{R}$ since this is true of each of its terms. Moreover, on each bounded interval $[a, b] \subseteq \mathbb{R}$, we have

$$\int_a^b F(t)\, dt = \sum_{n=1}^{\infty} \frac{1}{n^3} \int_a^b \sin nt\, dt.$$

To consider whether F is differentiable, we observe that the series of derivatives

$$\sum_{n=1}^{\infty} \frac{d}{dt}\left(\frac{\sin nt}{n^3}\right) = \sum_{n=1}^{\infty} \frac{\cos nt}{n^2}$$

also converges uniformly on $\mathbb{R}$ since $\Sigma_{n=1}^{\infty}(1/n^2) < +\infty$. Therefore, F is differentiable and its derivative

$$F'(t) = \sum_{n=1}^{\infty} \frac{\cos nt}{n^2}$$

is continuous, that is, $F \in C^1(\mathbb{R})$. However, if we differentiate the last series termwise, we obtain the series

$$\sum_{n=1}^{\infty} \frac{d}{dt}\left(\frac{\cos nt}{n^2}\right) = -\sum_{n=1}^{\infty} \frac{\sin nt}{n},$$

and although

$$\left|\frac{\sin nt}{n}\right| \leq \frac{1}{n}, \qquad n = 1, 2, \ldots,$$

we *cannot* use such estimates in the M-test since $\Sigma_{n=1}^{\infty}(1/n) = +\infty$. In Section 1.8, we will prove that the series $\Sigma_{n=1}^{\infty}(\sin nt/n)$ does converge pointwise (for each value of t), but it does not converge uniformly and its sum is *not* continuous.

To see why such assertions involve an interchange of limits, observe that in part A of Theorem 0.5 the continuity of $F = \Sigma_{n=1}^{\infty} f_n$ at $v_0 \in V$ would follow if we demonstrate that

$$\begin{aligned}\lim_{v \to v_0} F(v) &= \lim_{v \to v_0} \lim_{N \to \infty} F_N(v) = \lim_{N \to \infty} \lim_{v \to v_0} F_N(v) \\ &= \lim_{N \to \infty} F_N(v_0) = \sum_{n=1}^{\infty} f_n(v_0) = F(v_0),\end{aligned}$$

which requires both the continuity of the f_n and the indicated reversal of limits.

Now, let $F(u, v)$ be a real-valued function on $U \times V$ that has the separate finite limits

$$F_1(v) \stackrel{\text{def}}{=} \lim_{u \to u_0} F(u, v), \quad \text{for each } v \in V,$$

$$\text{and} \qquad F_2(u) \stackrel{\text{def}}{=} \lim_{v \to v_0} F(u, v), \quad \text{for each } u \in U.$$

Then we have the promised result.

(0.6) Limit Interchange Principle: *If either of these separate limits is approached uniformly with respect to the* **other** *variable, then*

$$\lim_{u \to u_0} \lim_{v \to v_0} F(u, v) = \lim_{v \to v_0} \lim_{u \to u_0} F(u, v),$$

provided that either of the iterated limits exists.

Proof: Suppose that $A = \lim_{v \to v_0} F_1(v)$ exists; then we wish to prove that $A = \lim_{u \to u_0} F_2(u)$. Now, for each $\epsilon > 0$:

$$|F_2(u) - A| \le |F(u, v) - F_1(v)| + |F(u, v) - F_2(u)| + |F_1(v) - A|,$$

and the last term can be made less than the given $\epsilon > 0$ if v is sufficiently "near v_0." The analysis of the remaining terms on the right side of this inequality depends on which variable supplies the hypothesized uniform convergence.

Case 1: If $|F(u, v) - F_1(v)| < \epsilon$ *uniformly in* v when u is sufficiently near u_0, then for *each* such u, we can choose v so near v_0 that also $|F(u, v) - F_2(u)| < \epsilon$.

Case 2: If $|F(u, v) - F_2(u)| < \epsilon$ *uniformly in* u when v is sufficiently near v_0, then for *one* such v we can make $|F(u, v) - F_1(v)| < \epsilon$ if u is sufficiently near u_0.

In both cases, $|F_2(u) - A| < 3\epsilon$, if u is sufficiently near u_0, and since $\epsilon > 0$ is arbitrary, we have $\lim_{u \to u_0} F_2(u) = A$, as desired. ■

We have already seen how this principle is used in establishing one important result. However, in the remainder of this text we shall simply appeal to the principle as necessary, since its actual applications may still be quite technical. In certain cases, especially those involving Lebesgue integration, the desired interchange can be effected in the presence of **nonuniform** convergence, but such considerations are beyond the scope of this text. (See [Rud] or any similar work on the fundamentals of analysis.)

PART
I

Basic Theory

A boundary value problem usually involves a partial differential equation in a region of euclidean space. To solve such a problem, we must find certain solutions of the equation. We can solve simple problems using Fourier series, as we shall see through many examples. However, there is growing recognition that it is easier to examine Fourier series independently of this association and that these series have other important applications, especially in signal processing. Therefore, in Chapter 1 we will look only at Fourier series and their behavior. Then, in Chapter 2, we will lay out the ingredients of boundary value problems. There we discover how linearity permits solutions to be obtained from superposition of simpler problems and how separation methods introduce Fourier components.

In Chapters 3–5, we will examine three principal types of physical process — diffusion described by the heat equation, steady-state described by the potential equation, and propagation described by the wave equation — together with boundary value problems related to each. Some of these problems yield to solution through Fourier series; others require methods that are discussed in later chapters.

1

Fourier Series of Periodic Functions

Many physical phenomena are effectively recurrent in time, especially those associated with the rotation of the earth and its movement in our solar system. On a smaller scale, we recognize the repetition inherent in describing a few swings of a pendulum and incorporate this regularity in the design of simple mechanical clocks to measure the passage of time. Also, we carry within us the potential biological regularity of a heartbeat (when resting) that can be revealed graphically in an EKG (electrocardiogram) and the presence of alpha rhythms while dreaming. Thus it is clearly of value to have an effective method of providing mathematical descriptions for those phenomena that exhibit natural recurrence after a given period.

In the present chapter we introduce Fourier series as a means of accomplishing this and investigate the associated questions of convergence and approximation. Convergence results are presented and applied in Section 1.5; their proofs are deferred until Section 1.8. We discuss both theoretical and numerical aspects of approximation. In Section 1.7 we introduce the fast Fourier transform to speed our calculation of the approximations. This device has made possible the widespread application of Fourier analysis in processing the many signals of our electronic existence.

1.1 Periodicity

An EKG is obtained by monitoring the electrochemically generated voltage differences that occur during the action of a heart and cause the heart muscles to contract. Two time histories of such voltage differences from the human heart are shown in Figure 1.1.

How do we compare them? The distinguishing feature of each is its apparent

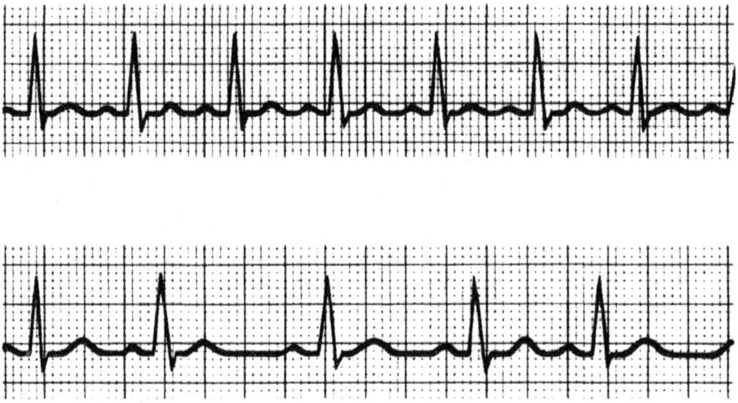

Figure 1.1

repetition over successive blocks of time. Within each block there is one sharp peak (corresponding to the onset of ventricular contractions that pump blood out of the heart) along with various other peaks or valleys.

We would like a means of assigning numerical characteristics to these electrical signals so that they can be compared. One such characteristic, the mean value per cycle, is unaffected by where we locate the cycle, but two very different signals might easily have the same mean value. To find characteristics that reflect more detailed structure of the signal, let's examine the most familiar periodic functions, the trigonometric sines and cosines.

One source of regularly recurrent behavior is that of a point fixed to the rim of a flywheel of radius r, which is rotating at constant angular velocity ω about a fixed central axis, as indicated in Figure 1.2.

At time t, the horizontal and vertical coordinates measured from the center of the wheel are $x = r\cos\omega t$ and $y = r\sin\omega t$, respectively, if we assume that at $t = 0$ their values are r and 0. For this reason, an electrical generator produces an alternating current that is purely sinusoidal in time with periods corresponding to frequencies such as 60 cycles per second. Indeed, it is this pure sine wave that is shown on an oscilloscope.

In the simplest model, a generator flywheel rotating at ω rad/s produces a signal proportional to $\sin\omega t$ (or $\cos\omega t$) that returns to its starting configuration first after one revolution in $L = 2\pi/\omega$ seconds, next after two revolutions in $2L$

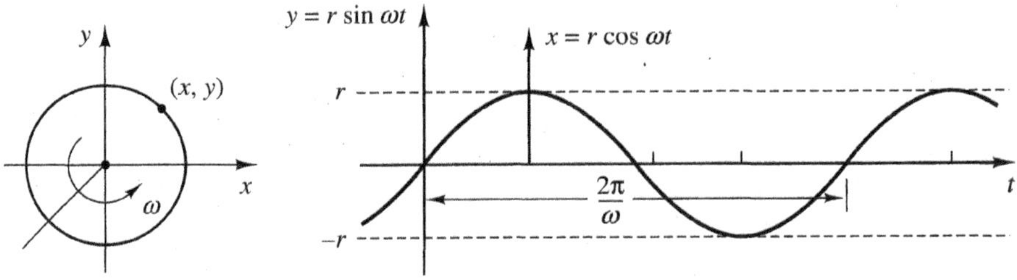

Figure 1.2

seconds, and so on. We say that $\sin \omega t$ (or $\cos \omega t$) has *base* period $L = 2\pi/\omega$ and periods L, $2L$, $3L$, and so on. In particular, when $\omega = 1, 2, 3, \ldots, n$, we see that the functions $\sin t$, $\cos t$, $\sin 2t$, $\cos 2t, \ldots, \sin nt$, $\cos nt$, *all* have period 2π even though their base periods may differ. (However, as n increases, $\sin nt$ and $\cos nt$ oscillate more rapidly over any given time interval.)

Trigonometric functions have natural periods, but most other elementary functions, including t, e^t, $t \sin t$, $t/(1+t^2)$, do not exhibit periodic behavior.

In general, a periodic function f is understood to be defined on $\mathbb{R}$, and f has period $L > 0$ when

$$f(t+L) = f(t), \qquad t \in \mathbb{R}. \tag{1}$$

Note that the constant function $f(t) = 1$, $t \in \mathbb{R}$, has period L for *every* $L > 0$, and it can also be regarded as a trigonometric function since $\cos 0t = 1$. Sums, products, and derivatives of functions of period L again have period L, but the indefinite integral of a periodic function might not be periodic. (See Problems 1.1F and 1.1G.)

A function f of given period may easily be converted into one of period 2π by a scale change. (See Section 1.6.) Since analysis is simpler for this period, we will usually consider an f which has period 2π, that is, one for which

$$f(t+2\pi) = f(t), \qquad t \in \mathbb{R}. \tag{1'}$$

(f may have periods less than 2π.) Alternatively, if f is given on any half-open interval $[a, a+2\pi)$ or $(a, a+2\pi]$, it can be extended with period 2π into each successive adjoining interval of length 2π. Again, for technical reasons, it is best to suppose that f is given on $[-\pi, \pi)$ or on $(-\pi, \pi]$. For example, $f(t) = t$ on $[-\pi, \pi)$ extends periodically to the sawtooth function of Figure 1.3a, but this is not the periodic extension of $f(t) = t$ on $(0, 2\pi]$ or for this same function on $(-\pi, \pi]$. On the other hand, the periodic extension of $f(t) = |t|$ on $[-\pi, \pi)$ is identical with that on $(-\pi, \pi]$, as we can see from Figure 1.3b.

Figure 1.3

Problem Set 1.1

1.1A Determine whether each of the following functions is periodic. If so, find its base period and sketch the graph.

1. $\sin \dfrac{2\pi t}{L}$ ***2.*** $\sin \dfrac{\pi t}{L}$

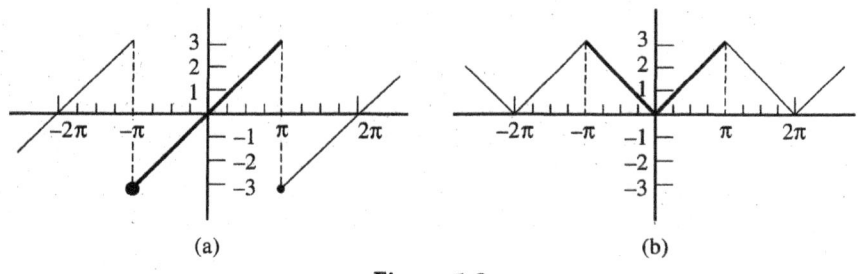

3. $\sin 3t$ **4.** $\cos mt$

5. $\sinh 2t$ **6.** e^t

7. $|\sin t|$ **8.** $\tan \pi t$

9. t^2 **10.** $\cos^2 t$

1.1B Sketch two or more periods of the following functions.

1. $f(t) = t^2, \quad -\pi < t \leq \pi, \quad f(t+2\pi) = f(t)$

2. $f(t) = |t-1|, \quad -\pi \leq t < \pi, \quad f(t+2\pi) = f(t)$

3. $f(t) = e^t, \quad -1 < t \leq 1, \quad f(t+2) = f(t)$

4. $f(t) = 1 - 2|t|, \quad -1 \leq t < 1, \quad f(t+2) = f(t)$

1.1C **1.** Find a formula for the function $f(t)$ in Figure 1.3a that is valid (a) in the interval $(\pi, 2\pi)$ and (b) in the interval $(-2\pi, -\pi)$.

2. Find a single formula for the function $f(t)$ in Figure 1.3b that is valid in the intervals $(-2\pi, -\pi)$ and $(\pi, 2\pi)$.

1.1D Suppose that $f(t)$ has period L. What is the corresponding period of $f(at)$ where a is a positive constant?

1.1E Show that if a function f has period L, then for $n = 1, 2, 3, \ldots$

1. $f(t + nL) = f(t)$ **2.** $f_n(t) \stackrel{\text{def}}{=} f(nt)$ also has period L.

1.1F If f and g are functions with period L, show that $\alpha f, f + g$, and $f \cdot g$ also have period L. (α is a constant.)

1.1G **1.** If f is differentiable on $\mathbb{R}$ with period L, how can we see graphically that the derivative f' also has period L?

2. Give a simple example of a periodic function f on $\mathbb{R}$ for which $F(x) = \int_0^x f(t)\, dt$ is not periodic.

1.1H **1.** If $f(x) = \cos \pi x$ and $g(x) = \cos \sqrt{2}\pi x$, is the sum $f + g$ periodic?

2. If f_1 is a periodic function with period L_1 and f_2 is a periodic function with period L_2, under what conditions on L_1 and L_2 is the sum $f_1 + f_2$ a periodic function?

1.2

Finite Representation; Orthogonality

For $n = 0, 1, 2, \ldots$, the trigonometric functions $\cos nt$ and $\sin nt$ have period 2π. Therefore, their *finite* linear combinations, including such expressions as

$$1 + \cos t, \qquad \sin 3t - \sqrt{2}\cos 17t, \quad \text{or} \quad \sqrt{3}\cos 2t - 4\sin 3t - \tfrac{2}{3}\cos 4t,$$

will also have period 2π, and in this section, we will briefly consider such functions. Each must have the form

$$f(t) = \frac{c_0}{2} + \sum_{n=1}^{N} (c_n \cos nt + s_n \sin nt) \tag{2}$$

for some choice of real coefficients $c_0, c_1, \ldots, c_N, s_1, \ldots, s_N$, where N is *finite*. ($c_0/2$ is just a real constant, but having the factor 1/2 simplifies presentation of later results.)

The function $\sin^2 t$ has period 2π, and through the identity

$$\sin^2 t = \tfrac{1}{2} - \tfrac{1}{2}\cos 2t$$

we see that it can be expressed in the form (2) with $c_0 = 1$, $c_2 = -1/2$, and all other coefficients equal to zero. Does every function f of period 2π admit finite representation in the form (2)? Perhaps not, but for each f, there is only *one* possibility; indeed, the coefficients c_n and s_n in (2) can be determined from f by the following procedure:

Suppose f is given by (2) for some coefficients c_n, s_n, when $n \le N$. For *fixed m*, multiply both sides by $\cos mt$ and integrate over $[-\pi, \pi]$. After obvious rearrangement the following equation results.

$$\int_{-\pi}^{\pi} f(t)\cos mt\,dt = \frac{c_0}{2}\int_{-\pi}^{\pi}\cos mt\,dt + \sum_{n=1}^{N} c_n \int_{-\pi}^{\pi}\cos nt\,\cos mt\,dt + \sum_{n=1}^{N} s_n \int_{-\pi}^{\pi}\sin nt\,\cos mt\,dt. \tag{2'}$$

Now, each integral on the right side of this equation is zero except that multiplying c_m because the relevant trigonometric functions are **orthogonal**[†] on $[-\pi, \pi]$; in fact, using standard trigonometric identities, it is easy to show that

$$\begin{aligned}\int_{-\pi}^{\pi}\sin nt\,\sin mt\,dt = \int_{-\pi}^{\pi}\cos nt\,\cos mt\,dt = 0, &\qquad m \ne n\\ \int_{-\pi}^{\pi}\sin nt\,\cos mt\,dt = 0, &\qquad \text{for all } m, n.\end{aligned} \tag{3}$$

(See Problem 1.2A.) Hence, when $m = 0$, since $\cos 0t = 1$, equation (2′) reduces to

$$\int_{-\pi}^{\pi} f(t)\,dt = \frac{c_0}{2}\int_{-\pi}^{\pi} dt = c_0\pi.$$

When $m = 1, 2, \ldots, N$, then (2′) reduces to

$$\int_{-\pi}^{\pi} f(t)\cos mt\,dt = c_m\int_{-\pi}^{\pi}\cos^2 mt\,dt = \frac{c_m}{2}\int_{-\pi}^{\pi}(1 + \cos 2mt)\,dt = c_m\pi.$$

In a similar manner we can show that

$$\int_{-\pi}^{\pi} f(t)\sin mt\,dt = s_m\pi. \tag{3'}$$

(See Problem 1.2C.)

When we combine these results and replace m by n, we obtain the following.

(1.1) Proposition: *If* $f(t) = (c_0/2) + \Sigma_{n=1}^{N}(c_n \cos nt + s_n \sin nt)$, $|t| \le \pi$, *then for* $n \le N$:

$$\left.\begin{matrix} c_n \\ s_n \end{matrix}\right\} = \frac{1}{\pi}\int_{-\pi}^{\pi} f(t)\left\{\begin{matrix}\cos nt\\ \sin nt\end{matrix}\right\}dt. \tag{4}$$

(continued)

[†] See the discussion in Section 0.1d.

Also,

$$\frac{1}{\pi}\int_{-\pi}^{\pi} f^2(t)\,dt = \frac{c_0^2}{2} + \sum_{n=1}^{N}(c_n^2 + s_n^2). \qquad \textbf{(4')}$$

Proof: For (4′), see Problem 1.2G. ■

Remarks: The coefficient formulas (4) are credited to Euler (1777) and Fourier (1807) who derived them in a related context. These formulas are important in several respects.

First, they establish our assertion that a given f has at most one representation of the form (2). For example, the right side of the identity $\sin^2 t = \frac{1}{2} - \frac{1}{2}\cos 2t$ cannot be replaced by any other linear combination of $\cos nt$ and $\sin nt$.

Second, they show that for each N, the $2N+1$ functions $1, \cos t, \cos 2t, \ldots, \cos Nt, \sin t, \ldots, \sin Nt$ are linearly independent. According to (4), their only linear combination f which is identically zero is that for which *all* coefficients are zero.

Third, for each given f, the coefficients can be determined *independently* and they characterize f. If both heart-action functions of Figure 1.1 admit finite representation in the form (2), then they can be compared through their respective coefficients.

Fourth, for each f, the values of the coefficients determined from (4) do not change with N. Therefore, f can be represented in form (2) only if the number of its coefficients with *nonzero values is finite.*

Example 1:

Figure 1.3b presents the graph of the periodic extension of $f(t) = |t|, |t| \le \pi$. Can this f be expressed in the form (2) for some finite integer N?

Consider the c_n integrals in (4).

$$c_0 = \frac{1}{\pi}\int_{-\pi}^{\pi} f(t)\,dt = \frac{1}{\pi}\left\{\int_{-\pi}^{0} -t\,dt + \int_0^{\pi} t\,dt\right\} = \pi. \qquad \textbf{(5a)}$$

However, since $\sin n\pi = 0$ and $\cos n\pi = (-1)^n$, then when $n \ge 1$:

$$c_n = \frac{1}{\pi}\int_{-\pi}^{\pi} f(t)\cos nt\,dt = \frac{1}{\pi}\left\{\int_{-\pi}^{0}(-t)\cos nt\,dt + \int_0^{\pi} t\cos nt\,dt\right\}$$

$$= \frac{1}{\pi}\left\{(-t)\frac{\sin nt}{n}\bigg|_{-\pi}^{0} + \int_{-\pi}^{0}\frac{\sin nt}{n}\,dt + t\frac{\sin nt}{n}\bigg|_0^{\pi} - \int_0^{\pi}\frac{\sin nt}{n}\,dt\right\}$$

$$= \frac{1}{\pi}\left\{\frac{-\cos nt}{n^2}\bigg|_{-\pi}^{0} + \frac{\cos nt}{n^2}\bigg|_0^{\pi}\right\} = \frac{2}{\pi}\frac{(-1)^n - 1}{n^2}, \qquad n = 1, 2, \ldots,$$

so that

$$c_n = \frac{2}{\pi}\frac{(-1)^n - 1}{n^2} = -\frac{4}{\pi n^2}\begin{cases} 0, & n = 2, 4, \ldots \\ 1, & n \text{ odd} \end{cases} \qquad \textbf{(5b)}$$

Since c_n is nonvanishing for every odd n, it follows that this function *cannot* have a finite representation in the form (2) for any integer N.

This conclusion could have been anticipated: A linear combination of the form

$$F_N(t) = \frac{c_0}{2} + \sum_{n=1}^{N}(c_n \cos nt + s_n \sin nt) \tag{6}$$

is smooth, even infinitely differentiable, as a function of t. Its graph must reflect this smoothness and cannot have corners or breaks such as those in Figure 1.3.

In Example 1, the coefficients $s_n = 0$, $n = 1, 2, \ldots$ (see Problem 1.2D), and if we insert the values for c_n found in (5a,b) into (6), we obtain the *sequence*

$$F_0(t) = \frac{\pi}{2}$$

$$F_1(t) = \frac{\pi}{2} - \frac{4}{\pi}\cos t$$

$$F_2(t) = \frac{\pi}{2} - \frac{4}{\pi}\cos t$$

$$F_3(t) = \frac{\pi}{2} - \frac{4}{\pi}\cos t - \frac{4}{\pi(3)^2}\cos 3t$$

$$\vdots$$

$$F_N(t) = \frac{\pi}{2} - \frac{4}{\pi}\sum_{n=1,3,\ldots}^{N}\frac{\cos nt}{n^2}.$$

The once laborious task of calculating enough values for such expressions to make an accurate graph can now be accomplished on desktop computers.† These graphs for $N = 3$, 7, and 9 are presented in Figure 1.4 (p. 24) in comparison to that of f. We see that F_3, F_7, and F_9 give successively better graphical approximations to f. We might conjecture that

$$f(t) = \lim_{N\to\infty} F_N(t),$$

but can this be true? Can smooth functions have jagged limits? Yes. In subsequent sections, we will see that when a periodic function f cannot be represented as some F_N, it can usually be *approximated* by these F_N in various theoretical senses.

Problem Set 1.2

1.2A *1.* Verify that formulas (3) are valid when $m = 0$ or $n = 0$.

2. Show that they hold otherwise by using trigonometric identities such as

$$2 \sin nt \sin mt = \cos (n - m)t - \cos (n + m)t.$$

1.2B When $n \neq 0$, use partial integrations to obtain formulas (3).

1.2C Replace $\cos mt$ by $\sin mt$ in (2′) and obtain (3′).

(*continued*)

† See the supplement provided with this text.

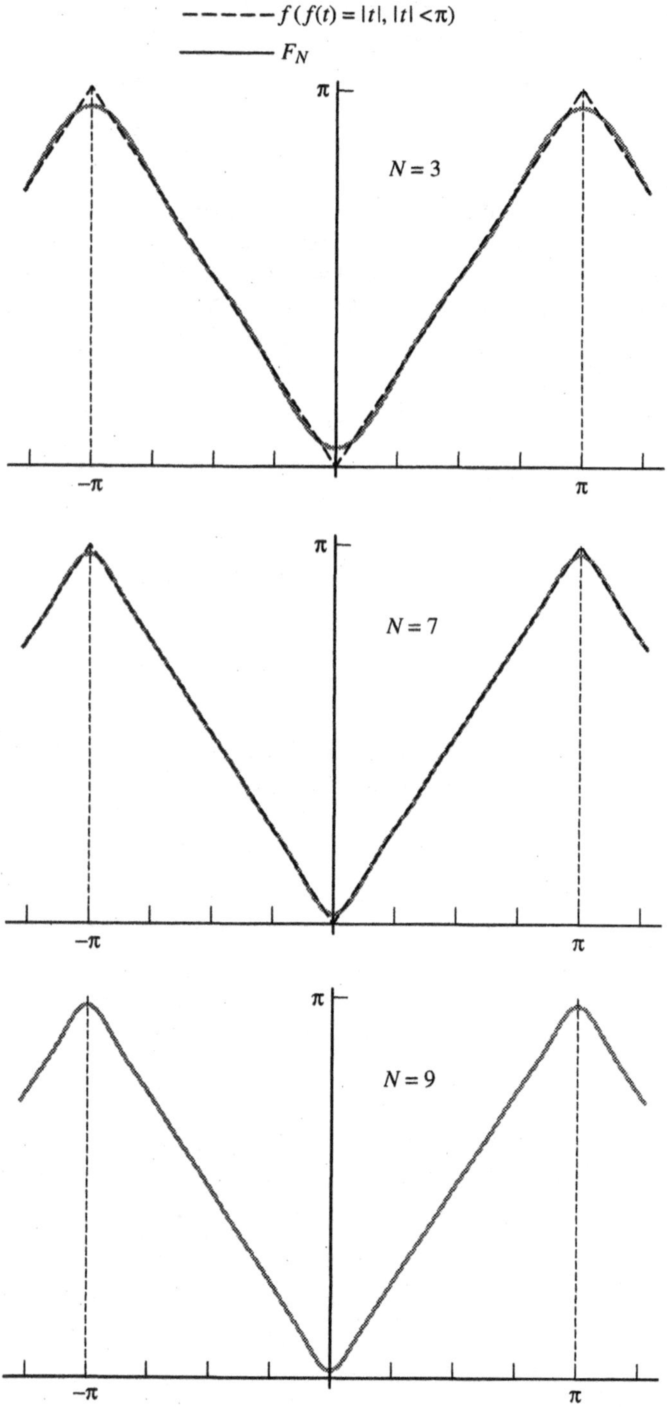

Figure 1.4

1.2D *1.* Verify that for $f(t) = |t|$ on $(-\pi, \pi)$, $s_n = 0$, $n = 1, 2, \ldots$

2. Give a graphical argument to explain this conclusion.

1.2E Could $t^2 = (c_0/2) + \Sigma_{n=1}^{N}(c_n \cos nt + s_n \sin nt)$ on $(-\pi, \pi]$ for any coefficients c_n, s_n, and any $N = 1, 2, \ldots$? Explain.

1.2F *1.* Verify that $\int_0^\pi \sin(nt/2)\sin(mt/2)\,dt = 0$ if m, n are distinct *even* integers or distinct *odd* integers.

2. Find $\int_0^\pi \sin^2(nt/2)\,dt$.

1.2G Multiply equation (2) by $f(t)$ and integrate to obtain (4′) in Proposition 1.1.

1.2H *1.* Represent $\sin^4 t$ in the form (2).

2. State whether the following is true or false: If functions f and $\tilde{f}$ both have finite representation, then their product $f\tilde{f}$ also has finite representation.

3. Represent $\sin^3 t$ in the form (2).

1.3

Fourier Series

Example 1 shows that representation of periodic functions by finite trigonometric sums need not be possible. At best, such smooth expressions as F_N in (6) approximate the sawtooth functions of Figure 1.3. We ask whether a given function f might admit representation in the *limiting series form*

$$f(t) = \lim_{N\to\infty} F_N(t) = \frac{c_0}{2} + \sum_{n=1}^{\infty}(c_n \cos nt + s_n \sin nt) \tag{7}$$

for the coefficients c_n, s_n, defined as before by

$$\left.\begin{matrix} c_n \\ s_n \end{matrix}\right\} = \frac{1}{\pi}\int_{-\pi}^{\pi} f(t)\left\{\begin{matrix} \cos nt \\ \sin nt \end{matrix}\right\}dt, \qquad n = 0, 1, 2, \ldots . \tag{8}$$

We assume that f is integrable on $[-\pi, \pi]$ and observe that the values of the coefficients are not affected by changing the values of f at a *finite* set of points in this interval.

With these coefficients, the formal series in (7) is known as the **Fourier series**† (of period 2π) **generated by** f, and each of its terms is called a **Fourier component** of f. Since we do not yet know whether this series converges and if so, in what sense it

† Although some cases of these series had been examined previously by Euler and by D. Bernoulli, it was the French mathematician Joseph Fourier (1768–1830) who first made systematic use of the general series in his efforts to solve problems in heat conduction. (See Chapter 3.) However, Fourier and his predecessors lacked the analytical techniques (then being developed by Cauchy) to resolve the questions of convergence of such series, and his methods were initially criticized. In fact, the efforts to understand the behavior of these series motivated the introduction and/or clarification of many important mathematical concepts including function, continuity and limit, convergence, integration, and measure.

represents f, we denote this relationship as follows:

$$f(t) \sim F(t) = \frac{c_0}{2} + \sum_{n=1}^{\infty}(c_n \cos nt + s_n \sin nt).$$

Functions f exist for which the formal series $F(t)$ fails to converge at many values of t, but fortunately such functions are rather complicated and do not arise in simple applications. (But see Problem 1.8E.) On the other hand, the series can converge at t, with a sum $F(t) \neq f(t)$, and this can occur with elementary functions.

Example 2:

When $f(t) = e^t$, $-\pi < t < \pi$, then from (8), f has the Fourier coefficients

$$c_0 = \frac{1}{\pi}\int_{-\pi}^{\pi} e^t\,dt = \frac{e^{\pi} - e^{-\pi}}{\pi} = \frac{2\sinh\pi}{\pi},$$

and, for $n = 1, 2, \ldots$,

$$\left.\begin{matrix} c_n \\ s_n \end{matrix}\right\} = \frac{1}{\pi}\int_{-\pi}^{\pi} e^t \begin{Bmatrix} \cos nt \\ \sin nt \end{Bmatrix} dt = \frac{2\sinh\pi(-1)^n}{\pi(1+n^2)}\left\{\begin{matrix} 1 \\ -n \end{matrix}\right.$$

These results can be obtained from partial integrations or with the aid of any integral table.

In this case, we see that *on the interval* $(-\pi, \pi)$,

$$e^t \sim F(t) = \frac{2}{\pi}\sinh\pi\left\{\tfrac{1}{2} + \sum_{n=1}^{\infty}(-1)^n\frac{(\cos nt - n\sin nt)}{1+n^2}\right\}, \qquad |t| < \pi.$$

Now, if e^t and $F(t)$ agree on this interval, they *cannot* agree for $|t| > \pi$ since e^t is strictly increasing, while F has period 2π. In fact, there are only a few values of t (such as 0, $\pm\pi$) for which it is evident that the last series converges and none for which the sum is easily computed. Moreover, since $F(-\pi) = F(\pi)$, while $e^{-\pi} \neq e^{\pi}$, it is difficult to believe that this series could represent e^t. We can appreciate the misgivings of some earlier mathematicians. However, we get some encouragement from the graph of the partial sum F_{12} shown in Figure 1.5 because as you see, it does resemble the graph of $f(t) = e^t$ on $(-\pi, \pi)$, except for the strange oscillations near $\pm\pi$. Should those oscillations be there, or is this evidence of computer malfunction? ■

When f is a polynomial, the calculation of its Fourier coefficients is facilitated by use of the following result (attributed to *Kronecker*):

If p_m is a polynomial of degree $\leq m$, and $g \in C[a, b]$, then

$$\int_a^b p_m(t)g(t)\,dt = [p_m G - p_m' G_1 + p_m'' G_2 + \cdots + (-1)^m p_m^{(m)} G_m](t)\,\Big|_a^b, \qquad \textbf{(8')}$$

where G is an integral of g, G_1 is an integral of G, and so on, and the formula is obtained by successive partial integrations.

For example, if $f(t) = t^2$, $|t| \leq \pi$, then from (8) its nth sine coefficient is

Graph of F_{12} ($f(t) = e^t$, $|t| < \pi$)

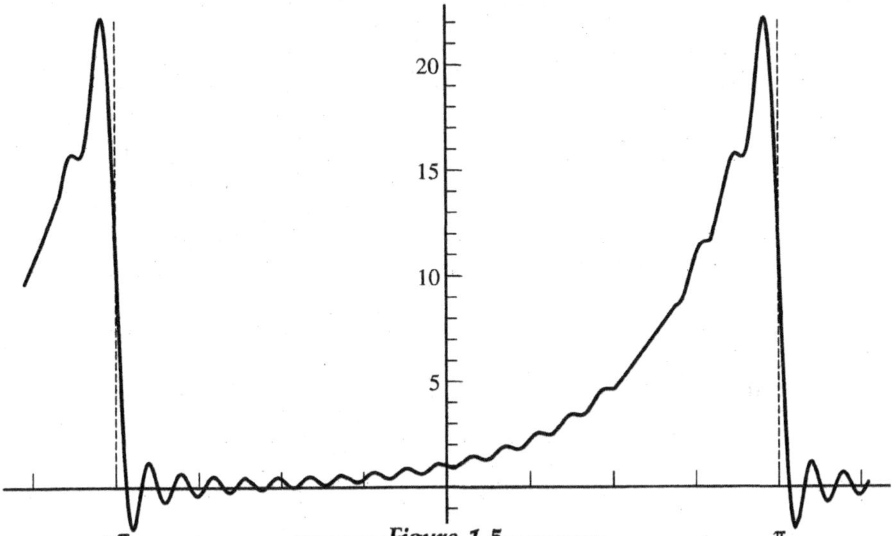

Figure 1.5

$s_n = (1/\pi)\int_{-\pi}^{\pi} t^2 \sin nt\, dt$. To use (8′) we take $p_2(t) = t^2$ and $g(t) = \sin nt$. Then

Repeat　　　　Differentiate

$$s_n = \frac{1}{\pi}\int_{-\pi}^{\pi} t^2(\sin nt)\, dt = \frac{1}{\pi}\left[t^2\left(-\frac{\cos nt}{n}\right) - 2t\left(-\frac{\sin nt}{n^2}\right) + 2\left(\frac{\cos nt}{n^3}\right)\right]_{-\pi}^{\pi},$$

Integrate

where the arrows indicate the preferred order of computation. If we now substitute the values $t = \pm\pi$ and subtract, we find that $s_n = 0$. In the next section we will see that this outcome was predictable without any calculation. However, $c_n \neq 0$. (See Problem 1.3C.)

In this text, unless specified otherwise, the Fourier coefficients of a function f on $(-\pi, \pi)$ will be c_n and s_n as defined in (8), while those of another function $\tilde{f}$ on $(-\pi, \pi)$ will be denoted $\tilde{c}_n$ and $\tilde{s}_n$, respectively. The generated Fourier series will have period 2π. Corresponding series with periods other than 2π are examined in Section 1.6.

Convergence and the M-Test

Since Fourier series are just a special type of series of functions, questions about convergence naturally arise. When we say that a given series of real- or complex-valued functions $\Sigma_{n=1}^{\infty} f_n(t)$ converges on a set T, we mean that for each $t \in T$ its sequence of partial sums

$$F_N(t) = \sum_{n=1}^{N} f_n(t), \qquad N = 1, 2, \ldots$$

has a *finite* limit $F(t)$ as $N \to \infty$. For this to occur at each $t \in T$ the error made in approximating $F(t)$ by $F_N(t)$ — namely,

$$E_N(t) \stackrel{\text{def}}{=} |F(t) - F_N(t)|$$

must be as small as we wish provided that N is large enough. In this case, we say that the series converges *pointwise* on T and has sum F. Comparison provides a simple test for convergence. For example, the series above converges at t, if each $|f_n(t)| < 1/n^2$ since $\Sigma_{n=1}^{\infty}(1/n^2) < +\infty$.

Now in principle, we can use the known partial sums F_N of a convergent series of functions to make inferences about the sum F, if we have suitable control of the error function E_N defined above. For useful control, the simplest conditions are those that make the error function E_N *uniformly* small on T, so that F_N is *uniformly* close to F. In this case, we say that the series converges **uniformly** on the set T, and, as you might expect, we get such convergence when the terms of the series are uniformly small enough. Uniform comparison is suggested and it results in the **M-test:**

If on the set T each $|f_n(\cdot)| \le M_n$, $n = 1, 2, \ldots$ and $\Sigma_{n=1}^{\infty} M_n < +\infty$, then the series $\Sigma_{n=1}^{\infty} f_n(t)$ converges uniformly on T.

To understand why a uniformly convergent series of functions is nicer than a nonuniformly convergent one, consider the following facts.

(1) *A uniformly convergent series of continuous functions is continuous; that is,* Limit(Σ) = Σ(Limit).

(2) *On a bounded set T, a uniformly convergent series of integrable functions is integrable, and* Integral(Σ) = Σ(Integral).

(3) *A uniformly convergent series of differentiable functions is differentiable when the series of derivatives also converges uniformly, and then* Derivative(Σ) = Σ(Derivative).

Each of these assertions involves an interchange of limiting operations (indicated symbolically) that might not be permissible without uniform convergence. For further discussion of this limit interchange principle of uniform convergence along with a proof of the M-test, see Section 0.3.

Problem Set 1.3

1.3A For each of the following functions f given on $(-\pi, \pi)$, use formula (8) to calculate the Fourier coefficients and write the resulting Fourier series F:

1. $f(t) = 1$
2. $f(t) = \begin{cases} 0, & \text{for } t < 0 \\ 1, & \text{for } t \ge 0 \end{cases}$
3. $f(t) = t(\pi - t)$
4. $f(t) = \frac{1}{2}t^2 - 1$
5. $f(t) = \begin{cases} 0, & \text{for } -\pi < t < 0 \\ t, & \text{for } 0 \le t < \pi \end{cases}$
6. $f(t) = \sin(t - \pi/6)$

7. $f(t) = \sin t \cos 2t$
8. $f(t) = |\sin t|$
9. $f(t) = e^t$ (Verify the integration results obtained in Example 2.)
10. $f(t) = e^{|t|}$
11. $f(t) = \cos^2 t$
12. $f(t) = \cos^3 t$. (*Hint:* Use a trigonometric identity.)

1.3B 1. If f is integrable on $(-\pi, \pi)$ with Fourier coefficients c_n, s_n and if $\tilde{f}$ is integrable on $(-\pi, \pi)$ with Fourier coefficients $\tilde{c}_n$, $\tilde{s}_n$, show that for any constants α and β, the Fourier coefficients for $\alpha f + \beta\tilde{f}$ are $\alpha c_n + \beta\tilde{c}_n$ and $\alpha s_n + \beta\tilde{s}_n$, respectively.

2. Conclude that
   a. the Fourier series of αf is that of f multiplied by the constant α.
   b. the Fourier series of $f + \tilde{f}$ is that of f plus that of $\tilde{f}$.

1.3C Use Kronecker's formula (8′) to compute c_n for $f(t) = t^2$ and $f(t) = t^3$.

1.4

Symmetry Considerations

Before examining the general situation for Fourier series, let's look at two special cases — those in which only cosine terms are present and those in which only sine terms are present. These arise from functions which have the same parity (or symmetry) as the cosine and sine, respectively.

(a) Even and odd functions

(1.2) Definition: *Let f be a real-valued function on $(-a, a)$. We say that f is an* **even function** *if $f(-t) = f(t)$ for all $t \in (-a, a)$, and f is an* **odd function** *if $f(-t) = -f(t)$ for all $t \in (-a, a)$, or in either case, if the conditions are met except at a finite set of values of t.* ∎

For example, it is easy to verify that on any symmetric interval, the functions

$$f(t) = 3,\ |t|,\ t^4 - 2t^2,\ \cosh t,\ \text{and}\ \cos nt, \qquad n = 0, 1, 2, \ldots, \text{are } \textit{even},$$

while the functions

$$f(t) = t,\ \pi t,\ t^5 - 3t^3,\ \sinh t,\ \text{and}\ \sin nt, \qquad n = 0, 1, 2, \ldots, \text{are } \textit{odd}.$$

($f(t) = 0$ is both even and odd.)

In general, the graphs of even functions are symmetric with respect to the vertical axis, while those of odd functions are symmetric with respect to the origin. On $(-\pi, \pi)$, a sample of each type is sketched in Figure 1.6. Observe that an extension of period 2π will retain the type of the original function, independently of the values assigned at $+\pi$ or $-\pi$.

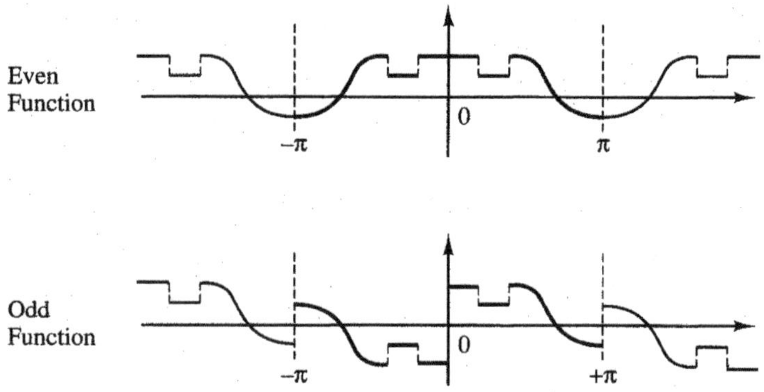

Figure 1.6

Products of even and odd functions obey the following rules.

Even × *Even*	and	*Odd* × *Odd*	are	*Even*.
Odd × *Even*	and	*Even* × *Odd*	are	*Odd*.

With the substitutions (+) for even and (−) for odd, these rules are easily remembered through the laws governing products of signs.

(1.3) Lemma: *Suppose that f is integrable on $[-a, a]$, for some $a > 0$.*

$$\text{If } f \text{ is even, then} \qquad \int_{-a}^{a} f(t)\,dt = 2\int_0^a f(t)\,dt.$$

$$\text{If } f \text{ is odd, then} \qquad \int_{-a}^{a} f(t)\,dt = 0.$$

Proof: These assertions follow from the fact that in either case

$$\int_{-a}^{a} f(t)\,dt = \int_{-a}^{0} f(t)\,dt + \int_0^a f(t)\,dt.$$

When f is even, the signed area represented by the integral over $[-a, 0]$ equals that over $[0, a]$, while when f is odd these areas have unlike signs and cancel. (See Figure 1.6.) ■

For example, the product $\sin nt \cos mt$ is odd and this gives a simple proof that

$$\int_{-\pi}^{\pi} \sin nt \cos mt\,dt = 0, \qquad \text{for all } m, n.$$

Similarly, $|t| \cos nt$ is even, hence,

$$\frac{1}{\pi}\int_{-\pi}^{\pi} |t| \cos nt\,dt = \frac{2}{\pi}\int_0^{\pi} t \cos nt\,dt.$$

Obviously, "most" functions are neither even nor odd, as evidenced by $f(t) = e^t$, among others. However,

$$e^t = \frac{e^t + e^{-t}}{2} + \frac{e^t - e^{-t}}{2},$$

$$\text{or} \qquad e^t = \cosh t + \sinh t,$$

in terms of the standard hyperbolic functions. Thus e^t can be expressed as the **sum of the even function** $\cosh t$ **and the odd function** $\sinh t$. A similar decomposition is possible for *any* real-valued function f on $(-a, a)$; viz.,

$$f(t) = f_E(t) + f_O(t), \tag{9}$$

where

$$f_E(t) \stackrel{\text{def}}{=} \frac{f(t)+f(-t)}{2} \text{ is even,} \qquad \text{and} \qquad f_O(t) \stackrel{\text{def}}{=} \frac{f(t)-f(-t)}{2} \text{ is odd.} \tag{9'}$$

Indeed,

$$f_E(-t) = \frac{f(-t)+f(t)}{2} = f_E(t) \qquad \text{and} \qquad f_O(-t) = \frac{f(-t)-f(t)}{2} = -f_O(t).$$

f_E and f_O are called the **even** and **odd parts** of f, respectively.

(b) Cosine and sine series: Now we can obtain the desired results.

(1.4) Proposition: *Let f be integrable over $(-\pi, \pi)$.*

If f is even, then $$f(t) \sim \frac{c_0}{2} + \sum_{n=1}^{\infty} c_n \cos nt,$$

where $$c_n = \frac{2}{\pi}\int_0^{\pi} f(t)\cos nt\, dt, \qquad n = 0, 1, 2, \ldots .$$

If f is odd, then $$f(t) \sim \sum_{n=1}^{\infty} s_n \sin nt,$$

where $$s_n = \frac{2}{\pi}\int_0^{\pi} f(t)\sin nt\, dt, \qquad n = 1, 2, \ldots .$$

Proof: Both assertions are immediate consequences of Lemma 1.3 and we establish only the second.

If f is odd, then the product $f(t)\cos nt$ is also odd, and hence by (8),

$$c_n = \frac{1}{\pi}\int_{-\pi}^{\pi} f(t)\cos nt\, dt = 0, \qquad n = 0, 1, 2, \ldots .$$

Similarly, $f(t)\sin nt$ is even, so that

$$s_n = \frac{1}{\pi}\int_{-\pi}^{\pi} f(t)\sin nt\, dt = \frac{2}{\pi}\int_0^{\pi} f(t)\sin nt\, dt, \qquad n = 1, 2, \ldots . \quad \blacksquare$$

Example 3:

$f(t) = |t|$ on $(-\pi, \pi)$ is even so that each sine coefficient s_n is zero. Moreover, we have

$$c_0 = \frac{2}{\pi}\int_0^{\pi} t\, dt = \frac{2}{\pi}\frac{\pi^2}{2} = \pi,$$

while from (5b), for $n = 1, 2, \ldots$

$$c_n = \frac{2[(-1)^n - 1]}{\pi n^2} = \begin{cases} 0, & n \text{ even} \\ -\dfrac{4}{\pi n^2}, & n \text{ odd.} \end{cases} \tag{10}$$

Thus on $(-\pi, \pi)$

$$|t| \sim \frac{\pi}{2} + \sum_{n=1}^{\infty} \frac{2}{\pi} \frac{(-1)^n - 1}{n^2} \cos nt$$

$$= \frac{\pi}{2} - \frac{4}{\pi} \sum_{n=1,3,}^{\infty} \frac{1}{n^2} \cos nt.$$

(Some authors make the substitution $n = 2k - 1$ and rewrite the last series in the form $\sum_{k=1}^{\infty} \frac{\cos(2k-1)t}{(2k-1)^2}$. Others replace $n = 1, 3,$ by n_{odd}.)

Example 4:

$f(t) = t$ on $(-\pi, \pi)$ is odd and it generates a sine series with the coefficients

$$s_n = \frac{2}{\pi} \int_0^{\pi} t \sin nt \, dt = \frac{2}{\pi} \left\{ -t \frac{\cos nt}{n} \Big|_0^{\pi} + \int_0^{\pi} \frac{\cos nt}{n} dt \right\}$$

$$= \frac{2}{\pi} \left(-\frac{\pi}{n} (-1)^n + \frac{\sin nt}{n^2} \Big|_0^{\pi} \right) = \frac{2(-1)^{n+1}}{n}, \qquad n = 1, 2, \ldots . \tag{11}$$

Thus on $(-\pi, \pi)$: $t \sim 2 \sum_{n=1}^{\infty} \frac{(-1)^{n+1}}{n} \sin nt.$

If f is given on $(-\pi, \pi)$, then its even part f_E generates the cosine series terms in the Fourier series of f, while its odd part f_O generates the sine series terms. This follows immediately when f is replaced by

$$f(t) = f_E(t) + f_O(t)$$

in the coefficient formulas. (See Problem 1.4M.)

(c) Even and odd extensions: When a function f on $(-\pi, \pi)$ is either even or odd, then to calculate its surviving Fourier coefficients, it is necessary only to know the values of the function on $(0, \pi)$. Conversely, if a function f is given initially only on $(0, \pi)$, we can consider its extension to $(-\pi, 0)$ in many ways, and each extension will produce its own Fourier series of period 2π.

If we make $f(-t) = f(t)$ for $-\pi < t < 0$, we obtain the **even extension** of f, which generates a cosine series. If we make $f(-t) = -f(t)$ for $-\pi < t < 0$, we obtain the **odd extension** of f, which generates a sine series. Other extensions will generate Fourier series that have both sine and cosine terms.

Certain applications require either a cosine series or a sine series; these series are associated with the even or odd periodic extension of f and that which is selected affects the behavior and value of the series at $0, \pi$. This is already evident from our previous examples, each of which represents one extension of $f(t) = t$ on $(0, \pi)$. The even extension was treated in Example 3, and we now see that the resulting series,

$$F(t) = \frac{\pi}{2} - \frac{4}{\pi} \sum_{n=1,3,}^{\infty} \frac{\cos nt}{n^2},$$

converges uniformly for all t by the M-test since

$$\left| \frac{\cos nt}{n^2} \right| \leq \frac{1}{n^2} \quad \text{and} \quad \sum_{n=1}^{\infty} \frac{1}{n^2} < +\infty.$$

Hence its sum is a continuous function by the limit interchange arguments of the previous section.

The odd extension of the same function was considered in Example 4. In this case, it is not obvious that the associated series,

$$F(t) = 2 \sum_{n=1}^{\infty} \frac{(-1)^{n+1}}{n} \sin nt,$$

actually converges except at $t = 0, \pm\pi, \pm 2\pi, \ldots$ where its value must be zero. If it does converge to $f(t) = t$ on $(-\pi, \pi)$, it will provide the odd extension of this f by automatically assigning the endpoint values $f(0) = f(\pi) = 0$. In particular, it cannot represent either of the limiting values of f at $\pm\pi$ considered previously but tries to average them. As we shall discover in Section 1.5, this behavior is characteristic near a simple discontinuity of the periodic extension.

Example 5:

$f(t) = t$ on $(0, \pi)$ can be extended to $(-\pi, 0)$ by defining $f(t) = 0$ there. When further extended periodically, the graph is as shown in Figure 1.7.

This extension is neither even nor odd, and as anticipated it generates the full Fourier series with coefficients

$$\left. \begin{matrix} c_n \\ s_n \end{matrix} \right\} = \frac{1}{\pi} \int_0^{\pi} t \left\{ \begin{matrix} \cos nt \\ \sin nt \end{matrix} \right\} dt,$$

which are precisely one-half of the corresponding ones for the previous extensions. This is not accidental since

$$f(t) = \tfrac{1}{2}(|t| + t), \qquad |t| < \pi.$$

Each of these extensions has base period 2π. However, if we define $f(t) = t + \pi$ on $(-\pi, 0)$, then the resulting extension has period π.

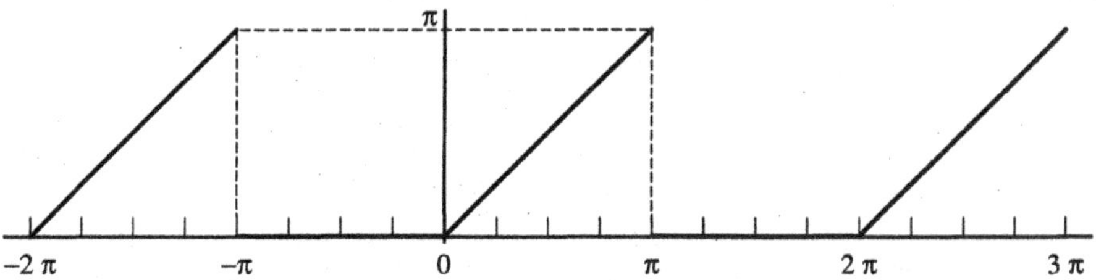

Figure 1.7

(d) Continuity at 0: When a function f defined on $(-a, a)$ is continuous at 0, it follows easily that both its even and odd parts are also continuous at 0. Indeed,

$$\lim_{t\to 0} f_E(t) = f(0), \qquad \text{while} \qquad \lim_{t\to 0} f_O(t) = 0.$$

However, suppose f has a simple (or "jump") discontinuity at 0 with distinct limiting values $f(0+)$ and $f(0-)$ from the right and left, respectively. Then f_O will exhibit a similar discontinuity, but f_E will be **effectively** continuous at 0 with limiting value

$$f_E(0) = \lim_{t\to 0} \frac{f(t) + f(-t)}{2} = \frac{f(0+) + f(0-)}{2}.$$

This observation will aid our discussion of convergence in Section 1.8.

Example 6:

The function

$$f(t) = \begin{cases} \pi, & -\pi < t \le 0 \\ t, & 0 < t \le \pi, \end{cases}$$

whose 2π-periodic extension is shown in Figure 1.8, has a simple discontinuity at 0 with $f(0+) = 0$ and $f(0-) = \pi$. Its even part $f_E(t) = \dfrac{\pi + |t|}{2}$ is effectively continuous there with limiting value $f_E(0) = \pi/2 = \dfrac{f(0+) + f(0-)}{2}$, as is evident graphically.

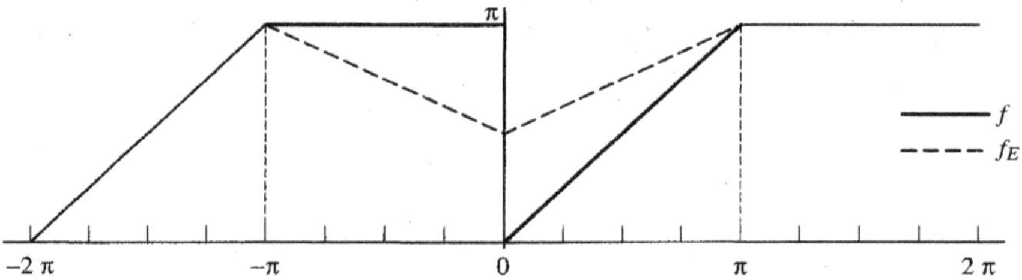

Figure 1.8

Problem Set 1.4

1.4A Classify each of the following functions on $\mathbb{R}$ as even, odd, or neither.

1. $f(t) = 1$ *2.* $g(t) = \sin^3 t$ *3.* $h(t) = t^3 - 2t$

4. $\phi(t) = t^2 + 3t$ *5.* $\psi(t) = (t^2 - 1)\sin t$ *6.* $G(t) = (e^t + 1)(e^{-t} + 1)$

7. $H(t) = \begin{cases} -1, & \text{for } t \le 0 \\ 1, & \text{for } t > 0 \end{cases}$ *8.* $\Phi(t) = \sin t + \cos t$

9. $\Psi(t) = \begin{cases} \arcsin t, & |t| < 1 \\ 0, & \text{otherwise} \end{cases}$ *10.* $\Upsilon(t) = \begin{cases} \dfrac{1}{1+t}, & t \ne -1 \\ 0, & t = -1 \end{cases}$

1.4B Which of the functions in Problem 1.4A, restricted to the interval $(-\pi, \pi)$, generates a sine series? a cosine series?

1.4C Find the even and odd parts of each of the functions in Problem 1.4A.

1.4D Prove that on a common interval $(-a, a)$

1. the product of two odd functions is an even function.
2. the product of two even functions is an even function.
3. the product of an even function and an odd function is an odd function.

1.4E Obtain the Fourier series F generated by f defined on $(-\pi, \pi)$ when

1. $f(t) = t^3$
2. $f(t) = t^4$
3. $f(t) = \begin{cases} -1, & t < 0 \\ 1, & t \geq 0 \end{cases}$
4. $f(t) = \begin{cases} 1, & |t| \leq \pi/2 \\ 0, & |t| > \pi/2 \end{cases}$
5. $f(t) = \begin{cases} t^2, & t \geq 0 \\ 0, & t < 0 \end{cases}$
6. $f(t) = \cos^2 t$

1.4F Graph the 2π-periodic extension of each of the functions in Problem 1.4E, assumed defined on $(-\pi, \pi]$, and indicate all points of discontinuity.

1.4G Assume each function in Problem 1.4E is defined only on $(0, \pi)$. For each:

1. Graph the odd periodic extension and obtain a sine series of period 2π.
2. Graph the even periodic extension and obtain a cosine series of period 2π.

1.4H For $f(t) = \pi - t$, $0 < t < \pi$, find

1. a sine series (with period 2π).
2. a cosine series (with period 2π).

1.4I For $f(t) = t(\pi - t)$, $0 < t < \pi$, find

1. a sine series (with period 2π).
2. a cosine series (with period 2π).

1.4J 1. Find a sine series for $f(t) = \cos t$ on $0 < t < \pi$.
2. Find a cosine series for $g(t) = \sin t$ on $0 < t < \pi$.

1.4K Suppose that $f(t)$ is 2π periodic and also satisfies $f(t + \pi) = -f(t)$.

1. Show that the Fourier coefficients c_n, s_n are zero if n is even.
2. Show that when n is odd, then

$$c_n = \frac{2}{\pi}\int_0^{\pi} f(t) \cos nt \, dt, \qquad s_n = \frac{2}{\pi}\int_0^{\pi} f(t) \sin nt \, dt.$$

1.4L 1. If g is integrable on $(0, \pi)$ where $g(t) = g(\pi - t)$, show that g generates a sine series of the form $\Sigma_{n=1,3,}^{\infty} s_n \sin nt$, with $s_n = (4/\pi) \int_0^{\pi/2} g(t) \sin nt \, dt$, $n = 1, 3, 5, \ldots$.
(*Hint:* Make a substitution to establish that $\int_{\pi/2}^{\pi} g(\pi - t) \sin nt \, dt = (-1)^{n+1} \int_0^{\pi/2} g(t) \sin nt \, dt$.)
2. What similar condition on g assumed integrable on $(0, \pi)$ ensures that g generates a cosine series of the form $\Sigma_{n=1,3,}^{\infty} c_n \cos nt$, with $c_n = (4/\pi) \int_0^{\pi/2} g(t) \cos nt \, dt$, $n = 1, 3, 5, \ldots$?

3. Observe that each of these conditions provides a means of extending g defined only on $(0, \pi/2)$.

1.4M Verify that f_E generates the cosine series terms of the Fourier series for an integrable f on $(-\pi, \pi)$. Make a similar deduction for f_O and use the results of Example 2 to find the Fourier series of $\cosh t$ and $\sinh t$.

1.4N 1. Show that f is even on a given interval $(-a, a)$ if and only if $2f(t) = f(t) + f(-t)$ on this interval.

2. Using part 1, show that an even polynomial p can contain only even powers of t.

1.4O 1. When $f \in C^1(-a, a)$, show that f is even if and only if f' is odd.

2. Is the preceding statement true after interchange of the words *odd* and *even*? Justify your assertions.

1.5

Convergence Results and Applications

Each function f that is (Riemann) integrable on $[-\pi, \pi]$ generates the formal Fourier series

$$F(t) = \frac{c_0}{2} + \sum_{n=1}^{\infty}(c_n \cos nt + s_n \sin nt),$$

with coefficients given by (8). This series may or may not converge and, as we have seen, need not have the value $f(t)$. In fact, total recovery of a function f from its *discrete sequence* of Fourier coefficients in this manner permits us to characterize the function with less information than is required to store its set of values. For irregular functions, such as those describing heart activity or seasonal weather patterns, this informational storage aspect of convergence is important and will be pursued in Section 1.7 through numerical approximations. Then, in Section 1.8 we will develop enough theoretical tools to supply conditions for convergence of a Fourier series, both at a single point and uniformly on the period interval. However, before embarking on the theoretical journey, let's use this section to examine its chief ports of call indicated in Theorem I (page 38) and Theorem II (page 41). A review of the discussion of convergence in Section 1.3 may clarify the issues.

Sectionally Smooth Functions

The presentation of convergence results is simpler for periodic functions f that are "smooth" except possibly at a finite set of points in each closed and bounded interval $[a, b]$. To identify such functions, we must describe their class more accurately.

(1.5) Definition: *If f is continuously differentiable (C^1) on a bounded open interval (a, b), we say that $f \in C^1[a, b]$ when* **both** *f and its derivative f' have continuous extensions to $[a, b]$; that is, each function has a finite one-sided limit from within (a, b) at a and at b. The endpoint limits of f are denoted $f(a+)$ and $f(b-)$.*

A function f is **sectionally smooth** *on $[a, b]$ when there is a* **finite partition** $a = t_0 < t_1 < \cdots < t_K = b$, *for which $f \in C^1[t_{k-1}, t_k]$, $k = 1, 2, \ldots, K$. If f is* **also continuous** *on $[a, b]$, we say that $f \in \hat{C}^1[a, b]$; then $f(a+) = f(a)$ and $f(b-) = f(b)$.* ■

If $f \in C^1[a, b]$, then at each $t \in (a, b)$, the graph of f has a tangent line with a well-defined slope that varies smoothly with t. Moreover, at a and b, this slope has limiting values that provide f with corresponding one-sided *derivatives*. These are given by l'Hôpital's rule as

$$f'_+(a) \stackrel{\text{def}}{=} \lim_{t \searrow a} \frac{f(t) - f(a+)}{t - a} = \lim_{t \searrow a} \frac{f'(t)}{1}$$

$$\text{and} \qquad f'_-(b) \stackrel{\text{def}}{=} \lim_{t \nearrow b} \frac{f(t) - f(b-)}{t - b} = \lim_{t \nearrow b} \frac{f'(t)}{1}.$$

Each function considered in the examples in this chapter is sectionally smooth on each interval $[a, b]$ of definition, and Figure 1.9a presents the graph of a sectionally smooth function exhibiting most of the features permitted.

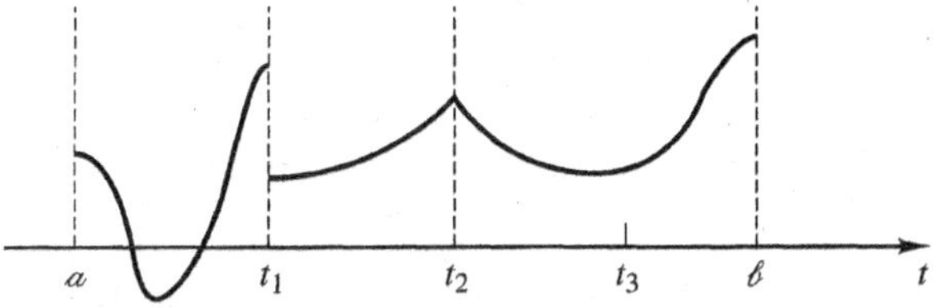

Figure 1.9(a)

This function f is not in $C^1[a, b]$ because it has a simple jump discontinuity at t_1 and because its derivative has such a discontinuity at t_2, as evidenced by the "corner" in the graph. However, f is in $C^1[a, t_1]$, $C^1[t_1, t_2]$, and $C^1[t_2, b]$, and thus f is sectionally smooth on $[a, b]$. Addition of the partition point t_3 would not affect this conclusion. This f is not continuous on $[a, b]$, but it is continuous on $[t_1, b]$; therefore, $f \in \hat{C}^1[t_1, b]$.

In general, if f is sectionally smooth on $[a, b]$, we can suppose that at each intermediate partition point t_k, either f or f' has a jump discontinuity; that is, each function has limits from the right and from the left, and these limits are distinct for f or f' (or both). At the endpoints a and b, both functions have one-sided limits from within (a, b) as in Definition 1.5. At the partition points, including the endpoints, values of f and f' need not be assigned, and if present, they can be disregarded. However, at all other points both f and f' are well defined and both functions are Riemann integrable over $[a, b]$. Finite sums and products of sectionally smooth functions are sectionally smooth on the same interval.

Now, let's summarize the principal results of this section and see how they are applied.

(1.6) Theorem I: *If f has period 2π and is sectionally smooth on $[-\pi, \pi]$ with Fourier coefficients $c_n, s_n, n = 0, 1, 2, \ldots$, then*

(A) $$F(t) \stackrel{\text{def}}{=} \frac{c_0}{2} + \sum_{n=1}^{\infty}(c_n \cos nt + s_n \sin nt) = \frac{f(t+) + f(t-)}{2}, \qquad t \in \mathbb{R}$$

$$(= f(t), \text{ if } f \text{ is continuous at } t).$$

(B) On each closed interval that excludes the points of jump discontinuity in f, the series in part A converges uniformly.

(C) $$\frac{c_0^2}{2} + \sum_{n=1}^{\infty}(c_n^2 + s_n^2) = \frac{1}{\pi}\int_{-\pi}^{\pi} f^2(t)\, dt. \qquad \textit{(Parseval's formula)}$$

■

Example 7:

If we set $a = -\pi$ and $b = \pi$, then the sectionally smooth function of Figure 1.9a has a periodic extension f fulfilling the hypotheses of Theorem I. From part A, we see that the Fourier series of f converges everywhere and the graph of its sum F coincides with that of f *except* possibly at the partition points and their periodic counterparts. At each of these exceptional points, F has the value (shown by a cross in Figure 1.9b) that *averages the one-sided limiting values of f.* At $t = -\pi$ and at $t = \pi$, F has the value $[f(-\pi) + f(\pi)]/2$ determined by considering the periodic extension of f. Since F has the same jump discontinuities that this periodic extension has, the series cannot converge uniformly on the interval $[t_2, \pi]$; however, by part B of Theorem I, it must converge uniformly on the interval $[t_2, t_3]$ shown in Figure 1.9b.

Thus, if we can graph a periodic sectionally smooth function f, we can graph F, the sum of its Fourier series, and distinguish intervals of uniform (and non-uniform) convergence.

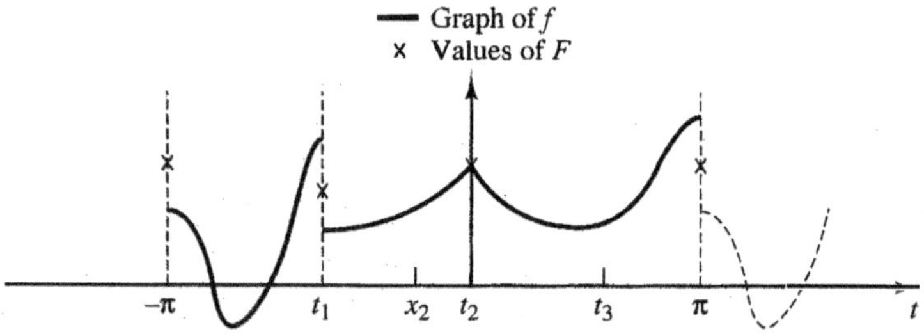

Figure 1.9(b)

Example 8:

The function

$$f(t) = \begin{cases} \pi, & \text{on } (-\pi, 0) \\ t, & \text{on } (0, \pi) \end{cases}$$

considered in Example 6 extends periodically as shown in Figure 1.8 (page 34). It is sectionally smooth on $[-\pi, \pi]$ and so its Fourier series converges everywhere, with sum F, where

$$F(t) = f(t), \qquad \text{for } 0 < |t| \le \pi, \qquad \text{but} \qquad F(0) = \pi/2.$$

Since this function is neither odd nor even we must expect a full series with both sine and cosine terms present.

The coefficients are calculated as follows utilizing (10) and (11) as needed.

$$c_0 = \frac{1}{\pi}\int_{-\pi}^{0} \pi\,dt + \frac{1}{\pi}\int_{0}^{\pi} t\,dt = \frac{3\pi}{2}$$

$$c_n = \frac{1}{\pi}\int_{-\pi}^{0} \pi \cos nt\,dt + \frac{1}{\pi}\int_{0}^{\pi} t\cos nt\,dt = \frac{\sin nt}{n}\Big|_{-\pi}^{0} + \frac{(-1)^n - 1}{\pi n^2}$$

$$= \frac{(-1)^n - 1}{\pi n^2}, \qquad n = 1, 2, \ldots$$

$$s_n = \frac{1}{\pi}\int_{-\pi}^{0} \pi \sin nt\,dt + \frac{1}{\pi}\int_{0}^{\pi} t\sin nt\,dt = -\frac{\cos nt}{n}\Big|_{-\pi}^{0} + \frac{(-1)^{n+1}}{n}$$

$$= \left(\frac{-1 + (-1)^n}{n} + \frac{(-1)^{n+1}}{n}\right) = -\frac{1}{n}.$$

Thus, we have that

$$\frac{3\pi}{4} + \sum_{n=1}^{\infty}\left(\left[\frac{(-1)^n - 1}{\pi n^2}\right]\cos nt - \frac{1}{n}\sin nt\right) = \begin{cases} \pi, & \text{on } [-\pi, 0) \\ \pi/2, & \text{at } 0 \\ t, & \text{on } (0, \pi] \end{cases}$$

Graphs of the corresponding finite sums

$$F_N(t) = \frac{3\pi}{4} + \sum_{n=1}^{N}\left(\left[\frac{(-1)^n - 1}{\pi n^2}\right]\cos nt - \frac{1}{n}\sin nt\right)$$

are presented in Figure 1.10 for values of $N = 5$, 11, and 31. As expected, the

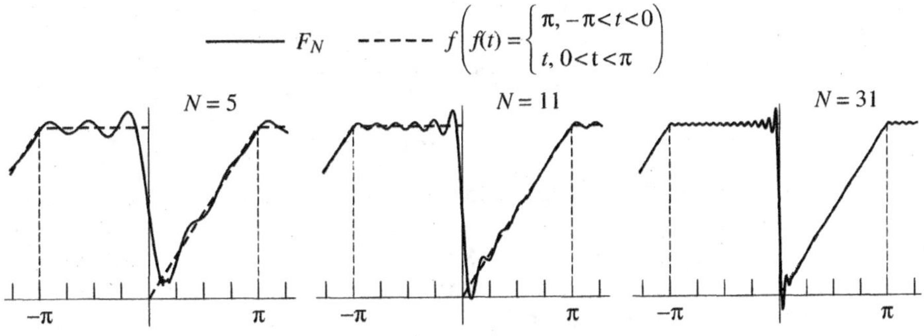

Figure 1.10

approximation improves with increasing N, and when $N = 31$, the corner point behavior at $t = \pi$ is modeled very well, while $F_{31}(0) \approx \pi/2$ as required. However, the most striking feature of the graph of F_{31} is the manner in which it oscillates near $t = 0$. As we shall see in Section 1.8d, this behavior is characteristic of Fourier series approximations for a function near a point of jump discontinuity; it *cannot* be eliminated by increasing the value of N. This behavior, called Gibbs' phenomenon, is a consequence of the fact that the Fourier series converges *nonuniformly* near such points.

Parseval's Formula

We have just seen how to use parts A and B of Theorem I. Now let's look at part C. Parseval's formula represents the limiting form of the result in (4′), and it holds for a large class of functions, including all those that are Riemann integrable on $[-\pi, \pi]$. To better appreciate its significance, suppose that f represents the time history of a signal of period 2π. Then $(1/2\pi)\int_{-\pi}^{\pi} f^2(t)\,dt$ can be interpreted as the **mean energy per cycle contained in the signal**. In this sense, the signal $s_n \sin nt$ has mean energy $(s_n^2/2\pi)\int_{-\pi}^{\pi} \sin^2 nt\,dt = s_n^2/2$. Similarly, $c_n \cos nt$ has mean energy $c_n^2/2$, when $n = 1, 2, \ldots$, while the constant signal $c_0/2$ has mean energy $c_0^2/4$. Thus, on division by 2, Parseval's formula indicates that the energy of f can be obtained by simply adding the energies of the separate Fourier components of f. This is another consequence of orthogonality of the trigonometric functions over the period interval. The basis for a geometric interpretation of Parseval's formula is discussed in Section 0.1e.

From Parseval's formula, we see that if *all* of the Fourier coefficients of some function f vanish, then

$$0 \le P(x) \stackrel{\text{def}}{=} \int_{-\pi}^{x} f^2(t)\,dt \le \int_{-\pi}^{\pi} f^2(t)\,dt = 0, \qquad x \in [-\pi, \pi].$$

Hence $P = 0$, and near each point x where f is continuous, $P'(x) = f^2(x) = 0$, (by the Fundamental Theorem of Calculus). Thus we conclude that

If f is continuous on $[-\pi, \pi]$, then $f \equiv 0$ when all of its Fourier coefficients vanish.

This conclusion would not hold if we eliminate any of the coefficients from consideration. For this reason, the trigonometric functions $\{1, \sin nt, \cos nt, n = 1, 2, \ldots\}$ are said to be **complete** on $[-\pi, \pi]$ (or on any translate of this interval).

Integration

Termwise integration of the Fourier series of a sectionally smooth periodic f is always possible over any bounded interval $[a, b]$, and the resulting sum is $\int_a^b f(t)\,dt$. This assertion follows from Theorem 0.5 if the series converges uniformly on $[a, b]$, but using Parseval's formula it can be extended to any periodic f that is Riemann integrable over $[-\pi, \pi]$. (See Problem 1.5F.) For example, $f(t) = t$ on $(-\pi, \pi)$ extended periodically gives the odd function of Example 4. Its Fourier series is

$$2\sum_{n=1}^{\infty}\frac{(-1)^{n+1}}{n}\sin nt = t, \qquad |t| < \pi, \tag{12}$$

and we can integrate this series termwise over $[0,\pi]$ to get

$$2\sum_{n=1}^{\infty}\frac{(-1)^n}{n}\frac{\cos nt}{n}\bigg|_0^{\pi} = 4\sum_{n=1,3,}^{\infty}\frac{1}{n^2} = \int_0^{\pi} t\,dt = \frac{\pi^2}{2} \qquad \text{or} \qquad \sum_{n=1,3,}^{\infty}\frac{1}{n^2} = \frac{\pi^2}{8}.$$

Differentiation

Differentiation is trickier than integration because termwise differentiation of a convergent series results in a series that need not converge! In fact, if we differentiate the series in (12) termwise, we get the new series $2\Sigma_{n=1}^{\infty}(-1)^{n+1}\cos nt$, which diverges at $t = 0$ even though $f(t) = t$ is infinitely differentiable on $(-\pi,\pi)$. Observe that f does *not* have a continuous extension of period 2π.

(1.7) Theorem II: *If f of period 2π is continuous on $\mathbb{R}$ and sectionally smooth on $[-\pi,\pi]$, then*

(D) *$f'(t) = \Sigma_{n=1}^{\infty} n(-c_n \sin nt + s_n \cos nt)$, at each t where $f''(t)$ exists.*

(E) *The Fourier series for f converges absolutely and uniformly, and we have the error estimate*

$$|f(t) - F_N(t)| \leq \sum_{n=N+1}^{\infty}\sqrt{c_n^2 + s_n^2} \leq \frac{1}{\sqrt{N\pi}}\left(\int_{-\pi}^{\pi}[f'(\tau)]^2\,d\tau\right)^{1/2}, \qquad t \in \mathbb{R},$$

where $$F_N(t) = \frac{c_0}{2} + \sum_{n=1}^{N}(c_n\cos nt + s_n \sin nt), \qquad N = 1,2,\ldots .$$ ■

Remarks: An f both continuous and sectionally smooth on $[-\pi,\pi]$ is in the class $\hat{C}^1[-\pi,\pi]$, and it has a continuous periodic extension iff $f(-\pi) = f(\pi)$. For $f \in \hat{C}^1[-\pi,\pi]$ with $f(-\pi) \neq f(\pi)$, a modified version of part D exists. (See Proposition 1.13.)

Example 9:

The function $f(t) = t^2$, $|t| \leq \pi$, has a continuous periodic extension, whose graph is shown in Figure 1.11. $f \in C^1[-\pi,\pi]$ and it fulfills the hypothesis of Theorem II (and so of Theorem I). Furthermore, f is even so that by part A,

$$f(t) = \frac{c_0}{2} + \sum_{n=1}^{\infty} c_n \cos nt = t^2, \qquad |t| \leq \pi.$$

When $|t| < \pi$, $f'(t) = 2t$ and $f''(t) = 2$. Hence, by part D,

$$2t = \sum_{n=1}^{\infty} -nc_n \sin nt, \qquad |t| < \pi.$$

However, from (12), we also have

$$\sum_{n=1}^{\infty}\frac{4(-1)^{n+1}}{n}\sin nt = 2t, \qquad |t| < \pi.$$

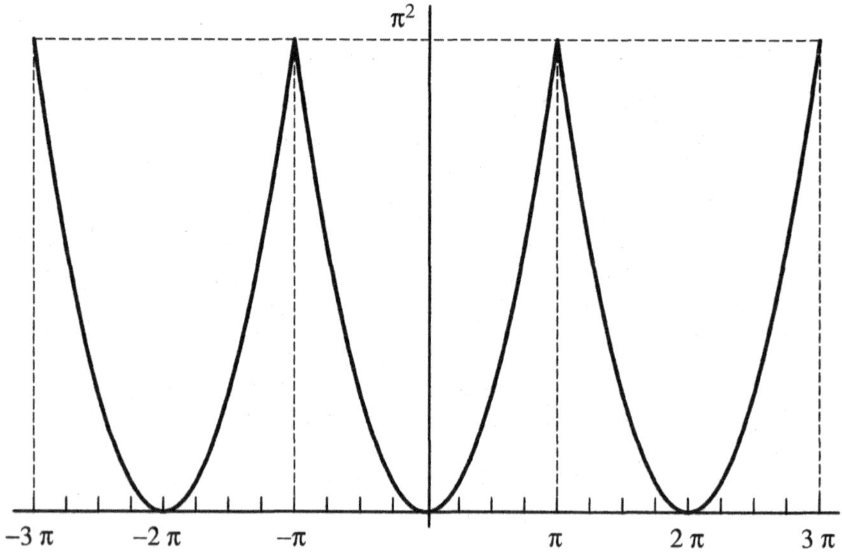

Figure 1.11

Therefore, we conjecture that $c_n = 4(-1)^n/n^2$, $n = 1, 2, \ldots$ and this can be verified. Clearly,

$$c_0 = \frac{2}{\pi}\int_0^{\pi} t^2\, dt = \frac{2}{3}\pi^2,$$

and when these coefficients are inserted, we obtain the Fourier series for f. Therefore,

$$f(t) = \frac{\pi^2}{3} + 4\sum_{n=1}^{\infty}\frac{(-1)^n}{n^2}\cos nt = t^2, \qquad |t| \leq \pi. \tag{13}$$

Evaluation at $t = \pi$ where $\cos nt = (-1)^n$ gives the series

$$4\sum_{n=1}^{\infty}\frac{1}{n^2} = \pi^2 - \frac{\pi^2}{3} \qquad \text{or} \qquad \sum_{n=1}^{\infty}\frac{1}{n^2} = \frac{\pi^2}{6}, \tag{13$'$}$$

whose partial sums may be used to approximate π^2 (and thus π). For this purpose, we have the error estimate

$$\left|\frac{\pi^2}{6} - \sum_{n=1}^{N}\frac{1}{n^2}\right| = \sum_{n=N+1}^{\infty}\frac{1}{n^2} \leq \int_N^{\infty}\frac{dt}{t^2} = \frac{1}{N}, \qquad N = 1, 2, \ldots, \tag{13$''$}$$

which reveals that we must sum $N = 100$ terms to be within $1/100$ of $\pi^2/6$. (A more efficient approximation is afforded through Parseval's formula. See Problem 1.5C).

Similar estimates are available at other values of t in the form

$$\left|f(t) - \left(\frac{\pi^2}{3} + 4\sum_{n=1}^{N}\frac{(-1)^n}{n^2}\cos nt\right)\right| = \left|4\sum_{n=N+1}^{\infty}\frac{(-1)^n}{n^2}\cos nt\right|$$

$$\leq 4\sum_{n=N+1}^{\infty}\frac{1}{n^2} \leq \frac{4}{N}, \qquad N = 1, 2 \ldots .$$

Since the right side is independent of t, we have a *uniform estimate* of the error made in approximating this f by a *finite* partial sum of its Fourier series. This error approaches zero uniformly as $N \to \infty$, which means that the Fourier series converges uniformly. Part E of Theorem II provides similar uniform estimates for any $f \in \hat{C}^1[-\pi, \pi]$ with $f(-\pi) = f(\pi)$.

Problem Set 1.5

1.5A The following functions f defined on $(-\pi, \pi)$ are extended to $\mathbb{R}$ with period 2π.

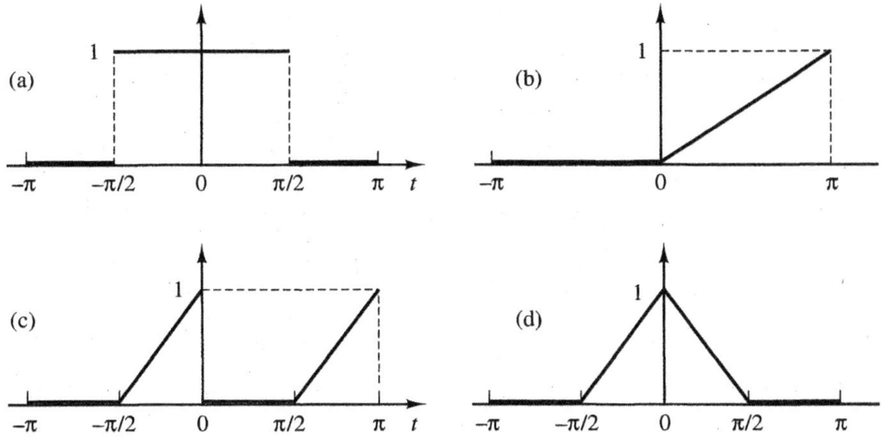

1. For each of the above graphs, give the value of the Fourier series F at each point $t \in [-\pi, \pi]$ where $F(t) \neq f(t)$.
2. Graph the Fourier series generated by the above functions.

1.5B Where, and to which values, will the Fourier series for each of the functions in Problem 1.4E converge on $[-\pi, \pi]$?

1.5C 1. Evaluate $\Sigma_{n=1}^{\infty}(-1)^n/n^2$. (*Hint:* See Example 9 and (13').)

2. Show that $f(t) = (t^3 - \pi^2 t)$ defined on $(-\pi, \pi)$ generates the Fourier sine series $12\Sigma_{n=1}^{\infty}(-1)^n(\sin nt/n^3)$. Evaluate it at $t = \pi/2$; at $t = \pi/4$.
3. Show that $\Sigma_{n=1}^{\infty} 1/n^4 = \pi^4/90$. (*Hint:* Use Parseval's formula.)

1.5D 1. According to part B of Theorem I, for which functions in Problem 1.4E will the Fourier series converge uniformly on $[0, \pi/2]$?

2. Give the same information for each of the functions in Problem 1.3A but on the interval $[-\pi, \pi]$.
3. Apply Parseval's formula (part C of Theorem I) to the functions of Examples 1, 2, and 4 and write the results.

1.5E 1. Use the Fourier series for $f(t) = \cosh t$ obtained in Problem 1.4M to derive that of $\sinh t$ on $(-\pi, \pi)$. Can the series for $\sinh t$ be differentiated to recover that of $\cosh t$? Explain.

2. Give Parseval's formula for each function.

1.5F *1.* Suppose that both f and $\tilde{f}$ of Problem 1.3B satisfy the conditions of Theorem I. Prove that

$$\frac{1}{\pi}\int_{-\pi}^{\pi} f(t)\tilde{f}(t)\,dt = \frac{c_0\tilde{c}_0}{2} + \sum_{n=1}^{\infty}(c_n\tilde{c}_n + s_n\tilde{s}_n).$$

(*Hint:* $(f+\tilde{f})^2 - (f-\tilde{f})^2 = 4f\tilde{f}$.)

2. The formula in part 1 holds when f and $\tilde{f}$ are just Riemann integrable on $[-\pi, \pi]$, and in particular, when

$$\tilde{f}(t) = \begin{cases} 1, & a \le t \le x \\ 0, & \text{otherwise} \end{cases}$$

for $[a, x] \subseteq [-\pi, \pi]$. Conclude that for such f,

$$g(x) \stackrel{\text{def}}{=} \int_a^x f(t)\,dt = \frac{c_0}{2}(x-a) + \sum_{n=1}^{\infty}\left(c_n\int_a^x \cos nt\,dt + s_n\int_a^x \sin nt\,dt\right),$$

whether or not the Fourier series for f converges.

1.5G *1.* Apply part D of Theorem II to the result of Example 3 to obtain

$$\frac{2}{\pi}\sum_{n=1}^{\infty}\frac{1-(-1)^n}{n}\sin nt = \begin{cases} -1, & \text{if } -\pi < t < 0 \\ 0, & \text{if } t = 0 \\ 1, & \text{if } 0 < t < \pi \end{cases}.$$

2. Do the same for the result of Problem 1.4I 2 to obtain

$$2\sum_{n=1}^{\infty}\frac{1+(-1)^n}{n}\sin nt = \begin{cases} -\pi - 2t, & \text{if } -\pi < t < 0 \\ 0, & \text{if } t = 0 \\ \pi - 2t, & \text{if } 0 < t < \pi \end{cases}.$$

1.5H Show that $f(t) = |t|^{3/4}$ is continuous at $t = 0$ but is not sectionally smooth in any interval containing $t = 0$.

1.6

Scale Change and Translation

The results just obtained for Fourier series of functions with period 2π have corresponding versions for functions of period $L = 2\ell$, relative to any interval $[a, a + L]$.

Suppose first that $f = f(x)$ is defined and integrable on $[-\ell, \ell]$. Then the new function

$$g(t) \stackrel{\text{def}}{=} f\left(\frac{\ell}{\pi}t\right)$$

is defined and integrable on $[-\pi, \pi]$. Moreover, g has the same continuity and differentiability properties as f, at corresponding points t and $x = (\ell/\pi)t$.

Thus as a function of t,

$$f\left(\frac{\ell}{\pi}t\right) \sim \frac{c_0}{2} + \sum_{n=1}^{\infty}(c_n \cos nt + s_n \sin nt)$$

or as a function of x,

$$f(x) \sim \frac{c_0}{2} + \sum_{n=1}^{\infty}\left(c_n \cos\frac{n\pi x}{\ell} + s_n \sin\frac{n\pi x}{\ell}\right). \tag{14}$$

The coefficient integrals of (8) are transformed by the substitution

$$t = \frac{\pi}{\ell}x \qquad \left(dt = \frac{\pi}{\ell}dx\right),$$

and for $n = 0, 1, 2, \ldots$, we find that

$$\left.\begin{matrix} c_n \\ s_n \end{matrix}\right\} = \frac{1}{\pi}\int_{-\pi}^{\pi} f\left(\frac{\ell}{\pi}t\right)\left\{\begin{matrix} \cos nt \\ \sin nt \end{matrix}\right\}dt = \frac{1}{\ell}\int_{-\ell}^{\ell} f(x)\left\{\begin{matrix} \cos\dfrac{n\pi x}{\ell} \\ \sin\dfrac{n\pi x}{\ell} \end{matrix}\right\}dx. \tag{15}$$

Convergence results, approximations, and so on, have corresponding formulations. For example, Parseval's formula becomes

$$\frac{c_0^2}{2} + \sum_{n=1}^{\infty}(c_n^2 + s_n^2) = \frac{1}{\ell}\int_{-\ell}^{\ell} f^2(x)\,dx. \tag{16}$$

Note that both $\cos(n\pi x/\ell)$ and $\sin(n\pi x/\ell)$ have period 2ℓ, when $n = 0, 1, 2, \ldots$.

In specific applications, either set of integrals in (15) may be used to evaluate the coefficients, and any known results may be incorporated.

Example 10:

For $f(x) = |x|$, $|x| \le \ell$,

$$f\left(\frac{\ell}{\pi}t\right) = \frac{\ell}{\pi}|t|, \qquad |t| \le \pi.$$

However, from Example 3,

$$\frac{\ell}{\pi}|t| = \frac{\ell}{\pi}\left[\frac{\pi}{2} - \frac{4}{\pi}\sum_{n=1,3,}^{\infty}\frac{1}{n^2}\cos nt\right], \qquad |t| \le \pi.$$

Hence $$|x| = \frac{\ell}{2} - \frac{4\ell}{\pi^2}\sum_{n=1,3,}^{\infty}\frac{1}{n^2}\cos\frac{n\pi x}{\ell}, \qquad |x| \le \ell,$$

where we have used part A of Theorem I to obtain actual representation in the first series. The M-test ensures uniform convergence of either series.

Translation

Each function f on $(-\ell, \ell)$ has an extension to $\mathbb{R}$ of period $L = 2\ell$ such that

$$f(x + L) = f(x), \qquad x \in \mathbb{R}. \tag{17}$$

The previous results are readily translated to other intervals of length $L = 2\ell$ through the following.

(1.8) Lemma: *If f has period L and is integrable over $[0, L]$, then for each $a \in \mathbb{R}$:*

$$\int_0^L f(a + x)\,dx = \int_a^{a+L} f(s)\,ds = \int_0^L f(x)\,dx. \tag{18}$$

Proof: The substitution $s = a + x$ with $ds = dx$ transforms the first *definite* integral to the second. The second may be represented as the sum

$$\int_L^{a+L} f(s)\,ds + \int_a^L f(x)\,dx$$

since s, x are dummy variables. Then the substitution $s = x + L$ allows the first integral in this sum to be expressed as

$$\int_0^a f(x + L)\,dx = \int_0^a f(x)\,dx,$$

in view of (17). Finally, the sum in question becomes

$$\int_0^a f(x)\,dx + \int_a^L f(x)\,dx = \int_0^L f(x)\,dx. \quad \blacksquare$$

If we divide each integral in (18) by L, we conclude that the mean value per cycle of a periodic function is unaffected by the location of the period interval, or equivalently, by translation of the function.

Observe that when f has period $L = 2\ell$, then each x-integrand in (15) and (16) also has period L, and Lemma 1.8 applies. Therefore, we can move the limits of integration from $-\ell$ and ℓ to a and $b = a + 2\ell$, respectively, to get results such as the following.

(1.9) Proposition: *If f is sectionally smooth on $[a, b]$, and $\ell = (b - a)/2 > 0$, then*

$$f(x) \sim \frac{c_0}{2} + \sum_{n=1}^{\infty}\left(c_n \cos\frac{n\pi x}{\ell} + s_n \sin\frac{n\pi x}{\ell}\right)$$

where

$$\left.\begin{matrix} c_n \\ s_n \end{matrix}\right\} = \frac{1}{\ell}\int_a^b f(x)\left\{\begin{matrix} \cos\dfrac{n\pi x}{\ell} \\ \sin\dfrac{n\pi x}{\ell} \end{matrix}\right\}dx, \qquad n = 0, 1, 2, \ldots . \tag{19}$$

The Fourier series has sum $[f(x+) + f(x-)]/2$ at each $x \in [a, b]$ if $f(a-) \stackrel{def}{=} f(b-)$ and $f(b+) \stackrel{def}{=} f(a+)$. If $f \in \hat{C}^1[a, b]$ and $f(a) = f(b)$, then the Fourier series converges uniformly on $[a, b]$ with sum f.

Proof: The convergence assertions are restatements from part A of Theorem I and part E of Theorem II. Recall that $f \in \hat{C}^1[a, b]$ when f is *both* continuous and sectionally smooth on $[a, b]$. $\blacksquare$

Warning: Unless $[a, b] = [-\ell, \ell]$, symmetry of f with respect to the midpoint of the interval $[a, b]$ does not imply that all $c_n = 0$ or that all $s_n = 0$ in (19).

Problem Set 1.6

1.6A Obtain the Fourier series of period 2 for each of the following functions defined on $(-1, 1)$ and indicate its sum on $[-1, 1]$ by a graph.

1. $f(x) = x$
2. $f(x) = |\sin x|$
3. $f(x) = \begin{cases} 0, & \text{if } x > 0 \\ -1, & \text{if } x \leq 0 \end{cases}$
4. $f(x) = x - x^2$
5. $f(x) = \cosh x$
6. $f(x) = \cos x$, $x > 0$ extended as an odd function.
7. $f(x) = x^3$, $x > 0$ extended as an even function.

1.6B Assume that each function in Problem 1.6A is defined on $(-1, 2)$. Indicate the sum of its Fourier series of period 3 by a graph.

1.6C For $f(x) = x(\pi - x)$, $0 < x < 1$, find

1. a sine series with period 2.
2. a cosine series with period 2.
3. a full Fourier series with period 1.

1.6D For $f(x) = \pi - x$, $0 < x < 2$, find

1. a sine series with period 4.
2. a cosine series with period 4.
3. a full Fourier series with period 2.

1.6E 1. Show that the function $f(x) = x$, $0 \leq x \leq \pi$ has an *odd-half-sine* series representation of the form

$$f(x) = \sum_{n=1,3,}^{\infty} s_n \sin\frac{nx}{2}, \qquad 0 \leq x \leq \pi$$

where $$s_n = \frac{2}{\pi}\int_0^{\pi} f(x) \sin\frac{nx}{2}\,dx, \qquad n = 1, 3, 5, \ldots .$$

(*Hint:* Let $x = 2t$, define $g(t) = f(2t)$, $0 \leq t \leq \pi/2$, and extend g as in Problem 1.4L.)

2. Obtain an *odd-half-cosine* series representation for $f(x) = \sin x$, $0 \leq x \leq \pi$.

1.7

Numerical Approximation

If f is integrable on $[a, b]$, then the set of coefficients $\{c_n, s_n\}_{n=0}^{\infty}$ defined in (19) provides a **spectral analysis** of f considered as a signal. The process of recovering

f from these coefficients in the form of a convergent Fourier series is called **spectral synthesis**.

When a periodic function is obtained experimentally, perhaps from an electrocardiogram, its coefficients cannot be determined exactly. Even if f can be precisely specified, it is seldom possible to determine the coefficients exactly from (19) since most functions do not have elementary integrals. Hence, we must resort to numerical approximation of the coefficient integrals, and in this text we consider only the simplest equally spaced rectangular approximations. Whatever method is used to determine the coefficient integrals, we still cannot calculate more than a finite number of coefficients, while the estimate in part E of Theorem 1.7 shows that uniform approximation of f within 1/100 requires roughly 10,000 exact coefficients. Thus we have two tasks, one theoretical and the other practical. The first is to determine how many approximate coefficients are required to achieve results of specified accuracy. The second is to expedite the calculation of large numbers of approximate coefficients. In this section, we present a few relevant facts that lead to the fast Fourier transform.

Approximation

Suppose f of period 2π is sectionally smooth on $[0, 2\pi]$. If N is given and we divide $[0, 2\pi]$ into N parts of equal length $t_1 = 2\pi/N$, by points $t_k = kt_1$, $k = 0, 1, 2, \ldots, N$ (as shown in Figure 1.12), then we have the following rectangular approximations to the coefficient integrals.†

$$\left.\begin{matrix} c_n \\ s_n \end{matrix}\right\} = \frac{1}{\pi}\int_0^{2\pi} f(t)\left\{\begin{matrix} \cos nt \\ \sin nt \end{matrix}\right\}dt \approx \frac{2}{N}\sum_{k=0}^{N-1} f(kt_1)\left\{\begin{matrix} \cos nkt_1 \\ \sin nkt_1 \end{matrix}\right\} = \left\{\begin{matrix} \tilde{c}_n \\ \tilde{s}_n \end{matrix}\right. . \qquad \textbf{(20)}$$

We take $n \le N' \le N$, where N' is selected later. There is redundancy in the sums defining the approximate coefficients $\tilde{c}_n$ and $\tilde{s}_n$ since

$$\tilde{c}_{N-n} = \tilde{c}_n \qquad \text{while} \qquad \tilde{s}_{N-n} = -\tilde{s}_n, \qquad n \le N. \qquad \textbf{(20')}$$

(See Problem 1.7B.) Therefore, we should not expect the approximations in (20) to be equally good for all values of $n \le N$. In fact, they favor the lesser values of n.

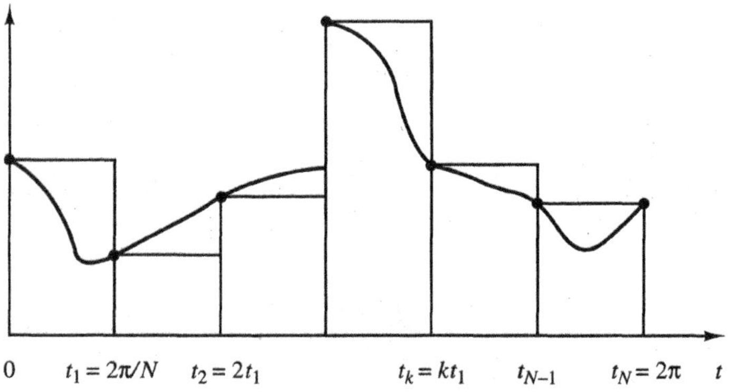

Figure 1.12

† At a point t where f is discontinuous, f is evaluated as $f(t+)$.

Example 11:

$$f(t) = \begin{cases} t, & 0 \le t \le \pi \\ 2\pi - t, & \pi \le t \le 2\pi \end{cases}$$

extended with period 2π is the function examined in Example 3 of Section 1.4. The exact nonvanishing coefficients are

$$c_0 = \pi \qquad \text{and} \qquad c_n = \frac{-4}{\pi n^2}, \qquad \text{if } n = 1, 3, 5, \ldots .$$

If we use (20) with $N = 2$, then

$$t_0 = 0 \qquad \text{and} \qquad t_1 = \pi$$

and the corresponding values $f_k = f(t_k)$ are

$$f_0 = 0 \qquad \text{and} \qquad f_1 = \pi.$$

Thus $\quad \tilde{c}_0 = \frac{2}{2}(f_0 + f_1) = \pi = \tilde{c}_2,$

and $\quad \tilde{c}_1 = (f_0 - f_1) = -\pi$

while $\quad \tilde{s}_1 = (f_0 \cdot 0 + f_1 \cdot 0) = 0.$

Of these, only $\tilde{c}_0$ and $\tilde{s}_1$ can be considered useful approximations.

When $N = 4$: then $\quad t_0 = 0,\ t_1 = \frac{\pi}{2},\ t_2 = \pi, \quad$ and $\quad t_3 = \frac{3\pi}{2},$

while $\quad f_0 = 0,\ f_1 = \frac{\pi}{2},\ f_2 = \pi, \quad$ and $\quad f_3 = \frac{\pi}{2}.$

Hence

$$\tilde{c}_0 = \tfrac{2}{4}(f_0 + f_1 + f_2 + f_3) = \pi\ (= \tilde{c}_4)$$

$$\tilde{c}_1 = \tfrac{1}{2}(f_0 + f_1 \cdot 0 - f_2 + f_3 \cdot 0) = -\frac{\pi}{2}\ (= \tilde{c}_3)$$

$$\tilde{c}_2 = \tfrac{1}{2}(f_0 - f_1 + f_2 - f_3) = 0$$

while $\quad \tilde{s}_1 = \tfrac{1}{2}(f_0 \cdot 0 + f_1 + f_2 \cdot 0 - f_3) = 0\ (= -\tilde{s}_3)$

and $\quad \tilde{s}_2 = \tfrac{1}{2}(f_0 \cdot 0 + f_1 \cdot 0 + f_2 \cdot 0 + f_3 \cdot 0) = 0.$

Of these, we see that $\tilde{c}_4$ and $\tilde{c}_3$ are poor approximations.

The situation encountered with this simple example is not unusual. In general, when $n > N/2$ formulas (20) do not yield useful approximations. For this reason, we shall take $N' \le N/2$ and consider the approximation to $F_{N'}$ given by

$$\tilde{F}_{N'}(t) = \frac{\tilde{c}_0}{2} + \sum_{n=1}^{N'} (\tilde{c}_n \cos nt + \tilde{s}_n \sin nt). \tag{21}$$

When $N = 2N' + 1$, it is not hard to show that

$$\tilde{F}_{N'}(t_j) = f(t_j) = f(jt_1), \qquad j \le N, \tag{22}$$

so that the infinitely differentiable trigonometric function $\tilde{F}_{N'}$ can be used to interpolate values for f between those "sampled" to form the approximation. Alternatively, we see that this last formula recovers the $2N'+1$ sampled values of f from the $2N'+1$ components of the approximate coefficients $\tilde{c}_0, \tilde{c}_1, \tilde{s}_1, \ldots, \tilde{c}_{N'}, \tilde{s}_{N'}$. (When N is even, the correspondence is less balanced.) (See Problem 1.7E.)

Example 12:

The function $f(t) = 1/(8-t)$, $t \in [0, 2\pi)$ does not admit exact calculation of its Fourier coefficients. However, if we use formula (20) with $N = 3$ (so $N' = 1$), then $t_0 = 0$, $t_1 = 2\pi/3$, $t_2 = 4\pi/3$, and the corresponding values $f_k = f(kt_1)$ are

$$f_0 = .125, \qquad f_1 = .169, \qquad f_2 = .262.$$

Thus

$$\tilde{c}_0 = \tfrac{2}{3}(f_0 + f_1 + f_2) = .371$$

$$\tilde{c}_1 = \tfrac{2}{3}(f_0 - \tfrac{1}{2} f_1 - \tfrac{1}{2} f_2) = -.061$$

$$\tilde{s}_1 = \tfrac{2}{3}\left(f_0 \cdot 0 + \frac{\sqrt{3}}{2} f_1 - \frac{\sqrt{3}}{2} f_2\right) = -.0537.$$

Hence,

$$\tilde{F}_1(t) = (.37/2) + (-.06)\cos t + (-.05)\sin t.$$

When $N = 5$, the corresponding values are calculated in Problem 1.7C, and the results for $N = 3$, 5, and 51 are shown in Figure 1.13.

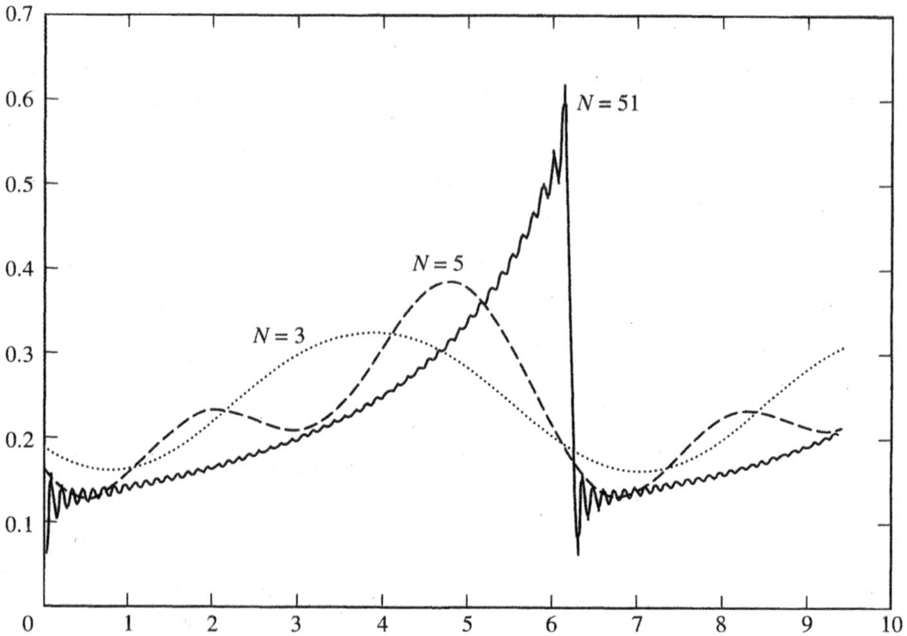

Figure 1.13

Graphs of $f(t) = \dfrac{1}{8-t}$, $0 < t < 2\pi$ and F_N for $N = 3, 5$, and 51.

The figure shows that these approximations also exhibit Gibbs' behavior at the discontinuities of the *periodic extension* of f. Moreover, we still want theoretical assurance that as $N \to +\infty$, the approximations $\tilde{F}_N$ do converge to f. (See Problem 1.7G.)

FFT: The Fast Fourier Transform

The practical problem of evaluating approximate Fourier coefficients has been of concern since the time of Fourier, and by 1900 there was a mechanical device for approximating the coefficients when $n \leq 100$. (To explain an apparent malfunction of this early computer, Gibbs was led to recognize how Fourier series behave near a jump discontinuity.) However, good approximation to irregular data requires coefficients for $n > 1000$, and it was not until 1960 that Cooley and Tukey found an effective means of computing them. The resulting "fast Fourier transforms," or FFTs, have revolutionized the applications of Fourier analysis. To appreciate why, let's look at the underlying algorithm (which has been traced back to Gauss in 1805).

It will be simpler if we first use Euler's formula, $\cos x + i \sin x = e^{ix}$, to combine the approximating sums of (20) into the equivalent *complex* form

$$\tilde{e}_n = \tilde{c}_n + i\tilde{s}_n = \frac{2}{N}\sum_{k=0}^{N-1} f(kt_1)e^{inkt_1}, \qquad n \leq N. \tag{23}$$

f is real valued so that $\tilde{e}_{N-n} = \bar{\tilde{e}}_n$ is the complex conjugate of $\tilde{e}_n$.

Now, if N is replaced by $2N$, then t_1 is replaced by $T_1 = t_1/2$ and the associated approximate coefficients $\tilde{E}_n$ are given by

$$\tilde{E}_n = \frac{2}{2N}\sum_{j=0}^{2N-1} f(jT_1)e^{injT_1}, \qquad n \leq 2N.$$

If we collect the terms where $j = 2k$ and those where $j = 2k + 1$, we see that

$$2\tilde{E}_n = \frac{2}{N}\sum_{k=0}^{N-1} f(kt_1)e^{inkt_1} + \frac{W^n 2}{N}\sum_{k=0}^{N-1} f(kt_1 + T_1)e^{inkt_1}.$$

Here $W = e^{i\pi/N}$ so that $W^N = -1$. For $n \leq N$, the first sum is just $\tilde{e}_n$ and the last,

$$\tilde{e}'_n = \frac{2}{N}\sum_{k=0}^{N-1} f\left(kt_1 + \frac{t_1}{2}\right)e^{inkt_1}, \tag{24}$$

is similar to $\tilde{e}_n$. Also, $e^{ikNt_1} = (e^{i2\pi})^k = 1$ when $k \leq N$, and we conclude that

$$2\tilde{E}_n = \tilde{e}_n + W^n\tilde{e}'_n \qquad \text{and} \qquad 2\tilde{E}_{N+n} = \tilde{e}_n - W^n\tilde{e}'_n, \qquad n \leq N. \tag{25}$$

These equations provide the computer algorithm for an FFT.

We can assume that the amount of computer time required to perform a complex addition (or subtraction) is negligible compared to that required for a complex multiplication. If (23) is used, then $(N + 1)N$ multiplications are required to handle N data points, but (25) offers substantial improvement. To understand how this follows, observe that if at most m complex multiplications are required to compute $\tilde{e}_n$ for all $n \leq N$, then according to (25) we require no more than $M = 2m + 2N$ multiplications to obtain $2\tilde{E}_n$ for all $n \leq 2N$. When $N = 2$, only additions are required, and when $N = 2^r$, it follows (by induction on r) that we can obtain all $\tilde{e}_n$ for $n \le N$ with no more than $(r+1)N$ multiplications. (See Problem 1.7F.) Now, $r = \log_2 N$ increases slowly with N, and for large $N = 2^r$ the FFT requires only $(\log_2 N)/N$ as much computer time as (23). By this assessment, when $r = 10$ so $N = 2^{10} = 1024$, the FFT is approximately 100 times faster!

For each N, the set of approximate coefficients $\tilde{e}_n$, $n = 0, 1, 2, \ldots, N$, is called the Nth order **discrete Fourier transform of** f. Conversely, if these $\tilde{e}_n$ are known, then formula (23) can be inverted to recover the sampled values

$$f(jt_1) = \frac{1}{2}\sum_{n=0}^{N-1} \tilde{e}_n e^{-injt_1}, \qquad j = 1, 2, \ldots, N \tag{26}$$

from which they were obtained. [Indeed, the last sum, computed from (23), is, after interchange,

$$\frac{1}{N}\sum_{k=0}^{N-1} f(kt_1) \sum_{n=0}^{N-1} e^{in(k-j)t_1};$$

but when $k = j$, the *inner* sum is $\Sigma_{n=0}^{N-1} 1 = N$, while when $k \neq j$, this inner sum is zero!]

Clearly the sums in (26) can again be calculated from an algorithm similar to (25), and for $N' \le N/2$, the $\tilde{e}_n$ can be used directly to provide the trigonometric approximations

$$\tilde{F}_{N'}(t) = \frac{1}{2}\sum_{n=-N'}^{N'} \tilde{e}_n e^{-int} \tag{27}$$

where $\tilde{e}_{-n} = \tilde{c}_n - i\tilde{s}_n = \bar{\tilde{e}}_n$ for $n \le N'$. (See Problem 1.7H.)

If a function is sampled on an interval $[0, L]$, then through the scale change $t = 2\pi x/L$ as in Section 1.6, each of these formulas is converted into a correspondent. Then (25) holds without change but, for example, (23) becomes

$$\tilde{e}_n = \frac{2}{N}\sum_{k=1}^{N} f(kx_1)e^{ink\pi x_1}, \qquad \text{where } x_1 = \frac{L}{N}.$$

As L varies, we have access to the coefficients $\tilde{e}_n$ associated with different frequencies $(n\pi L)/N$. In this manner, we can make an approximate spectral analysis of a signal of uncertain periodicity.

Problem Set 1.7

1.7A **1.** When $N = 8$ and $N' = 4$, calculate the approximate coefficients for the function of Example 11 and compare with the exact values.

2. Repeat this exercise for the function

$$f(t) = \begin{cases} t, & 0 \le t \le \pi \\ t - 2\pi, & \pi < t < 2\pi \end{cases}.$$

1.7B Derive the relations in (20′).

1.7C When $N = 5$ (and $N' = 2$) in Example 12, obtain the numerical coefficients required to express $\tilde{F}_2$.

1.7D Find $\tilde{F}_1$ and $\tilde{F}_2$ for the following functions defined on $(-\pi, \pi)$.

1. $f(t) = \sin\frac{t}{2}$
2. $f(t) = \sqrt{|t|}$
3. $f(t) = t^2$

1.7E 1. For $N = 2N' + 1$, substitute (20) in (21) to obtain

$$\tilde{F}_{N'}(t) = \frac{2}{N}\sum_{k=0}^{N-1} f(t_k)\left[\frac{1}{2} + \sum_{n=1}^{N'} \cos n(t - t_k)\right].$$

2. In deriving (33) we will show that the bracketed term in part 1 equals

$$\left[\frac{\sin N'\left(\frac{t - t_k}{2}\right)}{2\sin\frac{t - t_k}{2}}\right]$$

when $t \neq t_k$. When $t = t_j$, show that the bracketed term in part 1 becomes $N'/2$ if $k = j$ and is zero when $k \neq j$, so that (22) follows. (*Hint:* When $k \neq j$, use the fact that $t_k - t_j = (k - j)t_1$.)

3.* For $N' = N$, use the same reasoning to conclude that

$$\frac{1}{2}\left[\tilde{F}_N(t_j) - \frac{\tilde{c}_0}{2}\right] = f(t_j), \qquad j = 0, 1, 2, \ldots, N.$$

1.7F 1. Show how the operational counting assertion $M \leq 2m + 2N$ follows from (25), where m is the maximum number of multiplications required to obtain all $\tilde{e}_n$ (or $\tilde{e}_n'$) for $n \leq N$.

2. If $m \leq Nr$ multiplications are required to obtain all $2^{r-1}\tilde{e}_n$ when $n \leq N = 2^r$ (which is true when $r = 1$), conclude that $M \leq (2N)(r + 1)$, so that inductively we need at most $Nr + N$ multiplications to obtain the $\tilde{e}_n$, $n \leq N$.

1.7G 1. Suppose that $f \in C^1[0, 2\pi]$ and f has period 2π. Then, by the *law of the mean*: $|f(t) - f(t_k)| \leq M'|t - t_k|$ where $M' = \max|f'(t)|$. Let $M = M' + \max|f(t)|$, and show that

$$|c_n - \tilde{c}_n| + |s_n - \tilde{s}_n| \leq \frac{4\pi Mn}{N}, \qquad \text{if } n \leq N.$$

2. Conclude that if $N = N'^3$ in (21), then

$$|F_{N'}(t) - \tilde{F}_{N'}(t)| \leq \frac{8\pi M}{N'}, \qquad \text{which} \to 0 \quad \text{as} \quad N' \to +\infty,$$

so that *these* approximate Fourier sums $\tilde{F}_{N'} \to f$ uniformly on $[0, 2\pi]$, as $N' \to +\infty$ by Theorem 1.7.

1.7H 1. Use formulas (23) to transform (21) to (27).

2. Verify the interchanges and cancellations required to obtain (26) from (23).

1.7I 1. Suppose f is even and of period 2π. Show that $\tilde{s}_n = 0$, $n = 1, 2, \ldots$.

2. Suppose that f is odd and of period 2π. Must $\tilde{c}_n = 0$? Explain.

3. Suppose that f is odd and of period 2π with $f(-\pi) = f(\pi) = 0$. Show that $\tilde{c}_n = 0$.

1.8

Convergence Theorems for Fourier Series

In Section 1.5 we stated five major results about convergence of Fourier series of sectionally smooth functions. We now want to prove all of them. But why prove any of them, especially since mathematicians are sometimes accused of proving the obvious? Well, in the case of Fourier series, not much is obvious, nor are there many examples where we can watch a nonfinite Fourier series F represent the function f that generated it. So, without proof, we have little evidence for believing that F and f are even related, much less that they are almost identical. The proof is not simple, but the techniques used have far-reaching implications. Bear with it, and you will be introduced to some of the most significant results in mathematical analysis.

We need to prove parts A, B, and C of Theorem I and parts D and E of Theorem II. Here is our plan of attack: in Section 1.8a, we motivate C; in Section 1.8b, we establish A; in Section 1.8c, we establish D and E; and in Section 1.8d, we establish B and C. Whenever it is reasonable, we obtain more general results and consider their application, but far more general results are known and more specialized ones are produced each year. The most recent major contribution is Carleson's proof of 1966 that the Fourier series of a Lebesgue square-integrable f must converge pointwise "almost everywhere." (See [Ca].)

(a) Mean-square convergence; Bessel's inequality: Suppose f has period 2π and is Riemann integrable on $[-\pi, \pi]$. Then f generates the formal Fourier series

$$F(t) = \frac{c_0}{2} + \sum_{n=1}^{\infty} (c_n \cos nt + s_n \sin nt)$$

with coefficients

$$\left.\begin{matrix} c_n \\ s_n \end{matrix}\right\} = \frac{1}{\pi}\int_{-\pi}^{\pi} f(t) \left\{\begin{matrix} \cos nt \\ \sin nt \end{matrix}\right\} dt, \qquad n = 0, 1, 2, \ldots . \tag{28}$$

In what senses does the partial sum F_N of this series approximate f for large N? Here, for $N = 1, 2, \ldots$

$$F_N(t) = \frac{c_0}{2} + \sum_{n=1}^{N} (c_n \cos nt + s_n \sin nt). \tag{28'}$$

In order that the Fourier series of f converges to $f(t)$ at one point t, it is necessary that

$$|f(t) - F_N(t)| \to 0 \quad \text{as} \quad N \to \infty. \tag{28''}$$

Since (28″) need not hold at every t, we consider weaker types of convergence. Mean convergence of F_N to f occurs if

$$\int_{-\pi}^{\pi} |f(t) - F_N(t)|\, dt \to 0 \quad \text{as} \quad N \to \infty,$$

indicating that for sufficiently large N, the areal difference between the graphs of f and F_N can be made as small as desired. However, it is easier to utilize orthogonality in examining **mean-square convergence** of F_N to f, which occurs if

$$\int_{-\pi}^{\pi} [f(t) - F_N(t)]^2\, dt \to 0 \quad \text{as} \quad N \to \infty.$$

Moreover, we can give mean-square convergence a useful geometrical interpretation. (See the discussion in Section 0.1e.)

Our first reward from mean-square considerations is an important inequality obtained as follows. Note that

$$\begin{aligned}\int_{-\pi}^{\pi} [f(t) - F_N(t)]^2\, dt = \int_{-\pi}^{\pi} f^2(t)\, dt - \int_{-\pi}^{\pi} f(t)F_N(t)\, dt \\ + \int_{-\pi}^{\pi} F_N^2(t)\, dt - \int_{-\pi}^{\pi} f(t)F_N(t)\, dt.\end{aligned} \tag{29}$$

However, from (28′),

$$\begin{aligned}\int_{-\pi}^{\pi} f(t)F_N(t)\, dt &= \frac{c_0}{2}\int_{-\pi}^{\pi} f(t)\, dt + \sum_{n=1}^{N}\left[c_n\int_{-\pi}^{\pi} f(t)\cos nt\, dt + s_n\int_{-\pi}^{\pi} f(t)\sin nt\, dt\right] \\ &= \pi\left[\frac{c_0^2}{2} + \sum_{n=1}^{N}(c_n^2 + s_n^2)\right], \qquad \textit{by (28)}, \\ &= \int_{-\pi}^{\pi} F_N^2(t)\, dt, \qquad \textit{when (4′) of Proposition 1.1 is applied to } F_N.\end{aligned}$$

Thus, we find that for each $N = 1, 2, \ldots$

$$0 \le \frac{1}{\pi}\int_{-\pi}^{\pi} [f(t) - F_N(t)]^2\, dt = \frac{1}{\pi}\int_{-\pi}^{\pi} f^2(t)\, dt - \left(\frac{c_0^2}{2} + \sum_{n=1}^{N}(c_n^2 + s_n^2)\right). \tag{30}$$

$$\begin{aligned}\text{Hence,} \quad &\int_{-\pi}^{\pi} [f(t) - F_N(t)]^2\, dt \to 0, \quad \text{as} \quad N \to \infty \\ \text{iff} \quad &\frac{c_0^2}{2} + \sum_{n=1}^{\infty}(c_n^2 + s_n^2) = \frac{1}{\pi}\int_{-\pi}^{\pi} f^2(t)\, dt.\end{aligned} \tag{31}$$

In the last line, we see Parseval's formula, but it remains unproved; however, the inequality in (30) can be written as follows:

$$\frac{c_0^2}{2} + \sum_{n=1}^{N}(c_n^2 + s_n^2) \le \frac{1}{\pi}\int_{-\pi}^{\pi} f^2(t)\, dt$$

for each $N = 1, 2, \ldots$; as $N \to \infty$, this yields **Bessel's inequality** (of 1828)

$$\frac{c_0^2}{2} + \sum_{n=1}^{\infty} (c_n^2 + s_n^2) \le \frac{1}{\pi} \int_{-\pi}^{\pi} f^2(t)\, dt. \tag{32}$$

Since f is Riemann integrable on $[-\pi, \pi]$, then $\int_{-\pi}^{\pi} f^2(t)\, dt < +\infty$. It follows from (32) that the series $\Sigma_{n=1}^{\infty}(c_n^2 + s_n^2)$ converges; hence its nth term $\to 0$ as $n \to \infty$, which means that both c_n and s_n approach zero.

(1.10) Proposition: *If f is integrable on $[-\pi, \pi]$, then as $n \to \infty$*

$$c_n = \frac{1}{\pi} \int_{-\pi}^{\pi} f(t) \cos nt\, dt \to 0, \qquad \textit{and} \qquad s_n = \frac{1}{\pi} \int_{-\pi}^{\pi} f(t) \sin nt\, dt \to 0. \qquad \blacksquare$$

Thus the Fourier coefficients c_n, s_n of any integrable f must die out as $n \to \infty$. However, $\sum_{n=1}^{\infty} \frac{\sin nt}{\sqrt{n}}$ could not be the Fourier series of any Riemann integrable function f, since as $n \to \infty$ the coefficients $s_n = 1/\sqrt{n}$ do not approach zero fast enough to satisfy (32).

By similar analysis, we can also establish that for each N, F_N gives the best mean-square approximation to f among all choices

$$\tilde{F}_N(t) = \frac{\tilde{c}_0}{2} + \sum_{n=1}^{N} (\tilde{c}_n \cos nt + \tilde{s}_n \sin nt).$$

(See Section 9.2.)

(b) Pointwise convergence: Next, let's see how the Fourier series of an integrable f behaves at a point t where f has limits from either side. For convenience, we take $t = 0$, and our first goal is to find conditions that make

$$F_N(0) \to f_E(0) \stackrel{\text{def}}{=} \frac{f(0+) + f(0-)}{2}, \qquad \text{as} \qquad N \to \infty.$$

(Recall the discussion in Section 1.4d.)

Now when $t = 0$ in (28′), the sine terms vanish, $\cos n0 = 1$, $n = 0, 1, \ldots$, and on substitution of (28), we have

$$F_N(0) = \frac{c_0}{2} + \sum_{n=1}^{N} c_n = \frac{1}{\pi} \int_{-\pi}^{\pi} f(t) \left[\frac{1}{2} + \sum_{n=1}^{N} \cos nt \right] dt.$$

We want to know what happens as $N \to +\infty$, and in the integral, the dependence on N is confined to the bracketed term. We can sum this term by means of the following trigonometric identities.

$$2\left(\sin\frac{t}{2}\right) \left[\frac{1}{2} + \sum_{n=1}^{N} \cos nt \right] = \sin\frac{t}{2} + \sum_{n=1}^{N} 2 \cos nt \sin\frac{t}{2} \tag{32′}$$

$$= \sin\frac{t}{2} + \sum_{n=1}^{N} \left[\sin\left(n + \frac{1}{2}\right)t - \sin\left(n - \frac{1}{2}\right)t \right] = \sin\left(N + \frac{1}{2}\right)t,$$

(because this sum telescopes).

Now divide by $2\sin(t/2)$ and substitute the result to get **Dirichlet's formula**

$$F_N(0) = \frac{1}{2\pi} \int_{-\pi}^{\pi} f(t) \frac{\sin(N + \frac{1}{2})t}{\sin(t/2)}\, dt, \qquad N = 1, 2, \ldots . \tag{33}$$

This formula permits us to compute $F_N(0)$ directly by integrating f against the *Dirichlet kernel*

$$D_N(t) = \frac{1}{2\pi}\frac{\sin(N+\frac{1}{2})t}{\sin(t/2)}, \qquad 0 < |t| \le \pi \tag{33'}$$

whose graph for $N = 1, 3, 5$, and 10 is shown in Figure 1.14.

As you see, for $N = 5$ and 10, D_N peaks sharply near $t = 0$ and dies out rapidly as $|t|$ increases on $(0, \pi]$. Consequently, for large N we might expect the integral in (33) to be affected most by how f behaves near $t = 0$. These observations are encouraging and we could pursue them analytically (see, for example, [Wein]), but Proposition 1.10 offers us a much simpler *independent* approach. In (33), we can use the identity

$$\sin\left(N+\frac{1}{2}\right)t = \sin Nt \cos\frac{t}{2} + \cos Nt \sin\frac{t}{2}$$

to show that

$$F_N(0) = \frac{1}{2\pi}\int_{-\pi}^{\pi} f(t)\left\{\cos(t/2)\frac{\sin Nt}{\sin(t/2)}\right\}dt + \frac{1}{2\pi}\int_{-\pi}^{\pi} f(t)\cos Nt\,dt.$$

The term in braces is an *even* function of t so that its integral against f_O, the *odd* part of f, is zero, and if we use the decomposition $f = f_E + f_O$ of (9), we conclude that

$$F_N(0) = \frac{1}{2\pi}\int_{-\pi}^{\pi} f_E(t)\frac{\cos(t/2)}{\sin(t/2)}\sin Nt\,dt + \frac{1}{2\pi}\int_{-\pi}^{\pi} f(t)\cos Nt\,dt. \tag{33''}$$

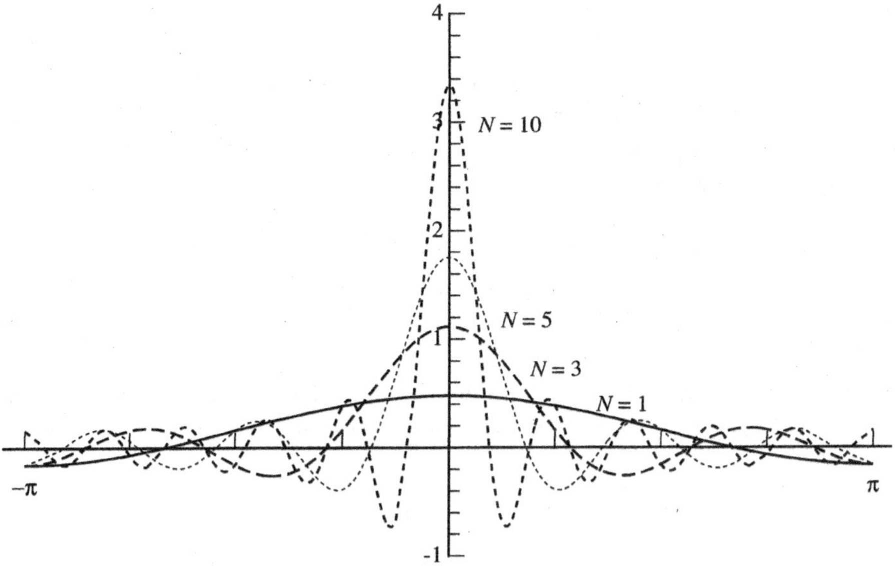

Figure 1.14

Graph of D_N

In (33″), the second integral is just $c_N/2$, which approaches zero as $N \to \infty$ by 1.10. We could also infer that the first integral approaches zero as $N \to \infty$, provided that the *odd* function g defined by

$$g(t) = f_E(t)\frac{\cos(t/2)}{\sin(t/2)} = \frac{f_E(t)}{t}\left[\frac{2(t/2)}{\sin(t/2)}\cos(t/2)\right], \qquad t \neq 0,$$

is integrable over $[0, \pi]$ (and so over $[-\pi, \pi]$). The bracketed term has limit 2 as $t \searrow 0$ so that it is well behaved on $[0, \pi]$. Consequently, g is integrable over $[0, \pi]$ if $f_E(t)/t$ has a limit as $t \searrow 0$, or equivalently,

if $f_E(0) = 0$ *and* $f_E(t)$ *has a derivative from the right at* 0.

Under these conditions on f_E, it follows that

$$F_N(0) \to 0 \qquad \text{as} \qquad N \to \infty,$$

and the real work is over. All that remains is to remove restrictions on f and on t, and we can accomplish this through simple modifications in f.

(1) If $c \stackrel{\text{def}}{=} f_E(0) \neq 0$, then we subtract c from f, which has the effect of subtracting the same constant c from the even part of f and from each partial sum F_N.

So, if we define $\quad \tilde{f}(t) = f(t) - c,$

then we see that $\quad \tilde{f}_E(t) = f_E(t) - c \quad$ vanishes at $t = 0$

and that with obvious notation $\quad \tilde{F}_N(t) = F_N(t) - c, \qquad$ for $N = 1, 2, \ldots$.

Moreover, *if* f *has derivatives from the right and from the left at* 0, then f_E (and so $\tilde{f}_E$) has a derivative from the right at 0 so that, as above,

$$\tilde{F}_N(0) \to 0 \qquad \text{as} \qquad N \to +\infty.$$

Therefore, as $N \to +\infty$: $\qquad F_N(0) = \tilde{F}_N(0) + c \to 0 + c = f_E(0).$

We have reached our first goal with a result worth stating.

Suppose that f *has possibly distinct limiting values* $f(0+)$ *and* $f(0-)$ *and associated one-sided derivatives*[†]

$$f'_+(0) \stackrel{\text{def}}{=} \lim_{t\searrow 0}\frac{f(t) - f(0+)}{t}, \qquad f'_-(0) \stackrel{\text{def}}{=} \lim_{t\nearrow 0}\frac{f(t) - f(0-)}{t}.$$

Then

$$\frac{c_0}{2} + \sum_{n=1}^{\infty} c_n = \frac{f(0+) + f(0-)}{2}. \tag{34}$$

(2) To derive corresponding results at a *fixed* $t \neq 0$, we introduce the translate $\tilde{f}(\tau) \stackrel{\text{def}}{=} f(\tau + t)$, $\tau \in \mathbb{R}$. The cosine coefficients of $\tilde{f}$ are given by

$$\begin{aligned}\tilde{c}_n &= \frac{1}{\pi}\int_{-\pi}^{\pi} \tilde{f}(\tau)\cos n\tau\, d\tau = \frac{1}{\pi}\int_{-\pi}^{\pi} f(\tau + t)\cos n\tau\, d\tau \\ &= \frac{1}{\pi}\int_{-\pi}^{\pi} f(\tau)\cos n(\tau - t)\, d\tau,\end{aligned}$$

if we note that the integrand is periodic and use Lemma 1.8. Then from the addition formula for $\cos n(\tau - t)$, it follows that

---

† Such derivatives are not essential: (34) is valid when the *square* of g defined above has a finite (possibly improper) integral over $[-\pi, \pi]$, in particular, it holds for $f(t) = |t|^{2/3}$.

$$\tilde{c}_n = \frac{1}{\pi}\left(\int_{-\pi}^{\pi} f(\tau)\cos n\tau\, d\tau\right)\cos nt + \frac{1}{\pi}\left(\int_{-\pi}^{\pi} f(\tau)\sin n\tau\, d\tau\right)\sin nt$$

or, by (28),

$$\tilde{c}_n = c_n \cos nt + s_n \sin nt, \qquad n = 0, 1, 2, \ldots . \tag{35}$$

Thus we obtain our principal result (which is similar to one first established by Dirichlet in 1829).

(1.11) Theorem: *If f has period 2π and is integrable over $[-\pi, \pi]$, then its Fourier series converges with sum $[f(t+) + f(t-)]/2$ at each t where f has derivatives from the right and from the left. If f is also continuous at t,* the sum is $f(t)$.

Proof: For fixed t, the translate $\tilde{f}(\tau) = f(\tau + t)$ has the same behavior at $\tau = 0$ as f does at t. Hence, under the stated conditions at t, by (34),

$$\frac{\tilde{c}_0}{2} + \sum_{n=1}^{\infty} \tilde{c}_n = \frac{\tilde{f}(0+) + \tilde{f}(0-)}{2} = \frac{f(t+) + f(t-)}{2}.$$

In view of (35), we can replace this equation by

$$\frac{c_0}{2} + \sum_{n=1}^{\infty}(c_n \cos nt + s_n \sin nt) = \frac{f(t+) + f(t-)}{2}, \tag{36}$$

and the last term reduces to $f(t)$ when f is continuous at t. ■

Remarks: Convergence at a point is guaranteed under weaker conditions (see [Z]); however, simple continuity of f is unfortunately not sufficient to ensure convergence. Since this theorem applies to sectionally smooth functions on $[-\pi, \pi]$, it establishes part A of Theorem I in Section 1.5. It also applies to some functions that are **not** sectionally smooth. (See Problem 1.8B.)

(c) Uniform convergence; differentiation: If $f \in \hat{C}^1[-\pi, \pi]$ and $f(-\pi) = f(\pi)$, then f has a continuous extension of period 2π, and (by Theorem 1.11) a Fourier series *representation*,

$$f(t) = \frac{c_0}{2} + \sum_{n=1}^{\infty}(c_n \cos nt + s_n \sin nt), \qquad |t| \le \pi. \tag{37}$$

For such f, Theorem II holds. Let's begin by establishing differentiability of the Fourier series under appropriate conditions.

(1.12) Theorem: *If $f \in \hat{C}^1[-\pi, \pi]$ with $f(-\pi) = f(\pi)$ is extended with period 2π, then*

$$f'(t) = \sum_{n=1}^{\infty} n(-c_n \sin nt + s_n \cos nt),$$

at each t where $f''(t)$ exists.

Proof: To establish this result, we relate the Fourier coefficients of f', denoted c_n' and s_n', to those of f.

Indeed, when $n \geq 1$, through partial integrations, the coefficient formulas (28) become

$$c_n = \frac{1}{\pi}\int_{-\pi}^{\pi} f(t)\cos nt\,dt = f(t)\frac{\sin nt}{n\pi}\bigg|_{-\pi}^{\pi} - \frac{1}{n\pi}\int_{-\pi}^{\pi} f'(t)\sin nt\,dt = -\frac{s_n'}{n};$$

and $$s_n = \frac{1}{\pi}\int_{-\pi}^{\pi} f(t)\sin nt\,dt = -f(t)\frac{\cos nt}{n\pi}\bigg|_{-\pi}^{\pi} + \frac{1}{n\pi}\int_{-\pi}^{\pi} f'(t)\cos nt\,dt = \frac{c_n'}{n}$$

(38)

since $f(-\pi) = f(\pi)$ and $\cos n(-\pi) = \cos n\pi$. (The required partial integrations can be justified by partitioning the interval.)

If $f''(t)$ exists, then f' is necessarily continuous at t so that by Theorem 1.11, its Fourier series converges at this t with sum $f'(t)$. However, from (38), we know that the Fourier coefficients for f' are $c_n' = ns_n$ and $s_n' = -nc_n$ when $n = 1, 2, \ldots,$ while

$$c_0' = \frac{1}{\pi}\int_{-\pi}^{\pi} f'(t)\,dt = \frac{f(\pi) - f(-\pi)}{\pi} = 0.$$

Thus $f'(t)$ is represented by the series given above. ■

Remarks: If $f \in \hat{C}^1[-\pi, \pi]$, the condition $f(-\pi) = f(\pi)$ ensures that f has a continuous extension of period 2π. When f is even, this condition holds, but when f is *odd*, it holds iff $f(\pi) = 0$. (Why?) However, in case f is odd and $f(\pi) \neq 0$, then $\tilde{f}$ defined by

$$\tilde{f}(t) \stackrel{\text{def}}{=} f(t) - \frac{t}{\pi}f(\pi), \qquad |t| \leq \pi,$$

is odd and in $\hat{C}^1[-\pi, \pi]$ with $\tilde{f}(\pi) = 0$. With the aid of this device, we can establish results such as the following.

(1.13) Proposition: *If $f \in \hat{C}^1[-\pi, \pi]$ is odd, then at each $t \in (-\pi, \pi)$ where $f''(t)$ exists,*

$$f'(t) = \frac{f(\pi)}{\pi} + \sum_{n=1}^{\infty}\left[ns_n + \frac{2f(\pi)(-1)^n}{\pi}\right]\cos nt.$$

Proof: See Problem 1.8I. ■

The coefficient relations (38) yield other important consequences for this class of functions. Observe first that by the M-test, series (37) converges *absolutely* and *uniformly everywhere* provided that

$$\sum_{n=1}^{\infty}(|c_n| + |s_n|) < +\infty \qquad \text{or} \qquad \sum_{n=1}^{\infty}\sqrt{c_n^2 + s_n^2} < +\infty, \tag{39}$$

since by Cauchy's inequality (Section 0.1a),

$$|c_n \cos nt + s_n \sin nt| \leq \sqrt{c_n^2 + s_n^2}\sqrt{\cos^2 nt + \sin^2 nt} = \sqrt{c_n^2 + s_n^2},$$

and $$\sqrt{c_n^2 + s_n^2} \leq |c_n| + |s_n| \leq 2\sqrt{c_n^2 + s_n^2}.$$

(1.14) Theorem: *If $f \in \hat{C}^1[-\pi, \pi]$ with $f(-\pi) = f(\pi)$, then its Fourier series (of period 2π) converges absolutely and uniformly on $[-\pi, \pi]$. When $N \geq 1$, we have the uniform error estimate*

$$|f(t) - F_N(t)| \leq \sum_{n=N+1}^{\infty} \sqrt{c_n^2 + s_n^2} \leq \frac{1}{\sqrt{N\pi}} \left(\int_{-\pi}^{\pi} (f'(\tau))^2 \, d\tau \right)^{1/2}, \qquad |t| \leq \pi. \tag{40}$$

Moreover,

(A) $\dfrac{c_0^2}{2} + \displaystyle\sum_{n=1}^{\infty} (c_n^2 + s_n^2) = \frac{1}{\pi} \int_{-\pi}^{\pi} f^2(t) \, dt;$ *(Parseval's formula)*

(B) *If* $c_0 = 0$, *then* $\displaystyle\int_{-\pi}^{\pi} f^2(t) \, dt \leq \int_{-\pi}^{\pi} (f'(t))^2 \, dt.$ *(Wirtinger's inequality)*

Proof: To establish (40) (and thus the uniform convergence of the series), we will need Bessel's inequality for f' in the form

$$\sum_{n=1}^{\infty} (c_n'^2 + s_n'^2) \leq \frac{1}{\pi} \int_{-\pi}^{\pi} (f'(\tau))^2 \, d\tau. \tag{40'}$$

When we subtract

$$F_N(t) = \frac{c_0}{2} + \sum_{n=1}^{N} (c_n \cos nt + s_n \sin nt)$$

from (37) and estimate as above, we see that

$$|f(t) - F_N(t)| \leq \sum_{n=N+1}^{\infty} |(c_n \cos nt + s_n \sin nt)| \leq \sum_{n=N+1}^{\infty} \sqrt{c_n^2 + s_n^2}.$$

But from (38), $\sqrt{c_n^2 + s_n^2} = (1/n)\sqrt{c_n'^2 + s_n'^2}$. Hence, if we use the extended Cauchy inequality from Section 0.1a (with $x_n = 1/n$, $y_n = \sqrt{c_n'^2 + s_n'^2}$, $n = 1, 2, \ldots$), we find that

$$|f(t) - F_N(t)| \leq \sum_{n=N+1}^{\infty} \sqrt{c_n^2 + s_n^2} = \sum_{n=N+1}^{\infty} \frac{1}{n} \sqrt{c_n'^2 + s_n'^2}$$

$$\leq \left(\sum_{m=N+1}^{\infty} \frac{1}{m^2} \right)^{1/2} \left(\sum_{n=N+1}^{\infty} (c_n'^2 + s_n'^2) \right)^{1/2}$$

Then, from (13″) and (40′) we obtain the estimate in (40),

$$|f(t) - F_N(t)| \leq \left(\frac{1}{N} \right)^{1/2} \left(\frac{1}{\pi} \int_{-\pi}^{\pi} (f'(\tau))^2 \, d\tau \right)^{1/2}$$

These estimates hold independently of t, which means that the series converges absolutely and uniformly. Moreover, if we square both sides of the last inequality and integrate the result over $[-\pi, \pi]$, we see that

$$\int_{-\pi}^{\pi} [f(t) - F_N(t)]^2 \, dt \leq \frac{2}{N} \int_{-\pi}^{\pi} f'(\tau)^2 \, d\tau, \qquad \text{which} \to 0 \qquad \text{as} \qquad N \to \infty.$$

Then, from (31), it follows that Parseval's formula of part A holds for f. Wirtinger's inequality of part B is a consequence of this result and Bessel's inequality for f'. (See Problem 1.8D.) ■

When f has higher derivatives, then in general its Fourier coefficients are smaller, and we can improve estimate (40). (See Problem 1.8K.) There is also the following companion result.

(1.15) Proposition: *If for some real numbers c_n, s_n, and integer $k > 0$ the series $\Sigma_{n=1}^{\infty} n^k(|c_n| + |s_n|) < +\infty$, then the formal series*

$$\frac{c_0}{2} + \sum_{n=1}^{\infty}(c_n \cos nt + s_n \sin nt)$$

is the Fourier series of a function f of period 2π in class C^k.

Proof: Since $|c_n| + |s_n| \le n^k(|c_n| + |s_n|)$, the formal series converges absolutely and uniformly with a continuous periodic sum f, say. Moreover, it must be the Fourier series generated by f, because uniform convergence permits us to interchange integration with summation when calculating the coefficient integrals (28) and then use orthogonality. Similarly, if $k = 1$ so that $\Sigma_{n=1}^{\infty} n(|c_n| + |s_n|) < \infty$, then the differentiated series

$$\sum_{n=1}^{\infty} n(-c_n \sin nt + s_n \cos nt)$$

converges absolutely and uniformly to a continuous function g such that $f(t) - \int_{-\pi}^{t} g(\tau)\,d\tau = (c_0/2) + \Sigma_{n=1}^{\infty} c_n(-1)^n = c$, say, so that $f'(t) = g(t)$ is continuous. Thus $f \in C^1$, and the argument admits repetition as needed. ■

Example 13:

$\Sigma_{n=1}^{\infty}(\sin nt)/2n^{7/2}$ is the Fourier series of an odd function f of period 2π with (sine) coefficients $s_n = 1/2n^{7/2}$, for which

$$\sum_{n=1}^{\infty} n^2|s_n| = \sum_{n=1}^{\infty} \frac{1}{2} n^{-3/2} < +\infty$$

Hence this $f \in C^2[a, b]$ for each interval $[a, b]$. From Theorem 1.12, we know that

$$f'(t) = \sum_{n=1}^{\infty} \frac{\cos nt}{2n^{5/2}}$$

moreover, since the further differentiated series

$$-\sum_{n=1}^{\infty} \frac{\sin nt}{2n^{3/2}}$$

also converges uniformly, we conclude by Theorem 0.5 that its value is $f''(t)$.

(d)* Nonuniform convergence; Gibbs' phenomenon: The function of Example 4, $f(t) = t$, $|t| < \pi$, has discontinuities at $t = \pm\pi$ in its extensions of period 2π. Therefore, its Fourier series

$$F(t) = \sum_{n=1}^{\infty} \frac{2(-1)^{n+1}}{n} \sin nt = t, \qquad |t| < \pi$$

cannot converge uniformly on $[-\pi, \pi]$. However, on each subinterval $[-b, b]$ where $|t| \le b < \pi$, this series *does* converge uniformly as a result of the following nontrivial identity for its partial sums. (See Problem 1.8F.)

$$F_N(t) = 2\sum_{n=1}^{N} \frac{(-1)^{n+1}}{n} \sin nt$$

$$= \tan(t/2) - \sec(t/2)\left[\frac{(-1)^N}{N+1}\sin(N+\tfrac{1}{2})t + \sum_{n=1}^{N} \frac{(-1)^n \sin(n+\frac{1}{2})t}{n(n+1)}\right]. \qquad \textbf{(41)}$$

Observe that as $N \to \infty$, the last series in (41) converges absolutely and uniformly by the M-test, so that the bracketed term converges uniformly for all $t \in \mathbb{R}$. Moreover, when $|t| \le b < \pi$, then $|\sec(t/2)| \le \sec(b/2)$, and consequently, the *sequence* $F_N(t)$ converges uniformly as $N \to \infty$.

When we study these expressions more carefully, we find that the maximum value of $|F_N(t)|$ occurs at some t_N near π and approaches a value $\approx 1.18\pi$ as $N \to \infty$. This is illustrated in Figure 1.15 with supporting analysis given in Problem 1.8G. In effect, to force the smooth oscillatory functions $\sin nt$ to approximate a discontinuity at a point, we must combine them so that the finite sums F_N deviate more from F near this point than the discontinuity might seem to require. In 1899, Gibbs proved that such "phenomenal" convergence behavior is characteristic of a Fourier series near a point of simple discontinuity in its sum.†

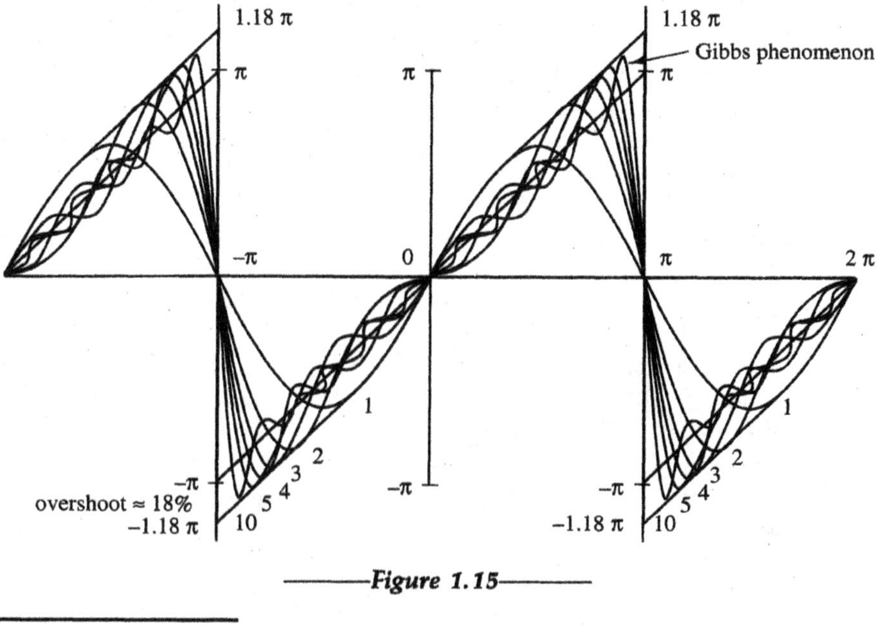

Figure 1.15

† Gibbs actually only rediscovered results already known to Wilbraham (c. 1850), but Gibbs' name is firmly attached to this behavior of a Fourier series.

If we divide this periodic sawtooth function by 2π and denote the result by $\sigma(t)$, then σ has as its only discontinuities unit jumps at the points $\pm\pi, \pm 3\pi, \ldots$, and its translate $\sigma(t - t_0 - \pi)$ has unit jumps only at $t_0, t_0 \pm 2\pi, \ldots, t_0 \pm 2n\pi$. With these functions, we can establish the convergence behavior of Fourier series of sectionally smooth functions.

For example, the function $f(t) = e^t$ on $(-\pi, \pi)$, extended periodically, has jumps of magnitude $a = e^{\pi} - e^{-\pi}$ at $\pm\pi, \pm 3\pi, \ldots$. Hence the new function

$$\tilde{f}(t) = f(t) - a\sigma(t)$$

is in $\hat{C}^1[-\pi, \pi]$ and $\tilde{f}(-\pi) = \tilde{f}(\pi)$, since we have subtracted a function (with the same jumps at the same points) that is otherwise continuous. By Theorem 1.14, the Fourier series for $\tilde{f}$ converges to $\tilde{f}$ uniformly. But then, the Fourier series for f is just that of $\tilde{f}$ plus a times that of σ, *both* of which converge uniformly on $[-\ell, \ell]$ when $\ell < \pi$. Therefore, the Fourier series for f also converges uniformly on this interval.

Periodic functions with a finite set of simple discontinuities in $(-\pi, \pi]$ can be analyzed similarly by subtracting appropriate multiples of translates of σ, one for each discontinuity in $(-\pi, \pi]$. These observations lead to the following, which validates parts B and C of Theorem I.

(1.16) Theorem: *If f has period 2π and is sectionally smooth on $[-\pi, \pi]$, then the Fourier series for f converges uniformly on each interval $[a, \ell]$ of continuity of f. Near each point where f has a jump discontinuity, the sequence F_N will exhibit Gibbs' behavior. Moreover, Parseval's formula holds for f.*

Proof: The first assertions have already been established. To show that Parseval's formula holds, we use the facts that by (13′), it holds for σ, and that

$$f = \tilde{f} + [\text{finite sum of scalar multiples of translates of } \sigma],$$

where $\tilde{f}$ satisfies the conditions of Theorem 1.14, and (by (13′) and Lemma 1.8) that Parseval's formula holds for each function in the bracketed sum on the right. It follows that Parseval's formula holds for f. (See Problem 1.8J.) ∎

Problem Set 1.8

1.8A For $N = 2$, write out in detail the computations leading from (29) to (30) in the derivation of Bessel's inequality.

1.8B The infinite staircase function $f(t) = 1/k$, $\pi/(k+1) < t \le \pi/k$, $k = 1, 2, \ldots$ with $f(0) = 0$, has an even extension that is Riemann integrable on $[-\pi, \pi]$, but it is not sectionally smooth on this interval. Explain, and graph this extension.

1. Show that Theorem 1.11 applies to the Fourier series for this function at each $t \in (0, \pi]$ and give the resulting sum, especially at $t_k = \pi/k$, $k = 1, 2, \ldots$.
2. This f is continuous at 0, but it does not have a one-sided derivative there. However, $f_1(t) = f(t)/t$ is integrable over $[0, \pi]$. (Why?) Examine carefully the arguments in Section 1.8b leading to (34) to show that (34) is valid for this f.

1.8C $\Sigma_{n=1}^{\infty}(1/n^5)\cos nt$ is the Fourier series for a certain function f. Give as much information as possible about this function.

1.8D 1. If f is as in Theorem 1.14, with $c_0 = (1/\pi)\int_{-\pi}^{\pi} f(t)\,dt = 0$, obtain part B from part A in the form

$$\frac{1}{\pi}\int_{-\pi}^{\pi} f^2(t)\,dt = \sum_{n=1}^{\infty}(c_n^2+s_n^2) \le \sum_{n=1}^{\infty} n^2(c_n^2+s_n^2) \le \frac{1}{\pi}\int_{-\pi}^{\pi}(f'(t))^2\,dt.$$

(*Hint:* Use Bessel's inequality for f'.)

2. Show that equality holds iff $f(t) = c_1\cos t + s_1 \sin t$.

1.8E* $f(t) = -\log[2\sin(t/2)]$ on $(0,\pi)$ has an even extension of period 2π for which the improper integrals

$$\int_{-\pi}^{\pi} |f(t)|\,dt \qquad \text{and} \qquad \int_{-\pi}^{\pi} f^2(t)\,dt$$

are finite.

1. It follows that this f generates a cosine series for which the arguments used in proving Theorem 1.11 still apply (except at $t=0$). Explain why this might be possible.
2. Show that $c_0 = 0$ (or $\int_0^{\pi}\log\sin(t/2)\,dt = -\pi\log 2$) by simple substitutions such as $t = 2x$, $x = \pi - \tau$.
3. Show that $c_n = 1/n$, $n = 1, 2, \ldots$ by partial integration using (33) for $f(t) = 1$.
4. Conclude that

$$\sum_{n=1}^{\infty}\frac{\cos nt}{n} = -\log\left(2\sin\frac{t}{2}\right), \qquad 0 < t \le \pi.$$

This Fourier series diverges at $t=0$, but evaluate it at $t=\pi$.

5. Through analysis similar to that associated with (41), it can be shown that the series in part 4 converges uniformly on each interval $[\tau,\pi]$ for $\tau > 0$. Use this fact to establish that for $0 < \tau \le \pi$,

$$\sum_{n=1}^{\infty}\frac{\sin n\tau}{n^2} = -\int_{\tau}^{\pi} f(t)\,dt.$$

6. Make the substitution $x = \pi - t$ in part 4 to obtain a Fourier cosine series representing $\log[2\cos(t/2)]$, and combine these results to produce a Fourier series for $\log\tan(t/2)$ on $(0,\pi)$.

1.8F 1. Let $S_N(t) = \Sigma_{k=0}^{N}(-1)^k \sin kt$. Verify that

$$2\cos\frac{t}{2}S_N(t) = (-1)^N \sin\left(N+\frac{1}{2}\right)t - \sin\frac{t}{2}.$$

2. Show that

$$\sum_{n=1}^{N}\frac{(-1)^n\sin nt}{n} = \sum_{n=1}^{N}\frac{S_n(t)-S_{n-1}(t)}{n}$$
$$= \frac{S_N(t)}{N+1} + \sum_{n=1}^{N}\frac{S_n(t)}{n(n+1)},$$

and combine this result with that of part 1 to get (41).

3. Use the identity $2\sin(t/2)\sin(n+\frac{1}{2})t = \cos nt - \cos(n+1)t$ to prove that

$$\sum_{n=0}^{N-1} \sin(n+\tfrac{1}{2})t = \frac{\sin^2(Nt/2)}{\sin(t/2)}, \qquad N = 1, 2, 3, \ldots .$$

*1.8G** (*Gibbs' phenomenon*):

1. Let $F_N(t) = 2\Sigma_{n=1}^{N}[(-1)^{n+1}\sin nt]/n$, as in (41), and argue that at $t_N = \pi[1-(1/N)]$, as $N \to \infty$:

$$F_N(t_N) = 2\sum_{n=1}^{N} \frac{\pi}{N}\left(\frac{N}{n\pi}\sin\frac{n\pi}{N}\right) \to 2\int_0^{\pi} \frac{\sin x}{x}\,dx = S, \qquad \text{say.}$$

2. Use the Maclaurin series representation for $\sin x$ and integrate termwise to obtain the alternating series

$$S = 2\sum_{m=1,3,\ldots}^{\infty} (-1)^{(m-1)/2}\frac{\pi^m}{m \cdot m!} = 2\left(\pi - \frac{\pi^3}{3\cdot 3!} + \frac{\pi^5}{5\cdot 5!} - \cdots\right).$$

3. Explain why

$$\frac{S}{\pi} > 2\left(1 - \frac{\pi^2}{3\cdot 3!} + \cdots - \frac{\pi^{10}}{11\cdot 11!}\right) \approx 1.18$$

and why these facts guarantee that the partial sums in part 1 must have Gibbs' behavior near π.

4. From the preceding analysis alone, could the limiting overshoot be less than 1.18π? Could it be more?
5. Use the square wave function

$$f(x) = \begin{cases} -1, & (2n-1)\pi < x \le 2n\pi \\ +1, & 2n\pi < x \le (2n+1)\pi \end{cases}, \qquad n = 0, \pm 1, \pm 2, \ldots$$

to illustrate Gibbs' phenomenon by plotting a few partial sums of the Fourier series for f. Use these partial sums to estimate the limiting overshoot of 1.18π.

1.8H 1. Suppose that f defined on $[-\pi, \pi]$ has a twice-continuously differentiable 2π-periodic extension. Show that its Fourier coefficients must satisfy

$$|c_n| \le \frac{C}{n^2} \quad \text{and} \quad |s_n| \le \frac{C}{n^2}, \qquad n = 1, 2, 3, \ldots$$

for some constant $C > 0$ independent of n. Conclude that its Fourier series converges uniformly *without* using Theorem 1.14.

2. Show that if a Fourier series converges with a discontinuous sum, then the series $\Sigma_{n=1}^{\infty}(|c_n| + |s_n|)$ cannot converge.

1.8I Prove Proposition 1.13 using the fact that Theorem 1.12 applies to the *odd* function $\tilde{f}(t) \stackrel{\text{def}}{=} f(t) - (t/\pi)f(\pi)$. Express the result in terms of f and its Fourier coefficients.

1.8J If Parseval's formula holds for f and for $\tilde{f}$, use (31) to show that

1. it holds for af for each constant a.
2. it holds for $f+\tilde{f}$, since $(u+v)^2 \le 2u^2 + 2v^2$.

Explain why the last assertion in Theorem 1.16 follows.

1.8K If f has period 2π and $f^{(k-1)} \in \hat{C}^1[-\pi, \pi]$ for some $k = 2, 3, \ldots$,

1. show that the Fourier coefficients of f are $\pm n^{-k}$ times those of $\tilde{f} = f^{(k)}$ so that $c_n^2 + s_n^2 = n^{-2k}(\tilde{c}_n^2 + \tilde{s}_n^2)$;
2. now use Cauchy's inequality and Bessel's inequality (for $\tilde{f}$) to get the error estimate

$$|f(t) - F_N(t)| \le \frac{\alpha_k}{N^{k-1/2}} \left[\int_{-\pi}^{\pi} (f^{(k)}(\tau))^2 \, d\tau \right]^{1/2}$$

where $\quad \alpha_k = [(2k-1)\pi]^{-1/2}, \qquad k = 1, 2, 3, \ldots .$

$$\left(\textit{Hint}: \sum_{n=N+1}^{\infty} n^{-2k} \le \int_N^{\infty} x^{-2k} \, dx. \right)$$

Why is the estimate in part 2 better than that in Theorem 1.14?

1.9

Additional Topics

The convergence theorems for Fourier series established in the last section were hard-won, both historically and in this text. However, their many applications justify the effort. In this section we shall examine a few items of interest directly related to the methods and results of this chapter, including the isoperimetric inequality and polynomial approximation.

The Isoperimetric Inequality

Can we increase the storage capacity of a silo by altering the base shape of its vertical walls while retaining their height and perimeter ℓ? It is easy to conclude that the traditional circular planform is better than that for a square (which in turn is superior to that for any other rectangle), but perhaps a less regular shape such as that indicated in Figure 1.16 might offer improvement.

That it cannot is one of the oldest geometrical conjectures (c. 850 BCE), and one that is surprisingly deep. It has received satisfactory analytical proofs only comparatively recently, one of which employs Wirtinger's inequality (part B of Theorem 1.14). We wish to prove that among all possible closed planar perimetral curves of length ℓ, the circle encloses the greatest area, namely, $\ell^2/4\pi$.

For convenience, assume that $\ell = 2\pi$ and that the perimeter S of a floor plan D

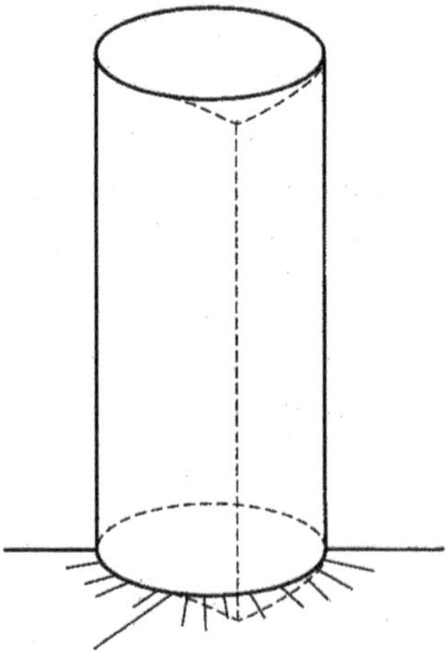

Figure 1.16

admits arc-length parameterization as shown by functions $x, y \in \hat{C}^1[-\pi, \pi]$, with $x(-\pi) = x(\pi)$ and $y(-\pi) = y(\pi)$. (See Figure 1.17.) Then $x'(t)^2 + y'(t)^2 = 1$ (except at possible corner points) and for this case we wish to establish that the enclosed area A cannot exceed π.

Assuming that D lies (only) to the left of the parametrized boundary S, as shown in Figure 1.17, we can use Green's theorem† to express the quantity A as a line integral. Thus

$$A = \int_D 1\,dx\,dy = -\int_{-\pi}^{\pi} y(t)x'(t)\,dt,$$

while $$2\pi = \int_{-\pi}^{\pi} [x'(t)^2 + y'(t)^2]\,dt \qquad \text{since } x'^2 + y'^2 = 1.$$

Hence $A \le \pi$, provided that $2\pi - 2A \ge 0$ or, with integral substitutions, that

$$\int_{-\pi}^{\pi} [x'(t)^2 + y'(t)^2]\,dt + 2\int_{-\pi}^{\pi} x'(t)y(t)\,dt \ge 0.$$

Upon rearranging terms, this integral inequality may be expressed in the form

$$\int_{-\pi}^{\pi} [x'(t) + y(t)]^2\,dt + \int_{-\pi}^{\pi} [y'(t)^2 - y^2(t)]\,dt \ge 0. \tag{42}$$

In (42), the first integral is nonnegative. The second suggests part B of Theorem 1.14, Wirtinger's inequality for y, and we may have $c_0 = (1/\pi)\int_{-\pi}^{\pi} y(t)\,dt = 0$ by arranging that the origin lies on the centroid of the perimetral curve. We conclude that when $\ell = 2\pi$, then $A \le \pi$ and that, in general, the **isoperimetric inequality**

$$\text{Area} \le (\text{Perimeter})^2/4\pi$$

holds (with equality iff the perimeter is a circle). (See Problem 1.9A.)

---

† See, for example, [Ap]. This also follows from the divergence theorem of Section 0.2 with $\mathbf{f} = (0, y)$ and $\mathbf{n} = (y', -x')$.

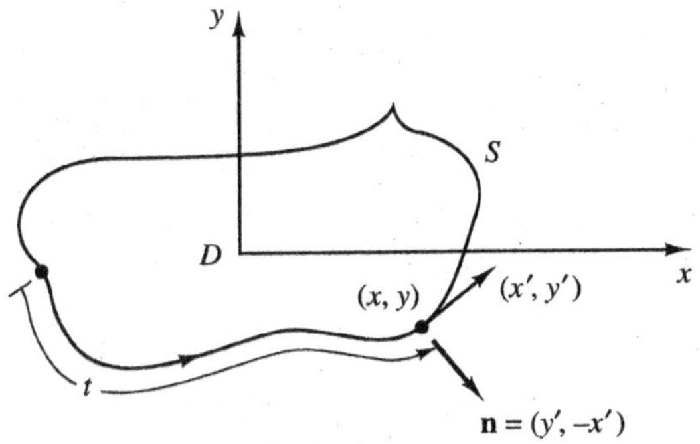

Figure 1.17

Polynomial Approximation

According to Theorem 1.14, we can approximate certain periodic continuous functions uniformly by trigonometric functions of the form

$$\frac{c_0}{2} + \sum_{n=1}^{N} (c_n \cos nt + s_n \sin nt).$$

We can also use other classes of functions to do the approximating, and in this section we will see how far we can go using just polynomials.

First, observe that from Euler's formula and the binomial theorem, it follows that for each $n = 0, 1, 2, \ldots$

$$\begin{aligned}\cos nt + i \sin nt = e^{int} &= (\cos t + i \sin t)^n \\ &= \cos^n t + n(i \sin t)\cos^{n-1} t + \frac{n(n-1)}{2}(i \sin t)^2 \cos^{n-2} t \\ &\quad + \cdots + (i \sin t)^n.\end{aligned}$$

In this sum, the terms involving $(i \sin t)^k$ are real, precisely when $k = 2m$, and then $(i \sin t)^k = (-1)^m (1 - \cos^2 t)^m$. Thus equating real parts, we see that

$$\cos nt = p_n(\cos t) \tag{43}$$

for a unique **polynomial** p_n of degree n. For example, $p_0(x) = 1$, $p_1(x) = x$, and $p_2(x) = 2x^2 - 1$. (When properly normalized, these p_n are known as the Tchebycheff polynomials; see Problem 1.9I.)

Now suppose that $\phi \in \hat{C}^1[-1, 1]$; then $f(t) \stackrel{\text{def}}{=} \phi(\cos t)$ has an *even* continuous extension of period 2π, which has derivatives from the right and left at each point. Using Theorem 1.14, we can approximate f uniformly by the partial sums of its *cosine* series

$$F_N(t) = \frac{c_0}{2} + \sum_{n=1}^{N} c_n \cos nt$$

and, by (43), replace each F_N by a polynomial in $\cos t$, say, $F_N(t) = P_N(\cos t)$. Thus given $\epsilon > 0$, we have for $t \in [0, \pi]$:

$$|\phi(\cos t) - P_N(\cos t)| < \epsilon, \qquad \text{if } N \geq N_\epsilon$$

and so, for $x = \cos t \in [-1, 1]$:

$$|\phi(x) - P_N(x)| < \epsilon, \qquad \text{when } N \geq N_\epsilon$$

This argument gives the desired uniform polynomial approximation for each $\phi \in \hat{C}^1[-\pi, \pi]$, and in particular, for a continuous ϕ which is *piecewise linear* on $[-\pi, \pi]$. But any $f \in C[-\pi, \pi]$ is uniformly continuous (see Appendix A.1) and so it can be approximated as closely as we please by a piecewise-linear ϕ that agrees with it at equally spaced points. Hence f can be uniformly approximated by a polynomial that approximates ϕ. Through translation and a scale change, we arrive at the following results.

(1.17) Theorem (Weierstrass): *If $f \in C[a, b]$, then it can be approximated uniformly on $[a, b]$ within a given $\epsilon > 0$ by a polynomial P_ϵ. If also*

$$\int_a^b f(x)x^n\,dx = 0, \qquad n = 0, 1, 2, \ldots, \qquad \text{then } f(x) = 0 \text{ on } [a, b].$$

Proof: It remains only to establish the last assertion. Its hypotheses imply that if $P(x) = a_0 + a_1x + a_2x^2 + \cdots + a_Nx^N$ is any polynomial, then

$$\int_a^b f(x)P(x)\,dx = 0, \qquad \text{so that} \qquad \int_a^b f^2(x)\,dx = \int_a^b f(x)(f(x) - P(x))\,dx.$$

However, $M = \int_a^b |f(x)|\,dx$ is finite, and we know that for each $\epsilon > 0$, $|f(x) - P_\epsilon(x)| < \epsilon$ for some polynomial P_ϵ. Thus

$$0 \leq \int_a^b f^2(x)\,dx \leq \int_a^b |f(x)||f(x) - P_\epsilon(x)|\,dx \leq M\epsilon$$

for arbitrary $\epsilon > 0$. Since f is continuous, it follows that $f^2(x) = 0$ when $x \in [a, b]$ so that $f = 0$ on $[a, b]$. ∎

Partial Summation

To establish uniform convergence of the sequence

$$\sum_{n=1}^N \frac{(-1)^{n+1}\sin nt}{n}, \qquad N = 1, 2, \ldots, \qquad |t| \leq b < \pi$$

in (41), we rearranged its terms to take advantage of the fact that

$$0 \leq \frac{1}{n} - \frac{1}{n+1} = \frac{1}{n(n+1)} \leq \frac{1}{n^2}, \qquad n = 1, 2, \ldots.$$

(See Problem 1.8F.) To accomplish this we used Abel's method of **partial summation,** which is based on the observation that

$$F_n \stackrel{\text{def}}{=} f_1 + f_2 + \cdots + f_n \quad \Leftrightarrow \quad f_n = F_n - F_{n-1}, \qquad n = 2, 3, \ldots.$$

Thus, for any real or complex numbers p_n, f_n, $n = 1, 2, \ldots$

$$\sum_{n=N+1}^M p_n f_n = \sum_{n=N+1}^M p_n(F_n - F_{n-1}) = \sum_{n=N+1}^M p_n F_n - \sum_{n=N}^{M-1} p_{n+1}F_n$$

or

$$\sum_{n=N+1}^M p_n f_n = p_{M+1}F_M - p_{N+1}F_N + \sum_{n=N+1}^M (p_n - p_{n+1})F_n. \tag{44}$$

This formula of partial summation bears a suggestive resemblance to one of partial integration for $\int_{N+1}^{M} p(x)f(x)\,dx$, but its elements are less easily remembered and we shall not pursue the analogy. For the above application, we should take

$$p_n = \frac{1}{n}, \qquad \text{and for fixed } t, \qquad f_n = f_n(t) = (-1)^{n+1}\sin nt, \qquad n = 1,2,\ldots;$$

the consequences were developed in Section 1.8d. In particular, from (41) we concluded that the series $\Sigma_{n=1}^{\infty}[(-1)^{n+1}/n]\sin nt$ converges uniformly (but not absolutely) on $T = [-\pi/2, \pi/2]$.

We can use uniform convergence on a set T of the series

$$F(t) = \sum_{n=1}^{\infty} f_n(t)$$

of real- or complex-valued functions f_n to establish uniform convergence of related series of the form $\Sigma_{n=1}^{\infty} p_n f_n(t)$ where $0 \le p_{n+1} \le p_n \le P < +\infty$, $n = 1,2,\ldots$. To prove this, we introduce the partial sums $F_n = f_1 + f_2 + \cdots + f_n$ for $n = 1,2,\ldots$ and modify formula (44) to obtain the following identity for positive integers N and $M > N$.

$$\sum_{n=N+1}^{M} p_n f_n = p_{M+1}(F_M - F) - p_{N+1}(F_N - F) + \sum_{n=N+1}^{M} (p_n - p_{n+1})(F_n - F). \qquad \mathbf{(44')}$$

Observe that the telescoping positive-term series $\Sigma_{n=N+1}^{\infty}(p_n - p_{n+1})$ converges with sum $p_{N+1} - \lim_{M\to\infty} p_M \le P$.

(1.18) Abel's Test:† *If $\Sigma_{n=1}^{\infty} f_n(t)$ converges uniformly for $t \in T$, and if $0 \le p_{n+1}(x) \le p_n(x) \le P$, $n = 1,2,\ldots$, $x \in X$, then the series $\Sigma_{n=1}^{\infty} p_n(x) f_n(t)$ converges uniformly on $X \times T$.*

Remarks: Absolute convergence is neither hypothesized nor implied. 0 and P may be replaced by any real constants A and B. The result also holds when the inequalities involving the p_n are reversed.

Proof:* Given $\epsilon > 0$, we can find $N(\epsilon)$ so large that $n \ge N(\epsilon) \Rightarrow |F_n - F| < \epsilon$ uniformly on T. (Why?) Thus if $n \ge N = N(\epsilon)$, then

$$0 \le (p_n - p_{n+1})\,|F_n - F| \le \epsilon(p_n - p_{n+1}).$$

As $M \to \infty$, the series on the right in (44′) converges absolutely by comparison, and its sum has absolute value $\le P\epsilon$. Similarly,

$$0 \le p_{M+1}|F_M - F| \le P|F_M - F|, \text{ which} \to 0 \text{ as } M \to \infty$$
$$\text{while} \qquad 0 \le p_{N+1}|F_N - F| \le P\epsilon.$$

It follows that the series $\Sigma_{n=N+1}^{\infty} p_n(x) f_n(t)$ converges at each $(x,t) \in X \times T$, and its sum is bounded (in absolute value) by $2P\epsilon$ *uniformly* on $X \times T$, when $N \ge N(\epsilon)$. This establishes the desired convergence. ■

† There are several tests for series convergence based on Abel's method of partial summation, but we require only this result.

Example 14:

The series $F(t) = \Sigma_{n=1}^{\infty}[(-1)^n/n] \sin nt$ converges uniformly on $T = [-\pi/2, \pi/2]$, and for $p_n(x) = x^n$, we see that if $x \in X = [0, 1]$, then

$$0 \le p_{n+1}(x) = x^{n+1} \le x^n = p_n(x) \le 1.$$

Therefore by Abel's test, the series

$$U(x, t) = \sum_{n=1}^{\infty} \frac{(-1)^n x^n}{n} \sin nt$$

converges uniformly on $X \times T$. Consequently U is continuous on $X \times T$ so that, in particular,

$$\lim_{x \nearrow 1} U(x, t) = U(1, t) = F(t), \qquad t \in T.$$

Problem Set 1.9

1.9A When $\ell = 2\pi$, show that the isoperimetric equality $A = \pi$ is possible only for functions $x(t)$, $y(t)$ in (42) that parametrize a circle. (*Hint:* See Problem 1.8D.)

1.9B *1.* For (43), find p_3, p_4, and p_5.

2. Use (43) to infer that for each $n = 0, 1, 2, \ldots$

$$\cos^n t = \frac{c_0}{2} + c_1 \cos t + c_2 \cos 2t + \cdots + c_n \cos nt,$$

for appropriate coefficients (which vary with n).

3. Find the coefficients required to express $\cos^3 t$ and $\cos^5 t$ in this form.

1.9C In (43) replace t by $t - (\pi/2)$, and show that when n is odd, $\sin nt = \sin(n\pi/2) p_n(\sin t)$.

1.9D *1.* Use Problem 1.9C to infer that when n is odd, $\sin^n t = s_1 \sin t + s_3 \sin 3t + \cdots + s_n \sin nt$, for appropriate coefficients $s_1, s_3, \ldots$ (which vary with n).

2. Express $\sin^3 t$ and $\sin^5 t$ in this form.

3. Could $\sin^2 t$ be so expressed? Explain.

1.9E In Theorem 1.17, show that the polynomial P_ϵ may be assumed to have the same endpoint values as f. (*Hint:* Subtract a linear function from P_ϵ.)

1.9F (For this problem, we assume that a computer graphics terminal is available.) Let

$$f(t) = \begin{cases} 1, & 0 < t < \pi, \\ -1, & -\pi < t < 0. \end{cases}$$

1. For $N = 10$, calculate the Fourier sums $F_N(t)$ where the coefficients are given by (28). Graph the error $e_{10}(t) = f(t) - F_{10}(t)$.

2. Compute the arithmetic mean approximation

$$\sigma_N(t) = \frac{1}{N}[F_0(t) + F_1(t) + \cdots + F_{N-1}(t)]$$

for $N = 10$ and graph the error $\epsilon_{10}(t) = f(t) - \sigma_{10}(t)$.

1.9G 1. Derive the following version of the summation by parts formula.

$$\sum_{n=1}^{M} p_n f_n = \sum_{n=1}^{M-1} F_n(p_n - p_{n+1}) + F_M p_M.$$

2. Use the formula in part 1 to solve Problem 1.8F2.

1.9H Consider the divergent series

$$\tfrac{1}{2} + \cos t + \cos 2t + \cdots + \cos(n-1)t + \cdots.$$

Show that this series generates the sequence of arithmetic means

$$\sigma_N(t) = \frac{\sin^2(Nt/2)}{2N\sin^2(t/2)}, \qquad N = 1, 2, \ldots .$$

(See Problems 1.8F and 1.9F.)

1.9I (*The Tchebycheff polynomials*).

1. Use the trigonometric identity

$$\cos nt + \cos(n-2)t = 2\cos t\cos(n-1)t$$

with (43) to obtain the recurrence relation

$$p_n(x) = 2xp_{n-1}(x) - p_{n-2}(x), \qquad n = 2, 3, \ldots .$$

2. Use the result in part 1 to obtain p_2, p_3, and p_4 from $p_0(x) = 1$ and $p_1(x) = x$. Then explain why $p_n(x) = 2^{n-1}x^n + \cdots$ for $n = 1, 2, \ldots$.
3. Verify that the nth Tchebycheff polynomial $T_n(x) \stackrel{\text{def}}{=} 2^{1-n}p_n(x) = x^n + \cdots$ satisfies $\max_{|x|\le 1} |T_n(x)| = 2^{1-n}$.
4.* Suppose that for some polynomial

$$P_n(x) = x^n + a_{n-1}x^{n-1} + \cdots + a_0$$

we have $\max_{|x|\le 1} |P_n(x)| < 2^{1-n}$. Obtain a contradiction by noting that $Q(x) = T_n(x) - P_n(x)$ is a polynomial of degree $< n$ that has the same sign as T_n at the $(n+1)$ values of $x = \cos t \in [-1, 1]$ where $\cos nt = \pm 1$.
5. Conclude that the best uniform approximation to x^n on $[-1, 1]$ by a polynomial Q of degree $< n$ is given by $Q(x) = T_n(x) - x^n$.

2

Linear Boundary Value Problems

The surprising fact about applied mathematics is that it works! Why should our universe let itself be so well-described mathematically that we can make mathematical predictions about its behavior? Perhaps partly because mathematics is the study of logical patterns, and to the extent that any orderly description is possible, it must use — or invent — mathematical language. Moreover, from antiquity science and mathematics have emerged hand in hand, with each supporting the development of the other. This definitely applies to boundary value problems, where the description of physical processes in terms of their rates of change led to a study of differential equations that in turn permitted more elaborate descriptions.

As a result, we now know how to model many physical processes mathematically using *ordinary differential equations* (pointwise relations between functions of one variable and their derivatives) or *partial differential equations* (pointwise relations between functions of several variables and their partial derivatives). In each case the equation is assigned the *order* of the highest-ordered derivative required to express it.

You probably remember finding solutions to ordinary differential equations on an interval that meet given conditions at the endpoint(s) of the interval. Similarly, with partial differential equations, we seek solutions in a given domain that meet one or more prescribed conditions at the boundary of the domain. The resulting pair consisting of

1. a partial differential equation in a domain
2. a set of prescribed auxiliary conditions

is referred to as a **boundary value problem** for the domain. The problem is **linear** when each of these ingredients can be specified by means of **linear operators** $\mathscr{L}: \mathscr{U} \to \mathscr{W}$, where $\mathscr{U}$ and $\mathscr{W}$ are appropriate function spaces. (See Section 0.1c.)

In this text, we concentrate on linear boundary value problems involving differential equations of the **second order** that arise in mathematical physics and engineering. However, some considerations are common to all linear boundary value problems. They are presented in this chapter (Section 2.3) together with separation methods for solving some of the simpler problems (Section 2.4), methods which we preview in Section 2.2 using Fourier series. But first, let's clarify some of the relevant concepts.

2.1

Linear Operators and Equations

Differentiation is a process that transforms a function u into another function w, and it does so linearly. In general, a transformation $\mathscr{L}: \mathscr{U} \to \mathscr{W}$ between real linear spaces $\mathscr{U}$ and $\mathscr{W}$ is **linear** when

$$\mathscr{L}(au + bv) = a\mathscr{L}(u) + b\mathscr{L}(v) \tag{1}$$

for all $u, v \in \mathscr{U}$ and constants $a, b \in \mathbb{R}$. Then, by repetition, if $u_1, u_2, \ldots, u_N \in \mathscr{U}$, it follows that

$$\mathscr{L}(a_1u_1 + a_2u_2 + \cdots + a_Nu_N) = a_1\mathscr{L}(u_1) + a_2\mathscr{L}(u_2) + \cdots + a_N\mathscr{L}(u_N), \tag{1$'$}$$

for any real coefficients $a_1, a_2, \ldots, a_N$. Hence

$$\mathscr{L}(u_n) = 0, \qquad n \le N \Rightarrow \mathscr{L}(U_N) = 0, \qquad \text{where } U_N \stackrel{\text{def}}{=} \sum_{n=1}^{N} a_nu_n, \tag{1$''$}$$

is called a *linear combination* of $u_1, u_2, \ldots,$ and u_N.

If $\mathscr{L}$ is linear as above, then for each *fixed* $F \in \mathscr{W}$, the *equation*

$$\mathscr{L}(u) = F \text{ is } \textit{linear} \text{ and } \begin{cases} \textit{homogeneous}, & \text{if } F = 0 \\ \textit{inhomogeneous}, & \text{if } F \neq 0. \end{cases}$$

Each $u \in \mathscr{U}$ for which $\mathscr{L}(u) = F$ is called a *solution* of this equation. A homogeneous equation always has the trivial solution $u = 0$ but might not have a nontrivial solution. An inhomogeneous equation might not have *any* solution; however, observe that if $\mathscr{L}(u_1) = F$ and $\mathscr{L}(u_2) = F$, then

$$\mathscr{L}(u_1 - u_2) = \mathscr{L}(u_1) - \mathscr{L}(u_2) = F - F = 0.$$

If these circumstances imply that $u_1 = u_2$, then equation $\mathscr{L}(u) = F$ has at most one solution and we say that its solution (if any) is *unique*. Consequently, we have the following results.

(2.1) Proposition: *The* **difference** *u of two solutions u_1 and u_2 of the* **same** *inhomogeneous linear equation is a solution of the corresponding* **homogeneous** *equation. Therefore, the solution to the inhomogeneous equation is unique when the homogeneous equation has zero as its only solution.* ■

Eigenvalue Problems

In case $\mathscr{U} \subseteq \mathscr{W}$, where $\mathscr{W}$ is a linear space of functions, there may be special *nonzero* $u \in \mathscr{U}$ such that

$$\mathscr{L}(u) = \lambda u,$$

for some number λ, real or complex. When this occurs, u is said to be an **eigenfunction** to the **eigenvalue** λ.† The search for such λ and u is referred to as the **eigenvalue problem** for $\mathscr{L}$ *on* $\mathscr{U}$.

A Little Conservation

Boundary value problems emerged from mathematical descriptions of the physical processes that relate conditions at the boundaries of a region to occurrences within the region. For example, heat generated inside a room affects the temperature at the walls, which in turn can raise or lower the temperature within. Electrostatic charge on a thin spherical shell produces a field with effects both inside and outside the shell. During flow through a pipe, material that enters one end must either leave the other or be accounted for by some occurrence within the pipe.

Such observations have resulted in the formulation of conservation laws expressed in terms of various physical quantities, including charge, mass, temperature, and momentum. If these quantities are assumed to be distributed over space and time, their local averages provide them pointwise definitions. Then, as functions of position and time, they are subject to the operations of calculus, and the principal operations — differentiation and integration — are *linear*.

There are also mathematical conservation laws relating the boundary values of a function to values of its derivatives within. The best-known of these is the Fundamental Theorem of Calculus in the form

$$\int_a^b f'(x)\,dx = f(x)\Big|_a^b = f(b) - f(a).$$

Here we suppose that f is continuously differentiable on a bounded open interval (a, b) and that f extends continuously to the closure $[a, b]$ (denoted $f \in D^1[a, b]$). The possibly improper Riemann integral on the left has the (limiting) value on the right.

In $\mathbb{R}^d$ (d-dimensional euclidean space), one version of this familiar result is the **divergence theorem** expressed as follows (see Section 0.2).

$$\int_V \nabla \cdot \mathbf{f}\,dV = \int_S \mathbf{f} \cdot \mathbf{n}\,dS, \qquad (d > 1), \tag{2}$$

where the integral over V may be improper.

Here we suppose that V is a **simple domain** of $\mathbb{R}^d$ with boundary $S = \partial V$ and closure $\bar{V} = V \cup S$, as defined in Section 0.1b. V is bounded and $\mathbf{n}$ is the unit

† Sometimes the English words *proper* or *characteristic* are used instead of the German adjective *eigen*, which denotes a property that is special to an individual.

normal vector on S directed *outward* from V. $\mathbf{f}$ is a vector field on V with d components denoted $f_1, f_2, \dots, f_d$, and each $f_j \in D^1(\bar{V})$ (i.e., f_j is continuously differentiable in V, and f_j has a continuous extension to $\bar{V}$). Finally, $\nabla \cdot \mathbf{f}$ denotes the **divergence** of the vector field $\mathbf{f}$. In Cartesian coordinates

$$\mathbf{x} = (x_1, x_2, \dots, x_d), \qquad \text{where } dV = dx_1\, dx_2 \,\cdots\, dx_d,$$

it is expresed as follows.

$$\nabla \cdot \mathbf{f} = \frac{\partial f_1}{\partial x_1} + \frac{\partial f_2}{\partial x_2} + \cdots + \frac{\partial f_d}{\partial x_d}. \tag{3}$$

We can regard $\mathbf{f}$ as the **flux** (density) **vector** of some scalar quantity Q, which is spread over V and considered to be in motion. That is, $\mathbf{f}$ directs the local flow of Q at a point in V at time t, and $|\mathbf{f}|$ prescribes its intensity as measured by the amount of Q per unit time that would cross a (d-1)-dimensional surface element S of unit area oriented *normal* to the flow. Then

$$\mathscr{N} \overset{\text{def}}{=} \int_S \mathbf{f} \cdot \mathbf{n}\, dS$$

represents the total flux of Q out of V across S. The divergence theorem (2) shows that $\mathscr{N} = 0$, when in V, the field $\mathbf{f}$ is **divergence free** in that $\nabla \cdot \mathbf{f} = 0$. Then the amount of Q within V is not affected by its flux across S, and a kind of conservation is achieved.

2.2

Harmonic Preview

We have just seen that divergence-free fields have a useful physical interpretation, but where can we find these mathematical objects?

If V is a domain in $\mathbb{R}^d$, then for any real-valued function $u \in C^2(V)$, the **gradient** $\mathbf{f} = \nabla u$ is a vector field in V with **potential** u. In Cartesian coordinates, $\mathbf{f} = \nabla u$ has components $u_{x_j} = \partial u / \partial x_j, j = 1, 2, \dots, d$, and if $u \in C^2(V)$, then the divergence of $\mathbf{f}$ is given by the **Laplacian**

$$\nabla^2 u \overset{\text{def}}{=} \nabla \cdot (\nabla u) = u_{x_1 x_1} + u_{x_2 x_2} + \cdots + u_{x_d x_d}. \tag{4}$$

If $u \in C^2(V)$, then $\mathbf{f} = \nabla u$ is divergence free in V when u satisfies **Laplace's equation**

$$\nabla^2 u = 0 \qquad \text{in} \qquad V. \tag{4$'$}$$

Real-valued solutions $u \in C^2(V)$ of this *partial differential equation* are said to be **harmonic** in V. The equation is *linear* since if also $v \in C^2(V)$, then

$$\nabla^2(au + bv) = a\nabla^2 u + b\nabla^2 v, \qquad \text{for any constants } a \text{ and } b. \tag{4$''$}$$

Consequently, linear combinations of harmonic functions are again harmonic. (See Problem 2.2A.)

Harmonic functions provide divergence-free fields to represent the flux of many physical quantities including heat, electrostatic charge, and mass. In $\mathbb{R}^2$, they are essential to the study of analytic functions of a complex variable, and in every dimension, they have become an important mathematical tool. In this text, we shall discover some of the properties that make these functions both natural and attractive, as well as the extent to which these properties are shared by solutions of other partial differential equations.

It is easy to see that a linear function

$$u(\mathbf{x}) = a_0 + a_1x_1 + a_2x_2 + \cdots + a_dx_d$$

is harmonic in $\mathbb{R}^d$, for any constants a_j, $j = 0, 1, 2, \ldots, d$. Are there any nonlinear examples? Not in $\mathbb{R}^1$, where $\nabla^2 u = u'' = 0$ on the interval (a, b) iff $u(x) = a_0 + a_1x$ for some constants a_0 and a_1. However, in $\mathbb{R}^2$, with $\mathbf{x} = (x, y)$, we have such harmonic functions as $x^2 - y^2$ and $e^x \cos y$. Moreover, the real and imaginary parts of the complex polynomial $(x + iy)^n$ are harmonic polynomials for each $n = 1, 2, 3, \ldots$ since

$$\frac{\partial^2}{\partial x^2}(x + iy)^n = -\frac{\partial^2}{\partial y^2}(x + iy)^n.$$

(See Problem 2.2B.) Although these harmonic polynomials can be obtained through the binomial formula, they can be expressed most easily in polar coordinates with the aid of Euler's formula, as follows.

$$(x + iy)^n = [r(\cos\theta + i\sin\theta)]^n = (re^{i\theta})^n = r^n e^{in\theta} = r^n \cos n\theta + ir^n \sin n\theta.$$

In $\mathbb{R}^1$, we can choose a_0 and a_1 so that the harmonic function $u(x) = a_0 + a_1x$ takes preassigned values at the endpoints of a bounded interval. Is this true in higher dimensions? In a given domain V of $\mathbb{R}^d$ can we find a harmonic function that extends smoothly (continuously) to $\bar{V}$ and takes prescribed boundary values on $S = \partial V$? Is there more than one such function?

In posing such questions we are fomulating boundary value problems for Laplace's equation in V and asking whether they have unique solutions. Before considering more general problems, let's see how Fourier series provide solutions in one nontrivial case.

Example 1 (*Harmonic functions in a disk*):

In polar coordinates (r, θ) for $\mathbb{R}^2$, when $n = 1, 2, \ldots$, each function $r^n \cos n\theta$ or $r^n \sin n\theta$ is harmonic everywhere, since these are just the real and imaginary parts of the complex polynomial $(x + iy)^n$. Thus in the *unit disk* $V = \{(r, \theta) : r < 1\}$, we have harmonic functions $u(r, \theta) = r^n \cos n\theta$ or $r^n \sin n\theta$, with the respective *boundary values* $u(1, \theta) = \cos n\theta$ or $\sin n\theta$, $n = 0, 1, 2, \ldots$. (See Figure 2.1.) Linear combinations of these functions are also harmonic with corresponding boundary values. For instance,

$$u(r, \theta) = \tfrac{1}{2} - \tfrac{1}{2} r^2 \cos 2\theta$$

is harmonic, and it has the boundary values

$$u(1, \theta) = \tfrac{1}{2} - \tfrac{1}{2} \cos 2\theta = \sin^2 \theta.$$

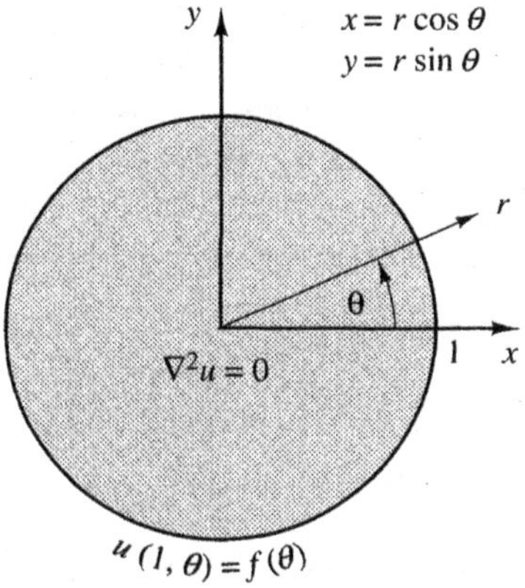

Figure 2.1

Now let's use this approach to *construct* a harmonic function $u = u(r, \theta)$ in V, with the boundary values

$$u(1, \theta) = |\theta|, \qquad -\pi \le \theta \le \pi.$$

(See Figure 2.2.) Note that $f(\theta) = |\theta|$ on $[-\pi, \pi]$ has a continuous extension of period 2π, and (from Example 3 in Section 1.4), the Fourier series *representation*

$$u(1, \theta) = |\theta| = \frac{\pi}{2} - \frac{4}{\pi} \sum_{n=1,3,}^{\infty} \frac{1}{n^2} \cos n\theta, \qquad -\pi \le \theta \le \pi.$$

Therefore, as a solution we try the *formal* series

$$U(r, \theta) \overset{\text{def}}{=} \frac{\pi}{2} - \frac{4}{\pi} \sum_{n=1,3,}^{\infty} \frac{1}{n^2} r^n \cos n\theta, \qquad r \le 1, \tag{5}$$

obtained from multiplying the nth term of the previous series by r^n to make it satisfy Laplace's equation. By the M-test this series converges absolutely and uniformly when $r \le 1$, since then

$$\left| \frac{1}{n^2} r^n \cos n\theta \right| \le \frac{1}{n^2}, \qquad n = 1, 2, \ldots \qquad \text{and} \qquad \sum_{n=1}^{\infty} \frac{1}{n^2} < +\infty.$$

Hence, U is continuous when $r \le 1$ (by the limit interchange principle). But is U harmonic when $r < 1$? If we can differentiate U by differentiating its series termwise, then

$$\nabla^2 U = -\frac{4}{\pi} \sum_{n=1,3,}^{\infty} \frac{1}{n^2} \nabla^2 (r^n \cos n\theta) = 0,$$

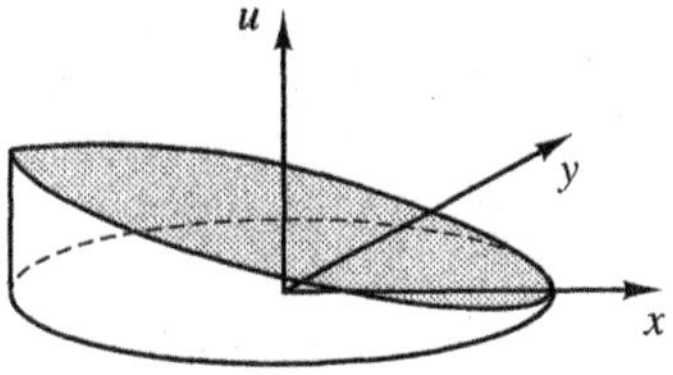

Figure 2.2

and in this case it can be shown that the limit interchange is valid. (We omit the details since we shall later examine them thoroughly.)

Now, this approach gives us a means of solving many similar problems for the unit disk. All that is necessary is a suitable Fourier series representation of the desired boundary values! It is also easy to see how to obtain corresponding results for a disk of arbitrary radius R, and we are led to formulate the following.

(2.2) Program: *To find a solution* $u = u(r,\theta)$ *of the* **boundary value problem**

$$\nabla^2 u = 0, \qquad r < R$$
$$u(R,\theta) = f(\theta), \qquad |\theta| \le \pi$$

where $f \in \hat{C}^1[-\pi,\pi]$ *and* $f(-\pi) = f(\pi)$ *(see* Figure 2.3):

1. *Find coefficients* c_n, s_n *to represent* f *by the Fourier series*

$$f(\theta) = \frac{c_0}{2} + \sum_{n=1}^{\infty} (c_n \cos n\theta + s_n \sin n\theta), \qquad |\theta| \le \pi.$$

2. *Multiply the* nth *term of the series by* $(r/R)^n$.
3. *Then the new series*

$$u(r,\theta) \stackrel{\text{def}}{=} \frac{c_0}{2} + \sum_{n=1}^{\infty} \left(\frac{r}{R}\right)^n (c_n \cos n\theta + s_n \sin n\theta), \qquad r \le R$$

gives the unique solution to the problem. ■

Simple Checks:

1. On the boundary, where $r = R$:

$$u(R,\theta) = \frac{c_0}{2} + \sum_{n=1}^{\infty} (1)^n (c_n \cos n\theta + s_n \sin n\theta) = f(\theta),$$

(from the first series)

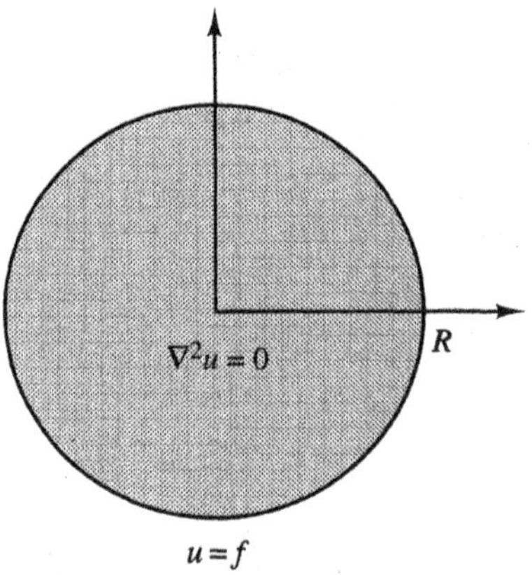

Figure 2.3

and in this manner, our solution meets the prescribed boundary condition.

2. Each term of the new series satisfies Laplace's equation since $r^n \cos n\theta$ and $r^n \sin n\theta$ do so, while c_n/R^n and s_n/R^n are just constants for each $n = 1, 2, \ldots$.

(2.3) Remarks:

1. We can approximate the solution $u = u(r, \theta)$ by the *everywhere*-harmonic functions

$$U_N(r, \theta) = \frac{c_0}{2} + \sum_{n=1}^{N} \left(\frac{r}{R}\right)^n (c_n \cos n\theta + s_n \sin n\theta), \qquad N = 1, 2, \ldots.$$

Using Theorem 1.7, we can obtain the following uniform error estimate: for $r \leq R$,

$$|\, u(r, \theta) - U_N(r, \theta)| \leq \frac{1}{\sqrt{N\pi}} \left(\int_{-\pi}^{\pi} (f'(\tau))^2 \, d\tau \right)^{1/2}, \qquad N = 1, 2, \ldots.$$

(See Problem 2.2L.) From the discussion in Section 1.3, we conclude that u must be continuous in the closed disk $\{r \leq R\}$ and in particular, as $r \nearrow R$, $u(r, \theta) \to u(R, \theta) = f(\theta)$. (Uniqueness will be established in Section 3.5.)

2. (*Poisson's formula*) Upon replacing the coefficients in the series by their integral definitions

$$\left.\begin{matrix} c_n \\ s_n \end{matrix}\right\} = \frac{1}{\pi} \int_{-\pi}^{\pi} f(t) \left\{\begin{matrix} \cos nt \\ \sin nt \end{matrix}\right\} dt, \qquad n = 0, 1, 2, \ldots$$

we see that

$$\begin{aligned} c_n \cos n\theta + s_n \sin n\theta &= \frac{1}{\pi} \int_{-\pi}^{\pi} f(t)(\cos nt \cos n\theta + \sin nt \sin n\theta) \, dt \\ &= \frac{1}{\pi} \int_{-\pi}^{\pi} f(t) \cos n(t - \theta) \, dt, \end{aligned}$$

and after permissible rearrangement, we find that

$$u(r, \theta) = \frac{1}{2\pi} \int_{-\pi}^{\pi} f(t) \left\{ 1 + 2 \sum_{n=1}^{\infty} \left(\frac{r}{R}\right)^n \cos n(t - \theta) \right\} dt, \qquad r \leq R.$$

When $r < R$, the expression in braces can be converted to geometric series with known sums. (See Problem 2.2H.) The result is the **Poisson integral formula,**

$$u(r, \theta) = \frac{1}{2\pi} \int_{-\pi}^{\pi} f(t) \frac{(R^2 - r^2)}{|(R, t) - (r, \theta)|^2} \, dt, \qquad r < R. \tag{6}$$

(The denominator in the integrand is the square of the euclidean distance between points with the stated *polar coordinates*; it admits explicit expression as $R^2 + r^2 - 2rR\cos(t - \theta)$.)

Poisson's formula represents a substantial improvement over the series form of the solution because it enables the values of u to be determined directly through integration. Moreover, the integral is defined and harmonic when $r < R$ (or when $r > R$), for any continuous f or, indeed, any integrable f. However, it is more difficult to prove that as $r \nearrow R$, $u(r,\theta) \to f(\theta)$. (See Example 12 in Section 3.5.)

Before leaving this section, note that if we formally differentiate the series (5) with respect to θ, we have

$$U_\theta(r,\theta) = -\frac{4}{\pi}\sum_{n=1,3,}^{\infty}\frac{\partial}{\partial\theta}\left(\frac{1}{n^2}r^n\cos n\theta\right) = \frac{4}{\pi}\sum_{n=1,3,}^{\infty}\frac{r^n}{n}\sin n\theta, \qquad r < 1.$$

This series vanishes when $\theta = 0$ and $\theta = \pi$, that is, on the flat boundary of the semidisk shown in Figure 2.4, where U_θ is a derivative in the **normal** direction. Thus for the semidisk, U solves the **mixed** boundary value problem indicated in the sketch. In fact, it is the unique solution for this problem. (See Theorem 3.2.) Related applications are explored in Problems 2.2C–F.

Example 1 and its consequences indicate how boundary value problems for a given linear partial differential equation might be formulated and solved. We must find a supply of elementary solutions (in this case, $r^n\cos n\theta$ and $r^n\sin n\theta$) whose linear combinations can be assembled as convergent series with given boundary behavior. Although explicit integral formulas such as Poisson's are rarely achieved, related integral formulas do exist. [G–L].

This approach was developed by Joseph Fourier (1768–1830) in prize-winning investigations of heat conduction initiated in 1804. In this work, Fourier introduced much of the machinery responsible for the successful use of partial differential equations in applied mathematics. He properly distinguished interior effects from boundary effects in formulating linear boundary value problems and introduced separation methods in appropriate coordinates to reduce the problems to ones involving ordinary differential equations. In addition, he insisted, against considerable opposition, that representation of boundary functions by means of related series and integrals was possible.

Problem Set 2.2

2.2A Let $u, v \in C^2(V)$, where V is a domain of $\mathbb{R}^d$.

1. Verify that for constants a and b,

$$(au+bv)_{x_jx_j} = au_{x_jx_j} + bv_{x_jx_j}, \qquad j = 1,2,\ldots,d.$$

2. Show how equation (4″) follows.
3. If also $w \in C^2(V)$, and $c \in \mathbb{R}$, use (4″) to show that

$$\nabla^2(au+bv+cw) = a\nabla^2u + b\nabla^2v + c\nabla^2w.$$

Do *not* differentiate.

2.2B 1. Show by induction that for $n = 1,2,\ldots$

$$\frac{\partial}{\partial y}(x+iy)^n = i\frac{\partial}{\partial x}(x+iy)^n.$$

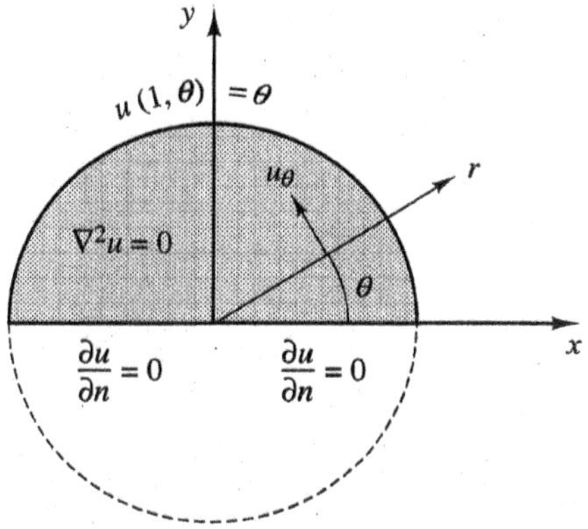

Figure 2.4

Use this formula to show that for each n, $\nabla^2(x+iy)^n = 0$.

2. Write $(x+iy)^n = u(x,y) + iv(x,y)$ and conclude that u and v are harmonic polynomials in $\mathbb{R}^2$
3. Use the binomial formula to obtain these polynomials when $n = 2, 3$, and 4. Show that if $(x+iy)^n = u_n + iv_n$, then

$$u_{n+1} = xu_n - yv_n \qquad \text{and} \qquad v_{n+1} = xv_n + yu_n, \qquad n = 1, 2, \ldots .$$

2.2C Use Program 2.2 to solve the following boundary value problem for the disk of unit radius: $\nabla^2 u = 0$, $r < 1$ where $u(1,\theta) = f(\theta)$, $|\theta| \le \pi$, if

1. $f(\theta) = \cos^2\theta$
2. $f(\theta) = \pi - |\theta|$
3. $f(\theta) = |\cos\theta|$
4. $f(\theta) = \theta(\pi - |\theta|)$

2.2D Repeat Problem 2.2C for the disk of radius 3, and use Poisson's integral formula to represent your answer.

2.2E Use the method suggested at the end of this section to find a harmonic function defined in the given region with the given boundary values.

1.

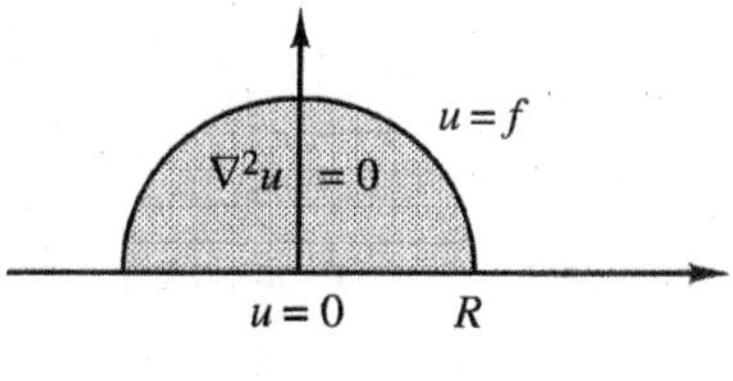

where $f(\theta) = \theta(\pi - \theta), 0 \le \theta \le \pi$.

2.

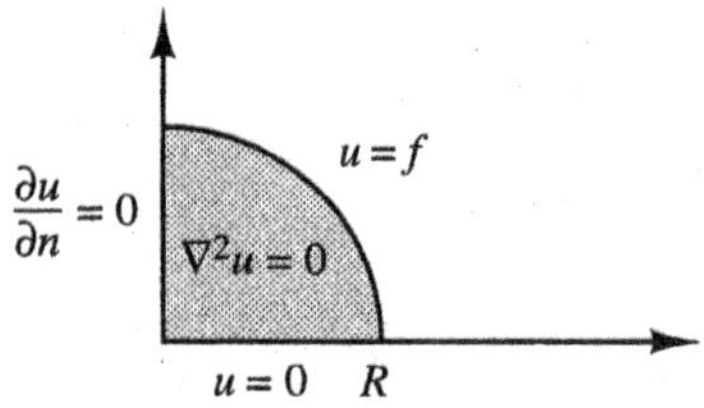

where $f(\theta)=\theta(\pi-\theta), 0\le\theta\le\pi/2$. (*Hint*: Differentiate previous series.)

2.2F Explain how the sine series $U(r,\theta)=\sum_{n=1}^{\infty} s_n r^n \sin n\theta$ could solve the *Dirichlet boundary value* problem indicated in the sketch

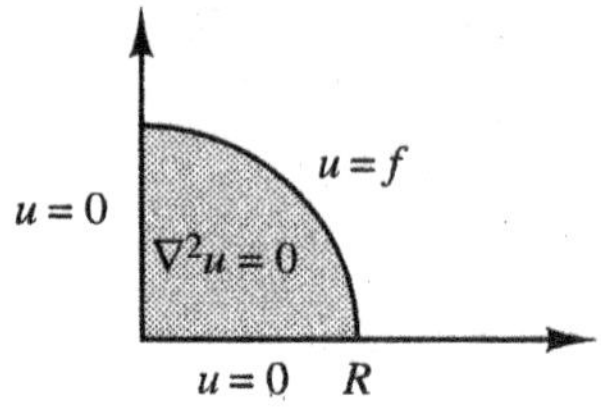

where $f(\theta)=\theta(\pi/2-\theta), 0\le\theta\le\pi/2$. (*Hint*: Consider an extension of f that makes $s_1=s_3=\cdots=0$; or, consider $f(\theta/2)$ and 2.2E1.)

2.2G *1.* Let $u=u(x,y)$ be C^2 in domain $D\subseteq\mathbb{R}^2$. Set $\tilde{u}(r,\theta)=u(r\cos\theta, r\sin\theta)$ and differentiate as needed to conclude that when $r>0$,

$$\tilde{u}_{rr}+\frac{1}{r}\tilde{u}_r+\frac{1}{r^2}\tilde{u}_{\theta\theta}=u_{xx}+u_{yy}=\nabla^2 u.$$

2. If $v(x,y)=u(x/R,y/R)$, show that $R^2\nabla^2 v=\nabla^2 u$.

3. Use part 1 to show directly that for each $n=0,1,2,\ldots$ both $r^n\cos n\theta$ and $r^n\sin n\theta$ satisfy Laplace's equation in $\{(r,\theta): r>0\}$.

2.2H *1.* For each complex number $z\ne 1$, verify that

$$\sum_{n=0}^{N} z^n=\frac{1-z^{N+1}}{1-z}$$

so that as $N\to\infty$, $\sum_{n=0}^{\infty} z^n=\dfrac{1}{1-z}$, $\quad |z|<1.$

2. Use Euler's formula $2\cos\tau=e^{i\tau}+e^{-i\tau}$ with part 1 to show that for $0\le r<R$

$$2\sum_{n=0}^{\infty}\left(\frac{r}{R}\right)^n\cos n\tau=\frac{R}{R-re^{i\tau}}+\frac{R}{R-re^{-i\tau}},$$

and simplify to obtain the result in (6).

2.2I *1.* When $R=1$, for fixed $t\in\mathbb{R}$, express the Poisson kernel

$$P(r,\theta-t)\overset{\text{def}}{=}\frac{1-r^2}{1+r^2-2r\cos(\theta-t)}$$

in the form

$$\tilde{P}(\xi, \eta) = -2\left\{\frac{1}{2} + \frac{\xi}{\xi^2 + \eta^2} \cos t + \frac{\eta}{\xi^2 + \eta^2} \sin t\right\}$$

in terms of the translated coordinates $\xi = x - \cos t$, $\eta = y - \sin t$ and differentiate as needed to show that $\nabla^2 \tilde{P} = \tilde{P}_{\xi\xi} + \tilde{P}_{\eta\eta} = 0$ provided $(\xi, \eta) \neq (0, 0)$. (*Hint*: Show that $\xi/(\xi^2 + \eta^2)$ is a harmonic function of its Cartesian variables.)

2. For this same fixed t, with $r \neq 1$, conclude that $P(r, \theta - t)$ is a harmonic function of the *translated* coordinates x and y.
3. If f is continuous on $[-\pi, \pi]$ with $f(-\pi) = f(\pi)$, explain why the function u defined by (6) is harmonic for $r < R = 1$ and for $r > R = 1$.

2.2J* If f is sectionally smooth on $[-\pi, \pi]$, show that the series U defined in Program 2.2 (with $R = 1$) satisfies $\lim_{r \nearrow 1} U(r, \theta) = [f(\theta+) + f(\theta-)]/2$. How is this related to the Poisson integral (6) with $R = 1$? (*Hint*: Use Theorem 1.18.)

2.2K Let $u, v \in C^2(V)$ where V is some domain of $\mathbb{R}^d$. Verify the identity

$$v\nabla^2 u = \nabla\cdot(v\nabla u) - (\nabla v)\cdot(\nabla u).$$

2.2L Obtain the error estimate given in part 1 of Remarks (2.3) by verifying that for $N = 1, 2, \ldots$ and $r \leq R$:

$$\sum_{n=N+1}^{\infty} \left|\left(\frac{r}{R}\right)^n (c_n \cos n\theta + s_n \sin n\theta)\right|$$

$$\leq \sum_{n=N+1}^{\infty} |c_n \cos n\theta + s_n \sin n\theta| \leq \sum_{n=N+1}^{\infty} \sqrt{c_n^2 + s_n^2}.$$

Then apply Theorem 1.7.

2.3

Linear Differential and Boundary Operators

Differential Operators

We have seen that linear equations are associated with linear operators, and that some nontrivial problems involving Laplace's *equation*, $\nabla^2 u = 0$, can be solved because the Laplace *operator*, ∇^2, is linear.

There are many other linear differential operators and, for example, if $a_0, a_1, a_2, \ldots, a_d$ are real-valued *functions* on a domain V of $\mathbb{R}^d$, then

$$L(u) = a_0 u + a_1 u_{x_1} + a_2 u_{x_2} + \cdots + a_d u_{x_d}$$

defines a linear differential operator in V. If each coefficient function a_j is continuous on V, then $L(u)$ is also continuous whenever $u \in C^1(V)$. We abbreviate this fact by writing $L : C^1(V) \to C(V)$. To show that L is linear, note that if $u, v \in C^1(V)$, then

$$L(u+v) = a_0(u+v) + a_1(u+v)_{x_1} + a_2(u+v)_{x_2} + \cdots + a_d(u+v)_{x_d}$$

$$= a_0 u + a_0 v + a_1 u_{x_1} + a_1 v_{x_1} + \cdots + a_d u_{x_d} + a_d v_{x_d}$$

(*where we have used the linearity of partial differentiation*)

$$= L(u) + L(v). \qquad \text{(after rearrangement)}$$

Similarly, if c is constant, then $L(cu) = cL(u)$. Consequently, if a and b are constants, then $L(au + bv) = aL(u) + bL(v)$, as desired. These operators are said to be **first order** when $a_1, a_2, \ldots, a_d$ do not vanish simultaneously at any point in V. For example,

$$L(u) = 3u_x - \sqrt{1 - x^2}\,u_y$$

defines a first-order operator L in the strip domain $V = \{|x| < 1\}$ of $\mathbb{R}^2$ expressed in terms of standard coordinates.

Similarly, in a domain V of $\mathbb{R}^2$, a **linear differential operator** of the **second order** is defined by

$$L(u) = a_0 u + a_1 u_x + a_2 u_y + a_3 u_{xx} + a_4 u_{xy} + a_5 u_{yy} \tag{7}$$

when the coefficient functions $a_0, a_1, \ldots, a_5$ are given in V and a_3, a_4, and a_5 do not vanish simultaneously. If each a_j is continuous on V, then $L : C^2(V) \to C(V)$, and it is straightforward to show that L is linear. Among the operators in (7), the simplest involving derivatives in both variables are those of the forms $u_{xy}, u_{xx} + au_y, u_{xx} + au_{yy}$, for some constant $a \neq 0$, and we will encounter these in many physical applications.

In every dimension, the **Laplacian** ∇^2 is linear and of second order, while $\nabla^4 = \nabla^2(\nabla^2)$ (the biharmonic operator) is linear and of order four. In physical applications involving space ($\mathbf{x}$) and time (t), ∇ and ∇^2 are restricted to operating on the spatial variable $\mathbf{x}$ only. Then, we have such standard second-order linear operators as

$$\begin{aligned} &\textit{1. } \textbf{heat operator:} && L(u) = u_t - \nabla^2 u \\ &\textit{2. } \textbf{wave operator:} && L(u) = u_{tt} - \nabla^2 u \end{aligned} \tag{8}$$

where $u = u(\mathbf{x}, t)$.

When V is a domain of $\mathbb{R}^d$, this last pair operates on $C^2(\mathscr{V})$ where $\mathscr{V} = V \times \mathbb{R} \subseteq \mathbb{R}^{d+1}$, but they are called **d-dimensional operators.**

Associated with each linear differential operator L are **linear differential equations** of the **same order**. The equation $L(u) = 0$ is **homogeneous**, and the equation $L(u) = F$ is **inhomogeneous** when $F \neq 0$. The homogeneous equation might have only the trivial solution $u = 0$, while the inhomogeneous equation need not have **any** solutions.

Equations involving products or quotients of u and/or its derivatives, or such algebraic functions of u as $\sqrt{1+u}$, or transcendental functions of u such as

e^u, $\sin u$, and so on are usually nonlinear. For example, the equation $K(u) \overset{\text{def}}{=} uu_x + u_t = 0$ is first order but it is *not* linear since

$$K(3u) = (3u)(3u)_x + (3u)_t = 3(3uu_x + u_t) \neq 3K(u).$$

Boundary Operators

Suppose that V is a domain with boundary S as in Section 0.1b. Then, for **second-order** linear partial differential equations, the most common auxiliary conditions are specified by means of the following three types of **boundary operators.**

1. **Dirichlet operator**: $B(u) = u|_S$,
   where the bar denotes restriction;
2. **Neumann operator**: $B(u) = (\partial u/\partial n)|_S = (\mathbf{n}\cdot\nabla u)|_S$
   where **n** is the unit normal vector to S directed outward from V;
3. **Robin operator**: $B(u) = ((\partial u/\partial n) + hu)|_S$
   where h is a given function on S.

For the last two operators, we assume that **n** is well defined at most points of S and that $u \in C^1(\bar{V})$. However, the Dirichlet operator requires only that $u \in C(\bar{V})$. Observe that each of these operators is also linear, in that for constants a, b and suitably defined functions u, v we have

$$B(au + bv) = aB(u) + bB(v). \tag{9}$$

By contrast, the boundary operator $B(u) = (\sin u)|_S$ is *nonlinear* since

$$B(3u) = (\sin 3u)|_S \neq 3(\sin u)|_S = 3B(u).$$

When the linear equation is of order higher than two, we require linear boundary operators involving correspondingly higher-order normal derivatives, operating on functions $u \in C^k(\bar{V})$ for $k > 1$. But for most of our applications, the above will suffice.

Associated with each linear boundary operator B are **boundary conditions** which are either

homogeneous : $B(u) = 0$

or

inhomogeneous : $B(u) = f$, a prescribed nonzero function on S.

B need not apply to the entire boundary, and we may have mixed boundary conditions[†] such as those for the quarter-disk problem shown in Figure 2.5a, where we have

the homogeneous Dirichlet condition $u = 0$ on S_1,

the homogeneous Neumann condition $\dfrac{\partial u}{\partial n} = 0$ on S_2,

and the inhomogeneous Dirichlet condition $u = f$ on S_3.

[†] A problem involving boundary conditions of only one type (Dirichlet, Neumann, or Robin) is often assigned the name of that type. For instance, the problem of Example 1 is a Dirichlet problem.

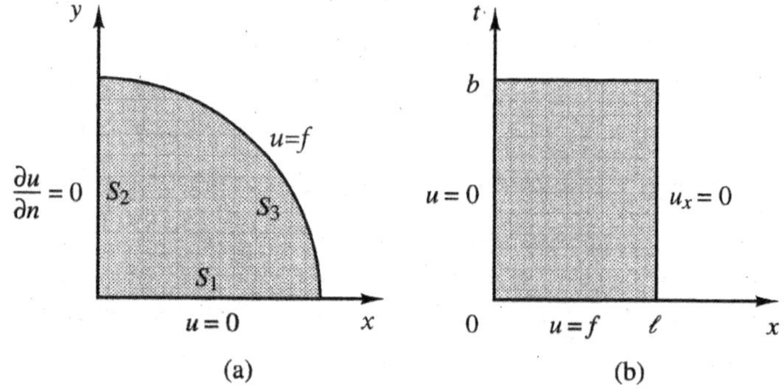

Figure 2.5

In a problem involving both space (**x**) and time (t), in a domain $\mathscr{V} = V \times (0, \mathscr{b})$, auxiliary conditions given at $t = 0$ (at $t = \mathscr{b}$) are called **initial conditions (final conditions)**. Figure 2.5b illustrates this situation for a problem involving $u = u(x,t)$ in $\mathscr{V} = (0, \ell) \times (0, \mathscr{b})$ with

the homogeneous Dirichlet condition $u(0,t) = 0, \quad 0 < t < \mathscr{b}$,

the homogeneous Neumann condition $\dfrac{\partial u}{\partial n}(\ell, t) = u_x(\ell, t) = 0, \quad 0 < t < \mathscr{b}$,

and the inhomogeneous *initial* condition $u(x, 0) = f(x), \quad 0 < x < \ell$.

In this problem we might also give final conditions; however, it may be neither necessary nor desirable to do so. Moreover, we might replace the *boundary* conditions on u specified at $x = 0$ and $x = \ell$ by **periodic conditions** of the type

$$u(0,t) = u(\ell, t) \qquad \text{or} \qquad u_x(0,t) = u_x(\ell, t).$$

In Part II of this text, we consider auxiliary conditions that place bounds or growth restrictions on the solution and/or its derivatives.

Superposition

In physical problems, homogeneous differential equations usually describe a system that is regulating itself internally, perhaps in response to *inhomogeneous auxiliary conditions*, but is not acted on by other external agents. When such agents are present, it is necessary to solve corresponding nonhomogeneous equations, and in the linear case, these take the form

$$L(v) = F.$$

A solution v to this equation represents a response to the external *distributed* "input" F. Observe that if v is any solution to this equation, then $L(av) = aL(v) = aF$, for each constant a; moreover, if $L(v_n) = F_n$, $n = 1, 2, \ldots$, then $L(v_1 + v_2) = F_1 + F_2$. Hence, for any constants, a and b, $av_1 + bv_2$ is a response to the input $aF_1 + bF_2$, and this solution is said to be obtained by **superposition** from v_1 and v_2.

Suppose a boundary value problem is **linear** in that both the differential operator L and *all* associated boundary operators B are linear. Then we can look for a solution formed by superposing solutions to simpler problems in which most of the features are homogeneous.

First, a solution w to a given linear nonhomogeneous boundary value problem indicated by

$$L(w) = F$$
$$B(w) = f,$$

may be represented as $w = u + v$, where u and v are, respectively, solutions to the problems

$$\begin{aligned} L(u) &= 0 \\ B(u) &= f \end{aligned} \qquad \text{and} \qquad \begin{aligned} L(v) &= F \\ B(v) &= 0. \end{aligned}$$

For example, a solution w to the boundary value problem for the annulus shown in Figure 2.6a involving the inhomogeneous equation $L(w) = F$ can be obtained by adding the solutions u and v of the simpler problems indicated in Figure 2.6b.

Moreover, u, a solution to the homogeneous equation $L(u) = 0$, is the sum of solutions to the simpler problems in Figure 2.6d, each of which involves only **one** inhomogeneous auxiliary condition. Such decompositions can be indicated with the aid of $+$ and $=$ symbols, as in these figures.

Finally, solutions to homogeneous equations meeting one or more homogeneous auxiliary conditions can sometimes be superposed to produce solutions that either match or approximate some given **inhomogeneous** condition.

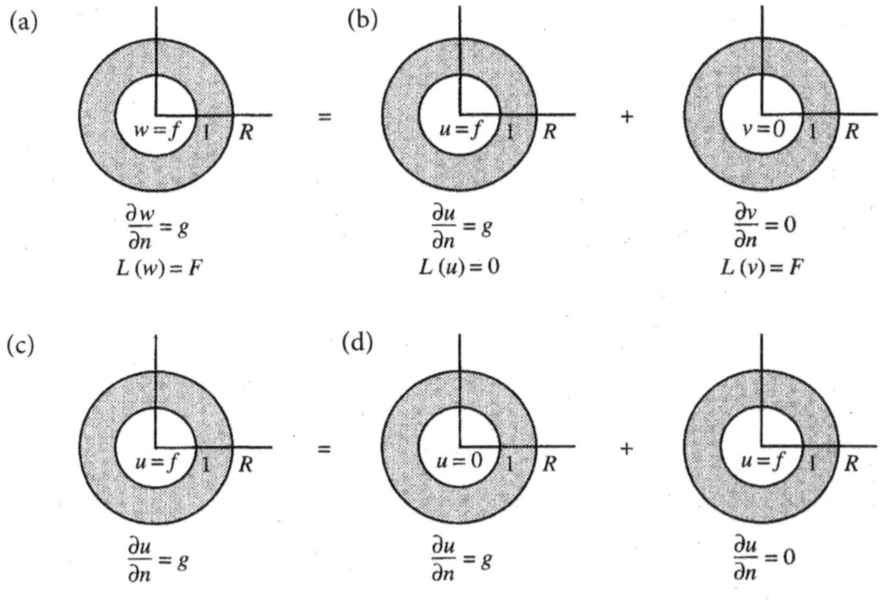

Figure 2.6

Solutions

(2.4) Definition: *We say that a function u is a solution of a given boundary value problem relative to a domain V when*

1. *Within the domain V, u has those continuous partial derivatives appearing in the differential equation, and they satisfy this equation;*
2. *u and those of its partial (or directional) derivatives set forth in the auxiliary conditions admit bounded continuous extensions from V as required, which meet these conditions.* ■

It is customary to assume that within the domain a solution u also has those continuous partial derivatives that can be formed without exceeding either the order of the equation or the separate orders of derivatives in its terms. For example, a solution u of the heat equation $u_t - u_{xx} = 0$ would be assumed to have continuous derivatives u_x, u_{xt}, and u_{xx} in V, while inferences about u_{tt} or u_{txx} are withheld. Similar conventions govern the extensions of u and its derivatives required for the auxiliary conditions. However, no assumptions are made about the solution outside $\bar{V}$, or indeed at boundary points not represented in the auxiliary conditions.

Such **actual** solutions (u), although desirable, are not necessarily available. In general, a candidate (U) obtained through series or integral methods (as in Example 1) should be regarded provisionally as a **formal** solution. Whether U is an actual solution depends in part on the smoothness and compatibility of the auxiliary conditions, both to each other and to the differential equations. To discover auxiliary conditions appropriate to a given problem, we require both experience and theoretical knowledge. Ideally, we would like to impose just enough conditions so that the problem has a unique actual solution.

Problem Set 2.3

2.3A Verify that the following operators are linear, where $\nabla^2 u = u_{xx} + u_{yy}$.

1. $L(u) \equiv \nabla^2 u - 3u_x$
2. $L(u) \equiv u_t - \nabla^2 u + 2(\sin x)u$
3. $L(u) \equiv u_{tt} - \nabla^2 u + 5u_x - 3(xy)u$
4. $L(u) \equiv u_{tt} - c\nabla^2 u$, where c is constant
5. $L(u) \equiv u_{tt} - k\nabla^2 u$ where $k = k(x, y, t)$
6. $L(u) \equiv \nabla \cdot (k\nabla u)$, where k is a C^1 function of the spatial variables
7. $L(u) \equiv \nabla^4 u \equiv \nabla^2(\nabla^2 u)$
8. $L(u) \equiv \nabla^4 u + 6\nabla^2 u - 8(\sin xt)u_t + 7(xy)u$

2.3B Which of the following operators are linear?

1. $L(u) \equiv u^2$
2. $L(u) \equiv 3u$
3. $L(u) \equiv \nabla \cdot (\nabla u)$
4. $L(u) \equiv |\nabla u|^2$
5. $L(u) \equiv uu_x + u$
6. $L(u) \equiv \nabla \cdot (\nabla u^2)$

2.3C Classify each of the following equations for $u = u(x, y, t)$ as (a) linear homogeneous, (b) linear nonhomogeneous, or (c) nonlinear.

1. $u_{xx} + u_y - (x^2 + y^2) = 0$
2. $u_x^2 - u_y = 0$
3. $u_{tt} = (x^2 + y^2)u + \nabla^2 u + \sin(xy)$
4. $u_t - \nabla\cdot((x^2 + y^2)\nabla u) - e^x e^y = 0$
5. $u^2 + 3u_x u_y - 6 = 0$
6. $\nabla^4 u + 6u = 3\nabla^2 u$
7. $\sin^2(xy)u_{xx} - \cos(xy)u_{xy} = 0$

2.3D Let S be the boundary of a spatial domain V in $\mathbb{R}^2$. Classify each of the following boundary conditions as (a) linear homogeneous, (b) linear nonhomogeneous, or (c) nonlinear.

1. $u|_S = 6$
2. $u^2|_S = 6$
3. $u|_S = \left.\dfrac{\partial u}{\partial n}\right|_S$
4. $\left.\dfrac{\partial u}{\partial n}\right|_S = (x \sin y)|_S$
5. $\left.\left(u\dfrac{\partial u}{\partial n}\right)\right|_S = 0$
6. $\left.\dfrac{\partial^2 u}{\partial n^2}\right|_S = 0$
7. $[(3 \sin x)u]|_S = \left.\left[(\cos y)\dfrac{\partial u}{\partial n}\right]\right|_S$

2.3E By trial and error, look for polynomial solutions $u(x, t)$ of degrees 2 and 3 in x for the following equations.

1. $u_t = u_{xx}$
2. $u_{tt} = u_{xx}$
3. $u_{tt} = -u_x$
4. $u_t = xu_x$

2.4

Separation Methods

Homogeneous Problems

A linear boundary value problem is **homogeneous** when the differential equation and *all* of the auxiliary conditions are homogeneous. Such problems **always** have the trivial solution $u \equiv 0$, but they may also have *linearly independent* nontrivial solutions $u_1, u_2, \ldots$.[†]

Then by superposition we see that each finite linear combination

$$U_N \stackrel{\text{def}}{=} a_1 u_1 + a_2 u_2 + \cdots + a_N u_N \tag{10}$$

is also a solution of the *same* homogeneous problem, for any constant coefficients $a_1, a_2, \ldots, a_N$. To produce a U_N in this manner that also satisfies some given inhomogeneous condition within specified accuracy, we probably need a large supply of linearly independent solutions $u_1, u_2, \ldots$. Where do we find it?

It is relatively easy to guess a few solutions to such basic linear equations as

[†] We want assurance that one u_n is **not** a linear combination of others, or equivalently, that if U_N in (10) is the zero function, then $a_1 = a_2 = a_3 = \cdots = a_N = 0$. **Functions** $u_1, u_2, \ldots$ are linearly independent if they are pairwise orthogonal on a common interval. (Recall the discussion leading to Proposition 1.1.)

Laplace's equation, the heat equation, or the wave equation. However, it is more difficult to guess solutions that meet given boundary conditions. Therefore, let's look at some standard situations where pattern recognition helps.

Simple Strip Problems

Suppose homogeneous boundary conditions are specified on a pair of parallel lines in $\mathbb{R}^2$ or parallel planes in $\mathbb{R}^3$. Then we can choose an orientation and scale so that these sets are defined by $x = 0$ and $x = \pi$, in some Cartesian system. For **second-order** equations, some common boundary conditions on these sets take one of the three forms indicated in Figure 2.7, labeled $\mathscr{D}$, $\mathscr{N}$, or $\mathscr{M}$ (for Dirichlet, Neumann, or Mixed).

By inspection we see that for

$$\left.\begin{aligned} &\mathscr{D}\colon X_n(x) = \sin nx, && n = 1, 2 \ldots \\ &\mathscr{N}\colon X_n(x) = \cos nx, && n = 0, 1, 2 \ldots \\ &\mathscr{M}\colon X_n(x) = \sin(nx/2), && n = 1, 3, 5 \ldots \end{aligned}\right\} \tag{11}$$

give functions $u(x, \cdot) = X_n(x)$ with derivatives $u_x(x, \cdot) = X_n'(x)$, which do satisfy the given boundary conditions. This is immediate for $\mathscr{D}$. For $\mathscr{N}$, note that $X_n'(x) = -n \sin nx = 0$ at $x = 0$ and $x = \pi$. For $\mathscr{M}$, we can verify that $X_n(x) = \sin(nx/2)$ does meet the conditions when n is odd, since $X_n'(\pi) = (n/2)\cos(n\pi/2) = 0$ if $n = 1, 3, \ldots$.

The graphs in Figure 2.8 for small integer values of n indicate what is happening in each case. Trigonometric functions have been chosen because they are elementary and because our experience with Fourier series in Example 1 suggests their utility. If possible, we wish to find solutions to the differential equation in each given case that use the appropriate X_n as factors. We will succeed for simple equations expressible in the separated form

$$u_{xx} = \tilde{L}(u),$$

where the differential operator $\tilde{L}$ does not involve x explicitly.

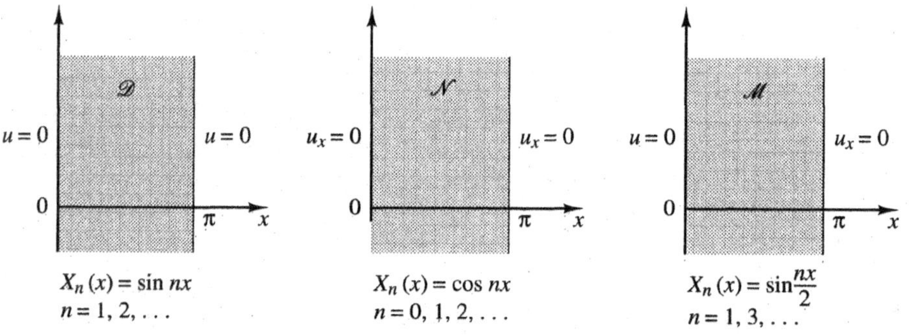

Figure 2.7

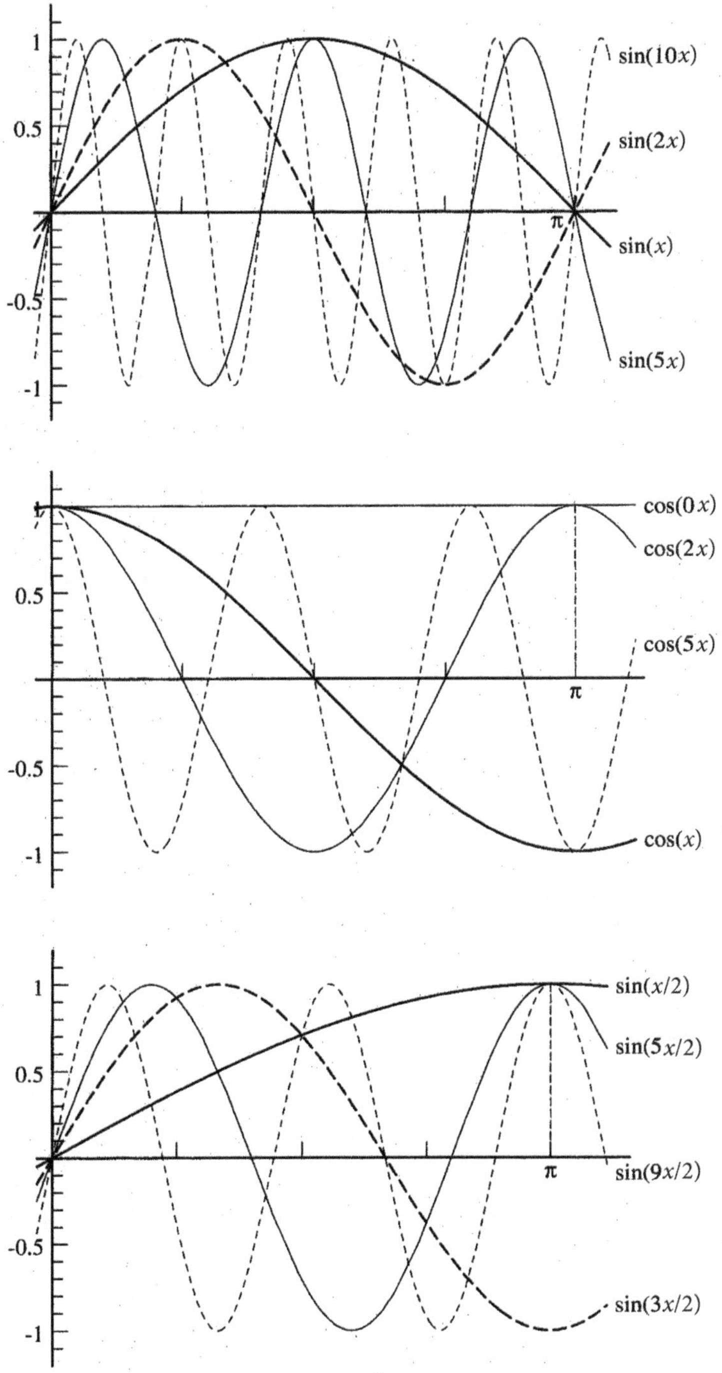

1
0.5
-0.5
-1
π
sin(10x)
sin(2x)
sin(x)
sin(5x)
cos(0x)
cos(2x)
cos(5x)
cos(x)
sin(x/2)
sin(5x/2)
sin(9x/2)
sin(3x/2)

Figure 2.8

Example 2:

Let us find solutions $u = u(x, t)$ to the *heat equation*

$$u_t - u_{xx} = 0, \qquad 0 < x < \pi, \qquad t > 0$$

that satisfy the boundary conditions

$$u(0, t) = u(\pi, t) = 0, \qquad t > 0.$$

(See Figure 2.9.) Here both the equation and the conditions are homogeneous, and the boundary conditions (of type $\mathscr{D}$) are satisfied by $X_n(x) = \sin nx$, $n = 1, 2, \ldots$. For each n,

$$u(x, t) \stackrel{\text{def}}{=} T(t) \sin nx$$

satisfies the same boundary conditions for any function T. u will satisfy the equation

$$u_t = u_{xx}$$

if

$$T'(t) \sin nx = -n^2 T(t) \sin nx.$$

Thus we set $T'(t) = -n^2 T(t)$, and for this ordinary first-order differential equation we have the well-known general solution

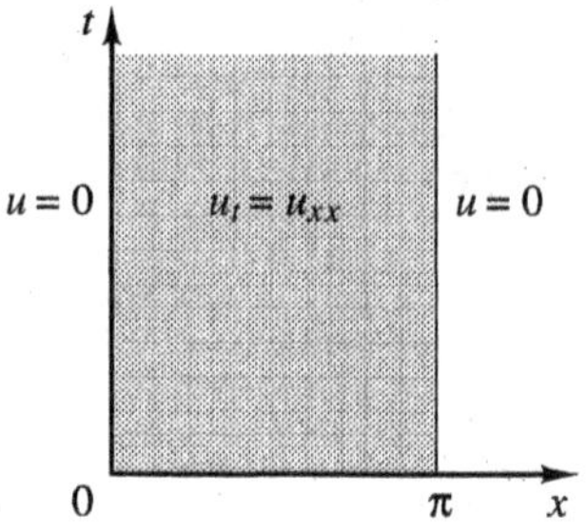

Figure 2.9

$$T_n(t) = b_n e^{-n^2 t},$$

where b_n is a constant. It follows that

$$u_n(x, t) = b_n e^{-n^2 t} \sin nx, \qquad n = 1, 2, \ldots \tag{12}$$

are solutions of the given homogeneous problem. Similarly, for constant a_n,

$$u_n(x, t) = a_n e^{-n^2 t} \cos nx, \qquad n = 0, 1, 2, \ldots \tag{12'}$$

are solutions of the same equation that satisfy boundary conditions of type $\mathscr{N}$. (See Problem 2.4A.)

Suppose that we want solutions to the more general heat equation

$$u_t = k u_{xx}, \qquad 0 < x < \ell, \qquad (k \text{ constant})$$

that satisfy the boundary conditions $u_x(0, t) = u_x(\ell, t) = 0$. These Neumann boundary conditions are of type $\mathscr{N}$, after the scale change that replaces x by $\pi x/\ell$. Thus for each $n = 0, 1, 2, \ldots$, we try a solution of the form

$$u(x, t) = T_n(t) \cos \frac{n\pi x}{\ell}$$

and find that we need to take

$$T_n(t) = a_n e^{-k(n\pi/\ell)^2 t}. \tag{13}$$

Example 3:

For constant $c > 0$, let's find solutions $u = u(x, t)$ of the *wave equation* $u_{tt} - c^2 u_{xx} = 0$, $0 < x < \pi$ with $u(0, t) = u_x(\pi, t) = 0$, $t \in \mathbb{R}$. Here, the boundary conditions are of type $\mathscr{M}$, and we seek product solutions in the form

$$u(x, t) = T(t) \sin \frac{nx}{2}, \qquad n = 1, 3, 5, \ldots$$

of the equation

$$u_{tt} = c^2 u_{xx}.$$

(See Figure 2.10.) Substituting and differentiating as required, we get

$$T''(t) \sin \frac{nx}{2} = -c^2 \left(\frac{n}{2}\right)^2 T(t) \sin \frac{nx}{2}.$$

We see that T should satisfy the equation

$$T'' + \left(\frac{nc}{2}\right)^2 T = 0, \qquad \text{for } n = 1, 3, \ldots,$$

and this equation has the general solution

$$T_n(t) = a_n \cos \frac{nc}{2} t + b_n \sin \frac{nc}{2} t,$$

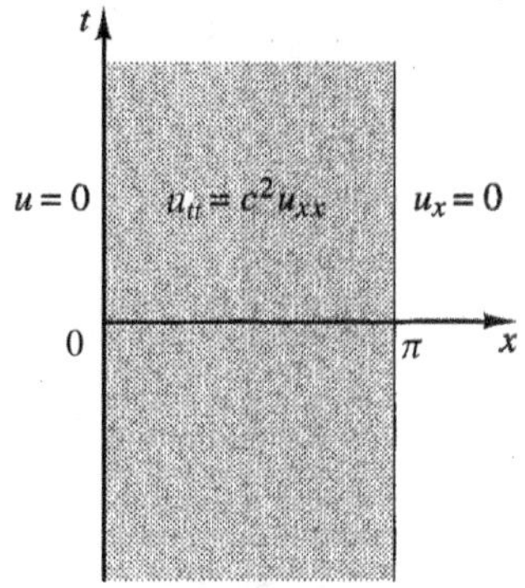

Figure 2.10

for constants a_n and b_n. Thus solutions of the given problem are

$$u_n(x, t) = \left(a_n \cos \frac{nc}{2} t + b_n \sin \frac{nc}{2} t\right) \sin \frac{nx}{2}, \qquad n = 1, 3, \ldots. \tag{14}$$

Observe that if we take $a_n = 0$, these u_n satisfy the additional homogeneous condition

$$u(x, 0) = 0.$$

If instead we take $b_n = 0$, they satisfy the condition

$$u_t(x, 0) = 0.$$

(See Figure 2.11.)

Other choices of a_n and b_n provide u_n that satisfy Robin conditions such as

$$u_t(x, 0) + 2u(x, 0) = 0.$$

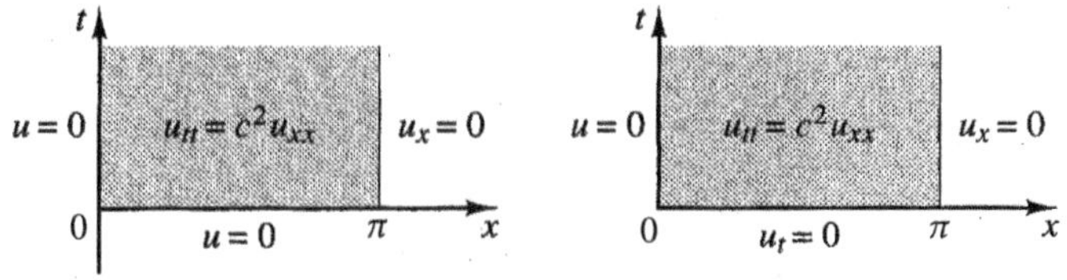

Figure 2.11

Example 4:

In polar coordinates r, θ, Laplace's equation takes the form (see Section 0.1g)

$$u_{rr} + \frac{1}{r}u_r + \frac{1}{r^2}u_{\theta\theta} = 0, \qquad r > 0. \tag{15}$$

To find solutions $u = u(r, \theta)$ satisfying the Neumann conditions

$$u_\theta(r, 0) = u_\theta(r, \pi) = 0, \qquad r > 0,$$

of Figure 2.12, we try

$$u(r, \theta) = R(r)\cos n\theta, \qquad n = 0, 1, 2, \ldots$$

and select $R = R(r)$ to make

$$\left(R'' + \frac{1}{r}R' - \frac{n^2}{r^2}R\right)\cos n\theta = 0.$$

Therefore, R should satisfy the equation

$$R'' + \frac{1}{r}R' - \frac{n^2}{r^2}R = 0,$$

which is **not** elementary. However, on multiplication by r^2, it becomes the Cauchy or equidimensional equation

$$r^2R'' + rR' - n^2R = 0. \tag{16}$$

The substitution $r = e^t$ replaces such equations by equations with constant coefficients in $T(t) \stackrel{\text{def}}{=} R(e^t)$. Indeed,

$$rR' = \dot{T} \qquad \text{and} \qquad r^2R'' = \ddot{T} - \dot{T}, \tag{17}$$

where the dot denotes differentiation with respect to t. In our case, (16) becomes

$$(\ddot{T} - \dot{T}) + \dot{T} - n^2T = \ddot{T} - n^2T = 0.$$

Thus, for $n = 1, 2, \ldots$ and constants a_n and $\tilde{a}_n$,

$$T_n(t) = a_ne^{nt} + \tilde{a}_ne^{-nt},$$

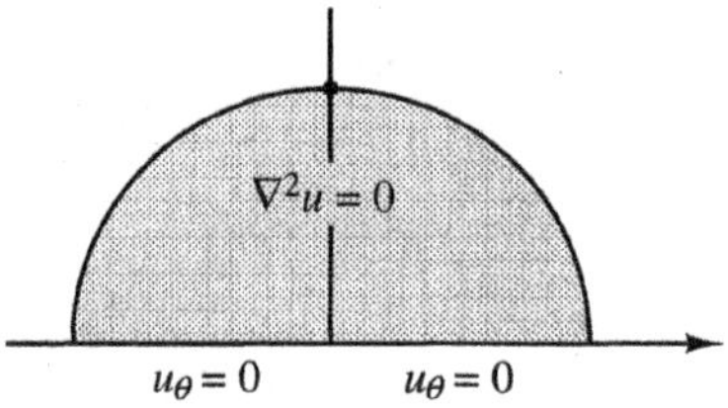

Figure 2.12

or, since $e^{nt} = (e^t)^n = r^n$ and $e^{-nt} = r^{-n}$,

$$R_n(r) = a_n r^n + \tilde{a}_n r^{-n}, \qquad n = 1, 2, \ldots. \tag{18}$$

As usual, the case $n = 0$ requires separate analysis. Here, $T_0(t) = a_0 + \tilde{a}_0 t$ so that

$$R_0(r) = a_0 + \tilde{a}_0 \log r. \tag{18'}$$

This gives the product solutions for $r > 0$

$$\begin{aligned} u_0(r, \theta) &= (a_0 + \tilde{a}_0 \log r) \\ u_n(r, \theta) &= (a_n r^n + \tilde{a}_n r^{-n}) \cos n\theta, \qquad n = 1, 2, \ldots \end{aligned} \tag{19}$$

for arbitrary coefficients a_n, $\tilde{a}_n$, $n = 0, 1, 2, \ldots$. Each of these functions is harmonic in $\mathbb{R}^2 \sim \{0\}$. However, when $\tilde{a}_n = 0$, the resulting function is one used in Example 1 and is harmonic in $\mathbb{R}^2$.

If we have conditions of type $\mathscr{D}$ instead of type $\mathscr{N}$ in this problem, we obtain (19) with $\cos n\theta$ replaced by $\sin n\theta$, for $n = 1, 2, \ldots$ only.

Separation of Variables

When it can be carried through, the strip approach just explored provides a sequence of nontrivial solutions to a given homogeneous problem. If this approach is inapplicable, there is a more basic technique with the same elements known as **separation of variables.** We again seek product solutions of functions of fewer variables, but now we do not preselect any of these functions.

Example 5:

To find solutions $u = u(x, t)$ to the heat equation $u_t = u_{xx}$, $0 < x < \pi$, with $u_x(0, t) = u(\pi, t) = 0$ (see Figure 2.13), we note that these boundary conditions are not of type $\mathscr{M}$. We have several alternatives. First, we might try to guess some trigonometric functions that would work. Second, we could try to convert the problem to one of type $\mathscr{M}$ through replacement of x by $\pi - x$. However, we can also take a more basic approach and seek product solutions

$$u(x, t) = X(x)T(t)$$

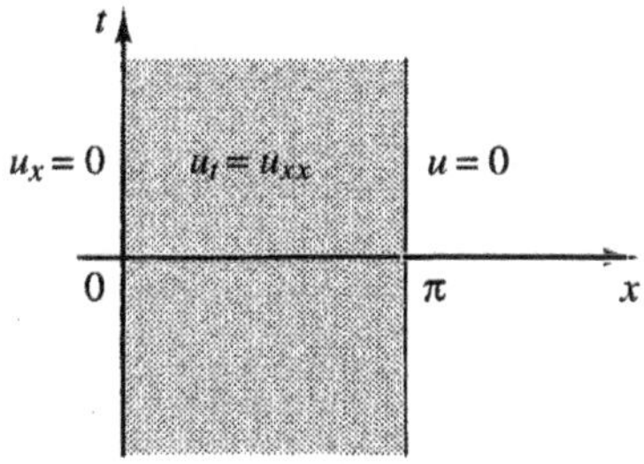

Figure 2.13

where now both functions X and T are unspecified. If we abbreviate as $u = XT$, then

$$u_t = XT', \qquad u_x = X'T, \qquad u_{xx} = X''T,$$

and on substitution in $u_t = u_{xx}$ we get

$$XT' = X''T.$$

Assuming that XT is nonvanishing, we can divide through by XT to **separate variables** in the form,

$$\frac{T'}{T} = \frac{X''}{X} = \lambda, \qquad \text{say.}$$

Now we argue that λ **must be constant**, since it is both a function only of x and a function only of t. Hence we are reduced to solving the *pair* of ordinary equations

$$T' = \lambda T \qquad X'' = \lambda X$$

involving the **same separation constant** λ. The first of these has the well-known general solution $T(t) = ae^{\lambda t}$. However, the form of solution of the second depends on whether λ is positive, negative, or zero. Moreover, in order that $u = XT$ satisfy the given boundary conditions

$$u_x(0, t) = 0 \qquad \text{and} \qquad u(\pi, t) = 0,$$

X must be selected to make

$$X'(0) = 0 \qquad \text{and} \qquad X(\pi) = 0.$$

Thus we must solve the **eigenvalue problem**†

$$\begin{cases} X'' - \lambda X = 0, & 0 < x < \pi \\ X'(0) = X(\pi) = 0 \end{cases} \tag{20}$$

for both λ and X.

Case 1: If $\lambda = 0$, then the resulting equation $X'' = 0$ has the general solution

$$X(x) = c + sx, \qquad \text{with } X'(x) = s,$$

for constants c and s.

$$X'(0) = 0 \Rightarrow s = 0, \qquad \text{so } X(x) = c.$$

$$\text{Then,} \qquad X(\pi) = 0 \Rightarrow c = 0, \qquad \text{so } X(x) \equiv 0.$$

Thus, $\lambda = 0$ gives only the trivial solution $X(x) \equiv 0$.

Case 2: If $\lambda = \omega^2 > 0$, then the resulting equation $X'' - \omega^2 X = 0$ has the general solution

$$X(x) = c \cosh \omega x + s \sinh \omega x,$$

---

† This is the eigenvalue problem for $L(X) \overset{\text{def}}{=} X''$ on the space $\mathscr{U} \overset{\text{def}}{=} \{X \in C^2[0, \pi] : X'(0) = X(\pi) = 0\}$ since $\mathscr{U} \subseteq \mathscr{W} = C[0, \pi]$.

with

$$X'(x) = c\omega \sinh \omega x + s\omega \cosh \omega x,$$

for some constants c and s.

$$\left.\begin{aligned} X'(0) &= 0 = s\omega \Rightarrow s = 0; \\ \text{then} \quad X(\pi) &= 0 = c \cosh \omega\pi \Rightarrow c = 0 \end{aligned}\right\} \Rightarrow X(x) \equiv 0.$$

Again, we have found only the trivial solution.

Case 3: If $\lambda = -\omega^2 < 0$, then the resulting equation $X'' + \omega^2 X = 0$ has the general solution

$$X(x) = c \cos \omega x + s \sin \omega x$$

with

$$X'(x) = -c\omega \sin \omega x + s\omega \cos \omega x,$$

and

$$X'(0) = 0 = s\omega \Rightarrow s = 0.$$

$$\text{Then} \qquad X(\pi) = 0 = c \cos \omega\pi,$$

and finally we have an alternative to $c = 0$, since

$$\cos \omega\pi = 0 \text{ iff } \omega\pi = \frac{n\pi}{2}, \qquad n = 1, 3, 5, \ldots .$$

Thus

$$\lambda_n = -\left(\frac{n}{2}\right)^2 \qquad \text{and} \qquad X_n(x) = \cos \frac{nx}{2}, \qquad n = 1, 3, 5, \ldots .$$

give nontrivial solutions to the eigenvalue problem (20).

The X_n are the eigenfunctions to this eigenvalue problem and these eigenfunctions X_n provide corresponding product solutions

$$u_n(x, t) = e^{-(n/2)^2 t} \cos \frac{nx}{2}, \qquad n = 1, 3, 5, \ldots \tag{21}$$

to the given boundary value problem.

Remarks:

1. Had the separation method been used in each previous example, the X_n previously guessed would have emerged as eigenfunctions to an eigenvalue problem for the equation $X'' = \lambda X$. More labor is required and sign errors are common, but this approach uncovers *all* possible eigenfunctions X_n and thereby *all* possible product solutions. In the next three chapters we will see what can be accomplished when, as in these examples, the eigenfunctions are just the elements for a Fourier series.

2. In some physical applications, we encounter the equation

$$(\tau u_x)_x - qu = -\rho \tilde{L}(u) \tag{22}$$

where τ, q, and ρ are suitably specified functions of x on an interval, and the linear differential operator $\tilde{L}$ does not involve x explicitly. Separation methods are usually effective, and values of the separation constant λ are found from eigenvalue problems of the **Sturm–Liouville** type involving the equation

$$(\tau X')' - qX = -\lambda\rho X \tag{22'}$$

and various homogeneous auxiliary conditions. We will return to these problems and their applications in Chapter 7 and examine them more thoroughly in Chapter 9, where we also consider some multidimensional analogues.

3. The separation method was introduced by Fourier and is used throughout applied mathematics. If the strip approach is inappropriate, then separation should be tried. However, it too can fail. Moreover, an equation may have simple solutions that cannot be expressed in elementary product form. For instance, Laplace's equation $u_{xx} + u_{yy} = 0$ has as solutions $x^2 - y^2$ and certain other polynomials. The one-dimensional wave equation $u_{tt} - u_{xx} = 0$ has as solutions $(x+t)^2$, $\sin(x-t)$, and indeed any C^2 function of $(x+t)$ or of $(x-t)$.

Example 6:

It may be useful to consider complex product solutions of a real equation. The heat equation $u_t = ku_{xx}$ does not admit a nontrivial oscillatory solution of the form

$$u(x,t) = X(x)\cos\omega t, \qquad \omega > 0.$$

However, we can find a solution of the form

$$u(x,t) = X(x)e^{i\omega t}$$

if $X'' = i(\omega/k)X$. (Why?)
With $\omega/k = 2a^2$, the latter equation has complex solutions

$$X(x) = e^{\alpha x} \qquad \text{when} \qquad \alpha^2 = i2a^2 \qquad \text{or} \qquad \alpha = \pm(1+i)a.$$

Therefore for each $a \in \mathbb{R}$,

$$u(x,t) = e^{\alpha x}e^{i\omega t} = e^{ax}e^{i(2a^2kt+ax)}$$

is a complex product solution of the heat equation, while its real part,

$$v(x,t) = e^{ax}\cos(\omega t + ax), \tag{23}$$

is a real-valued solution not readily discovered by direct means; its imaginary part is another.

*General Separation and Nonhomogeneous Problems**

We have just seen how to use separation methods to find solutions $u = u(x,t)$ of the homogeneous equation $u_{xx} = u_t$ on the domain $(0,\pi) \times (0,+\infty)$. More generally, suppose we have a domain of the form $V \times \tilde{V}$, where V and $\tilde{V}$ are also domains, and linear differential operators L and $\tilde{L}$ that involve only variables of

V and $\tilde{V}$, respectively. Then we can use separation methods to find solutions u to the homogeneous equation

$$L(u) = \tilde{L}(u), \qquad \text{in} \qquad V \times \tilde{V}. \tag{24}$$

In fact, functions X on V and $\tilde{X}$ on $\tilde{V}$ generate the product solution $u = X\tilde{X}$ of equation (24) in $V \times \tilde{V}$, provided that they satisfy, respectively, the reduced equations

$$L(X) = \lambda X \qquad \text{and} \qquad \tilde{L}(\tilde{X}) = \lambda \tilde{X} \tag{24'}$$

for some common separation constant λ. If X is also required to satisfy homogeneous auxiliary conditions relative to V, then the resulting *eigenvalue problem* yields values for λ and associated eigenfunctions X. In turn, these values of λ may permit complementary solutions $\tilde{X}$ of the second reduced equation $\tilde{L}(\tilde{X}) = \lambda \tilde{X}$ that meet certain homogeneous conditions relative to $\tilde{V}$.

Once primary separation has been achieved and eigenfunctions X_n to eigenvalues λ_n have been found for $n = 1, 2, \ldots$, it is possible to solve some nonhomogeneous problems. In Example 1, we saw how properly assembled solutions of a homogeneous problem can satisfy inhomogeneous boundary conditions. In addition, the eigenfunctions X_n provide corresponding solutions to the *nonhomogeneous equation*

$$\mathscr{L}(v) \overset{\text{def}}{=} \tilde{L}(v) - L(v) = F$$

for certain functions F prescribed on $V \times \tilde{V}$. Indeed, if $\tilde{X}_n$ is a solution of the reduced nonhomogeneous equation

$$\tilde{L}(\tilde{X}) - \lambda_n \tilde{X} = \tilde{f}_n \qquad \text{on } \tilde{V}, \tag{25}$$

then $v_n \overset{\text{def}}{=} X_n \tilde{X}_n$ is a solution to the nonhomogeneous equation

$$\mathscr{L}(v_n) = X_n \tilde{f}_n.$$

(See Problem 2.4G.) Therefore the series $\sum_{n=1}^{\infty} X_n \tilde{X}_n$ constitutes a *formal* solution to the equation

$$\mathscr{L}(v) = \sum_{n=1}^{\infty} X_n \tilde{f}_n. \tag{25'}$$

Finally, note that in solving any of the reduced problems on V or on $\tilde{V}$, further separation may be possible.

Problem Set 2.4

2.4A 1. Verify that for $n = 0, 1, \ldots$ $u_n(x, t) = a_n e^{-n^2 t} \cos nx$ is a solution to the boundary value problem $u_t = u_{xx}$, $0 < x < \pi$, $t > 0$, with $u_x(0, t) = u_x(\pi, t) = 0$, $t > 0$.

2. Find corresponding solutions to the same equation satisfying

$$u(0, t) = u_x(\pi, t) = 0, \qquad t > 0.$$

2.4B Find elementary product solutions for each of the following homogeneous problems.

1. $L(u) \equiv u_t - ku_{xx} = 0$ in $\{0 < x < \pi, t > 0\}$; (k is constant.)
   $u(0,t) = u(\pi,t) = 0,$ $t > 0.$

2. $L(u) \equiv u_t - u_{xx} = 0$ in $\{0 < x < \ell, t > 0\}.$
   $u(0,t) = u(\ell,t) = 0,$ $t > 0.$

3. $L(u) \equiv u_{xx} + u_{yy} = 0$ in $\{0 < x < \pi, 0 < y < b\}.$
   $u_x(0,y) = u_x(\pi,y) = 0,$ $0 < y < b.$
   $u_y(x,0) = 0,$ $0 < x < \pi.$

4. $L(u) \equiv u_{tt} - u_{xx} = 0$ in $\{0 < x < \pi, t > 0\}.$
   $u_x(0,t) = u_x(\pi,t) = 0,$ $t > 0.$
   $u(x,0) = 0,$ $0 < x < \pi.$

5. $L(u) \equiv u_{tt} - u_{xxt} = 0$ in $\{0 < x < \pi, t > 0\}.$
   $u(0,t) = u(\pi,t) = 0,$ $t > 0.$
   $u(x,0) = 0,$ $0 < x < \pi.$

2.4C Find product solutions for the problem

$$L(u) \equiv u_t - u_x = 0 \quad \text{in} \quad \{0 < x < \ell, t > 0\}$$

satisfying

1. $u(x,0) = 0,$ $0 < x < \ell$ or 2. $u(0,t) = u(\ell,t) = 0,$ $t > 0.$

2.4D* Consider the following problem involving the linearized Korteweg–deVries equation

$$L(u) \equiv u_t - u_{xxx} = 0 \quad \text{in} \quad \{0 < x < \ell, t > 0\}$$

and the boundary conditions $u(0,t) = u(\ell,t) = 0$, $u_x(0,t) = 0$, $t > 0$.

1. Show that nontrivial elementary product solutions must be of the form
$$u(x,t) = e^{\lambda t}X(x)$$
for some constant $\lambda \neq 0$ chosen to make X a solution to the eigenvalue problem
$$X''' = \lambda X,$$
with appropriate boundary conditions.
2. Verify that $X(x) = e^{\alpha x}$ will be a solution if $\alpha^3 = \lambda = 8\omega^3$, say, and conclude that
$$X(x) = be^{2\omega x} + e^{-\omega x}(c\cos(\sqrt{3}\omega x) + s\sin(\sqrt{3}\omega x))$$
for certain constants b, c, s.
3. Show that ω should be chosen so that
$$\sin\left(\sqrt{3}\omega\ell + \frac{\pi}{6}\right) = \frac{e^{3\omega\ell}}{2}.$$
4. Argue graphically that this last equation has an infinite set of solutions $\omega_n < 0$, $n = 1, 2, \ldots$. Is there a solution $\omega_0 \geq 0$?

5. Show directly that the eigenvalue problem in part 1 cannot have nontrivial solutions when $\lambda > 0$ since $\lambda \int_0^\ell X^2(x)dx \leq 0$. (*Hint*: Use partial integration.)

2.4E Show that product solutions for the problem

$$L(u) \equiv u_t - ku_{xx} = 0 \qquad \text{in} \qquad \{0 < x < \pi, t > 0\}$$

$$u(0,t) = u_x(\pi,t) + u(\pi,t) = 0, \qquad t > 0$$

are given by

$$u_n(x,t) = e^{-k\omega_n^2 t} \sin \omega_n x \qquad \text{where } \omega_n = -\tan \omega_n \pi, \qquad n = 1, 2, \ldots$$

2.4F Verify by substitution that $u(x,t) = (1/\sqrt{t})\, e^{-x^2/4kt}$ satisfies the equation $u_t = ku_{xx}$, for $t > 0$.

2.4G 1. Show that solutions X and $\tilde{X}$ of (24′) give product solutions $u = X\tilde{X}$ of $L(u) = \tilde{L}(u)$.

2. If $L(X_n) = \lambda_n X_n$ and $\tilde{X}_n$ satisfies (25), show that $v_n = X_n \tilde{X}_n$ satisfies $\mathscr{L}(v_n) = X_n \tilde{f}_n$.

2.4H Show that the equation $(x + y)u_x = u_y$ has only trivial product solutions of the form $u(x,y) = X(x)Y(y)$.

2.4I 1. Find nontrivial product solutions $u = u(x,y,t)$ to the boundary value problem

$$u_t = \nabla^2 u = u_{xx} + u_{yy} \qquad \text{in} \qquad \{0 < x < \pi, 0 < y < \pi, t > 0\}$$

$$\text{with} \qquad u_x(0,y,t) = u_x(\pi,y,t) = 0, \qquad 0 < y < \pi, t > 0$$

$$\text{and} \qquad u(x,0,t) = u(x,\pi,t) = 0, \qquad 0 < x < \pi, t > 0.$$

(*Hint*: Try a solution of the form $u(x,y,t) = T(t) \cos mx \sin ny$, where $m = 0, 1, 2, \ldots$ and $n = 1, 2, 3, \ldots$.)

2. How would you modify your solution to satisfy the same equation with the boundary conditions

$$u(0,y,t) = u_x(\pi,y,t) = 0, \qquad 0 < y < \pi, t > 0$$

$$u_y(x,0,t) = u_y(x,\pi,t) = 0, \qquad 0 < x < \pi, t > 0.$$

2.4J Find product solutions $u = u(x,y,z)$ for the boundary value problem

$$\nabla^2 u = 0 \qquad \text{in} \qquad \{0 < x < \ell, 0 < y < b, z > 0\}$$

$$\text{with} \qquad u_x(0,y,z) = u_x(\ell,y,z) = 0, \qquad 0 < y < b, z > 0$$

$$u(x,0,z) = u(x,b,z) = 0, \qquad 0 < x < \ell, z > 0.$$

2.5

Scale Changes and Dimensional Considerations

Laplace's equation expressed in polar coordinates shows how a change of coordinate system can affect the appearance of a partial differential equation. Moreover, the choice of coordinate system affects the feasibility of separation for a given problem. Within a given coordinate system, simple changes in scale can also be useful, but they must be applied throughout.

For example, the scale changes $\bar{x} = x/\ell$, $\bar{y} = y/b$ transform the rectangle $\{0 < x < \ell,\ 0 < y < b\}$ onto the unit square $\{0 < \bar{x} < 1,\ 0 < \bar{y} < 1\}$. However, they also transform a solution $u = u(x, y)$ of Laplace's equation $u_{xx} + u_{yy} = 0$ in the rectangle to a solution $\bar{u} = u(\ell\bar{x}, b\bar{y})$ of the equation

$$\bar{u}_{\bar{x}\bar{x}} + \frac{\ell^2}{b^2}\bar{u}_{\bar{y}\bar{y}} = 0.$$

$\bar{u}$ will be harmonic in the square iff $\ell = b$. Thus we can only preserve harmonicity through scale changes between geometrically similar domains. On the other hand, instead of seeking a solution $\bar{u}$ to the equation

$$\bar{u}_{\bar{x}\bar{x}} + 4\bar{u}_{\bar{y}\bar{y}} = 0$$

in the unit square, it might be easier to find a harmonic function u in the rectangle $\{0 < x < 2, 0 < y < 1\}$.

Similarly, the scale changes $\bar{x} = \pi x/\ell$, $\bar{t} = (\pi^2/\ell^2)kt$, for positive k and ℓ, transform a solution $u = u(x, t)$ of the heat equation $u_t = ku_{xx}$ in the strip $\{0 < x < \ell,\ t > 0\}$ into a solution $\bar{u} = u(\ell\bar{x}/\pi, \ell^2\bar{t}/k\pi^2)$ of the simpler equation $\bar{u}_{\bar{t}} = \bar{u}_{\bar{x}\bar{x}}$ in the strip $\{0 < \bar{x} < \pi, \bar{t} > 0\}$. In this case, problem constants have been absorbed by the variables.

There are also more profound aspects of change of scale associated with the underlying dimensions of the problem, as we will see in the following example.

Example 7:

Suppose that we seek solutions $u = u(x, t)$ of the equation

$$u_t = ku_{xx} - vu_x \qquad \text{for} \qquad 0 < x < \ell, \tag{26}$$

where k and v are positive constants. How do these solutions depend on x, t, v, k, and ℓ? Introduce new independent variables $\bar{x} = x/\ell$ and $\bar{t} = t/T$, say, and set $\bar{u}(\bar{x}, \bar{t}) = u(\ell\bar{x}, T\bar{t})$. Then

$$u_t = \frac{1}{T}\bar{u}_{\bar{t}}, \qquad u_x = \frac{1}{\ell}\bar{u}_{\bar{x}}, \qquad \text{and} \qquad u_{xx} = \frac{1}{\ell^2}\bar{u}_{\bar{x}\bar{x}}.$$

Thus the new equation becomes

$$\begin{aligned} \frac{1}{T}\bar{u}_{\bar{t}} &= \frac{k}{\ell^2}\bar{u}_{\bar{x}\bar{x}} - \frac{v}{\ell}\bar{u}_{\bar{x}}, \\ \text{or} \qquad \bar{u}_{\bar{t}} &= \frac{kT}{\ell^2}\bar{u}_{\bar{x}\bar{x}} - \frac{vT}{\ell}\bar{u}_{\bar{x}}. \end{aligned} \tag{26'}$$

We may make $vT/\ell = 1$ if we choose $T = \ell/v$. Then

$$\bar{t} = \frac{vt}{\ell}, \qquad \frac{kT}{\ell^2} = \frac{k}{\ell v},$$

and (26′) takes the simpler form

$$\bar{u}_{\bar{t}} = \frac{k}{\ell v}\bar{u}_{\bar{x}\bar{x}} - \bar{u}_{\bar{x}}.$$

For a specific value of $k/\ell v$, a solution $\bar{u} = \bar{u}(\bar{x}, \bar{t}, k/\ell v)$ of the last equation when $0 < \bar{x} < 1$ provides a corresponding solution

$$u(x, t) = \bar{u}\left(\frac{x}{\ell}, \frac{vt}{\ell}, \frac{k}{\ell v}\right)$$

of the original equation, when $0 < x < \ell$.

If (26) has physical origin, then each symbol appearing in it would have some assigned dimensions. For example, x might represent length, t time, u velocity, and so on. These assignments are related by the fact that each term in a sum must have the *same* dimensions, in this case, velocity/time, the dimensions of u_t. In particular, independently of the dimensions of u:

$$\text{the dimensions of } k \text{ must be } \frac{(\text{length})^2}{\text{time}},$$

$$\text{the dimensions of } v \text{ must be } \frac{\text{length}}{\text{time}}.$$

Thus if ℓ is a reference length, then

$$\pi_1 = \frac{x}{\ell}, \qquad \pi_2 = \frac{vt}{\ell}, \qquad \text{and} \qquad \pi_3 = \frac{k}{\ell v}$$

are *dimensionless*, and we have shown that the solution u of equation (26) for $0 < x < \ell$ is a function of π_1, π_2, and π_3 instead of the five separate quantities x, t, v, k, and ℓ. Here, the linearity of the equation assisted us in reaching this conclusion. In other situations, similar results are obtained through the π-Theorem of Buckingham [Log] [Bir].

Problem Set 2.5

2.5A Verify that for constant k, the scale changes

$$\bar{x} = \frac{x}{\ell} \qquad \text{and} \qquad \bar{t} = \frac{kt}{\ell^2}$$

transform a solution $\bar{u}$ of the normalized (heat) equation $\bar{u}_{\bar{t}} = \bar{u}_{\bar{x}\bar{x}}$ into a solution $u(x, t) = \bar{u}(x/\ell, kt/\ell^2)$ of the standard (one-dimensional) heat equation $u_t = ku_{xx}$.

2.5B With $\bar{x} = x/\ell$, make a scale change in t to normalize the following equations for $u = u(x, t)$ (that is, eliminate the positive constant k from the equation).

1. $u_{tt} = k^2 u_{xx}$
2. $u_t = k u_{xxx}$
3. $u_{tt} = k u_{xxx}$
4. $u_{tt} = k^2 u_{xx} - k u_{xt}$

Give a related normalized equation.

2.5C Consider the *nonlinear reaction–diffusion equation*

$$\frac{\partial u}{\partial t} = Au(B - u) + D\frac{\partial^2 u}{\partial x^2}.$$

Show by a suitable choice of dimensionless variables $\bar{u}$, $\bar{x}$, and $\bar{t}$ that it can be put in the normalized form

$$\frac{\partial \bar{u}}{\partial \bar{t}} = \bar{u}(1 - \bar{u}) + \frac{\partial^2 \bar{u}}{\partial \bar{x}^2}.$$

2.5D The *telegrapher's equation* has the form

$$u_{xx} = LCu_{tt} + (RC + GL)u_t + RGu$$

where R is resistance per unit length, L is inductance per unit length, C is capacitance per unit length, and G is conductance per unit length of cable.

1. Let $u(x, t) = v(x, t)e^{\alpha t}$ and choose α so that the differential equation for v has no v_t term.
2. Find the relation between R, L, C, and G so that the equation in part 1 reduces to $v_{xx} = LCv_{tt}$.
3. If instead, we make the scale changes

   $$t = \alpha\tau \qquad \text{and} \qquad x = \beta\xi,$$

   find α and β so that the equation becomes

   $$\bar{u}_{\tau\tau} + \gamma\bar{u}_\tau + \bar{u} = \bar{u}_{\xi\xi},$$

   which has only one free parameter γ. Find this γ.

3

Diffusion Problems: The Heat Equation

When a toxic dye is introduced into one end of a still pool of water, it spreads or *diffuses* so that its concentration changes both spatially and temporally. The process by which this occurs is somewhat mysterious, but it can be given a probabilistic explanation similar to that used to account for the spread of an epidemic through a population. In effect, a proportion of the "diseased" molecules infects the neighbors and forms a toxic plume that spreads throughout. Similarly, when a radiator is turned on at one end of a cold room, its heat diffuses into the room by conduction and changes the temperature. Flow through a porous medium is another process in which something diffuses in the manner suggested by these examples. This general class of phenomena will be explored in the present chapter, commencing with heat conduction, which is a familiar part of shared experience. In fact, the successful mathematical analysis of diffusion problems began with Fourier's work on heat conduction around 1804.

3.1 Heat Conduction

To describe the temperature distribution in a room as a function of position, $\mathbf{x}$, and time, t, we need an equation governing the behavior of heat under various conditions. The task is complicated by the passage of individuals in and out of the room, the presence of radiators, windows, lights, etc. Thus, we make our first idealization: that individuals and interior radiators or cooling units produce a combined prescribed distribution of heat "sources" of local strength $Q = Q(\mathbf{x}, t)$ per unit volume per unit time, and that they can in effect be replaced by a knowledge of Q. Other heat-producing processes, such as chemical reactions, radioactive fallout, etc., will be either neglected or assumed accounted for by Q. Air circulation (convection) by means of fans, wind, buoyancy, etc., will be excluded. These conditions could be achieved in a room of a space lab where convective currents arising from the buoyancy of air cannot be present. (Diffusion with convection is considered at the end of this section.) At this point, we do not need to know the mechanism of heat production — or indeed what heat is (!) — although our analysis will help define it operationally.

We select within the room a fixed simple domain V such as a cube or a sphere, which is called a *control volume* (see Figure 3.1). Relative to V and its boundary S, we postulate the following *energy conservation law*:

$$\begin{pmatrix}\text{The time rate of}\\ \text{change of total}\\ \text{heat within } V\end{pmatrix} = \begin{pmatrix}\text{The rate of heat}\\ \text{generated by the}\\ \text{sources, } Q\end{pmatrix} - \begin{pmatrix}\text{The heat flux}\\ \text{outward}\\ \text{through } S.\end{pmatrix} \quad \textbf{(1)}$$

Although we still need not know what heat is, we are requiring that it have this much regularity.

The first term on the right is represented mathematically by the integral

$$\int_V Q\,dV = \int_V Q(\mathbf{x}, t)\,dV,$$

since this is equivalent to the definition of Q. Similarly, the second term on the right may be expressed as the boundary integral $\int_S \mathbf{f}\cdot\mathbf{n}\,dS$ if we hypothesize a vector $\mathbf{f}$ of heat flux per unit area per unit time. By the divergence theorem of Section 0.2,

$$\int_S \mathbf{f}\cdot\mathbf{n}\,dS = \int_V \nabla\cdot\mathbf{f}\,dV;$$

hence the right side of (1) may be expressed as the integral

$$\int_V (Q - \nabla\cdot\mathbf{f})\,dV. \quad \textbf{(2)}$$

Next, we introduce thermometry and agree that locally the amount of heat present is proportional to the *absolute* temperature u where $u = u(\mathbf{x}, t)$. Then, since V is independent of time, the term on the left in (1) can be expressed as

$$\frac{d}{dt}\int_V \nu u\,dV = \int_V (\nu u)_t\,dV, \quad \textbf{(2')}$$

where $\nu = \nu(\mathbf{x}, t)$ is known as the *specific heat at constant volume*. For our purposes, ν is simply the local proportionality-dimensional-factor that converts

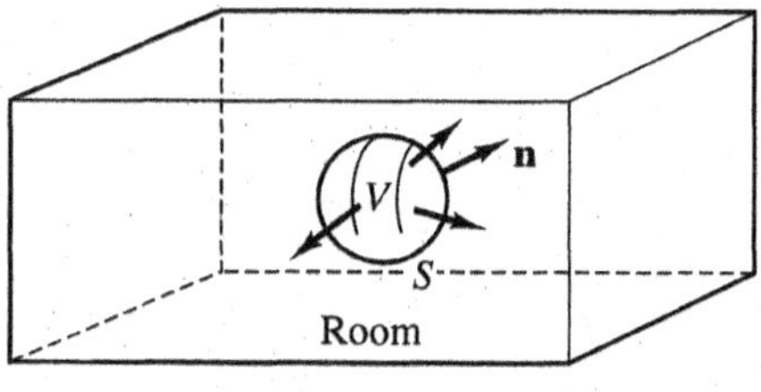

Figure 3.1

absolute temperature to heat per unit volume, and we suppose ν to be determined experimentally in air or in other media of interest.

Thus, substituting (2) and (2′) and transposing, we convert (1) to the integral equation

$$\int_V [(\nu u)_t - Q + \nabla \cdot \mathbf{f}]\, dV = 0, \tag{3}$$

where ν and Q are known, but $\mathbf{f}$ remains to be related to u. We have tacitly assumed that all quantities are continuous with as much differentiability as required to permit the indicated operations. In particular, we assume that the bracketed term in (3) is continuous. Then we obtain a remarkable result: This term must be identically zero in the room! for we have regarded V as being *any* control volume in the room, so that the quantity in brackets is specified everywhere in the room. If this quantity were positive at a point at some time t, then from its assumed continuity it would remain so in a small control box V containing that point—hence the integral over this V would be positive, contradicting (3). A similar argument precludes negativity at this point. Hence the bracketed term must be zero at this and every point at each time t; i.e., from (3) we deduce that

$$(\nu u)_t - Q + \nabla \cdot \mathbf{f} \equiv 0. \tag{4}$$

Equation (4) is a crude form of the heat equation or, with appropriate interpretation, the equation governing other diffusion processes.

(a) Heat flux; Fourier's law: What is heat flux? And perhaps more basically, what is heat? In the early development of thermodynamics, heat was conceived of as a kind of fluid substance (called caloric) that was stored in a hot body and "flowed" naturally toward colder ones. But in 1798 Rumford and Davy demonstrated that more heat could be produced by friction than is permitted by this caloric model. Nevertheless, its language still shapes our terminology and subjective feel. For example, we can conceive of heat flux as a vector in the direction of this hypothetical "flow."

A decisive step was taken in 1807 with Fourier's postulated *law of heat conduction* that in an *isotropic* medium (one whose properties are directionally independent at each point), the heat-flux vector is given by

$$\mathbf{f} = -K\nabla u, \tag{4'}$$

where $K = K(\mathbf{x}, t, u)$ is a positive scalar function that depends on the material being heated[†] and is called its **thermal conductivity**. To see why postulate (4′) is plausible, recall that a nonzero gradient ∇u is a vector in the direction of the maximum spatial rate of change of temperature u and of that magnitude. Moreover, ∇u is normal to the isothermal surfaces of constant temperature. Thus in formulating (4′), Fourier simply asserted that heat should "flow" along directions normal to isothermal surfaces, and with the negative sign, from hot surfaces to colder ones. K measures the local "speed" of flow, and we suppose that it can be determined experimentally for each material of interest.

[†] Although air and cast metals are thermally isotropic, many modern materials, including worked metals and composite ceramics, are not. For such materials, K should be regarded as a square matrix in (4′).

When we insert (4′) into (4) we obtain the **heat equation**

$$(\nu u)_t = \nabla \cdot (K \nabla u) + Q, \tag{5}$$

which gives us a partial differential equation for the unknown temperature function u.

Note that equation (5) and its integral form (3) fix relations between the units of the quantities involved. For example, if we agree to measure a quantity of heat in Btu (British thermal units) and temperature in degrees, then Q must have dimensions of Btu/unit volume-unit time. Hence, ν must have those of Btu/unit volume-degree and $\mathbf{f}$ those of Btu/unit area-unit time, which means that K must have those of Btu/unit time-unit length-degree. Thus, without knowing any of these quantities explicitly, we impose a dimensional consistency that helps design experiments to determine them.

(b) Simplifications:

As it stands, the heat equation (5) is rather complicated and is not even linear because of the possible dependence of K on u, but simplified versions are appropriate for particular applications. Here are some possibilities:

(1) $Q \equiv 0$, which occurs when heat is not being generated or lost internally but only redistributes itself.

(2) $\nu = \nu(\mathbf{x})$, which is reasonable over a small range of temperatures (and for a gas such as air). Then $(\nu u)_t = \nu u_t$.

(3) $K = K(u, t)$ or $K = K(u)$, which is reasonable in homogeneous media. Then $\nabla K = K_u \nabla u$ by the chain rule, and $\nabla \cdot (K \nabla u) = K \nabla^2 u + K_u |\nabla u|^2$.

(4) $\nu =$ constant and $K =$ constant, which is reasonable for a uniform medium over a small range of temperature. Then $\nabla K = \mathbf{0}$, and $\nabla \cdot (K \nabla u) = K \nabla^2 u$.

Under all four assumptions (5) can be replaced by the **standard heat equation**

$$u_t = k \nabla^2 u, \tag{6}$$

where $k = K/\nu$ is a positive material constant called the **thermal diffusivity**. Equation (6) is linear; moreover, it is unaffected by the subtraction of a reference temperature from u, or by a change in the units of u from degrees Celsius to degrees Fahrenheit, say.

We can also consider cases where u itself depends on fewer variables. For example, if $u = u(\mathbf{x})$ alone, then $u_t = 0$ and under assumptions (1) and (2), the heat equation (5) reduces to its *steady-state* form

$$\nabla \cdot (K \nabla u) = 0, \tag{6′}$$

while (6) reduces to Laplace's equation. Consequently, in the absence of sources, the steady-state temperature distribution in a homogeneous isotropic medium is governed by Laplace's equation; if sources are present, it is governed by **Poisson's equation,**

$$\nabla^2 u = -Q/K. \tag{6″}$$

If u depends on fewer spatial variables, then the vector terms on the right in (5) are reduced accordingly. The temperature distribution inside a long tunnel of constant cross section could be essentially two-dimensional away from the ends. The temperature distribution along the center line of an insulated long thin rod is essentially one-dimensional and is governed by the equation

$$u_t = ku_{xx} \qquad \text{for} \qquad u = u(x, t) \tag{7}$$

under the above simplifying assumptions.

In general, we justify simplifications through careful consideration of special circumstances or through the desire to have a solvable problem and accept the limited validity of the solution. In particular, unless a method of solution is apparent, we can simplify until one becomes apparent, keeping a record of the assumptions made at each stage.

For example, in the insulated rod above, the steady-state temperature u obeys the ordinary differential equation

$$u_{xx} = u''(x) = 0, \qquad \text{where} \qquad u = u(x);$$

thus, u must take the form $u(x) = ax + b$, for constants a, b, which can be found to match prescribed temperatures at the ends. Of course, this result gives no information about how heat diffuses along a cold rod that is suddenly heated at one end, but it does provide a simple test of our model, for it predicts that the steady-state temperature at the midpoint of the rod should be the average of those at the ends. Unless this also occurs experimentally, some of our assumptions are inappropriate.

Example 1 (*A geothermal heat pump*):

A heat pump can provide an energy-efficient means of both heating and cooling a house. One modern type of heat pump designed for year-round operation uses liquid circulated underground as its primary source (see Figure 3.2). To understand why this works, let's look at the effects of the sun's yearly cycle on the Earth's ground temperature at moderate depth x. If we neglect the heating from within (by taking $Q = 0$) and the curvature of the Earth's surface, then we can assume that at time t, the temperature $u = u(x, t)$ satisfies (7) for some mean value of k. At $x = 0$, the surface temperature,

$$u(0, t) = f(t),$$

say, is obtained from surface measurements as a function of time. For simplicity, let's assume that fluctuations of the surface temperature about a

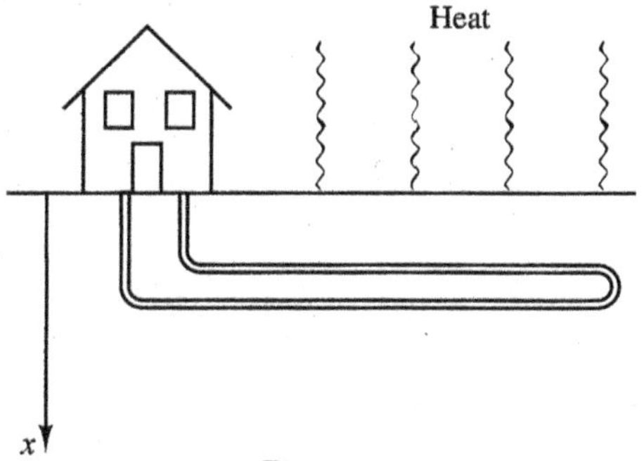

Figure 3.2

reference mean of zero can be represented as

$$f(t) = Y \cos \omega t,$$

which permits yearly variation at frequency ω and amplitude Y.

Then we recall from Example 6 of Section 2.4 that

$$u_a(x, t) = e^{-ax} \cos(\omega t - ax) \tag{8}$$

satisfies (7), and at $x = 0$, we see that

$$u_a(0, t) = \cos(\omega t).$$

Through linearity, we conclude that

$$u(x, t) = Yu_a(x, t) = Ye^{-ax} \cos(\omega t - ax), \tag{8'}$$

supplies a solution to our problem provided that we take

$$2a^2 k = \omega \qquad \text{or} \qquad a = \sqrt{\omega/2k}.$$

(We reject the *mathematically* valid possibility that $a < 0$ on the *physical* grounds that the temperature should not increase with depth.) Since ω and Y are known, we can approximate a from (8′) through simultaneous measurements of temperature at various depths. This also gives an approximate value of k, the ground thermal diffusivity, and comparative graphs of the temperature cycle at several depths are shown in Figure 3.3 when $k = 2 \times 10^{-7}\ \text{m}^2/\text{sec}$. These graphs illustrate interesting features of the solution. The temperature below the surface continues to fluctuate with the yearly cycle, but the amplitude diminishes with depth. Moreover, there is a phase shift with depth which suggests designing a heat pump to take advantage of temperatures at cellar depths (where $x = \pi/a$) that are warm in winter and cool in summer!

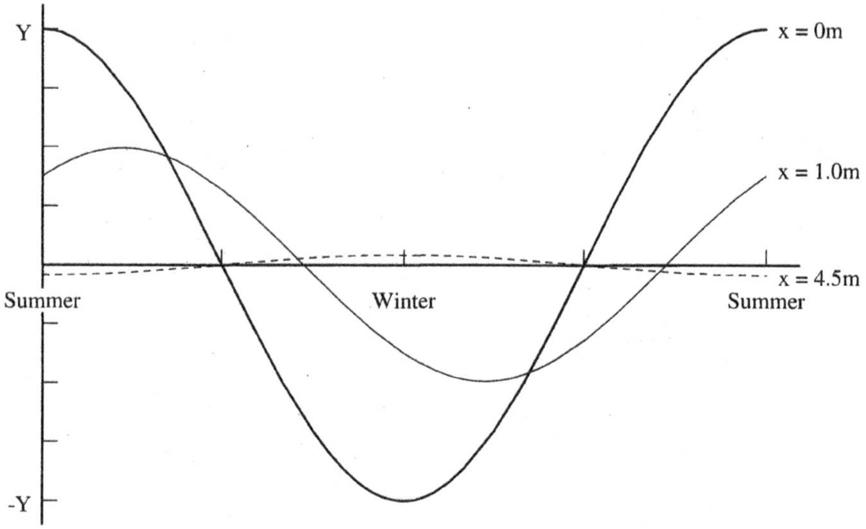

Figure 3.3

However, before we can make such predictions with confidence, we must exclude there being a different (bounded) solution to our boundary value problem. To do this, we require a nontrivial uniqueness argument, which we postpone until Section 8.5, but in Section 3.2 we shall learn how to establish uniqueness in simpler cases.

Diffusion with Convection

If dye is injected into a stream of flowing water, as shown in Figure 3.4, then the natural static diffusion of the dye in water is augmented by the convective effects of the flow. Let $u = u(\mathbf{x}, t)$ denote the mass concentration (per unit volume) of the dye, and $\mathbf{f} = \mathbf{f}(\mathbf{x}, t)$ be its associated flux (per unit area per unit time). Then conservation of mass of the dye for a *fixed* control volume V within the flow leads to the integral equation

$$\frac{d}{dt}\int_V u\,dV = \int_{S=\partial V} (-\mathbf{f}\cdot\mathbf{n})\,dS. \tag{9}$$

(We neglect the amount of dye being generated or lost internally, say through chemical reactions.) Thus u and $\mathbf{f}$ should be related through the equation

$$u_t = -\nabla\cdot\mathbf{f}.$$

For static diffusion, **Fick's law** of mass flux,

$$\mathbf{f} = -\kappa\nabla u,$$

is patterned after Fourier's law of heat flux, and it is found to give good experimental agreement for a properly chosen positive diffusivity coefficient $\kappa = \kappa(\mathbf{x}, t, u)$. Therefore, u satisfies the **diffusion equation**

$$u_t = \nabla\cdot(\kappa\nabla u),$$

whose resemblance to the heat equation (5) is apparent. However, if the water is flowing with local velocity $\mathbf{v} = \mathbf{v}(\mathbf{x}, t)$, then Fick's law is often modified as follows:

$$\mathbf{f} = -\kappa\nabla u + u\mathbf{v},$$

and the resulting **diffusion-convection equation** is

$$u_t = \nabla\cdot(\kappa\nabla u) - \nabla\cdot(u\mathbf{v}). \tag{9'}$$

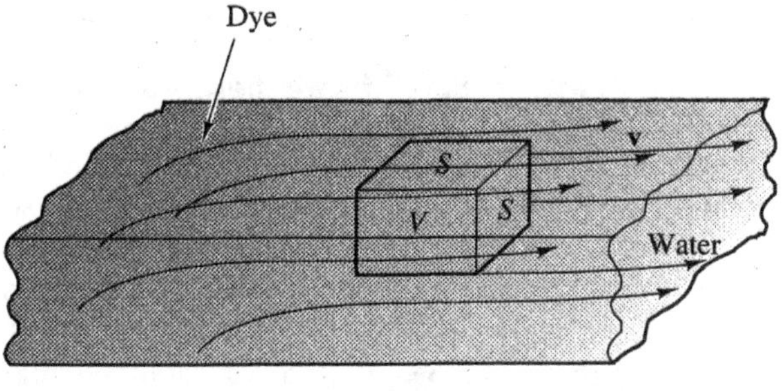

Figure 3.4

If $\nabla \cdot \mathbf{v} = 0$, which occurs if $\mathbf{v}$ is constant or if the flow is incompressible (see Section 6.4), then we have the simpler equation

$$u_t = \nabla \cdot (\kappa \nabla u) - \mathbf{v} \cdot \nabla u. \tag{9''}$$

Example 2 (*A transport-mixing process*):

To speed production in the petrochemical industry, materials in liquid form are mixed as they flow along a pipe. Suppose a new chemical (the solute) is introduced into a mixture (the solvent) flowing with constant velocity v inside a long straight circular pipe of diameter ℓ, as shown in Figure 3.5. It is important to know how much mixing has occurred at time t at distance x downstream from the point of introduction. If we assume that this is a one-dimensional diffusion-convection process, then the mass concentration of the solute $u = u(x, t)$ obeys (9″) with $\mathbf{v} = (v, 0, 0)$. The static diffusivity κ can be approximated experimentally and may be assumed constant. Hence (9′) reduces to

$$u_t = \kappa u_{xx} - v u_x,$$

the equation analyzed dimensionally in Section 2.5. The dimensionless ratio $v\ell/\kappa$, called the *Péclet number*, measures the relative importance of convection (v) to diffusion (κ) in the process. Two such processes with the same Péclet number and auxiliary conditions may be expected to exhibit the same behavior at corresponding points; they are said to be *similar* processes.

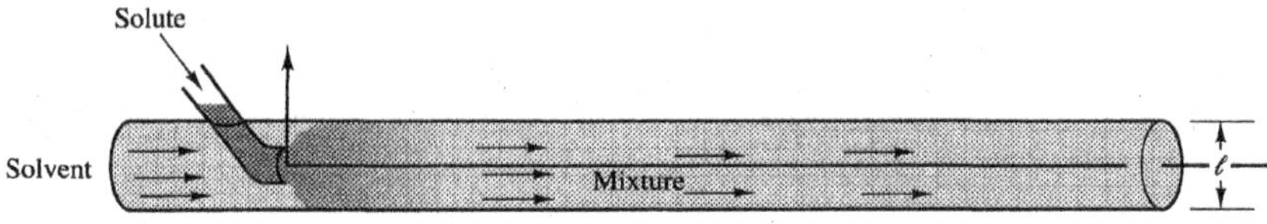

Figure 3.5

Problem Set 3.1

3.1A ***1.*** For $u = u(x, y, t)$, write the explicit two-dimensional version of equation (5).

2. Simplify your equation in part 1 by taking ν and K to be constants and introducing $k = K/\nu$ and $q = Q/\nu$ (of what units?).

3. If polar coordinates (r, θ) are introduced as in Problem 2.2G, what is the simplified equation for $\tilde{u}(r, \theta, t) = u(r\cos\theta, r\sin\theta, t)$?

3.1B Suppose that a long thin rod of length ℓ with constant circular cross section is made of heat-conducting metal such as iron and has perfectly insulated sides. In the absence of internal heat sources, we assume that the temperature along the center line is given by $u = u(x, t)$.

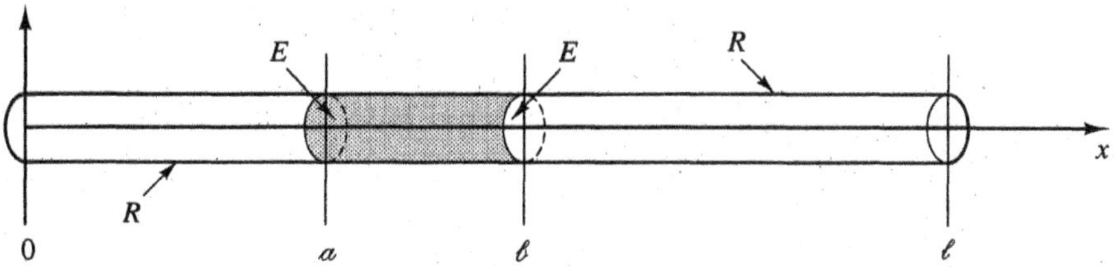

1. Let V be a section of the rod between a and b $(0 \leq a < b \leq \ell)$. Since $\mathbf{f}\cdot\mathbf{n}$ is required to vanish on the insulated sides, R, deduce that

$$\int_V (\nu u)_t \, dV = -\int_E \mathbf{f}\cdot\mathbf{n} \, dS$$

where E represents the end surfaces of the section.

2. Suppose $\nu = \nu(x)$ and that $f_1 = f_1(x, t)$ is the x-component of $\mathbf{f}$. Show that

$$\int_a^b \nu u_t \, dx = -[f_1(b, t) - f_1(a, t)] = -\int_a^b \frac{\partial f_1}{\partial x} \, dx,$$

so that for arbitrary a and b, we must have

$$\nu u_t + \frac{\partial f_1}{\partial x} = 0, \qquad 0 < x < \ell.$$

3. Conclude that under Fourier's law of heat conduction, with $K = K(x, t, u)$,

$$\nu u_t = (K u_x)_x.$$

4. Obtain a similar result when $Q = Q(x, t) \neq 0$. (This analysis shows that heat conduction in a long thin rod or bar with insulated sides is governed by the one-dimensional version of (5).)

3.1C When the sides, R, of the rod in Problem 3.1B are not perfectly insulated, the Robin condition $(\partial u/\partial n)|_R = -hu|_R$ is appropriate (for an outside temperature of zero) where $h = h(x) > 0$.

1. Under these conditions, use Fourier's law on R as in Problem 3.1A to conclude that the equation in Problem 3.1B should be replaced by

$$\int_a^b (\nu u)_t \, dx = -\int_a^b \frac{\partial f_1}{\partial x} \, dx - (2/r) \int_a^b (Khu) \, dx,$$

where r is the radius of the rod.

2. With arbitrary a and b, what is the resulting equation of heat conduction corresponding to that in Problem 3.1B?
3. What is the corresponding equation when the outside temperature is a constant $U \neq 0$, so that $[(\partial u/\partial n) + hu]|_R = hU|_R$?

3.1D Suppose that for a considerable period of time, the midnight temperature is

21°C and the midday high is 35°C. Approximate the daily surface temperature variation between successive highs at $t = 0, 24, 48, \ldots$ hours by a simple cosine function. Then find the temperature variation $u(x,t)$ at depth x at time t in a lake, assuming the thermal diffusivity of the water to be $k \approx 1.4 \times 10^{-3}\ \text{cm}^2/\text{sec}$. (*Hint*: See Example 1.)

3.1E Suppose that in Example 1 the surface temperature of the earth is represented by

$$u(0,t) = Y\cos\omega t + D\cos\delta t,$$

which permits both yearly and daily oscillations at different frequencies. Show that the resulting solution can be represented by $u(x,t) = Yu_1(x,t) + Du_2(x,t)$, where each u_j is of the form obtained. Write the solution explicitly.

3.1F Use the substitution $\xi = x - vt$, $\eta = t$ to transform the one-dimensional diffusion-convection equation

$$u_t = ku_{xx} - vu_x$$

into a heat equation for $\tilde{u}(\xi,\eta) = u(\xi + v\eta, \eta)$.

3.1G In the concluding exposition of diffusion with convection

1. Apply the divergence theorem to the integral balance equation (9), to obtain

$$\int_V [u_t + \nabla\cdot\mathbf{f}]dV = 0,$$

and explain why the (continuous) integrand must vanish.
2. In the case of static diffusion where $\mathbf{v} = \mathbf{0}$, give a physical argument motivating Fick's law and show how this law leads to the diffusion equation. (*Hint*: Compare Fick's law with Fourier's law of heat conduction.)
3. Show how the modified Fick's law for $\mathbf{f}$ gives equation (9″) when $\nabla\cdot\mathbf{v} = \mathbf{0}$.
4. Suppose that the dye can be generated (or lost) internally through chemical reactions. How should we modify the following?
   (a) the diffusion equation
   (b) the diffusion-convection equation

3.2

Auxiliary Conditions and Uniqueness

We indicated in Section 2.3 that a reasonable boundary value problem requires both a differential equation such as (6) and a compatible set of auxiliary conditions to ensure uniqueness of solution. In diffusion problems, we need to consider both initial conditions and boundary conditions.

Initial/Final Conditions

Consider a room V in which $Q = Q(\mathbf{x}, t)$ is specified. Suppose that at 6:00 A.M. the temperature in the room is a uniform 60°F or has some other prescribed distribution and that we wish to predict its subsequent behavior.

This would correspond to the (Dirichlet) *initial condition* in which

$$u|_{t=0} = u(\mathbf{x}, 0) = u_0(\mathbf{x}) \tag{10}$$

is given at time 0 (= 6:00 A.M.) at each point $\mathbf{x} \in V$.

Experience teaches us that (10) is not sufficient to determine the temperature distribution at the later time $T =$ 10:00 P.M., nor would it be appropriate to specify another initial condition such as

$$u_t|_{t=0} = u_t(\mathbf{x}, 0)$$

since this should be given by, say, equation (6), as

$$u_t|_{t=0} = k\nabla^2 u_0.$$

We can also admit a *final condition* where $u|_{t=T} = u(\mathbf{x}, T)$ is prescribed, but this cannot fix the temperature in the intervening period $0 < t < T$ since rooms with zero initial and final temperatures might have different temperatures at say, 11:00 A.M., as a result of the sun.

Boundary Conditions

To account for effects of sun, surface lighting, windows, floor coverings, etc., we must consider the boundary conditions at the walls, floor, and ceiling of a room. For example, by Newton's law of cooling, a body loses or gains heat at a rate proportional to the difference between its temperature, u, and that of its surroundings, U. At the boundary S of the room V, this law can be expressed through the condition

$$-\left.\frac{\partial u}{\partial n}\right|_S = [h(u - U)]|_S$$

for an appropriate boundary function $h > 0$. Consequently, we have

a *Robin condition,* where $\left.\left(\dfrac{\partial u}{\partial n} + hu\right)\right|_S$ is prescribed.

Experience suggests that knowledge of the initial temperature distribution at time $t = 0$ and of this boundary condition at each time $t \in (0, T)$ *is* sufficient to fix the temperature within the room at time T. Before considering mathematical arguments, note that experience is less persuasive when h is negative.

Other boundary conditions are possible and two special cases are worth identifying:

a *Dirichlet condition,* where $u|_S$ is prescribed;

i.e., the temperature of the walls, floor, and ceiling is known or controlled;

a *Neumann condition,* where $\left.\dfrac{\partial u}{\partial n}\right|_S$ is prescribed;

and with it $\mathbf{f}\cdot\mathbf{n}|_S = -[K(\partial u/\partial n)]|_S$, the normal component of the boundary heat-flux vector. Neumann conditions distinguish boundary surfaces with perfect insulation where $\mathbf{f}\cdot\mathbf{n}|_S = 0 = (\partial u/\partial n)|_S$, from those which permit heat flux.

It is plausible that specifying either Dirichlet or Neumann boundary conditions for all $t \in (0, T)$, together with the initial distribution (10), will fix the temperature at time T within a room of given Q. For example, if $Q = 0$ and the initial and boundary temperatures are held at 60°F, we would not expect to find a temperature other than 60°F within the room. We might have the same expectation if the initial temperature is 60°F and the boundary surfaces are perfectly insulated so that there is no heat flux through them. In Theorem 3.1 below we will find mathematical support for these conjectures.

For an actual room, the boundary conditions will vary from surface to surface, and indeed along the surfaces. For example, part of a wall containing a window might be heavily insulated and part of a ceiling might have surface lighting. Such situations can be handled by

$$\textit{Mixed conditions}, \text{ where } \left.\left(\beta \frac{\partial u}{\partial n} + hu\right)\right|_S \text{ is prescribed}$$

for appropriate boundary functions β and h not vanishing simultaneously at any point. Observe that with proper selection of β and h we can recover any of the previous conditions.

Finally, some applications involve external feedback mechanisms that transfer the heat lost at one wall to another. For example, in the one-dimensional case covered by (7), we might have the so-called

$$\textit{Periodic conditions}, \text{ where } \quad u(0,t) = u(\ell,t)$$
$$\text{or} \quad u_x(0,t) = u_x(\ell,t), \qquad t > 0.$$

This pair of conditions arises naturally when considering heat conduction in a ring. (See Problem 3.2F.)

Observe that the boundary operator associated with each of these auxiliary conditions *is* linear when K is independent of u.

Uniqueness Theorems

It is simpler to treat a boundary value problem that has only one solution. In the present case we just discussed auxiliary conditions that might provide uniqueness, and in the following theorems we demonstrate how mathematics supports physical intuition.

According to Proposition 2.1, uniqueness for an *inhomogeneous linear* problem is a consequence of uniqueness for the corresponding *homogeneous* problem, so we just have to show that $u = 0$ is the only solution to the homogeneous problem.

(3.1) Theorem (*Uniqueness*): *Let $\nu \in C(\bar{V})$ and $K \in D^1(\bar{V})$ be positive on V, a simple domain of $\mathbb{R}^d$ with boundary S. Then $u = u(\mathbf{x}, t) = 0$ is the only solution to the boundary value problem*

$$\nu u_t = \nabla\cdot(K\nabla u) \quad \text{in} \quad \mathscr{V} = V \times (0,\infty) \tag{11}$$
$$\text{with} \quad u(\mathbf{x},0) = 0, \quad \mathbf{x} \in V$$

if for all $t > 0$, either

(a) $u|_S = 0$, *or*

(b) $(\partial u/\partial n)|_S = 0$, *or*

(c) $(\partial u/\partial n)|_S = -(hu)|_S$ *for a positive function* $h \in C(S)$, *or*

(d) one of the preceding conditions is specified at each boundary point.

Proof: Observe that in all cases $[Ku(\partial u/\partial n)]|_S \leq 0$, $t > 0$. We will show that for $t > 0$, the "energy" integral

$$E(t) \stackrel{\text{def}}{=} \frac{1}{2}\int_V \nu u^2\, dV = \frac{1}{2}\int_V \nu(\mathbf{x})u^2(\mathbf{x}, t)\, dV,$$

is zero.

Clearly $E(t) \geq 0$ since $\nu > 0$, and $E(0) = 0$ since $u(\mathbf{x}, 0) \equiv 0$. Therefore, we try to establish that $E(t)$ cannot increase with t by proving that $E'(t) \leq 0$. From $E(t) = 0$, we conclude that the continuous nonnegative integrand $\nu(\mathbf{x})u^2(\mathbf{x}, t)$ vanishes so that $u(\mathbf{x}, t) = 0$, $\mathbf{x} \in V$.

We must still establish that $E'(t) \leq 0$. To see how we might achieve this, let's operate formally, using (11) and standard vector identities as required to get

$$\begin{aligned} E'(t) &= \frac{d}{dt}\int_V \frac{\nu u^2}{2}\, dV = \int_V u(\nu u_t)\, dV = \int_V u(\nabla\cdot(K\nabla u))\, dV \\ &= \int_V \nabla\cdot(uK\nabla u)\, dV - \int_V K|\nabla u|^2\, dV. \end{aligned} \tag{11'}$$

The last integral is nonnegative (with K), and we can use the divergence theorem on the previous one, if we set $\mathbf{f} = uK\nabla u$. Then, when $\mathbf{n}$ is defined on S,

$$\mathbf{f}\cdot\mathbf{n}|_S = uK(\nabla u\cdot\mathbf{n})|_S = \left(Ku\frac{\partial u}{\partial n}\right)\Big|_S \leq 0,$$

in view of our initial observation, and (11′) gives

$$E'(t) \leq \int_V \nabla\cdot\mathbf{f}\, dV = \int_S \mathbf{f}\cdot\mathbf{n}\, dS = \int_S Ku\frac{\partial u}{\partial n}\, dS \leq 0,$$

as desired.

Now, let's reexamine the ingredients of the formal argument. In general, such arguments are valid if u is so smooth that all integrals used exist and the operations are permissible. To guarantee this, we must impose more smoothness on u near the boundary of $\mathscr{V}$ than the auxiliary conditions require. However, in this case, the key integrands are nonnegative, and standard approximation arguments from within V show that the conclusion remains valid provided that each component of ∇u extends continuously to $\bar{V} \times (0, +\infty)$. This hypothesis is implied by boundary conditions of type (b) or (c). For condition (a), an independent approach to uniqueness will be presented in Section 3.5. ■

Under steady-state conditions $u = u(\mathbf{x})$ satisfies the nonhomogeneous **potential equation**

$$\nabla\cdot(K\nabla u) = -Q, \qquad \text{in } V,$$

where $Q = Q(\mathbf{x})$ is prescribed. By modifying the proof just given, we can derive analogous results.

(3.2) Corollary: *Let S be the boundary of a simple domain V of $\mathbb{R}^d$ and let $K \in D^1(\bar{V})$ be positive on V. Then $u = u(\mathbf{x}) =$ constant is the only solution to the boundary value problem*

$$\nabla\cdot(K\nabla u) = 0 \qquad \text{in } V$$

with boundary conditions of type (a), (b), (c), or (d). Moreover, $u(\mathbf{x}) \equiv 0$ unless $(\partial u/\partial n)|_S = 0$ is required.

Proof: If we set $\nu = 0$ in equation (11′) and use the divergence theorem as before, we find that

$$0 \leq \int_V K|\nabla u|^2\, dV = \int_V \nabla\cdot(uK\nabla u)\, dV = \int_S Ku\frac{\partial u}{\partial n}\, dS \leq 0.$$

Hence in V, $|\nabla u|^2 = 0$ or $\nabla u = \mathbf{0}$, which means that u cannot change value on a line segment in V or on any open ball in V. Then we invoke the ball-chain condition of the *domain* V (Section 0.1b) to conclude that u is constant in V, so that $(\partial u/\partial n)|_S = 0$. Under Neumann conditions on S, no stronger assertion about u can be made. However, if conditions of type (a) or type (c) hold at *any* boundary point $\mathbf{s} \in S$, then

$$u(\mathbf{x}) = u(\mathbf{s}) = 0, \qquad \mathbf{x} \in V.$$

Again, these conclusions can be validated under conditions (b) or (c), and for (a) an alternative proof is given in Section 3.5. ■

Remark: The uniqueness results just obtained hold for more general bounded domains V. However, if V is unbounded then additional auxiliary conditions are required. (See Section 8.5.)

Problem Set 3.2

3.2A Verify the vector identity $u\nabla\cdot(K\nabla u) = \nabla\cdot(uK\nabla u) - K|\nabla u|^2$ in $\mathbb{R}^3$, using rectangular Cartesian coordinates $\mathbf{x} = (x, y, z)$. (Assume that K is C^1 and u is C^2 in $\mathbf{x}$.)

3.2B If $f \in C[0, 1]$, let $u(x, t)$ satisfy

$$\begin{cases} u_t = u_{xx}, & 0 < x < 1, t > 0 \\ u(0,t) = u(1,t) = 0, & t \geq 0 \\ u(x,0) = f(x), & 0 \leq x \leq 1 \end{cases}$$

1. Derive the identity $2u(u_t - u_{xx}) = (u^2)_t - (2uu_x)_x + 2u_x^2$.
2. Use the result in part 1 to show that for any $T \geq 0$, $\int_0^1 u^2(x,T)\,dx \leq \int_0^1 f^2(x)\,dx$, and explain why this problem has at most one solution.

3.2C Give a physical interpretation in terms of heat conduction for the following problem.

$$\begin{cases} u_t = u_{xx} - u, & 0 < x < 1, t > 0 \\ u_x(0,t) = u(0,t), & t > 0 \\ u_x(1,t) = 0, & t > 0 \\ u(x,0) = 1, & 0 < x < 1 \end{cases}$$

Give a uniqueness argument for its solutions similar to that in Problem 3.2B.

3.2D 1. Write out the steps of the proof of Theorem 3.1 for the simple one-dimensional equation

$$\nu u_t = K u_{xx}, \qquad 0 < x < \ell, t > 0,$$

where ν and K are positive constants. Show why

$$\frac{d}{dt}\int_0^\ell \nu u^2\,dx \le 2\int_0^\ell \frac{\partial}{\partial x}(uKu_x)\,dx = 2uKu_x\Big|_{x=0}^{x=\ell}$$

2. Conclude that under the (Dirichlet) conditions $u(0,t) = u(\ell,t) = 0, t > 0$, we have

$$0 \le \int_0^\ell \nu u^2\,dx \le \int_0^\ell \nu u^2\Big|_{t=0} dx,$$

and the last energy integral vanishes if $u|_{t=0} = u(x,0) = 0,\ 0 < x < \ell$.

3. How would your arguments change if $\nu = \nu(x)$? if, in addition, $K = K(x,t)$? if u is added to the right side of the equation?

4. How would your arguments in part 2 change if

(a) $u_x(0,t) = u_x(\ell,t) = 0,\ t > 0$? (b) $u(0,t) = u_x(\ell,t) = 0,\ \ t > 0$?
(c) $u_x(0,t) - u(0,t) = u(\ell,t) = 0,\ t > 0$?

3.2E Show that the Korteweg–deVries problem from Problem 2.4D* has at most one solution with prescribed $u(x,0) = f(x), 0 < x < \ell$. (*Hint*: Show that the difference u of any two solutions satisfies

$$\frac{1}{2}\frac{d}{dt}\int_0^\ell u^2\,dx = \int_0^\ell uu_t\,dx \le 0.)$$

3.2F Suppose the homogeneous metal bar of Problem 3.1B is formed into a ring by joining its ends. Then the temperature $u = u(x,t)$ along the center line is governed by the heat equation $\nu u_t = (Ku_x)_x$ with the periodic conditions

$$u(0,t) = u(\ell,t), \qquad u_x(0,t) = u_x(\ell,t), \qquad t > 0,$$

where $0 < K = K(x)$ satisfies $K(0) = K(\ell)$, and $0 < \nu = \nu(x)$. Show that there is at most one solution to this problem with prescribed continuous initial temperature $u(x,0) = f(x),\ 0 < x < \ell$ (where $f(0) = f(\ell)$). (*Hint*: Guided by the proof of Problem 3.2D1, show that

$$\frac{d}{dt}\int_0^\ell \nu u^2\,dx \le 0.)$$

3.2G* Consider the diffusion-convection equation $u_t = \nabla\cdot(\kappa\nabla u + u\mathbf{v})$, in $\mathscr{V} = V \times (0, +\infty)$, where V is a simple domain with boundary S. Assume that κ is independent of u and that $u|_S = g$, $u|_{t=0} = f$, where f and g are given.

1. If $\nabla\cdot\mathbf{v} = 0$, show that the difference u of two solutions to the problem satisfies

$$u_t = \nabla\cdot(\kappa\nabla u) + \mathbf{v}\cdot\nabla u \qquad \text{with } u|_S = 0 \text{ and } u|_{t=0} = 0.$$

2. Establish that

$$uu_t \le \nabla\cdot(\kappa u\nabla u) + \nabla\cdot\left(\tfrac{1}{2}u^2\mathbf{v}\right).$$

3. Prove that

$$\frac{d}{dt}\int_V \frac{u^2}{2}\,dV \le \int_S \kappa u\frac{\partial u}{\partial n}\,dS + \int_S \frac{u^2}{2}\,\mathbf{v}\cdot\mathbf{n}\,dS$$

and conclude that $u \equiv 0$.

4. Are there any other conditions that would result in a unique solution for this problem? (*Hint*: Suppose S is the boundary of a swimming pool.)

3.3

One-Dimensional Problems

In this section we examine some diffusion problems that require only one spatial dimension to formulate. We begin with problems in heat conduction and then move into several other fields.

Example 3:

You have probably burned your fingers at least once by touching the handle of a metal spoon whose lower end is immersed in boiling water. Let's think of the handle as a long thin bar of length ℓ and see how heat flows along it (see Figure 3.6). At time t, the temperature u is usually constant over each cross section and can be represented by $u = u(x, t)$, $0 < x < \ell$. Our spoon handle doesn't have internal heat sources, but to keep it from losing heat along its length we might sheath it in asbestos and so insulate its lateral surface. Then u should satisfy the one-dimensional version of equation (6),

$$u_t = (ku_x)_x, \qquad 0 < x < \ell, t > 0. \tag{12}$$

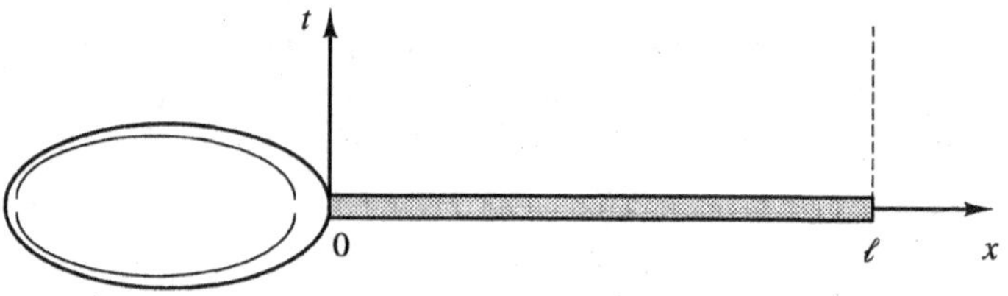

Figure 3.6

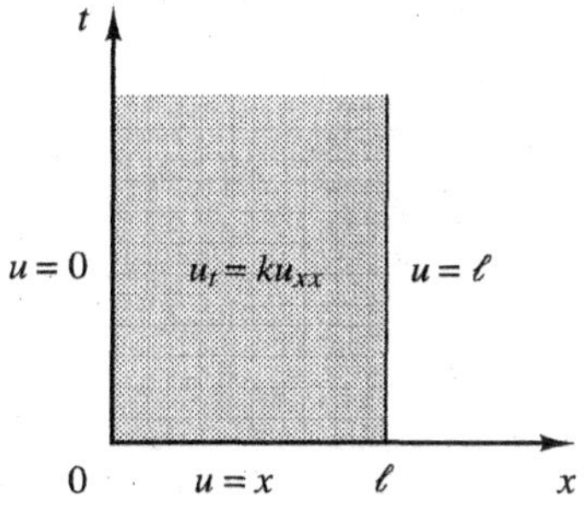

Figure 3.7

(See Problem 3.1B.) Here, $0 < k = k(x)$ is the mean thermal diffusivity at x of dimensions length^2/time, and it can vary along a bar of a given material of nonconstant cross section or a bar of nonuniform composition. When k is constant, we have the simpler equation

$$u_t = ku_{xx}, \tag{13}$$

which will challenge us enough for the present and has surprisingly many applications.

By inspection, $u(x, t) = x$ satisfies equation (13),

$$\text{the boundary conditions} \begin{cases} u(0, t) = 0 \\ u(\ell, t) = \ell \end{cases}$$

$$\text{and the initial condition} \quad u(x, 0) = x.$$

Therefore u is the *unique* solution to the boundary value problem indicated in Figure 3.7. Physically, u gives the temperature distribution in our asbestos-sheathed bar if we heat it initially so that $u(x, 0) = x$ and thereafter hold the end temperatures at their initial values.† Observe that this solution is time-independent (steady-state) and that along the bar the heat flux, which is proportional to $-ku_x = -k$, is constant.

Example 4:

Suppose instead that the same initially heated bar has length $\ell = \pi$ and perfectly insulated ends. Then its temperature function u must satisfy the Neumann conditions

$$u_x(0, t) = u_x(\pi, t) = 0, \qquad t > 0 \tag{13'}$$

since now heat cannot escape through the ends. In this case we are faced with the strip problem shown in Figure 3.8. Although we might try to guess a solution, it is better to use the separation methods of Section 2.4. There (in

† We can also think of u as giving the temperature distribution in the bar when we keep the ends at temperatures 0 and ℓ for eternity! In particular, at each time t, the temperature along the bar is distributed linearly, which suggests a means of generating the given initial profile.

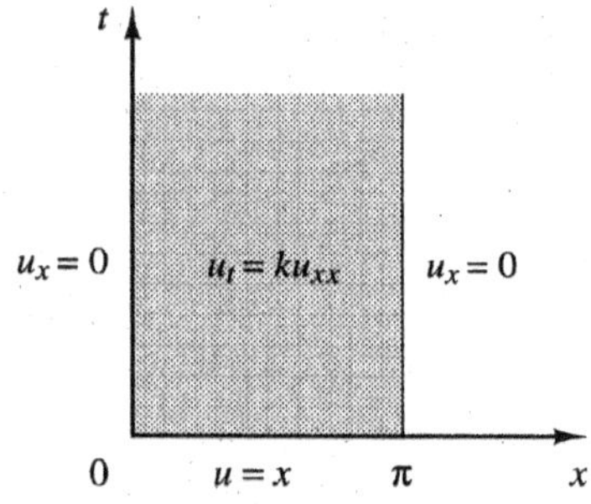

Figure 3.8

Example 2) we found that each product

$$u_n(x,t) = e^{-kn^2t}\cos nx, \qquad n = 0, 1, 2, \dots$$

solves the homogeneous part of this problem, and by linearity so does the finite sum

$$U_N(x,t) = \frac{c_0}{2} + \sum_{n=1}^{N} c_n e^{-kn^2t}\cos nx$$

for any $N = 1, 2, 3, \dots$, and constants $c_0, c_1, c_2, \dots, c_N$. In particular,

$$u(x,t) = 3e^{-kt}\cos x - 5e^{-9kt}\cos 3x$$

would solve our problem (uniquely) if we wanted the initial temperature distribution

$$u(x,0) = 3\cos x - 5\cos 3x.$$

But we want $u(x,0) = x$, and it is here that Fourier series come in. If we recall from Example 3 in Section 1.4 that

$$x = \frac{\pi}{2} - \frac{4}{\pi}\sum_{n=1,3,}^{\infty} \frac{1}{n^2}\cos nx, \qquad 0 \le x \le \pi,$$

then it is natural to try the corresponding series

$$U(x,t) = \frac{\pi}{2} - \frac{4}{\pi}\sum_{n=1,3,}^{\infty} \frac{1}{n^2} e^{-kn^2t}\cos nx \tag{14}$$

as a *formal* solution to our problem.

It is easy to generalize this approach and produce formal solutions to similar strip problems with the initial condition $u(x,0) = f(x)$ as follows.

For the boundary value problem indicated in Figure 3.9 with strip conditions of type $\mathscr{D}$, $\mathscr{N}$, or $\mathscr{M}$ at $x = 0, \ell$:

Represent f as a Fourier series of functions $X_n(x)$ that meet the strip

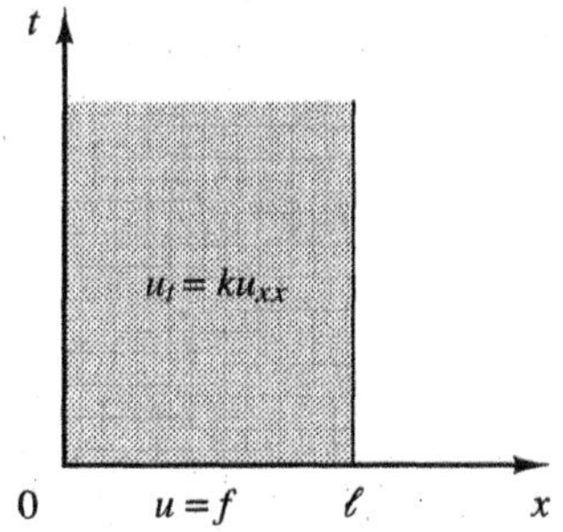

Figure 3.9

conditions; then replace each $X_n(x)$ by a product of the form

$$u_n(x,t) = \exp(-\lambda_n kt)X_n(x)$$

that satisfies the equation $u_t = ku_{xx}$. The resulting series $U(x,t)$ is a formal solution to the problem.

In this way we can convert a suitable Fourier series representation of the given function $f(x)$ into a formal series solution of the problem just as we did in Section 2.2.

Example 5:

For the problem shown in Figure 3.10, we choose functions

$$X_n(x) = \sin\frac{n\pi x}{2\ell}, \qquad n = 1, 3, \ldots$$

to satisfy the mixed strip conditions and use them to represent x^2 by the associated sine series

$$x^2 = \sum_{n=1,3,}^{\infty} s_n \sin\frac{n\pi x}{2\ell}, \qquad 0 < x < \ell;$$

this requires (see Problem 1.6E) that

$$s_n = \frac{2}{\ell}\int_0^\ell x^2 \sin\frac{n\pi x}{2\ell}\,dx = \frac{16\ell^2}{n^2\pi^2}\left[\sin\frac{n\pi}{2} - \frac{2}{n\pi}\right], \qquad n = 1, 3, \ldots .$$

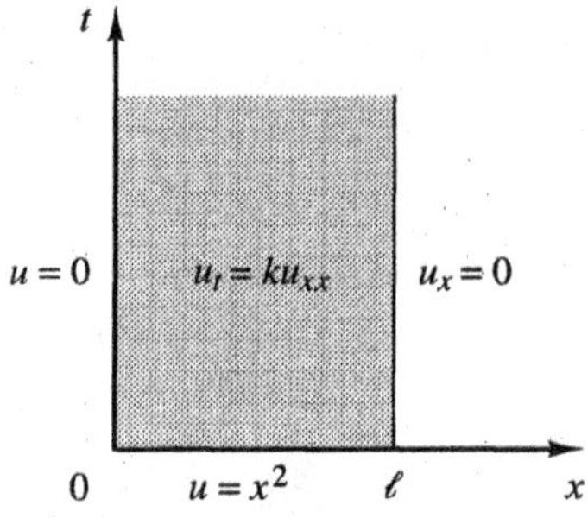

Figure 3.10

By substitution we see that

$$u_n(x,t) = \exp(-\lambda_n kt)\sin\frac{n\pi x}{2\ell}$$

satisfies the equation $u_t = ku_{xx}$, when $\lambda_n = (n\pi/2\ell)^2$, $n = 1, 3, \ldots$. Consequently, with the s_n just found,

$$U(x,t) = \sum_{n=1,3,}^{\infty} s_n \exp\left(-\left(\frac{n\pi}{2\ell}\right)^2 kt\right)\sin\frac{n\pi x}{2\ell} \tag{15}$$

provides a formal solution of our problem.

Anatomy of a Solution

What good is a *formal* solution to a *real* boundary value problem? In particular, what can we say about series (14) as being a solution to the boundary value problem of Example 4? First, note that each associated *approximate solution*

$$U_N(x,t) = \frac{\pi}{2} - \frac{4}{\pi}\sum_{n=1,3,}^{N}\frac{1}{n^2}e^{-kn^2t}\cos nx, \qquad N = 1, 2, \ldots$$

is a very smooth (infinitely differentiable) function of x and t that is defined everywhere.

U_N solves the homogeneous part of the problem *exactly* (why?), and at $t = 0$ its values

$$U_N(x,0) = \frac{\pi}{2} - \frac{4}{\pi}\sum_{n=1,3,}^{N}\frac{1}{n^2}\cos nx$$

give a Fourier approximation to the given initial condition. In this case, the approximation is uniform by the M-test. In fact, for $t \geq 0$: $e^{-kn^2t} \leq 1$ (since $k > 0$) so that

$$|c_n u_n(x,t)| = \left|\frac{4}{\pi n^2}e^{-kn^2t}\cos nx\right| \leq \frac{4}{\pi n^2}, \qquad n = 1, 2, \ldots,$$

and we have the following *uniform* error estimate

$$|U(x,t) - U_N(x,t)| \leq \sum_{n=N+1}^{\infty}|c_n u_n(x,t)| \leq \frac{4}{\pi}\sum_{n=N+1}^{\infty}\frac{1}{n^2} \leq \frac{4}{\pi N}.$$

Thus for $t \geq 0$, U is approximated uniformly by global solutions U_N of equation (13). But is U itself a solution? Let's examine series (14) more closely. When $t \geq 0$, the series converges uniformly so that its sum U is continuous. In particular,

$$\text{as } t \searrow 0: \quad U(x,t) \to U(x,0) = x, \qquad 0 \leq x \leq \pi.$$

Therefore, U satisfies the initial condition.

Also, if we differentiate the series termwise with respect to x, we have for $t \geq t_0 > 0$ a uniformly convergent series, since

$$\left|\frac{e^{-kn^2t}\sin nx}{n}\right| \leq e^{-knt_0} \quad \text{and} \quad \sum_{n=1}^{\infty} e^{-knt_0} < +\infty.$$

Thus for $t \geq t_0 > 0$ we have $U_x(x, t) = \frac{4}{\pi}\sum_{n=1,3,}^{\infty} \frac{1}{n} e^{-kn^2 t} \sin nx$, and

$$\begin{aligned} \text{as } x \searrow 0: \quad & U_x(x, t) \to U_x(0, t) = 0, \\ \text{as } x \nearrow \pi: \quad & U_x(x, t) \to U_x(\pi, t) = 0. \end{aligned}$$

Hence U satisfies the boundary conditions.

Finally, we see that termwise differentiation of (14) once with respect to t or twice with respect to x results in series with terms

$$\begin{aligned} U_t: \quad & \frac{4}{\pi} k e^{-kn^2 t} \cos nx \\ U_{xx}: \quad & \frac{4}{\pi} e^{-kn^2 t} \cos nx \end{aligned}$$

which can be estimated as above when $t \geq t_0 > 0$. We conclude that for $t > 0$,

$$L(U) \stackrel{\text{def}}{=} U_t - kU_{xx} = \lim_{N \to \infty} L(U_N) = 0.$$

Consequently, U is an actual solution to our boundary value problem, and by Theorem 3.1, it is *the* unique solution.

These arguments illustrate how uniform convergence is used to show that a formal solution has the properties required of an actual solution, but such arguments are rather technical so we will only sketch them hereafter.

Now let's look at some diffusion problems from other fields.

Example 6 (*Kiln-drying lumber*):

To build houses we need wood. Well-seasoned wood is best but natural seasoning (drying) is time-consuming. To accelerate the process, freshly cut timber is stacked in an oven or kiln and dried under controlled conditions. While the wood is drying, its state is monitored by comparing the weight w of a unit cube to that w_0 of a dry cube of the same kind of wood. The nondimensional quantity $u = (w - w_0)/w_0$, called the **moisture content**, is a measure of relative wetness that decreases to zero as the wood dries.

To model the drying lumber, assume that at time t the moisture content u "*transfuses*" across the grain from wetter to dryer regions *at a rate proportional to the slope of the moisture gradient*. Suppose that a long thin plank of wood from a tree is placed in a kiln at time $t = 0$. Consider a typical interior cross section as indicated in Figure 3.11 and neglect edge effects. Then at time t the moisture content $u = u(x, t)$ should obey a one-dimensional diffusion equation

$$u_t = k(u_x)_x = ku_{xx}, \qquad 0 < x < \ell, t > 0, \tag{16}$$

for some positive constant k that depends on kiln conditions and the type of wood being dried. For convenience, suppose that the plank is of width $\ell = \pi$. Then $u(0, t) = u(\pi, t) = 0$, $t > 0$ (since the outer faces remain dry), and this suggests use of the elementary products

$$u_n(x, t) = e^{-kn^2 t} \sin nx, \qquad \text{with } u_n(x, 0) = \sin nx, \qquad n = 1, 2, \ldots.$$

Let's assume that at time $t = 0$, the wood is uniformly wet, so that for some

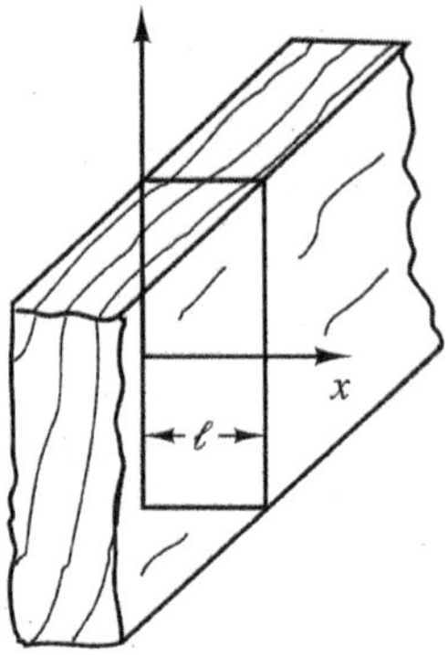

Figure 3.11

constant $W > 0$,

$$u(x,0) = f(x) = W, \qquad 0 < x < \pi.$$

Since this function has the *sine* series representation

$$W = \frac{4W}{\pi} \sum_{n=1,3,}^{\infty} \frac{\sin nx}{n}, \qquad 0 < x < \pi,$$

we consider as a formal solution the corresponding series

$$U(x,t) = \frac{4W}{\pi} \sum_{n=1,3,}^{\infty} \frac{e^{-kn^2t}}{n} \sin nx, \qquad t > 0. \tag{16'}$$

By examining the behavior for $t \geq t_0 > 0$, as in the previous discussion, we can establish that when $t > 0$, U satisfies the equation and meets the boundary conditions. This series does not converge uniformly for $t \geq 0$, and it is not clear that

$$\text{as } t \searrow 0: \quad U(x,t) \to W, \qquad 0 < x < \pi;$$

however, this limiting behavior can be established. The difficulty reflects the fact that our boundary conditions are incompatible with the initial condition at $x = 0$ and $x = \pi$. In effect, we have required the outer faces to be wet at time $t = 0$ and instantaneously dry when $t > 0$. Only a discontinuous function could behave in this manner, and such functions cannot be represented by uniformly convergent series. Instead, the partial sums of this series,

$$U_N(x,t) = \frac{4W}{\pi} \sum_{n=1,3,}^{N} \frac{e^{-kn^2t}}{n} \sin nx,$$

exhibit Gibbs' behavior near $(0,0)$ and $(\pi,0)$, as shown in Figure 3.12 (p. 129).

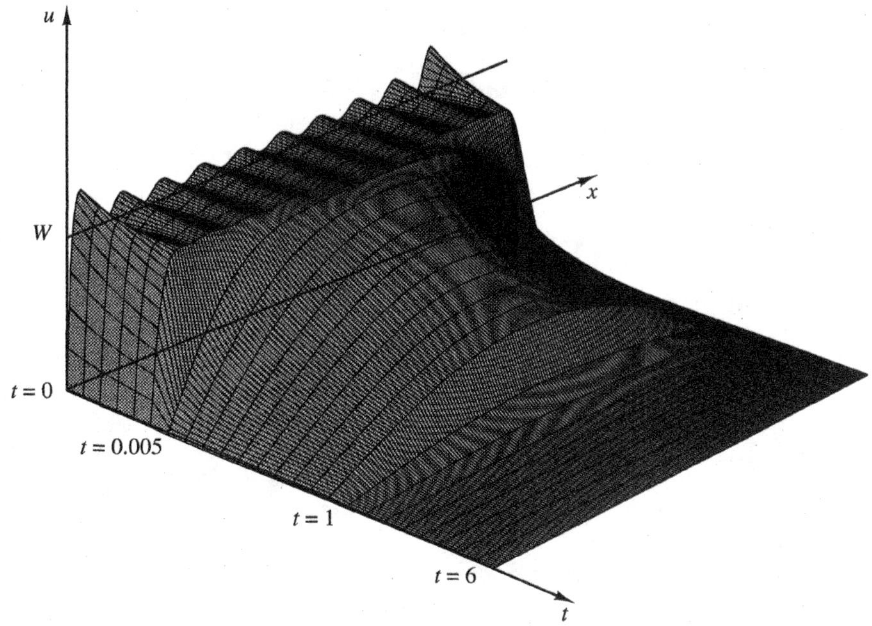

Figure 3.12

Observe that for $t > 0$ a geometric series estimate gives

$$|U(x,t)| \leq \frac{4W}{\pi} \sum_{n=1,3,}^{\infty} e^{-kn^2t} \leq \frac{4W}{\pi} e^{-kt} \sum_{n=0}^{\infty} (e^{-kt})^n = \frac{4W}{\pi} \frac{e^{-kt}}{1-e^{-kt}} = \frac{4W}{\pi} \frac{1}{e^{kt}-1}$$

so that, for example,

$$|U(x,t)| \leq .01W, \qquad \text{if } kt > \log\left(1 + \frac{400}{\pi}\right) \approx 2.11.$$

Once k is known, we can use this estimate to predict how much time is required for the lumber to become 99% dry. Moreover, through comparative measurements, we can use our solution to obtain approximate values for k under kiln conditions.

Example 7 (*Transistor design*):

When we switch on a television or stereo set today, we probably energize transistors, which are used instead of the older vacuum tubes to amplify and control electrical currents. The simplest transistor is a solid-state device formed by

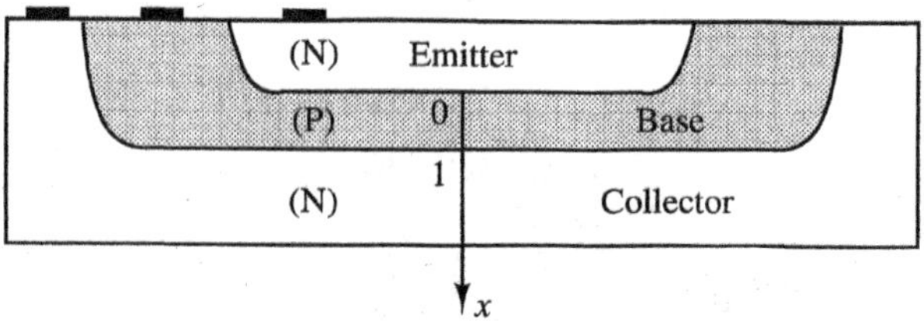

Figure 3.13

bonding three semiconductors of different types† in the order N–P–N, designated emitter–base–collector, respectively.

During normal operation, voltages applied to each component are such that the flow of excess electrons from the emitter into the base controls the current produced by the corresponding flow from the base into the collector. The significant electrical activity is confined to the base and, to improve efficiency, during base fabrication the net impurity strength is "graded" to produce a constant built-in field that causes a convective flow of electrons from emitter to collector.

What happens inside the base when we switch the transistor *off* at time $t = 0$? By neglecting some edge effects, we can imagine the base to be situated on an x-interval $[0, 1]$ (as shown in the cross section in Figure 3.13) and consider its electron-concentration density $\eta = \eta(x, t)$, which for $t > 0$ should vanish at $x = 0$ and $x = 1$. Moreover, it can be shown that a constant built-in field produces the initial concentration

$$\eta(x, 0) = 1 - \exp(-2a(1 - x)), \qquad 0 < x < 1,$$

for a positive drift constant a, and a transient response modeled by the diffusion-convection equation

$$\eta_t = D(\eta_{xx} - 2a\eta_x), \qquad 0 < x < 1, t > 0, \tag{17}$$

for a positive charge-diffusivity constant D. Both a and D depend on the materials and type of construction employed, and certain related constants have been suppressed to simplify the presentation. The details will be found in [L–W].

If we substitute $\eta(x, t) = e^{ax - Da^2 t} u(x, t)$ in (17), we can show that u satisfies the heat equation

$$u_t = Du_{xx}, \qquad 0 < x < 1, t > 0, \tag{17$'$}$$

the boundary conditions $u(0, t) = u(1, t) = 0$, $t > 0$, and the initial condition

$$u(x, 0) = f(x) \stackrel{\text{def}}{=} e^{-ax}(1 - e^{-2a(1-x)}) = e^{-ax} - e^{-2a}e^{ax}$$
$$= 2e^{-a} \sinh a(1 - x), \qquad 0 < x < 1.$$

† A semiconductor is made by adding impurities to the crystal-lattice of silicon to alter its natural low electrical conductivity. It is of type P or N, according to whether the resulting composite is electron-poor or electron-rich.

A formal solution to the latter boundary value problem is given by

$$U(x, t) = \sum_{n=1}^{\infty} s_n e^{-D\pi^2 n^2 t} \sin n\pi x$$

where the s_n are chosen to make

$$U(x, 0) = f(x), \qquad 0 < x < 1.$$

With f as above,

$$s_n = 2 \int_0^1 f(x) \sin n\pi x \, dx = \frac{\alpha n\pi}{n^2\pi^2 + a^2}, \qquad n = 1, 2, \ldots \tag{18}$$

where $\alpha = 4e^{-a} \sinh a$. (See Problem 3.3K.) Thus, the series

$$\mathfrak{N}(x, t) = \alpha e^{ax} \sum_{n=1}^{\infty} \frac{n\pi}{n^2\pi^2 + a^2} \, e^{-D(n^2\pi^2+a^2)t} \sin n\pi x \tag{18'}$$

provides a formal solution to the original boundary value problem. For $t > 0$, $\mathfrak{N}$ satisfies the given differential equation and boundary conditions properly. However, it is less easy to establish that as $t \searrow 0$,

$$\mathfrak{N}(x, t) \to 1 - e^{-2a(1-x)},$$

especially when x is near zero where the initial base requirement is not compatible with the boundary condition.

*Example 8:**

In transistor design, a physical quantity of interest is the *total charge reclaimable through the emitter*, defined within a constant factor by the improper integral $Q_0 = \int_0^\infty \mathfrak{N}_x(0, t)dt$. Differentiation of (18′) and evaluation at $x = 0$ gives

$$\begin{aligned} Q_0 &= \lim_{\substack{T \nearrow +\infty \\ \epsilon \searrow 0}} \alpha \int_\epsilon^T \sum_{n=1}^{\infty} \frac{(n\pi)^2}{n^2\pi^2 + a^2} \, e^{-D(n^2\pi^2+a^2)t} \, dt \\ &= \lim_{\substack{T \nearrow +\infty \\ \epsilon \searrow 0}} \frac{\alpha}{D} \sum_{n=1}^{\infty} \frac{-(n\pi)^2}{(n^2\pi^2 + a^2)^2} \, e^{-D(n^2\pi^2+a^2)t} \Big|_\epsilon^T . \end{aligned}$$

When $t \geq 0$, the uniform convergence of this last series permits the limit interchange required to conclude that

$$Q_0 = \frac{\alpha}{D} \sum_{n=1}^{\infty} \left(\frac{n\pi}{n^2\pi^2 + a^2} \right)^2 = \frac{1}{D\alpha} \sum_{n=1}^{\infty} s_n^2,$$

where the s_n are precisely the sine coefficients of $f(x) = 2e^{-a} \sinh a(1 - x)$ obtained above in (18).

Consequently, we can use Parseval's formula (part C of Theorem 1.6) to evaluate the last series as follows:

$$\sum_{n=1}^{\infty} s_n^2 = 2\int_0^1 f^2(x)dx = 8e^{-2a}\int_0^1 (\sinh^2 ax)\,dx$$

$$= 4e^{-2a}\int_0^1 (\cosh 2ax - 1)dx = 4e^{-2a}\left(\frac{\sinh 2a}{2a} - 1\right),$$

and we obtain for Q_0 the simple expression

$$Q_0 = \frac{e^{-a}}{D\sinh a}\left(\frac{\sinh 2a}{2a} - 1\right),$$

which indicates the relative importance of drift (a) and diffusion (D).

Problem Set 3.3

3.3A Verify that each of the following functions is a solution of (13) and give a boundary value problem that it solves in $\{0 < x < \ell, t > 0\}$ involving (a) Dirichlet conditions, (b) Neumann conditions, or (c) Robin conditions where $[(\partial u/\partial n) + u]|_{x=0,\ell}$ is prescribed.

1. $u(x, t) = e^{x+kt}$
2. $u(x, t) = x^2 t + kt^2 + \frac{1}{12k}x^4$
3. $u(x, t) = 3e^{-kt}\cos x + 2e^{-4kt}\sin 2x$
4.* $u(x, t) = \frac{1}{\sqrt{t}}\exp[-x^2/4kt], \quad t > 0$

3.3B Show that if u_0 is a constant, then

$$u(x, t) = u_0\left[1 - \frac{2}{\pi}\cos\left(\frac{\pi x}{\ell}\right)\exp\left(-\frac{\pi^2 kt}{\ell^2}\right)\right]$$

is the solution to a heat conduction problem for a bar of length ℓ with insulated ends. What initial conditions are satisfied with this solution?

3.3C Use Fourier series methods to obtain a formal solution, $U = U(x, t)$, to each of the following problems involving the equation $u_t = ku_{xx}$ in $\{0 < x < \pi, t > 0\}$. Comment briefly on the behavior of $U(x, t)$ as $t \searrow 0$ in relation to aspects of compatibility.

1. $u(0, t) = u(\pi, t) = 0; \quad u(x, 0) = x, \quad 0 < x < \pi.$
2. $u(0, t) = u(\pi, t) = 0; \quad u(x, 0) = \begin{cases} x, & 0 < x < \pi/2 \\ \pi - x, & \pi/2 < x < \pi. \end{cases}$
3. $u_x(0, t) = u_x(\pi, t) = 0; \quad u(x, 0) = x, \quad 0 < x < \pi.$
4. $u_x(0, t) = u_x(\pi, t) = 0; \quad u(x, 0) = \begin{cases} x, & 0 < x < \pi/2 \\ \frac{\pi}{2}, & \pi/2 < x < \pi. \end{cases}$
5. $u(0, t) = u_x(\pi, t) = 0; \quad u(x, 0) = x, \quad 0 < x < \pi.$ (*Hint*: See Example 5.)
6. $u_x(0, t) = u(\pi, t) = 0; \quad u(x, 0) = \begin{cases} 0, & 0 < x < \pi/2 \\ 1, & \pi/2 < x < \pi. \end{cases}$
7. $u(0, t) = u(\pi, t) = 0; \quad u(x, 0) = \sum_{n=1}^{\infty} \frac{1}{n^2}\sin nx.$

3.3D Use Fourier series methods to obtain a formal solution, $U = U(x,t)$, to each of the following problems involving the equation $u_t = ku_{xx}$ in $\{0 < x < \ell, t > 0\}$.

1. $u(0,t) = u(\ell,t) = 0; \quad u(x,0) = x, \quad 0 < x < \ell.$
2. $u(0,t) = u(\ell,t) = 0; \quad u(x,0) = \begin{cases} x, & 0 < x < \ell/2 \\ \ell - x, & \ell/2 < x < \ell. \end{cases}$
3. $u_x(0,t) = u_x(\ell,t) = 0; \quad u(x,0) = x, \quad 0 < x < \ell.$
4. $u_x(0,t) = u_x(\ell,t) = 0; \quad u(x,0) = \begin{cases} x, & 0 < x < \ell/2 \\ \frac{\ell}{2}, & \ell/2 < x < \ell. \end{cases}$
5. $u(0,t) = u_x(\ell,t) = 0; \quad u(x,0) = x, \quad 0 < x < \ell.$
   (*Hint*: See Example 5.)
6. $u_x(0,t) = u(\ell,t) = 0; \quad u(x,0) = \begin{cases} 0, & 0 < x < \ell/2 \\ 1, & \ell/2 < x < \ell. \end{cases}$

3.3E* 1. If $s_n = (2/\pi)\int_0^\pi f(x)\sin nx\,dx$, $n = 1, 2, \ldots$, for an integrable f satisfying $(2/\pi)\int_0^\pi |f(x)|dx = M < +\infty$, conclude that for $t \geq t_0 \geq 0$, the series $U(x,t) = \sum_{n=1}^{\infty} s_n u_n(x,t)$ converges absolutely and uniformly.

2. Verify that $\sum_{n=1}^{\infty} n^2 e^{-kn^2 t_0} < \infty$, and show that for $t \geq t_0 > 0$, the differentiated series for U_x, U_{xx}, and U_t also converge uniformly.

3.* If $f \in \hat{C}^1[0,\pi]$ with $f(0) = f(\pi) = 0$, show that the series for U converges uniformly and absolutely for $t \geq 0$ so that it gives the unique solution to the problem shown in the figure. (*Hint*: Use part E of Theorem 1.7 to conclude that $\sum_{n+1}^{\infty} |s_n| < +\infty$.)

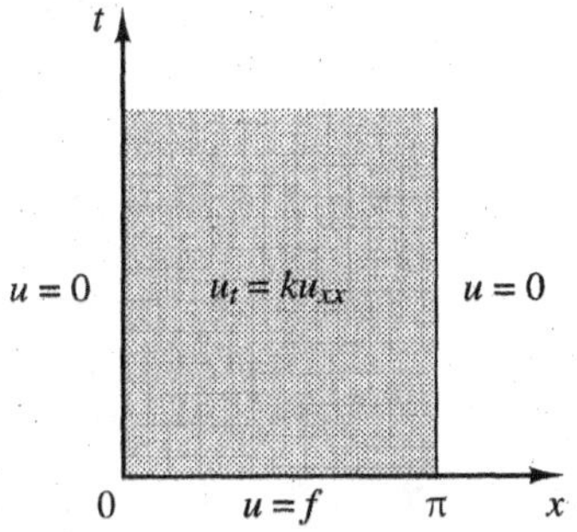

4.* If f is sectionally smooth on $[0,\pi]$ with $f(0+) = f(\pi-) = 0$, show that at each $x \in (0,\pi)$:

$$\lim_{t \searrow 0} U(x,t) = \frac{f(x+) + f(x-)}{2}.$$

(*Hint*: Use Abel's test with $p_n(t) = e^{-kn^2 t}$ as in Example 14 of Section 1.9.)

3.3F How would the arguments in Problem 3.3E change for an analysis of the series

$$U(x,t) = \frac{c_0}{2} + \sum_{n=1}^{\infty} c_n e^{-kn^2 t} \cos nx,$$

where $c_n = (2/\pi)\int_0^\pi f(x)\cos nx\,dx$, $n = 0, 1, 2, \ldots$?

3.3G Using Problem 2.4E, explain how one might attack the problem

$$\begin{cases} L(u) \equiv u_t - ku_{xx} = 0 & \text{in} \quad \{0 < x < \pi, t > 0\} \\ u(0,t) = u_x(\pi,t) + u(\pi,t) = 0, & t > 0 \\ u(x,0) = x^2, & 0 < x < \pi. \end{cases}$$

What prevents your having a complete explicit formal solution?

3.3H When the lateral surface of a long thin circular rod of uniform composition is not insulated, then (as in Problem 3.1C) the governing equation for the temperature distribution $u = u(x,t)$ is

$$u_t = (ku_x)_x - \gamma u, \qquad 0 < x < \ell, \qquad t > 0,$$

for suitable positive constants k and γ.

1. For simplicity, suppose $\ell = \pi$ and $\gamma = 1$. If the ends of the bar are insulated, show that elementary product solutions $u = X(x)T(t)$ are possible, provided that for some constant λ,

$$\begin{cases} kX'' - X = -\lambda X, & 0 < x < \pi \\ X'(0) = X'(\pi) = 0 \end{cases}$$

2. Solve the eigenvalue problem in part 1 and use your result to obtain a formal solution to the initial value problem for this bar when $u(x,0) = x^2$.
3. Discuss briefly whether your formal solution is an actual solution.
4. Suppose that the rod is *nonuniform*, requiring that k and γ be considered as positive functions of x. How will your conclusions change? What prevents your obtaining a complete formal solution to the initial value problem in part 2? Can you find approximate solutions?

3.3I Suppose that the plank of Example 6 initially has the parabolic moisture content $u(x,0) = x(\pi - x)$, $0 < x < \pi$. Obtain the corresponding series solution and discuss its convergence behavior. (Note that this initial condition *is* compatible with the given boundary conditions.) What is the solution when this parabolic distribution is approximated by $(\pi^2/4)\sin x$?

3.3J
1. Show that the nonlinear *Burger's equation* $v_t + vv_x = v_{xx}$ is satisfied by $v = -2u_x/u$ if $u_t = u_{xx}$.
2. Use this fact to find a solution $v = v(x,t)$ to Burger's equation for $0 < x < \pi$, $t > 0$, with $v(0,t) = v(\pi,t) = 0$, $t > 0$ and $v(x,0) = 1$, $0 < x < \pi$.
3. Is the solution of the boundary value problem in part 2 unique?

3.3K (Transistor design from Example 7)

1. Show that for constants a, b, and k the equation $v_t = k(v_{xx} - 2av_x + bv)$ can be replaced by $u_t = ku_{xx}$ under the substitution $v(x,t) = e^{ax+\beta t}u(x,t)$, provided $\beta = k(b - a^2)$.
2. Verify that on the interval $(0,1)$, the Fourier sine coefficients of $f(x) = e^{-ax} - e^{-2a}e^{ax}$ are given by (18).

3.4

Nonhomogeneous Problems

In the previous section we saw how to attack one-dimensional diffusion problems involving a homogeneous differential equation and a pair of homogeneous end conditions. Although certain inhomogeneous end conditions can be accommodated rather easily, it is considerably more difficult to obtain solutions to nonhomogeneous equations. However, the two types of problems are interrelated.

By inspection: $s(x, t) = ax + a_0$ is a global steady-state solution of the equation

$$L(u) \overset{\text{def}}{=} u_t - ku_{xx} = 0$$

and with $a = a_1 - a_0$ we have

$$s(0, t) = a_0; \qquad s(1, t) = a + a_0 = a_1,$$

for preassigned values a_0, a_1.

Thus if a u is a solution to the problem shown in Figure 3.14a, then $v = u + s$ is a solution to the problem shown in Figure 3.14b.

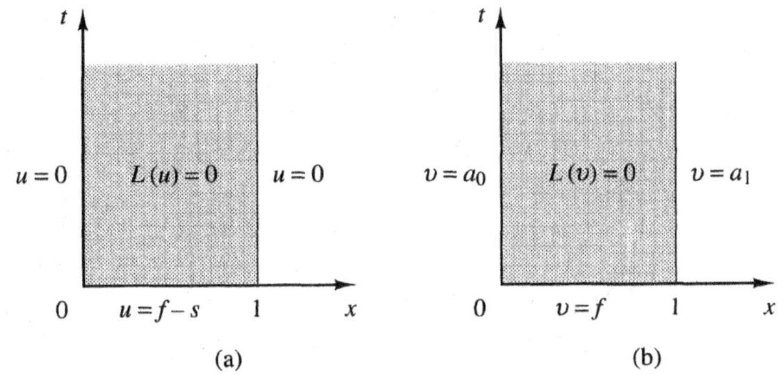

Figure 3.14

Example 9 (*Prescribed end temperatures*):

Suppose that heat sources of unequal temperatures a_0 and a_1 are applied to opposite ends of an iron bar of unit length and insulated sides, which is initially at the constant temperature $v(x, 0) = v_0$. Set $a = a_1 - a_0$. Then, the subsequent temperature distribution in the bar is given by

$$v(x, t) = ax + a_0 + u(x, t),$$

where $u(x, t)$ is a solution of the problem

$$L(u) \overset{\text{def}}{=} u_t - ku_{xx} = 0 \qquad \text{in} \qquad \{0 < x < 1, t > 0\},$$

$$\text{with} \qquad u(0, t) = u(1, t) = 0, \qquad t > 0,$$

and the initial distribution

$$u(x,0) = v_0 - [ax + a_0] = (v_0 - a_0) - ax, \qquad 0 < x < 1.$$

u can be found by the methods of the previous section and then

$$v(x,t) = ax + a_0 + \sum_{n=1}^{\infty} s_n e^{-kn^2\pi^2 t} \sin n\pi x, \tag{19}$$

where

$$s_n = \frac{2(v_0 - a_0)}{n\pi}(1 - (-1)^n) + \frac{2a(-1)^n}{n\pi}, \qquad n = 1, 2, \ldots . \tag{19'}$$

Since the exponential terms die out with increasing t, we see that $u(x,t)$ represents the transient part of the temperature distribution $v(x,t)$ as v approaches its steady-state values,

$$v_s(x,t) = ax + a_0, \qquad 0 < x < 1.$$

Other examples involving inhomogeneous boundary conditions are given in Problem Set 3.4.

Nonhomogeneous Equations

Suppose that we attack the problem shown in Figure 3.15, with $L(u) = u_t - ku_{xx}$, by observing that $s(x,t) = x^2$ satisfies the auxiliary conditions. Then we find that $v(x,t) = u(x,t) - x^2$ would solve the problem when

$$v(0,t) = v(\ell,t) = v(x,0) = 0, \tag{20}$$

provided that v is a solution of the nonhomogeneous equation

$$L(v) = L(u) - L(s) = 0 - L(s) = 2k.$$

(This equation would govern the temperature v in a bar containing a uniformly distributed heat source such as an electrical heating element.)

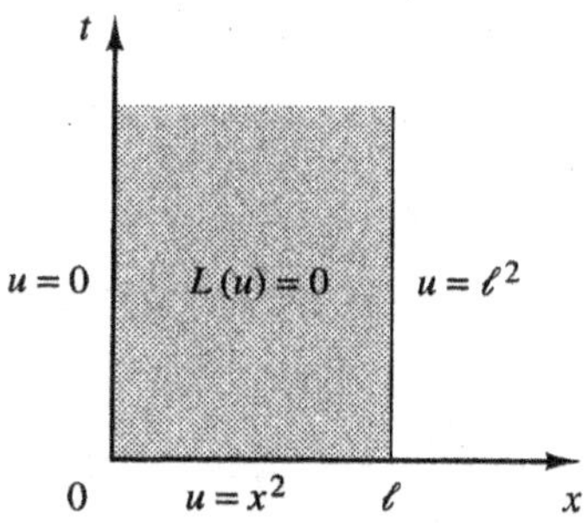

Figure 3.15

Example 10:

Let's find a solution $v = v(x, t)$ to the boundary value problem

$$L(v) \stackrel{\text{def}}{=} v_t - kv_{xx} = F \quad \text{in} \quad \{0 < x < \pi, t > 0\}$$

$$\text{where} \qquad F = F(x, t) = \begin{cases} x, & 0 \le x \le \pi/2 \\ \pi - x, & \pi/2 \le x \le \pi, \end{cases}$$

with the boundary conditions $v(0, t) = v(\pi, t) = 0$, $t > 0$ and the initial condition $v(x, 0) = 0$, $0 < x < \pi$ (see Figure 3.16). To do so we shall continue the analysis begun at the end of Section 2.4.

First, note that the strip conditions and the homogeneous equation require the eigenfunctions $X_n(x) = \sin nx$, $n = 1, 2, \ldots$. Moreover, the product $v_n = T_n X_n$ satisfies the particular nonhomogeneous equation

$$L(v_n) = L(T_n X_n) = (T_n' + kn^2 T_n) X_n = f_n X_n \tag{21}$$

for any given function $f_n = f_n(t)$, provided that $T_n = T_n(t)$ satisfies the equation

$$T_n' + kn^2 T_n = f_n. \tag{22}$$

To have $v_n(x, 0) = 0$, we should take $T_n(0) = 0$; then by Appendix B.1, T_n can be expressed in the integral form

$$T_n(t) = e^{-kn^2 t} \int_0^t e^{kn^2 \tau} f_n(\tau)\, d\tau. \tag{22'}$$

If possible, the f_n should be found to give the representation

$$F(x, t) = \sum_{n=1}^{\infty} f_n(t) X_n(x) = \sum_{n=1}^{\infty} f_n(t) \sin nx, \qquad 0 < x < \pi \tag{23}$$

so that for each $t > 0$, we need

$$\begin{aligned} f_n(t) &= \frac{2}{\pi} \int_0^{\pi} F(x, t) \sin nx\, dx \\ &= \frac{2}{\pi} \left\{ \int_0^{\pi/2} x \sin nx\, dx + \int_{\pi/2}^{\pi} (\pi - x) \sin nx\, dx \right\} \\ &= \frac{4}{\pi n^2} \sin \frac{n\pi}{2} = s_n, \qquad \text{say}, \qquad n = 1, 2, \ldots . \end{aligned} \tag{23'}$$

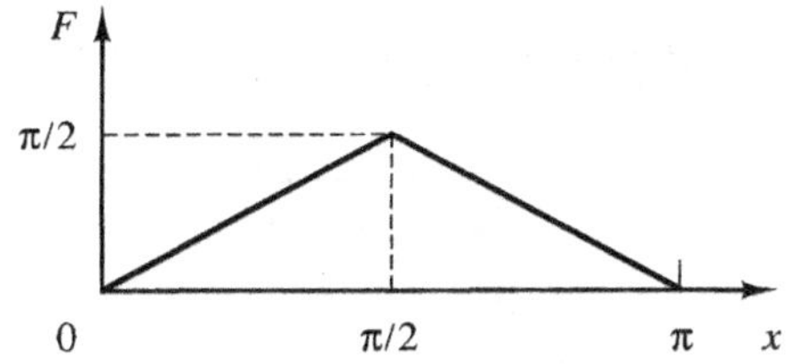

Figure 3.16

When we substitute these f_n in (22′), we obtain

$$T_n(t) = s_n \frac{1 - e^{-kn^2 t}}{kn^2}, \qquad n = 1, 2, \ldots$$

and thereby the formal series solution

$$V(x, t) = \sum_{n=1}^{\infty} T_n(t) X_n(x) = \sum_{n=1}^{\infty} s_n \frac{1 - e^{-kn^2 t}}{kn^2} \sin nx$$

$$= \frac{4}{\pi k} \sum_{n=1}^{\infty} \left(\sin \frac{n\pi}{2}\right) \frac{(1 - e^{-kn^2 t})}{n^4} \sin nx.$$

Because of the presence of the $1/n^4$ factors, it follows that for $t \geq 0$, this series converges absolutely and uniformly (by the M-test); hence, V is continuous and it meets the required auxiliary conditions. Moreover, the series obtained by differentiating termwise as required to consider $L(V) = V_t - kV_{xx}$ also converge uniformly and we see that

$$L(V) = \sum_{n=1}^{\infty} L(T_n X_n) = \sum_{n=1}^{\infty} f_n X_n.$$

By construction,

$$\sum_{n=1}^{\infty} f_n(t) X_n(x) = \frac{4}{\pi} \sum_{n=1}^{\infty} \frac{1}{n^2} \sin \frac{n\pi}{2} \sin nx = \begin{cases} x, & 0 \leq x \leq \pi/2 \\ \pi - x, & \pi/2 \leq x \leq \pi, \end{cases}$$

since from Theorem 1.6 this Fourier series representation is valid. We conclude that the series V is the unique solution to our problem.

If we replace F as given by $F(x, t) = x$, $0 \leq x \leq \pi$, then this same construction provides the formal solution

$$V(x, t) = \frac{2}{k} \sum_{n=1}^{\infty} \frac{(-1)^{n+1}}{n^3} \left(1 - e^{-kn^2 t}\right) \sin nx.$$

Now, the series itself converges uniformly as before, but attempted differentiations are less easy to validate. In this case, we note that each finite sum

$$V_N(x, t) = \frac{2}{k} \sum_{n=1}^{N} \frac{(-1)^{n+1}}{n^3} \left(1 - e^{-kn^2 t}\right) \sin nx$$

satisfies the given homogeneous conditions and the equation

$$L(V_N) = 2 \sum_{n=1}^{N} \frac{(-1)^{n+1}}{n} \sin nx,$$

the right side of which approximates the given F uniformly on each x-interval $[0, \ell]$ for $\ell < \pi$.

Finally, consider an arbitrary $F = F(x, t)$ that admits series representation in the form

$$F(x,t) = \sum_{n=1}^{\infty} f_n(t)X_n(x) = \sum_{n=1}^{\infty} f_n(t)\sin nx$$

for appropriate functions $f_n(t)$. Then for each fixed $t \geq 0$, the $f_n(t)$ are just the Fourier (sine) coefficients of $F(\cdot, t)$ and, assuming sufficient integrability, it follows that

$$f_n(t) = \frac{2}{\pi}\int_0^{\pi} F(x,t)\sin nx\,dx, \qquad n = 1, 2, \ldots .$$

If F is continuous, then each of the f_n so defined is also continuous, and we obtain the associated T_n by integration as before. Then

$$V(x,t) = \sum_{n=1}^{\infty}\left(\int_0^t e^{-kn^2(t-\tau)}f_n(\tau)\,d\tau\right)\sin nx \tag{24}$$

should supply a *formal* solution to the equation $L(v) = F$ that satisfies all of the auxiliary conditions (20) when $\ell = \pi$. See [Wein].

Of course, other boundary conditions result in different eigenfunctions X_n and associated representations.

Example 11:

To solve $L(v) \stackrel{\text{def}}{=} v_t - v_{xx} = xe^t$ with $v_x(0,t) = v_x(\pi,t) = 0,\ \ t > 0$ and $v(x,0) = 0,\ 0 < x < \pi$, we know from Example 4 that the appropriate eigenfunctions are $X_n(x) = \cos nx,\ n = 0, 1, 2, \ldots$. Hence we wish to represent

$$xe^t = \frac{f_0(t)}{2} + \sum_{n=1}^{\infty} f_n(t)\cos nx, \qquad 0 < x < \pi, \tag{25}$$

which gives, in analogy to (23′),

$$f_n(t) = \frac{2}{\pi}e^t\int_0^{\pi} x\cos nx\,dx = \begin{cases} \dfrac{-4}{\pi}\dfrac{e^t}{n^2}, & n \text{ odd} \\ 0, & n = 2, 4, \ldots \end{cases}$$

$$f_0(t) = \frac{2}{\pi}e^t\frac{\pi^2}{2} = e^t\pi.$$

We note that since xe^t has a continuous periodic even extension in x for each t, then (25) holds with these coefficients, and the $1/n^2$ factors guarantee uniform convergence when $0 \leq t \leq \tau$, for each $\tau > 0$.

Next, we substitute these f_n in (22′) with k=1 to obtain

$$T_n(t) = \frac{-4}{\pi n^2}\int_0^t e^{-n^2(t-\tau)}e^{\tau}\,d\tau = \frac{-4}{\pi n^2}e^{-n^2 t}\int_0^t e^{(n^2+1)\tau}\,d\tau$$

$$= \frac{-4}{\pi n^2}e^{-n^2 t}\frac{(e^{(n^2+1)t} - 1)}{n^2+1} = \frac{-4}{\pi n^2}\frac{(e^t - e^{-n^2 t})}{n^2+1}, \qquad n = 1, 3, 5, \ldots,$$

while $\quad T_0(t) = \pi\int_0^t e^{\tau}d\tau = \pi(e^t - 1).$

Thus, we have the formal series solution

$$\begin{aligned} V(x,t) &= \frac{T_0(t)}{2} + \sum_{n=1,3,}^{\infty} T_n(t)\cos nx \\ &= \frac{\pi}{2}(e^t - 1) - \frac{4}{\pi}\sum_{n=1,3,}^{\infty}\left(\frac{e^t - e^{-n^2 t}}{n^2(n^2+1)}\right)\cos nx, \end{aligned} \tag{26}$$

and, in view of the $1/n^2(n^2+1)$ factors, it will converge uniformly together with the necessary termwise differentiated series so that V is *the* solution of our problem.

Problem Set 3.4

3.4A Obtain a formal solution V to each of the following problems.

1. $L(v) \equiv v_t - 2v_{xx} = 0$ in $\{0 < x < 1, t > 0\}$, with $v(0,t) = 0, v(1,t) = 100, t > 0$ and $v(x,0) = x, 0 < x < 1$.
2. $L(v) \equiv v_t - v_{xx} = 0$ in $\{0 < x < \pi, t > 0\}$, with $v(0,t) = 10, v(\pi,t) = -10, t > 0$ and $v(x,0) = x^2, 0 < x < \pi$.
3. $L(v) \equiv v_t - v_{xx} = 0$ in $\{0 < x < \pi, t > 0\}$, with $v(0,t) = 0, v_x(\pi,t) = -e^{-t}, t > 0$ and $v(x,0) = \begin{cases} 1, & 0 < x < \frac{\pi}{2} \\ 0, & \frac{\pi}{2} < x < \pi. \end{cases}$ (*Hint*: Take $s(x,t) = e^{-t}\sin x$.)
4. $L(v) \equiv v_t - v_{xx} = 0$ in $\{0 < x < \pi, t > 0\}$, with $v(0,t) = 0, v_x(\pi,t) = -e^{-4t}, t > 0$ and $v(x,0) = x, 0 < x < \pi$.
5. $L(v) \equiv v_t - v_{xx} = x^2 e^t$ in $\{0 < x < \pi, t > 0\}$, with $v(0,t) = v(\pi,t) = 0, t > 0$ and $v(x,0) = 0, 0 < x < \pi$.
6. $L(v) \equiv v_t - kv_{xx} = xt$ in $\{0 < x < 1, t > 0\}$, with $v(0,t) = 0, v(1,t) = 0, t > 0$ and $v(x,0) = Ax + B, 0 < x < 1$.

3.4B Show that the four functions $u_0 = 1$, $u_1 = x$, $u_2 = x^2 + 2kt$, $u_3 = x^3 + 6kxt$ (known as heat polynomials) are solutions of (13). Find a linear combination of these functions that satisfies the boundary conditions $u(0,t) = 0$, $u(1,t) = t$, and explain how it could be used to solve a nonhomogeneous problem.

3.4C 1. If the lateral surface of a rod is not insulated, there is heat exchange by convection with the surrounding medium. When the surrounding medium has constant temperature T_0, the rate at which heat is lost from the rod is proportional to the difference $u - T_0$. The governing partial differential equation is $ku_{xx} = u_t + \gamma(u - T_0)$, $k, \gamma > 0$. Show that the change of variables $u(x,t) = T_0 + v(x,t)e^{-\gamma t}$ leads to the heat equation $v_t = kv_{xx}$.

2. Use the above technique to solve the problem

$$\begin{cases} u_{xx} = u_t + 4u - 20, & 0 < x < \pi, t > 0 \\ u(0,t) = 5 = u(\pi,t), & t > 0 \\ u(x,0) = 5 + 2x, & 0 < x < \pi. \end{cases}$$

3.4D The ends $x = 0$ and $x = 100$ of a rod 100 cm in length with an insulated lateral surface are held at temperatures 0°C and 100°C, respectively, until steady-state conditions prevail. Then at time $t = 0$, the temperatures of the two ends are interchanged. Find the subsequent temperature distribution throughout the rod.

3.4E The steady-state problem corresponding to

$$(\mathrm{P})\begin{cases} u_t = ku_{xx} + cu, & 0 < x < \pi, t > 0 \\ u_x(0,t) = T_1, & t > 0 \\ u_x(\pi,t) = T_2, & t > 0 \\ u(x,0) = f(x), & 0 < x < \pi \end{cases} \quad \text{is } (\mathrm{P_S})\begin{cases} kv_{xx} + cv = 0, & 0 < x < \pi \\ v_x(0) = T_1 \\ v_x(\pi) = T_2. \end{cases}$$

1. Solve the problem when $c < 0$.
2. Show that ($\mathrm{P_S}$) has a solution when $c = 0$ if and only if $T_1 = T_2$.
3. If $c > 0$, replace π in ($\mathrm{P_S}$) by ℓ. For which values of ℓ does ($\mathrm{P_S}$) have a unique solution for all T_1 and T_2?
4. Show that if u solves (P) and v solves ($\mathrm{P_S}$), then $w(x,t) = u(x,t) - v(x)$ solves

$$\begin{cases} w_t = kw_{xx} + cw, & 0 < x < \pi, t > 0 \\ w_x(0,t) = w_x(\pi,t) = 0, & t > 0 \\ w(x,0) = g(x), & 0 < x < \pi \end{cases}$$

where $g(x) = f(x) - v(x)$.

3.4F* (*Duhamel's principle*) Suppose that $F \in C^1(\bar{V} \times \mathbb{R})$ where V is a domain of $\mathbb{R}^d$ with boundary S. Suppose also that for each $\tau > 0$, $U(x, t, \tau)$ is a solution of the boundary value problem for the equation $u_t = u_{xx}$ in $\mathscr{V} = V \times (0, +\infty)$ with the boundary condition, $u|_S = 0$, $t > 0$ and the initial condition $u(\mathbf{x}, 0) = f(\mathbf{x}, \tau), \mathbf{x} \in V$.

1. Show that $v(\mathbf{x}, t) \overset{\text{def}}{=} \int_0^t U(\mathbf{x}, t - \tau, \tau)\, d\tau$ gives a formal solution to the boundary value problem

$$v_t = v_{xx} + F \qquad \text{in } \mathscr{V}$$

with $v|_S = 0$, $t > 0$ and $v(\mathbf{x}, 0) = 0$, $\mathbf{x} \in V$. (Assume that all integrands are continuous.)

2. Derive a similar result when U satisfies the Neumann condition $(\partial u/\partial n)|_S = 0$ instead of the Dirichlet condition.
3. Use this principle to solve the problem of Example 11, and compare your answer with that in (26).

3.5

Continuous Dependence of Solution on Auxiliary Data

In this chapter, we envision experimental verification of our mathematical results. Is this possible? For example, how can we verify that a bar has been heated to exactly 200°F when even the most accurate digital thermometer must round off at some point? In fact, we can never measure any real temperature exactly but a mathematical solution is determined by some hypothetically exact specification. Intuitively, we reason that approximation should suffice, but what does that mean?

Surprisingly, this question was not considered seriously until this century when J. Hadamard called attention to its importance. Our assumption that approximation of auxiliary conditions is permissible can be translated into the more precise statement that small variations in auxiliary data will be accompanied by small variations in the solution; i.e., the solution depends **continuously** or **stably** on the auxiliary data. Boundary value problems with unique solutions that exhibit this stability are called (after Hadamard) **well-posed** or **properly posed**: all other problems are said to be **ill-posed** or **improperly posed.** Although we will concentrate on well-posed problems, there are important ill-posed problems that have been investigated, especially in recent years. See [Pa].

The Maximum Principle

Physical intuition about heat conduction leads us to an important observation about the behavior of solutions u of the diffusion equation

$$\nu u_t = \nabla \cdot (K \nabla u) + Q, \tag{27}$$

$$\text{where} \quad Q(\mathbf{x}) \leq 0 < \nu(\mathbf{x}), \quad \text{and} \quad K = K(\mathbf{x}) \geq k_0 > 0 \text{ is } C^1, \quad \text{with } |\nabla K| \text{ bounded.} \tag{28}$$

Suppose that equation (27) describes the temperature u inside a room V in which heat is not being generated ($Q \leq 0$). Physical experience suggests that at time $\tau > 0$, the temperature u should be no higher than the largest temperature that occurred, either in V at time $t = 0$ or on the boundary surfaces S at any intervening time. If we set $\mathscr{V} = V \times [0, +\infty)$ and let $\mathscr{A}$ be its space-time boundary as indicated in Figure 3.17, then this physical conjecture takes the abbreviated global form

$$\max_{\mathscr{V}} u \leq \max_{\mathscr{A}} u,$$

an inequality known as the **maximum principle.**†

Before attempting its mathematical verification, note that *if* it holds, then the solutions of equation (27) depend continuously on the values that they take on

† In such inequalities, "max" or "min" should be interpreted as supremum or infimum, respectively. See [Ru].

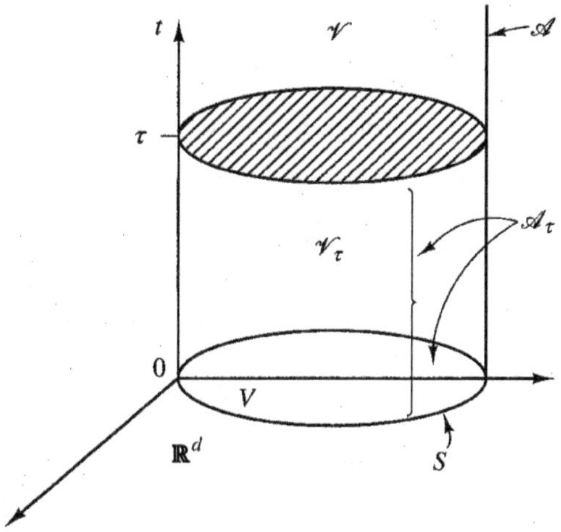

Figure 3.17

the auxiliary set $\mathscr{A}$. For if $\tilde{u}$ is another solution, then both $u - \tilde{u}$ and $\tilde{u} - u$ are solutions of equation (27) with $Q \equiv 0$. Hence if $|u - \tilde{u}| \leq \delta$ on $\mathscr{A}$, then in $\mathscr{V}$,

$$u - \tilde{u} \leq \max(u - \tilde{u})|_{\mathscr{A}} \leq \delta$$

and

$$\tilde{u} - u \leq \max(\tilde{u} - u)|_{\mathscr{A}} \leq \delta.$$

Thus in $\mathscr{V}$: $|u - \tilde{u}| \leq \delta$; that is, solutions of (27) which are close together (within δ) on $\mathscr{A}$ will be equally close together inside $\mathscr{V}$ for each $t > 0$. Consequently, and in this sense, the solution depends continuously on its Dirichlet data. Moreover, when $u = \tilde{u}$ on $\mathscr{A}$, then $u \equiv \tilde{u}$ inside $\mathscr{V}$, and this gives a proof of uniqueness of solution for the associated Dirichlet boundary value problem.

These arguments indicate how to use the maximum principle to show that a mathematical model for a diffusion problem is experimentally realistic. But we can't use it until we establish it, and this requires a little work.

Suppose that V is a bounded domain of $\mathbb{R}^d$ and for fixed $\tau > 0$, let $\mathscr{V}_\tau = \bar{V}x[0, \tau]$ and $\mathscr{A}_\tau = \mathscr{A} \cap \{(\mathbf{x}, t) : 0 \leq t < \tau\}$; i.e., $\mathscr{A}_\tau$ is obtained by excluding from the boundary of $\mathscr{V}_\tau$ the shaded region indicated in Figure 3.17.

(3.3) Theorem (*The maximum principle*): *Under assumptions* (28), *each continuous function u on $\mathscr{V}_\tau$ that satisfies* (27) *inside $\mathscr{V}_\tau$ obeys the following maximum principle:*

$$\max_{\mathscr{V}_\tau} u \leq \max_{\mathscr{A}_\tau} u \tag{29}$$

Proof:* According to a standard vector identity, each solution u of (27) also satisfies the equation

$$\nu u_t = K\nabla^2 u + \nabla K \cdot \nabla u + Q \tag{30}$$

Suppose first that Q *is negative everywhere* in $\mathscr{V}_\tau$. At a fixed time $t_0 \in (0, \tau]$, consider the values of the function $u(\cdot, t_0)$ on V. Let $\mathbf{x}_0 \in V$. If at $(\mathbf{x}_0, t_0)$, this function has a local maximum value, the first and second derivative tests give

$$\nabla u = \mathbf{0} \qquad \text{and} \qquad \nabla^2 u = u_{x_1x_1} + u_{x_2x_2} + \cdots + u_{x_dx_d} \leq 0,$$

so that $\nabla K \cdot \nabla u = 0$ and $K\nabla^2 u \leq 0$. Hence, from (30),

$$u_t(\mathbf{x}_0, t_0) \leq \left(\frac{Q}{\nu}\right)(\mathbf{x}_0, t_0) < 0,$$

which implies that a *larger* value of u will be found at $(\mathbf{x}_0, t)$ for some $t < t_0$. Therefore,wherever we stand on the graph of u over $V \times (0, \tau]$, we can look toward the auxiliary boundary $\mathscr{A}_\tau$ and find some other point that is at least as high as we are. Since u is continuous on $\bar{\mathscr{V}}_\tau$, it follows that the maximum principle holds† when $Q < 0$, but our argument breaks down if $Q \leq 0$. (Why?)

If $Q \leq 0$ and x is the "first" spatial coordinate of $\mathbf{x}$, then for each $\epsilon > 0$ and $\alpha > 0$,

$$v(\mathbf{x}, t) \stackrel{\text{def}}{=} u(\mathbf{x}, t) + \epsilon e^{\alpha x} \tag{31}$$

satisfies an equation similar to (30) having a new Q that *is* negative independently of ϵ with proper choice of α. Consequently, v *does* satisfy the maximum principle and as ϵ approaches zero, u differs from *some* v by as little as we please. Through this argument, whose details are presented in Problem 3.5F, we can show that u obeys the maximum principle in $\bar{\mathscr{V}}_\tau$. ■

If we apply the maximum principle in the manner indicated previously, we get the following result that partially extends Theorem 3.1.

(3.4) Corollary: *The solutions u of* (27) *inside* $\mathscr{V}_\tau$ *that extend continuously to* $\bar{\mathscr{V}}_\tau$ *depend continuously and uniquely on their values on* $\mathscr{A}_\tau$. ■

Now, if we set $\nu = 0$ in (27), the arguments used above simplify and yield corresponding results for steady-state solutions $u = u(\mathbf{x})$ of the potential equation

$$\nabla \cdot (K \nabla u) = -Q. \tag{32}$$

(3.5) Theorem: *Suppose assumptions* (28) *for Q and K hold in a bounded domain V of* $\mathbb{R}^d$.

1. *Then each* $u \in C(\bar{V})$ *that satisfies equation* (32) *in V obeys the following maximum principle:*

$$\max_{\bar{V}} u \leq \max_{S} u, \qquad \text{where } S = \partial V. \tag{33}$$

2. *The solutions u of (32) in V that extend continuously to* $\bar{V}$ *depend continuously and uniquely on their values on S.* ■

† A more satisfactory analytical use of continuity can be made by appealing to the compactness of $\bar{\mathscr{V}}_\tau$. (See Appendix A.1.)

(3.6) Remarks: The foregoing arguments have various extensions. For example, the solutions satisfy a minimum principle if the inequality on Q is reversed and both principles if $Q = 0$. In particular, harmonic functions satisfy both maximum and minimum principles in each bounded domain. Moreover, the maximum principle can be shown to hold even when ν, Q, or K varies with u, but continuous dependence cannot be inferred. See [P-W] and [Sp].

*Example 12:**

We can use the maximum principle for harmonic functions to extend the Poisson integral representation obtained in (6) of Section 2.2 to continuous boundary functions f as follows.

Each $f \in C[-\pi, \pi]$ with $f(-\pi) = f(\pi)$ can be approximated uniformly within $1/k$ by some $f_k \in C^1[-\pi, \pi]$ with $f_k(-\pi) = f_k(\pi), k = 1, 2, \ldots$. (This is an easy consequence of the Weierstrass approximation theorem (1.17).)

Now, each f_k considered as a function on the boundary of the disk $\{r \leq R\}$ has a continuous extension u_k, which is harmonic for $r < R$, given by the Poisson integral

$$u_k(r, \theta) = \frac{1}{2\pi} \int_{-\pi}^{\pi} P(r, \theta, t) f_k(t)\, dt$$

$$\text{where} \qquad P(r, \theta, t) = \frac{R^2 - r^2}{|(R, t) - (r, \theta)|^2}.$$

As $k \to \infty : f_k \to f$ uniformly, so that by the limit interchange principle, for fixed $r < R$ and θ

$$u_k(r, \theta) \to U(r, \theta) \stackrel{\text{def}}{=} \frac{1}{2\pi} \int_{-\pi}^{\pi} P(r, \theta, t) f(t)\, dt.$$

We know that U is harmonic for $r < R$, (see Problem 2.2I) so we only need to show that U provides a continuous extension of f on the closed disk $\{r \leq R\}$.

However, for each integer pair (k, ℓ) the function $u_k - u_\ell$ is harmonic, so that by the maximum and minimum principles,

$$|u_k(r, \theta) - u_\ell(r, \theta)| \leq \max_{t \in [-\pi, \pi]} |f_k(t) - f_\ell(t)|.$$

When $\ell \to \infty$, this gives, for $r < R$,

$$|u_k(r, \theta) - U(r, \theta)| \leq \max_{t \in [-\pi, \pi]} |f_k(t) - f(t)|.$$

Therefore, as $k \to \infty$, the u_k converge uniformly on the closed disk to the function U that is necessarily continuous and equal to f on the boundary. Thus U has a continuous extension to the closed disk, with boundary values f, as desired.

Problem Set 3.5

3.5A *1.* In proving the maximum principle, it was asserted that at an interior point where a C^2 function u attains its maximum value, we must have

$\nabla u = 0$ and $\nabla^2 u \leq 0$. Establish this assertion by considering separately each coordinate direction.

2. Conclude that a solution u of $\nabla^2 u = F$ in a domain V cannot have an interior maximum at a point where $F > 0$.

3.5B 1. Use the results from Problem 3.5A to prove Theorem 3.5 for $K = 1$ when $Q < 0$ in V.

2. If $Q \leq 0$ in V, modify the argument, using (31) to obtain a new equation for v with a negative $\tilde{Q}$.

3.5C 1. Explain how the global maximum principle $\max_{\mathscr{V}} u \leq \max_{\mathscr{A}} u$ follows from Theorem 3.3.

2. Does the converse assertion hold?

3.5D The Dirichlet problem

$$\begin{cases} u_t = -u_{xx} & \text{in} \quad \mathscr{V} = \{0 < x < \pi, t > 0\} \\ u(0,t) = u(\pi,t) = 0, & t > 0 \\ u(x,0) = f(x), & 0 < x < \pi, \end{cases}$$

is improperly posed.

1. Can you give a physical argument involving time reversal ($t \to -t$) that explains why this might occur?

2. Find elementary product solutions u_n of this problem satisfying

$$u_n(x,0) = f_n(x) = \frac{1}{n}\sin nx, \qquad n = 1, 2, \ldots .$$

3. As $n \to \infty$, show that $u_n \to 0$ on the space-time boundary $\mathscr{A}$ of $\mathscr{V}$, but $u_n(\pi/2, 1) \not\to 0$.

3.5E 1. When $\nabla \cdot \mathbf{v} = 0$ and $\kappa > 0$, explain why solutions of the diffusion-convection equation (9″) also satisfy a maximum principle. (Assume that $|\mathbf{v}|$ is bounded.)

2. If the dye of concentration u in (9′) cannot be generated internally but might be removed by chemical reactions, will a maximum principle still hold?

3. What assumptions would lead to a minimum principle?

4. What assumptions would lead to continuous dependence on Dirichlet data?

3.5F* (Completion of the proof of Theorem 3.3.)

1. Show that if u satisfies (30), then v defined in (31) satisfies the equation

$$\nu v_t - K\nabla^2 v - \nabla K \cdot \nabla v = -\epsilon\alpha(K\alpha + k)e^{\alpha x} + Q = \tilde{Q}, \qquad \text{say},$$

where $k = K_x$ is the "first" component of ∇K.

2. When $Q \leq 0$, show that $\tilde{Q}$ is negative if the constant α is taken so large that $(K\alpha + k) > 0$. Why does assumption (28) permit this choice?

3. Show that if $\tilde{Q} < 0$, then v satisfies the maximum principle in $\mathscr{V}_T$.
4. Assume that v satisfies the maximum principle in $\mathscr{V}_T$ and verify the following inequalities in $\mathscr{V}_T$:

$$u \leq v \leq \max_{\mathscr{A}_T} v = \max_{\mathscr{A}_T}(u + \epsilon e^{\alpha x}) \leq (\max_{\mathscr{A}_T} u) + \epsilon e^{\alpha \ell}$$

where ℓ is an upper bound for x when $\mathbf{x} = (x, \ldots) \in \bar{V}$. Now let $\epsilon \searrow 0$ to obtain the desired result.

3.5G If we neglect heating from the Earth's core, then at time t, the temperature $u = u(r, t)$ at a distance r from the center of a spherical Earth of radius R is described approximately by the equation

$$u_t = k\,(u_{rr} + (2/r)u_r) \qquad (r > 0)$$

for an appropriate diffusivity constant k. (See Problem 4.1D)

1. Show that $v = ru$ satisfies the equation $v_t = kv_{rr}$.
2. If the surface temperature can be described by $u(R, t) = Y\cos(\omega t + ar)$ where $a = (\omega/2k)^{1/2}$ and $Y > 0$, show that

$$u(r,\ t) = Y(R/r)\,[e^{r-R}]\cos(\omega t + aR)$$

and discuss this equation.

4

Steady-State Problems: The Potential Equation

4.1 Introduction

In the last chapter we used the term **steady-state** to describe solutions that are independent of time. Let's agree that a state of a system is *steady* when it can be specified through **macroscopic** conditions that are not affected by the passage of time. The microscopic conditions may be wildly erratic as long as their macroscopic averages are unchanging. For example, the absolute temperature of air as read from a thermometer may appear constant, even though, according to one model, we are actually measuring the mean random kinetic energy of the molecules.

To maintain steady-state conditions during a physical process, there must be a balance between what enters each fixed control volume V in the form of mass, momentum, energy, and so on, and what leaves it. When applied to energy in the form of heat, this observation results in the time-independent version of equation (4) of Section 3.1;

$$\nabla \cdot \mathbf{f} = Q, \qquad \text{in } V$$

Here, $\mathbf{f}$ is the heat-flux vector, Q represents the volume density of heat sources, and other means of heat production have been neglected. Similarly, for conservation of mass in steady-flow hydrodynamics, we need

$$\nabla \cdot (\rho \mathbf{v}) = \sigma, \qquad \text{in } V$$

where now $\mathbf{f} = \rho\mathbf{v}$ is the mass-flux vector defined by the product of the mass density ρ per unit volume and the velocity $\mathbf{v}$, and σ represents the volume density of possible mass sources in V. For example, water pumped into the interior of a tank at a constant rate is a source of mass production that must be compensated by corresponding outflow to produce steady-state conditions within the tank.

In electrostatics of an isotropic medium, we have Gauss' law:

$$\nabla \cdot (\epsilon \mathbf{E}) = 4\pi\rho,$$

expressed in terms of the dielectric parameter ϵ of the medium, the volume density of charge ρ, and the electrical field intensity vector $\mathbf{E}$. However, the corresponding law of magnetostatics is

$$\nabla \cdot (\mu \mathbf{H}) = 0$$

since magnetic "charges" or *poles* can only be distributed in locally opposing pairs.

We can see that each of these equations is of the divergence form

$$\nabla \cdot \mathbf{f} = F, \tag{1}$$

where it is assumed that the vector field $\mathbf{f}$ has C^1 components and F is continuous. When $\mathbf{f} = \nabla\phi$ for some potential $\phi \in C^2(V)$, (1) becomes **Poisson's equation**

$$\nabla^2\phi = F, \tag{1'}$$

which reduces to that of Laplace when $F \equiv 0$. This occurs, for example, in steady-state heat conduction with constant thermal conductivity, K.

In steady-flow hydrodynamics, it is usual to assume that the density ρ is constant, so that $\nabla \cdot (\rho\mathbf{v}) = \rho\nabla \cdot \mathbf{v}$. However, flows with velocity $\mathbf{v} = \nabla\phi$ relative to some C^2 potential ϕ have coordinate velocity components v_1, v_2, v_3, satisfying relations of the form

$$\frac{\partial v_i}{\partial x_j} = \frac{\partial v_j}{\partial x_i}\left(= \frac{\partial^2\phi}{\partial x_i\,\partial x_j}\right), \qquad i, j = 1, 2, 3 \tag{2}$$

in Cartesian coordinates. Such flows are said to be **irrotational** because around any smooth, closed curve $\mathscr{C}$ bounding a surface S in the flow, the **circulation** is zero; i.e.,

$$\int_{\mathscr{C}} \mathbf{v} \cdot \mathbf{T}\, ds = 0 \tag{2'}$$

where along $\mathscr{C}$, $\mathbf{T}$ is the unit tangent vector in the direction of increasing arc length s. This is a consequence of Stokes' theorem. [Ed].

Now, in electrostatics, the field $\mathbf{E} = (v_1, v_2, v_3)$ automatically satisfies (2) and therefore in each convex domain V, $\mathbf{E} = \nabla\phi$ for some $\phi \in C^2(V)$. (See Problem 6.3B.) Similarly, in magnetostatics of a *nonconducting* medium, the field $\mathbf{H} = \nabla\phi$ for some $\phi \in C^2(V)$.

In these cases, $\mathbf{f} = -K\nabla\phi$ for a positive function K and then (1) becomes

$$\nabla \cdot (K\nabla\phi) = -F. \tag{3}$$

We will call (3) the **potential equation** and show that it is associated with other steady-state situations. In two dimensions, it governs small deformations of an elastic membrane in static equilibrium, and this affords easy visualization of the solutions as well as a valuable thermomechanical analogy (see Section 4.2). This equation arises also in the study of soil mechanics, gravitational attraction, and many other fields.

When K and F are independent of ϕ then (3) is linear, and in Section 3.2 we found conditions for uniqueness of solution to associated boundary value problems. Moreover, from Theorem 3.5, we know that in a bounded region V solutions depend continuously on their boundary values and so admit possibility of experimental verification. Finally, when $F \leq 0$, the solutions to (3) should satisfy the maximum principle in V. Even with this much theoretical knowledge, it is still difficult to find explicit solutions for associated boundary value problems, and we make further simplifications.

When $F = 0$ and $K = 1$, the solutions of (3) are by definition harmonic functions (see Section 2.2). Consequently, harmonic functions provide solutions to *some* boundary value problems in each field governed by (3), and this explains their prevalence in applied mathematics.

Problem Set 4.1

4.1A Suppose that everywhere within a medium in $\mathbb{R}^2$, the heat-flux vector is given by $\mathbf{f} = 3\mathbf{i} + 4\mathbf{j}$ calories per square centimeter per second. Set $\mathbf{f} = -K\nabla\phi$, and

1. find the temperature $\phi(x, y)$ in degrees, assuming that $\phi(0, 0) = 0$ and that the thermal conductivity K of the medium is 0.1 calories per centimeter-degree-second.
2. sketch the equipotentials or isothermal curves on which ϕ equals 0, 1, -1, and 2.

4.1B A fluid is rotating about the z-axis with constant angular velocity ω. Every particle at a distance r from the z-axis traces a circle at speed $|\mathbf{v}| = \omega r$ in a plane perpendicular to the z-axis. Describe this field by writing an equation for the velocity vector. Show that there is *no* potential ϕ for this field.

4.1C (*Coulomb's law*). For electrostatics in $\mathbb{R}^3$

$$\mathbf{F} = \frac{\gamma q Q}{|\mathbf{x} - \mathbf{y}|^3}(\mathbf{x} - \mathbf{y})$$

describes the force exerted by a charge Q located at $\mathbf{y}$ on a charge q located at $\mathbf{x}$. Find a potential ϕ for this force. Show directly that for $\mathbf{x} \neq \mathbf{y}$, ϕ is harmonic in $\mathbf{x}$.

4.1D *1.* For $\mathbf{x} \in \mathbb{R}^d$, let $r = |\mathbf{x}| = (x_1^2 + x_2^2 + \cdots + x_d^2)^{1/2}$ and show that if $\psi = \psi(r)$, then

$$\nabla^2\psi = \psi'' + \frac{d-1}{r}\psi', \qquad r > 0.$$

2. Conclude that each radially symmetric **harmonic** function in $\mathbb{R}^d \sim \{0\}$ is of the form

$$\psi(r) = a + \frac{b}{r^{d-2}}, \qquad \text{if } d > 2,$$

where a and b are constants. What is the form when $d = 2$?

4.1E Let V be the unit square $\{0 < x < 1, 0 < y < 1\}$ and let $u(x, y)$ be twice continuously differentiable with respect to x and y in V, continuous on the closed square, and zero on the boundary of V. Show that

$$\int_0^1 \int_0^1 u u_{xx}\, dx\, dy = -\int_0^1 \int_0^1 u_x^2\, dx\, dy$$

$$\int_0^1 \int_0^1 u u_{yy}\, dx\, dy = -\int_0^1 \int_0^1 u_y^2\, dx\, dy.$$

Use these results to conclude that the only smooth solution to the problem

$$\begin{cases} \nabla^2 u = 0, & \text{in } V \\ u = 0, & \text{on } \partial V \end{cases}$$

is the zero function.

4.1F Suppose that $\mathbf{v}$ has C^1 components v_1, v_2, v_3 that satisfy conditions (2) in the ball $B_R(\mathbf{0})$ of $\mathbb{R}^3$.

1. Let $\mathscr{C}_r$ be a circle in the x_1, x_2 plane of radius $r < R$ centered at the origin, and show that

$$\int_{\mathscr{C}_r} \mathbf{v} \cdot \mathbf{T}\, ds = \int_{\mathscr{C}_r} \mathbf{f} \cdot \mathbf{n}\, ds = 0$$

where $\mathbf{n} \cdot \mathbf{T} = 0$ and $\mathbf{f} = (-v_2, v_1)$.
(*Hint:* Use the divergence theorem for $\mathbb{R}^2$.)

2. For what other closed curves $\mathscr{C}$ inside the ball will

$$\int_{\mathscr{C}} \mathbf{v} \cdot \mathbf{T}\, ds = 0?$$

Give a physical explanation as to why this line integral is called the circulation and why its vanishing might characterize irrotational flow.

4.2

Static Equilibrium of Membranes

A body is said to be in *static equilibrium* under a system of applied forces when its elements are at rest. In this chapter we characterize static equilibrium through (Daniell) **Bernoulli's**† **principle of minimum potential energy**, which expresses the conviction that the universe should operate efficiently. In fact, potential energy can only be defined within some additive reference constant, and then *minimum*

† During the years following the emergence of calculus and Newtonian mechanics, the Swiss brothers Jakob (1654–1705) and Johann (1667–1748) Bernoulli were among the foremost contributors to the development of mathematics and its applications. This tradition was continued by Johann's son Daniell (1700–1782) and above all by Johann's pupil Leonhard Euler (1707–1783). Among many other achievements, these natural philosophers created mathematical theories for elasticity and hydrodynamics and introduced variational methods to attack related problems.

potential energy is associated with systems in *stable equilibrium*. A system in unstable equilibrium such as a balanced house of cards usually exhibits maximal potential energy that can be determined by minimizing its negative. More general principles have been formulated and chief among them is Hamilton's principle, which is taken up in Section 5.1. (See [Tr, Ch 8] for further discussion.)

A drumhead is made by stretching a thin elastic skin tightly across a rigid circular frame. If we neglect the weight of the skin in comparison with the tensile force of stretching, we would expect the drumhead to be flat. However, a distorted frame will result in a warped drumhead whose shape is of interest. Let's generalize these observations slightly.

A **membrane** is a thin elastic two-dimensional surface, such as a drumhead, which is capable of sustaining tension under stretching but has negligible resistance to bending or torsion. When a thin isotropic membrane is stretched over curved (rigid) wires, then **Bernoulli's principle** asserts that the membrane will assume the shape that **minimizes the associated potential energy**.

In particular, suppose the wire(s) can be described by nearly planar curves that "lie over" plane curves S enclosing a simple domain V, as illustrated in Figure 4.1. Then the potential energy of strain is measured by the amount by which the actual surface area of the membrane exceeds that of V.

If the static vertical deflection of the membrane at $(x, y) \in V$ is given by $u(x, y)$, then from calculus, the surface area has element

$$\sqrt{1 + u_x^2 + u_y^2}\, dV = \sqrt{1 + |\nabla u|^2}\, dV$$

compared with that of $dV = dx\, dy$ so that the total potential energy of strain should be essentially

$$\int_V \tau\left(\sqrt{1 + |\nabla u|^2} - 1\right) dV.$$

Here, the *tension function* $\tau = \tau(x, y) > 0$ measures the local resistance to stretching; τ should be constant when the membrane is of uniform thickness and composition. The boundary deflection, $u|_S = g$, is fixed.

To simplify the analysis, we suppose that admissible deflections u are so nearly constant that the slopes $\nabla u = (u_x, u_y)$ are small. We write $|\nabla u| \ll 1$ to denote that

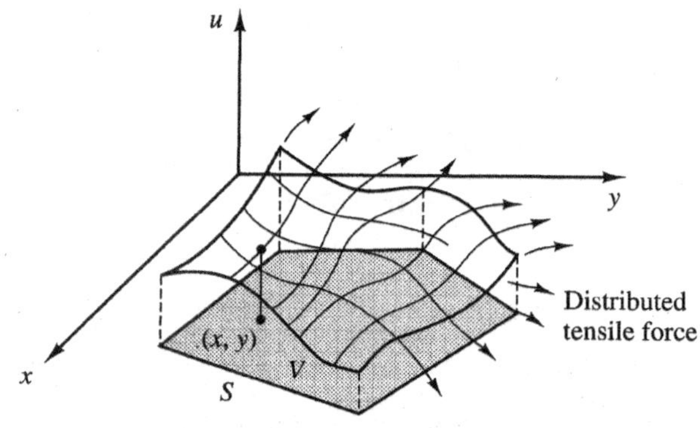

Figure 4.1

they are negligible in comparison to a unit magnitude. Then we see that

$$\sqrt{1+|\nabla u|^2}-1=(\sqrt{1+|\nabla u|^2}-1)\frac{(\sqrt{1+|\nabla u|^2}+1)}{(\sqrt{1+|\nabla u|^2}+1)}$$

$$=\frac{|\nabla u|^2}{\sqrt{1+|\nabla u|^2}+1}\approx\frac{|\nabla u|^2}{2};$$

hence, we wish to find the function $u \in C^1(\bar{V})$ that assumes given boundary values ($u|_S = g$) and minimizes the energy integral

$$E(u)=\frac{1}{2}\int_V \tau|\nabla u|^2\,dV, \tag{4}$$

neglecting provisionally the potential energy due to external loading.

This problem in variational calculus can be attacked directly as follows: Suppose that τ in $C^1(\bar{V})$ is positive on V.

We seek $\quad u \in C^1(\bar{V})$ with $u|_S = g$ for which

$$\begin{aligned} &E(u+v) \geq E(u), && \text{for all } v \in C^1(\bar{V}),\ \text{with } v|_S = 0, \\ \text{or}\quad &E(u+v)-E(u) \geq 0, && \text{for all such } v. \end{aligned} \tag{4$'$}$$

For if we had such a function u, then any other deflection function with the *same* boundary values has the form $u+v$, for some v with *zero* boundary values, and the inequality (4′) simply asserts the minimality of $E(u)$. We can also assume that $u \in C^2(V)$.

Substituting and subtracting, we obtain after cancellations that

$$|\nabla(u+v)|^2-|\nabla u|^2=(|\nabla u|^2+2\nabla u\cdot\nabla v+|\nabla v|^2)-|\nabla u|^2,$$

$$\text{so that}\qquad E(u+v)-E(u)=\int_V \tau(\nabla v\cdot\nabla u)\,dV+\underbrace{\frac{1}{2}\int_V \tau|\nabla v|^2\,dV}_{\geq 0},$$

$$\geq \int_V [\nabla\cdot(\tau v\nabla u)-v\nabla\cdot(\tau\nabla u)]\,dV \tag{4$''$}$$

$$=\int_S \tau v(\nabla u\cdot\mathbf{n})\,dS-\int_V v\nabla\cdot(\tau\nabla u)\,dV,$$

by the divergence theorem applied to $\mathbf{f}=\tau v\nabla u$. Now, since $v|_S = 0$, the first integral in the last line vanishes, and the second would also vanish provided that in V, u satisfies the partial differential equation $\nabla\cdot(\tau\nabla u)=0$. These arguments are valid in each dimension, and they can be made rigorous through approximations within V.

(4.1) Theorem: *Let $\tau \in D^1(\bar{V})$ be positive on V, a simple domain of $\mathbb{R}^d$. Then each $u_0 \in C^1(\bar{V})$ that is a solution of*

$$\nabla\cdot(\tau\nabla u)=0, \qquad \text{in } V \tag{5}$$

minimizes $E(u)=\frac{1}{2}\int_V \tau|\nabla u|^2\,dV$ uniquely among all functions in $C^1(\bar{V})$ that have the same boundary values as u_0.

Proof: Each function $u \in C^1(\bar{V})$ that matches u_0 on $S = \partial V$ can be expressed as $u = u_0 + v$, where $v = u - u_0$, so that $v|_S = 0$. But under the hypothesized conditions, we have just shown that

$$E(u_0 + v) - E(u_0) = E(v) = \frac{1}{2}\int_V \tau\,|\nabla v|^2\,dV \geq 0 \qquad (\text{since } \tau > 0)$$

with equality iff $\tau\,|\nabla v|^2 = 0$ in V; i.e., $\nabla v = \mathbf{0}$ in V (since $\tau > 0$), or

$$\begin{aligned} v &= \text{const. (since } V \text{ is a domain)} \\ &= v|_S = 0. \end{aligned}$$

Therefore $u = u_0$. ■

(4.2) Corollary: *In the preceding theorem, when $d = 2$ and $|\nabla u_0| \ll 1$, then u_0 describes the shape of an unloaded isotropic*† *membrane that is stretched over V with tensile resistance τ and boundary deflection $u_0|_S$.*

Proof: Recall Bernoulli's principle. ■

In this derivation, we neglected the potential energy arising from an external distributed loading, such as the weight of the membrane. When this is taken into account, the same conclusion holds for the *nonhomogeneous* potential equation

$$\nabla \cdot (\tau \nabla u) = -F \tag{6}$$

where the distributed loading F is taken to be positive in the direction of positive deflection u. (See Problem 4.2B.)

A Thermomechanical Analogy

If we identify u with ϕ and τ with K, and reinterpret F, (6) is identical with (3), and this provides remarkable analogies. Let's examine the thermomechanical connection in two dimensions.

Suppose that we want to "see" the steady-state temperature distribution ϕ in a metal plate having the shape of a planar domain V and insulated faces. We stretch a thin membrane in this shape that has local resistance τ proportional to K and give it boundary conditions corresponding to those of ϕ and external loading F proportional to Q. If $|\nabla u| \ll 1$, the measured deflections u will be proportional to the desired temperatures ϕ, and curves of constant height will be the isothermal curves. (These curves in turn define the equipotential surfaces for a related problem in two-dimensional electrostatics.) When the stretched membrane is not loaded ($F = 0$), its deflection at an interior point should not exceed the extremes of its boundary deflections. This observation provides a two-dimensional visualization of the maximum and minimum principles. See Theorem 3.5.

The thermomechanical analogy is also useful in explaining the *Neumann compatibility requirement*, which arises when the divergence theorem is used to integrate (6) over a simple domain V. Then, we see that

$$\int_V (-F)\,dV = \int_V \nabla \cdot (\tau \nabla u)\,dV = \int_S \tau \frac{\partial u}{\partial n}\,dS,$$

† For anisotropic membranes, τ must be replaced by an appropriate positive-definite matrix function.

since on S, $\tau(\nabla u \cdot \mathbf{n}) = \tau(\partial u/\partial n)$. Thus, there can be *no* solution to (6), with prescribed boundary slope $(\partial u/\partial n)|_S = g$, unless

$$\int_V F\, dV = -\int_S \tau g\, dS. \tag{7}$$

This puzzling requirement has an easy explanation when viewed as a problem in heat conduction with $\tau = K$, $F = Q$. It then says that for thermal equilibrium, the total amount of heat being generated inside V at the rate

$$\int_V Q\, dV = \int_V F\, dV$$

must equal that which flows outward through the boundary S at the rate

$$\int_S \mathbf{f} \cdot \mathbf{n}\, dS = -\int_S K \frac{\partial u}{\partial n}\, dS = -\int_S Kg\, dS.$$

For example, if we seek a solution to Laplace's equation ($K \equiv 1, F \equiv 0$), in a given region V with the prescribed Neumann condition $(\partial u/\partial n)|_S = g$, we *must* require that $\int_S g\, dS = 0$. At a "free" boundary point of a membrane where the deflection u is unspecified, we should have the homogeneous condition $(\partial u/\partial n) = 0$ since the associated normal (tensile) strain must vanish.

For the membrane, a Robin condition, in which $[(\partial u/\partial n + hu)]|_S$ is specified for some given boundary function h, corresponds to a kind of elastic support at the boundary but does not carry a similar implied compatibility requirement.

Problem Set 4.2

4.2A Verify the calculations leading to (4″) and its reformulation using the divergence theorem.

4.2B To take into account a static external distributed loading F on a bounded membrane, the potential energy function should be $\tilde{E}(u) = E(u) - \int_V Fu\, dV$, where $E(u)$ is defined in (4).

1. Show that if $v|_S = 0$, then

$$\tilde{E}(u+v) - \tilde{E}(u) = E(v) - \int_V v[\nabla \cdot (\tau \nabla u) + F]\, dV.$$

2. Conclude that $\tilde{E}$ is minimized by any u that meets given Dirichlet boundary conditions and satisfies equation (6).
3. If V is a simple domain, show that this u is the unique minimizing function.

4.2C*If the external loading F of a stretched bounded membrane is varied slightly, it is natural to expect that the resulting deflection u will also vary slightly. For a uniform membrane ($\tau = const.$) we wish to show that solutions u of Poisson's equation $\nabla^2 u = -F/\tau = f$, say, with $u|_S = g$ given, depend continuously on f.

1. By considering the difference u of solutions u_1 (when $f = f_1$) and u_2 (when $f = f_2$), prove that it suffices to show that the solution of $\nabla^2 u = f$, with $u|_S = 0$, is small when $|f|$ is small.

2. Assume that $f \in C(\bar{V})$ where V is a bounded domain so that $M = 4^{-1} \max_{\bar{V}} |f| < +\infty$. Set $v = u + Mr$ where $r(x, y) = x^2 + y^2$ and show that $\nabla^2 v \geq 0$.
3. Conclude that $u \leq v \leq \max_S v = \max_S M(x^2 + y^2) = MR^2$ for an appropriate R. (*Hint:* Use (3.5).)
4. Apply the same reasoning to $-u$ to obtain that, in V, $|u| \leq (R^2/4) \max_{\bar{V}} |f|$. Why does this give the conclusion desired in part 1?

4.3

Simple Two-Dimensional Problems

In the previous sections we have seen that various steady-state processes in an isotropic medium are governed by the potential equation (3). Moreover, when the medium is homogeneous and "source-free," (3) reduces to Laplace's equation, which in two-dimensional rectangular coordinates x, y is

$$\nabla^2 u = u_{xx} + u_{yy} = 0. \tag{8}$$

In addition, this equation governs the static deflection $u = u(x, y)$ of an unloaded stretched membrane of uniform properties. Its solutions in a domain V are harmonic in $V \times \mathbb{R}$ and so provide solutions to problems in $\mathbb{R}^3$ whose conditions are independent of z. In this and the next section we examine some of the simpler boundary value problems associated with equation (8) in domains V of rectangular or circular shapes. Corresponding problems for Poisson's equation are taken up in Section 4.5, but more general problems are reserved until the later chapters.

Rectangular Problems

A model for the roof of a theater can be made by stretching a thin sheet of mylar over a rectangular frame with an arch at one end as shown in Figure 4.2. For convenience, we use the dimensions and coordinate system indicated and neglect the weight of the mylar, which is assumed to behave like a membrane. The deflection

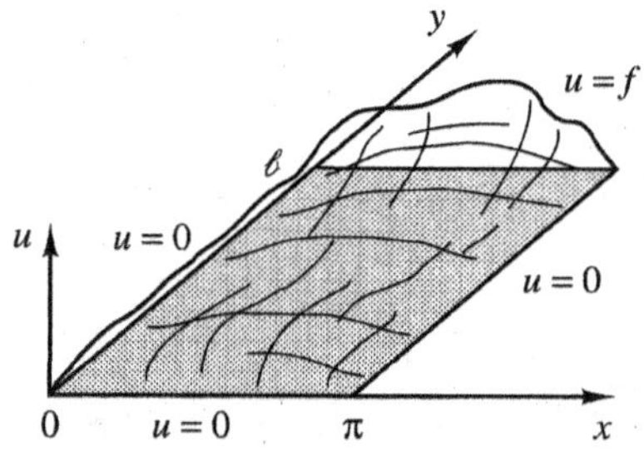

Figure 4.2

$u = u(x, y)$ at intermediate points can be found as a solution of (8) in the rectangle $V = \{0 < x < \pi, 0 < y < \ell\}$ with the boundary conditions

$$u(0, y) = u(\pi, y) = 0, \qquad 0 < y < \ell$$
$$u(x, 0) = 0; u(x, \ell) = f(x), \qquad 0 < x < \pi$$

where f describes the shape of the arch.

Example 1:

Suppose that the arch can be described by

$$f(x) = \sum_{n=1}^{N} s_n \sin nx, \qquad 0 < x < \pi,$$

for some constants $s_1, s_2, \ldots, s_N$. As in Section 2.4, observe that

$$u_n(x, y) = Y_n(y) \sin nx, \qquad n = 1, 2, \ldots$$

satisfies the first two boundary conditions (which are of strip-type $\mathscr{D}$). Also,

$$\nabla^2 u_n(x, y) = [Y_n''(y) - n^2 Y_n(y)] \sin nx = 0$$
$$\text{if} \qquad Y_n'' - n^2 Y_n = 0,$$
$$\text{or} \qquad Y_n(y) = a_n \cosh ny + b_n \sinh ny,$$
for some constants $a_n, b_n, n = 1, 2, \ldots$.

Finally, $u_n(x, 0) = 0$ iff $Y_n(0) = 0 = a_n$, and thus for each constant b_n,

$$u_n(x, y) = b_n \sinh ny \sin nx$$

solves the homogeneous part of the problem. In particular, $u_1(x, y) = b_1 \sinh y \sin x$ gives the deflection function for the simple sinusoidal arch described by $f(x) = b_1 \sinh \ell \sin x$. Similarly, we see that

$$U_N(x, y) = \sum_{n=1}^{N} b_n \sinh ny \sin nx$$

is the *unique* solution to the given problem when $s_n = b_n \sinh n\ell$, $n = 1, 2, \ldots, N$, so that

$$f(x) = U_N(x, \ell) = \sum_{n=1}^{N} (b_n \sinh n\ell) \sin nx = \sum_{n=1}^{N} s_n \sin nx, \qquad 0 < x < \pi,$$

as desired. This requires that

$$b_n \sinh n\ell = s_n = \frac{2}{\pi} \int_0^{\pi} f(x) \sin nx \, dx, \qquad n = 1, 2, \ldots, N$$

and gives us the unique solution

$$U_N(x, y) = \sum_{n=1}^{N} s_n \frac{\sinh ny}{\sinh n\ell} \sin nx. \tag{8'}$$

For a more general integrable f, we regard the U_N of (8′) as *approximate solutions* to our problem, and

$$U(x,y) = \sum_{n=1}^{\infty} s_n \frac{\sinh ny}{\sinh n\ell} \sin nx \tag{9}$$

as a **formal** solution. U *is* harmonic for $0 < y < \ell$, since

$$|s_n \sin nx| \le |s_n| \le \frac{2}{\pi}\int_0^{\pi} |f(x)|\,dx = M, \qquad \text{say,}$$

and when $0 \le y \le y_0 < \ell$:

$$\frac{\sinh ny}{\sinh n\ell} = \frac{e^{ny} - e^{-ny}}{e^{n\ell} - e^{-n\ell}} = \frac{e^{ny}(1 - e^{-2ny})}{e^{n\ell}(1 - e^{-2n\ell})} \le \frac{e^{-n(\ell-y)}}{1 - e^{-2\ell}} \le \frac{e^{-n(\ell-y_0)}}{1 - e^{-2\ell}}.$$

The negative exponential factors $e^{-n(\ell-y_0)}$ are so small that when $0 \le y \le y_0 < \ell$, series (9) converges uniformly by the M-test. Moreover, it can be differentiated as often as desired in this strip without loss of uniform convergence, since successive termwise differentiations only introduce factors of $n, n^2, \ldots$ into similar estimates.

It follows that U is continuous for $0 \le y < \ell$ and harmonic for $0 < y < \ell$. Therefore, U is an exact solution to the homogeneous part of the problem, but the behavior of $U(x,y)$ as $y \nearrow \ell$ depends on that of f.

Example 2:

For the roof model being considered, the triangular arch function of Figure 3.16

$$f(x) = \begin{cases} x, & 0 \le x \le \pi/2 \\ \pi - x, & \pi/2 \le x \le \pi \end{cases}$$

does not cause obvious physical difficulties. Let's see how this observation is reflected in the behavior of the mathematical solution U of (9).

First, note that $f \in \hat{C}^1[0,\pi]$, with $f(0) = f(\pi) = 0$. Therefore, according to Theorem 1.7, f has the Fourier sine series **representation**

$$f(x) = \sum_{n=1}^{\infty} s_n \sin nx, \qquad 0 \le x \le \pi, \qquad \text{with } \sum_{n=1}^{\infty} |s_n| < +\infty.$$

(Indeed, for the f given above,

$$s_n = \frac{4}{\pi n^2} \sin\frac{n\pi}{2}, \qquad n = 1, 2, \ldots$$

so that

$$U(x,y) = \frac{4}{\pi}\sum_{n=1}^{\infty} \frac{1}{n^2}\left(\sin\frac{n\pi}{2}\right)\frac{\sinh ny}{\sinh n\ell} \sin nx, \qquad 0 < y < \ell.)$$

Since

$$0 \le y \le \ell \Rightarrow \left| s_n \frac{\sinh ny}{\sinh n\ell} \sin nx \right| \le |s_n|,$$

it follows by the M-test that in this strip, the series for U converges uniformly and therefore U is continuous. Consequently, as $y \nearrow b$:

$$U(x,y) \to U(x,b) = \sum_{n=1}^{\infty} s_n \sin nx = f(x), \qquad 0 \le x \le \pi, \tag{9'}$$

and under these conditions on f, U is the *unique* solution to our boundary value problem. A computer-generated approximation to its graph is shown in Figure 4.3. (See the supplement to this text.)

When f is only sectionally smooth on $[0,\pi]$, it can be shown that (9′) holds at each x if $f(x)$ is replaced by $[f(x+)+f(x-)]/2$ where the endpoint limits are those of the *odd* periodic extension of f. (See Problem 3.3E.) In particular, the arch function $f(x) = 1$ has an odd periodic extension that is discontinuous at $x = 0$ and $x = \pi$, and the corresponding formal series solution U exhibits Gibbs' behavior near the points $(0, b)$ and (π, b) in its effort to accommodate the incompatibility in boundary conditions. And how does the mylar handle this situation? It tears!

If we recall that a uniform scale change preserves harmonicity, we see that

$$U(x,y) = \sum_{n=1}^{\infty} s_n \frac{\sinh(n\pi y/\ell)}{\sinh nb} \sin \frac{n\pi x}{\ell} \tag{10}$$

with $\quad s_n = \dfrac{2}{\ell}\displaystyle\int_0^{\ell} f(x) \sin \frac{n\pi x}{\ell}\, dx, \qquad n = 1, 2, \ldots$

gives us a formal solution to the same boundary value problem for the geometrically similar rectangular membrane with sides of lengths ℓ and $\tilde{b} = \ell b/\pi$.

Of course, different homogeneous boundary conditions result in different eigenvalue problems and associated series solutions.

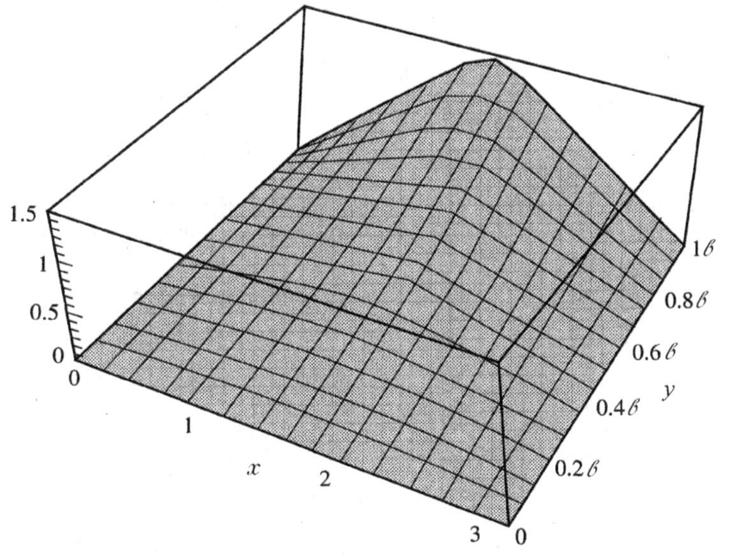

Figure 4.3

Example 3:

The steady-state temperature distribution $u = u(x, y)$ in a rectangular steel plate with insulated faces and a pair of adjacent insulated edges can be obtained from a solution to the following boundary value problem (see Figure 4.4).

$$\nabla^2 u = 0 \qquad \text{in } \{0 < x < \pi, \qquad 0 < y < \ell\}$$
$$\text{with} \qquad u_y(x, 0) = 0, \qquad 0 < x < \pi,$$
$$\text{and} \qquad u_x(\pi, y) = 0, \qquad 0 < y < \ell.$$

If the temperature on the other edges is prescribed by $u(0, y) = 0$, $0 < y < \ell$ and $u(x, \ell) = f(x)$, $0 < x < \pi$, then we get the formal solution

$$U(x, y) = \sum_{n=1,3,}^{\infty} b_n \cosh\frac{n}{2}y \sin\frac{n}{2}x, \qquad 0 < x < \pi,\ 0 < y < \ell,$$

$$\text{where} \qquad b_n \cosh\frac{n}{2}\ell = s_n = \frac{2}{\pi}\int_0^{\pi} f(x)\sin\frac{n}{2}x\,dx, \qquad n = 1, 3, \ldots .$$

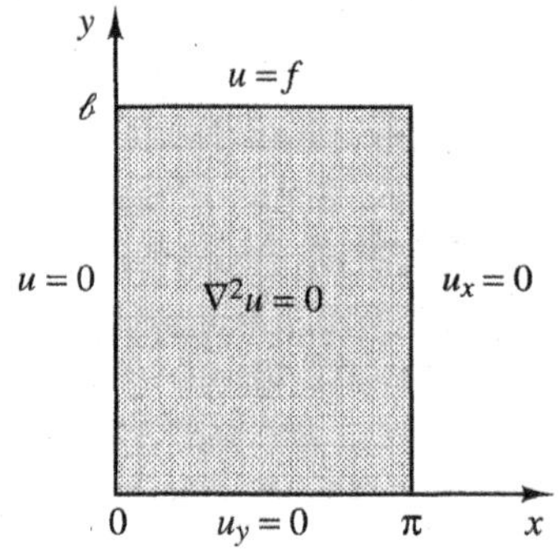

Figure 4.4

Circular and Annular Problems

When the planar region V has radial symmetry (i.e., V is a disk or a ring-shaped region [an annulus]), it is easier to specify boundary conditions in polar coordinates, (r, θ).

For a disk of unit radius centered at the origin, the Poisson integral (of Example 12 in Section 3.5),

$$u(r, \theta) = \frac{1}{2\pi}\int_{-\pi}^{\pi} f(t)\frac{1 - r^2}{|(1, t) - (r, \theta)|^2}\,dt$$

supplies the unique solution to the (Dirichlet) problem

$$\nabla^2 u = 0,\ 0 \le r < 1$$
$$\text{with} \qquad u(1, \theta) = f(\theta), \qquad \text{for } f \in C[-\pi, \pi] \qquad \text{with } f(-\pi) = f(\pi).$$

This integral also provides a *bounded* harmonic function in the *exterior* of the unit disk that has given boundary values. When f has appropriate symmetry, these harmonic functions solve certain problems relative to a semidisk and a quarter-disk.

For more general problems concerning Laplace's equation in such regions, we return to series methods for the equation in polar coordinates; viz.,

$$\nabla^2 u \equiv u_{rr} + \frac{1}{r}u_r + \frac{1}{r^2}u_{\theta\theta} = 0, \qquad (r > 0).$$

By Example 4 of Section 2.4, this equation has elementary product solutions

$$u_n(r,\theta) = r^n(c_n \cos n\theta + s_n \sin n\theta), \qquad n = 0, 1, 2, \ldots \tag{11a}$$

valid for $r \geq 0$, as well as those valid only for $r > 0$; viz.,

$$\begin{aligned} \tilde{u}_0(r,\theta) &= \log r \\ \tilde{u}_n(r,\theta) &= r^{-n}(\tilde{c}_n \cos n\theta + \tilde{s}_n \sin n\theta), \qquad n = 1, 2, \ldots \end{aligned} \tag{11b}$$

for arbitrary constants c_n, s_n, $\tilde{c}_n$, and $\tilde{s}_n$.

Example 4:

Many kinds of heat exchangers are available. In one model, illustrated cross-sectionally in Figure 4.5, air is used as an insulator between long concentric circular cylindrical walls of radii 1 and $R > 1$ (say) that operate at different constant temperatures, T_1 and T_0. Will the steady-state temperature $u = u(r, \theta)$ vary *linearly* along radii between these walls? To answer this question, we recall that in the chamber between the walls, u should be a radially symmetric solution of Laplace's equation; thus,

$$u_{rr} + \frac{1}{r}u_r = \frac{1}{r}(ru_r)_r = 0, \qquad 1 < r < R.$$

Hence ru_r is constant, so that $u = u(r) = \tilde{c}_0 \log r + c_0$ for constants $\tilde{c}_0$ and c_0, which can be determined from the wall temperatures. We see that the temperature profile is *not* linear (as it would be between straight parallel walls) but *logarithmic*. This behavior could not have been anticipated easily.

If the wall temperatures in the last example are not constant, we must determine more general solutions to Laplace's equation in the annulus $V = \{(r,\theta): 1 < r < R\}$.

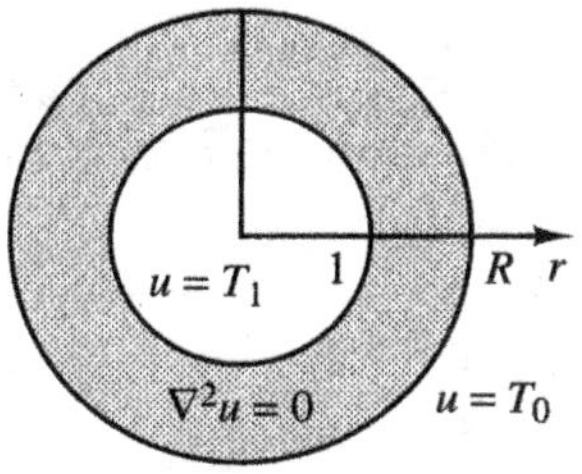

Figure 4.5

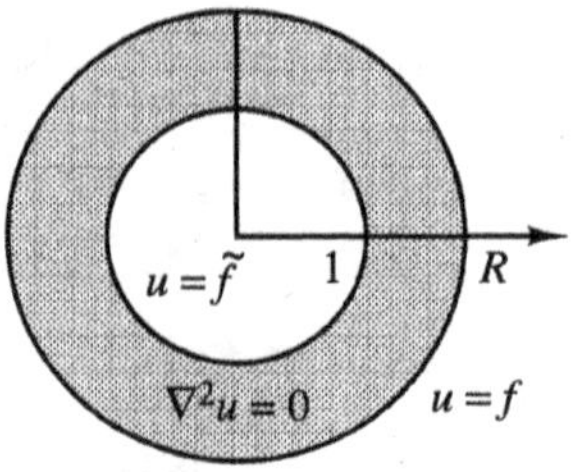

Figure 4.6

In view of formulas (11a) and (11b), it is natural to seek solutions of the form

$$u(r,\theta) = U(r,\theta) + \tilde{U}(r,\theta),$$

$$\text{where} \qquad U(r,\theta) = \frac{c_0}{2} + \sum_{n=1}^{\infty} r^n (c_n \cos n\theta + s_n \sin n\theta) \tag{12a}$$

$$\text{and} \qquad \tilde{U}(r,\theta) = \frac{\tilde{c}_0}{2} \log r + \sum_{n=1}^{\infty} r^{-n} (\tilde{c}_n \cos n\theta + \tilde{s}_n \sin n\theta). \tag{12b}$$

The two boundary circles ($r = 1, r = R$) are shown in Figure 4.6. At either one we may have a Dirichlet, Neumann, or Robin boundary condition (or a mixture). Unfortunately, this problem does not separate usefully into a pair of problems, one for each boundary, and coupling between the coefficients is always present. For example, to meet the Dirichlet conditions

$$u(1,\theta) = \tilde{f}(\theta), \qquad u(R,\theta) = f(\theta), \qquad -\pi < \theta < \pi,$$

we require through orthogonality that the coefficients be chosen to satisfy

$$c_0 = \frac{1}{\pi}\int_{-\pi}^{\pi} \tilde{f}(\theta)\, d\theta; \qquad c_0 + \tilde{c}_0 \log R = \frac{1}{\pi}\int_{-\pi}^{\pi} f(\theta)\, d\theta, \tag{13a}$$

and for $n = 1, 2, \ldots,$

$$\left.\begin{matrix} c_n + \tilde{c}_n \\ s_n + \tilde{s}_n \end{matrix}\right\} = \frac{1}{\pi}\int_{-\pi}^{\pi} \tilde{f}(\theta) \left\{\begin{matrix} \cos n\theta \\ \sin n\theta \end{matrix}\right\} d\theta; \quad \left.\begin{matrix} c_n R^n + \tilde{c}_n R^{-n} \\ s_n R^n + \tilde{s}_n R^{-n} \end{matrix}\right\} = \frac{1}{\pi}\int_{-\pi}^{\pi} f(\theta) \left\{\begin{matrix} \cos n\theta \\ \sin n\theta \end{matrix}\right\} d\theta. \tag{13b}$$

The individual coefficients can be found by simple algebra. In particular, *both coefficients in any line of equations* (13) *are zero when both integrals in that line vanish.* Then convergence can be analyzed in the usual way since at the boundaries only Fourier series are involved. Observe, though, that the terms of the series for $\tilde{U}$ should be estimated by their (absolute) values on the *inner* boundary, $r = 1$ since $r \geq 1 \Rightarrow r^{-n} \leq 1$, for $n = 1, 2, \ldots$.

(4.3) Proposition: *If* $f, \tilde{f} \in \hat{C}^1[-\pi, \pi]$ *and each function has the same value at* $\pm\pi$, *let* U *and* $\tilde{U}$ *be given by equations* (12) *with coefficients from* (13). *Then*

$$u(r,\theta) = U(r,\theta) + \tilde{U}(r,\theta)$$

provides the unique solution to the Dirichlet problem

$$\nabla^2 u = 0 \quad \text{in the annulus} \quad V = \{(r,\theta): \quad 1 < r < R\},^{\dagger}$$

with $u(1,\theta) = \tilde{f}(\theta)$ *and* $u(R,\theta) = f(\theta)$, $-\pi \le \theta \le \pi$.

Proof:* The conditions on f, $\tilde{f}$ guarantee (by Theorem 1.7) the absolute convergence of the series with terms

$$c_n R^n + \tilde{c}_n R^{-n} = \frac{1}{\pi}\int_{-\pi}^{\pi} f(\theta)\cos n\theta \, d\theta = f_n, \text{ say,}$$

and that with terms

$$c_n + \tilde{c}_n = \frac{1}{\pi}\int_{-\pi}^{\pi} \tilde{f}(\theta)\cos n\theta \, d\theta = \tilde{f}_n, \text{ say,} \qquad n = 1, 2, \ldots .$$

Then, $c_n = \tilde{f}_n - \tilde{c}_n$ so that the series with terms

$$\tilde{f}_n - \tilde{c}_n(1 - R^{-2n}) = f_n / R^n$$

also converges absolutely, as must that with terms $\tilde{c}_n(1 - R^{-2n})$. (Why?) But $1 - R^{-2n} \ge 1 - R^{-1}$ for $n \ge 1$, and we see that

$$\sum_{n=1}^{\infty} |\tilde{c}_n| \le \sum_{n=1}^{\infty} |\tilde{c}_n| \frac{(1 - R^{-2n})}{(1 - R^{-1})} < +\infty.$$

Consequently, $\Sigma_{n=1}^{\infty} |c_n| R^n < +\infty$ and corresponding arguments give $\Sigma_{n=1}^{\infty}(|\tilde{c}_n| + |\tilde{s}_n|) < +\infty$, $\Sigma_{n=1}^{\infty}(|c_n| + |s_n|)R^n < +\infty$. Hence, the series in equations (12) converge absolutely and uniformly for $1 \le r \le R$, and when $1 < r < R$, they can be differentiated as often as desired. Thus $u = U + \tilde{U}$ is harmonic in V, continuous on $\bar{V}$, and has the appropriate boundary values. Therefore it is the unique solution to our problem by Theorem 3.5. ■

Example 5:

When $\tilde{f}(\theta) = 0$ and $f(\theta) = |\theta|$, $0 \le |\theta| < \pi$, the conditions of Proposition 4.3 are satisfied. Moreover, all s_n and $\tilde{s}_n$ are zero. (Why?) From (13a) we have $c_0 = 0$, $c_0 + \tilde{c}_0 \log R = (2/\pi)\int_0^{\pi} \theta d\theta = \pi$, and the remaining part of (13b) gives

$$c_n + \tilde{c}_n = 0, \qquad c_n R^n + \tilde{c}_n R^{-n} = \frac{2}{\pi}\int_0^{\pi} \theta \cos n\theta d\theta = \frac{-4}{\pi n^2}\begin{cases} 0, & n = 2, 4, \ldots \\ 1, & n = 1, 3, \ldots \end{cases}$$

(See Example 3 in Section 1.4). Hence, $c_n = \tilde{c}_n = 0$, for $n = 2, 4, 6, \ldots$, but

$$c_n = -\tilde{c}_n = \frac{-4}{\pi n^2}\frac{1}{R^n - R^{-n}}, \qquad \text{for } n = 1, 3, 5, \ldots .$$

Therefore, we see that

$$u(r,\theta) = \frac{\pi}{2}\frac{\log r}{\log R} - \frac{4}{\pi}\sum_{n=1,3,}^{\infty} \frac{r^n - r^{-n}}{R^n - R^{-n}}\frac{\cos n\theta}{n^2}, \qquad 1 \le r \le R$$

† To solve the corresponding problem in the annulus $\{a < r < aR\}$ for $a > 0$, just replace r by r/a in the right sides of (12a, b).

is the solution to the corresponding Dirichlet problem for the annulus $\{1 < r < R\}$. If we replace r in this formula by $r/3$, say, then we get the solution of the same problem in the annulus $\{3 < r < 3R\}$, and R can still be specified.

Problem Set 4.3

4.3A Show that $u(x,y) = x^2 - y^2$ and $u(x,y) = xy$ are solutions of Laplace's equation. Sketch the surfaces $z = u(x,y)$. What boundary conditions do these functions satisfy on the lines $x = 0$, $x = a$, $y = 0$, $y = b$? Consider values of both u and $\partial u/\partial n$.

4.3B If a solution of Laplace's equation in the square $\{0 < x < 1, 0 < y < 1\}$ has the form $u(x,y) = Y(y)\sin \pi x$, of what form is the function Y? Find a function Y that makes $u(x,y)$ satisfy the boundary conditions $u(x,0) = 0$, $u(x,1) = \sin \pi x$.

4.3C 1. Find a nonzero polynomial of the form $p(x,y) = ax^2 + bxy + cy^2 + dx + ey + f$ that satisfies Laplace's equation and the boundary conditions

$$p(0,y) = 0 \qquad p(x,0) = 0.$$

2. Show that a harmonic polynomial of this form cannot have a strict local maximum point or a strict local minimum point.

4.3D Use Fourier series methods together with the information given to obtain formal solutions U to each of the following problems, and indicate the extent of validity of your solution when (i) $f(x) = x^2$ or (ii) $f(x) = x(\pi - x)$ or (iii)

$$f(x) = \begin{cases} 1, & 0 < x < \pi/2 \\ 0, & \pi/2 < x < \pi. \end{cases}$$

1. $\nabla^2 u = 0$ in $\{0 < x < \pi, 0 < y < b\}$,
$u(0,y) = u(\pi,y) = 0$, $0 < y < b$,
$u(x,0) = 0$, $u(x,b) = f(x)$, $0 < x < \pi$.
$(u_n(x,y) = b_n \sinh ny \sin nx$, $n = 1, 2, \ldots)$

2. $\nabla^2 u = 0$ in $\{0 < x < 2, 0 < y < 1\}$,
$u(0,y) = u(2,y) = 0$, $0 < y < 1$,
$u(x,0) = 0$, $u_y(x,1) = x^2$, $0 < x < 2$.
$\left(u_n(x,y) = b_n \sinh\dfrac{n\pi y}{2} \sin\dfrac{n\pi x}{2}, \quad n = 1, 2, \ldots\right)$

3. $\nabla^2 u = 0$ in $\{0 < x < \pi, 0 < y < b\}$,
$u_x(0,y) = u_x(\pi,y) = 0$, $0 < y < b$,
$u(x,0) = 0$, $u(x,b) = f(x)$, $0 < x < \pi$.
$(u_n(x,y) = b_n \sinh ny \cos nx$, $n = 1, 2, \ldots$, but $u_0(x,y) = b_0 y)$

4. $\nabla^2 u = 0$ in $\{0 < x < 2, 0 < y < 1\}$,
$u_x(0,y) = u_x(2,y) = 0$, $0 < y < 1$,
$u_y(x,0) = 0$, $u(x,1) = 6$, $0 < x < 2$.
$\left(u_n(x,y) = b_n \cosh\dfrac{n\pi y}{2} \cos\dfrac{n\pi x}{2}, \quad n = 0, 1, 2, \ldots\right)$ Can you also guess a solution?

5. $\nabla^2 u = 0 \quad \text{in} \quad \{0 < x < \pi, 0 < y < b\},$
$u(0,y) = u_x(\pi, y) = 0, \qquad 0 < y < b,$
$u(x,0) = 0, \qquad u(x,b) = x(\pi - x), \qquad 0 < x < \pi.$
$\left(u_n(x,y) = b_n \sinh\frac{ny}{2}\sin\frac{nx}{2}, \qquad n = 1,3,5,\ldots\right)$

4.3E Suppose that the rectangular plate of Example 3 is *not* thermally isotropic but instead has heat-flux vector $\mathbf{f} = -(u_x, 4u_y)$.

1. Show that the resulting steady-state temperature distribution $u = u(x,y)$ should obey the equation $u_{xx} + 4u_{yy} = 0$.
2. Explain how you could find a solution to the corresponding boundary value problem.

4.3F Suppose we have a straight coaxial cable of radius 1 whose outer sheath is maintained at a potential value of A and whose inner core of radius $\frac{1}{4}$ is maintained at a potential value of B. Find the planar electrostatic field in a plane perpendicular to this cable. (*Hint:* See Example 4.)

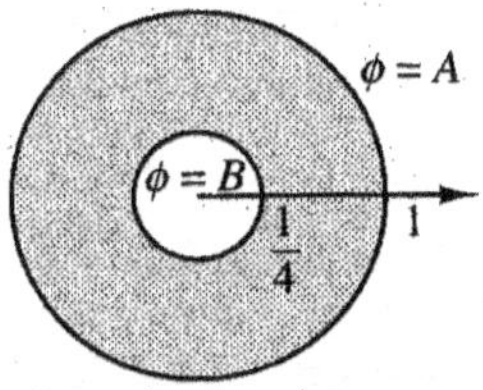

4.3G 1. Obtain a solution $U(r,\theta)$ to the Neumann problem $\nabla^2 u = 0$ in $\{0 \le r < 1\}$ with $u_r(1,\theta) = \theta(\pi - \theta)$, $0 \le \theta \le \pi$, extended as an *odd* function of period 2π.

2. Is there a solution when $u_r(1,\theta)$ is extended as an even function? Explain.

4.3H How could a solution be obtained for the (mixed) problem

$$\nabla^2 u = 0 \quad \text{in} \quad \{0 < r < 1, 0 < \theta < \pi\},$$
$$u(r,0) = u(r,\pi) = 0, \qquad 0 < r < 1, \qquad u_r(1,\theta) = f(\theta), \qquad 0 < \theta < \pi,$$

from a solution of a Neumann problem?

4.3I In the proof of Proposition 4.3, use $c_n R^n = f_n - \tilde{c}_n R^{-n}$ and established results to deduce the convergence of $\Sigma_{n=1}^{\infty} |c_n| R^n$.

4.3J Obtain a series solution for the Dirichlet problem

$$\nabla^2 u = 0 \quad \text{in} \quad \{(r,\theta):\ 1 < r < R\}$$
$$u(1,\theta) = \theta(\pi - |\theta|), \qquad u(R,\theta) = 1, \qquad |\theta| \le \pi.$$

State two different physical problems to which this solution applies.

4.3K Suppose that in Proposition 4.3, f and $\tilde{f}$ are both even functions so that from (13b), the sine coefficients $s_n = \tilde{s}_n = 0$, $n = 1, 2, \ldots$. Then the formal solution has the reduced form

$$u(r,\theta) = \frac{c_0 + \tilde{c}_0 \log r}{2} + \sum_{n=1}^{\infty}(c_n r^n + \tilde{c}_n r^{-n}) \cos n\theta.$$

1. Show that this gives a unique solution to the mixed problem indicated in the following figure.

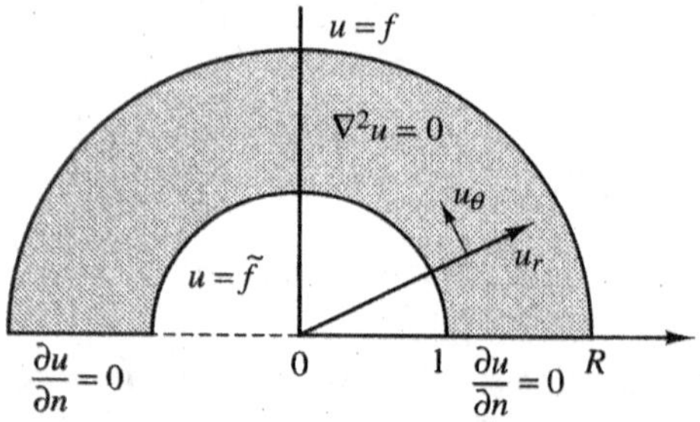

2. Give a condition on the coefficients c_n, $\tilde{c}_n$ which would ensure that $u_\theta(r, \pi/2) = 0$ and requirements on f and $\tilde{f}$ so that these conditions are satisfied. (*Hint:* $c_n = \tilde{c}_n = 0$ iff $f_n = \tilde{f}_n = 0$ where f_n and $\tilde{f}_n$ were used in proving Proposition 4.3.)
3. Show that $f(\theta) = \theta(\pi - \theta)$ and $\tilde{f}(\theta) = \theta^2(\pi - \theta)^2$ on $[0, \pi]$ with even extensions meet the requirements.

4.3L 1. Show that the substitution $r = e^x$ transforms Laplace's equation for $u(r, \theta)$ in the form $L(u) \equiv r^2 u_{rr} + r u_r + u_{\theta\theta} = 0$, $1 < r < R$ into

$$\tilde{L}(\tilde{u}) \equiv \tilde{u}_{xx} + \tilde{u}_{\theta\theta} = 0, \qquad 0 < x < \log R = \ell$$

where $\tilde{u}(x, \theta) = u(e^x, \theta)$ or, alternatively, $u(r, \theta) = \tilde{u}(\log r, \theta)$. Thus, each solution $\tilde{u}$ to $\tilde{L}(\tilde{u}) = 0$ in the "rectangle" $\tilde{V}_\beta = \{(x, \theta)\colon\ 0 < x < \ell,\ 0 < \theta < \beta\}$ provides a solution u to $L(u) = 0$ in the annular sector $V_\beta = \{(r, \theta)\colon\ 1 < r < R, 0 < \theta < \beta\}$.

2. Take $\beta = \pi/4$ and find the harmonic function in the sector that satisfies the boundary conditions

$$u(r, 0) = u\left(r, \frac{\pi}{4}\right) = 0, \qquad 1 < r < R$$

$$u_r(1, \theta) = 0, \qquad u_r(R, \theta) = \theta\left(\frac{\pi}{4} - \theta\right), \qquad 0 < \theta < \frac{\pi}{4}.$$

4.4*

Complex Methods of Solution

In two dimensions, there is an unbreakable bond between harmonic functions and complex analytic functions, and steady-flow hydrodynamics supplies a good physical basis for the connection. Assume that the flow velocity is given by $\mathbf{v} = \nabla\phi$ for some potential ϕ (so that the flow is *irrotational*) and let

$$u = \phi_x \qquad \text{and} \qquad v = \phi_y$$

denote the Cartesian components of $\mathbf{v}$. If the flow is *source-free*, then

$$\nabla \cdot \mathbf{v} = \nabla^2 \phi = 0 \qquad \text{or} \qquad u_x + v_y = 0,$$

and ϕ is necessarily harmonic. Now, in each disk the last condition guarantees the existence of a **stream function** ψ, with gradient $\nabla\psi = (-v, u)$, coupled to ϕ through the **Cauchy–Riemann relations**

$$\phi_x = \psi_y; \qquad \phi_y = -\psi_x. \tag{14}$$

It follows that $\nabla^2\psi = (-\phi_y)_x + (\phi_x)_y = 0$. ψ, called a **harmonic conjugate** to ϕ, is determined uniquely within an additive constant by a line integral of the first derivatives of ϕ, and a similar construction is possible in each simply connected domain. However, in other domains (those with "holes"), a harmonic function ϕ need not admit a *single-valued* harmonic conjugate. (See Problems 4.4E, G, and O.)

Complex Analytic Functions

If we introduce the **complex potential**

$$f = \phi + i\psi \qquad \text{with derivatives} \qquad f_x \stackrel{\text{def}}{=} \phi_x + i\psi_x \quad \text{and} \quad f_y \stackrel{\text{def}}{=} \phi_y + i\psi_y,$$

then equations (14) are equivalent to requiring that f be a solution of the first-order linear equation

$$f_y = if_x. \tag{14$'$}$$

To see this more clearly, note that

$$f_y = \phi_y + i\psi_y \qquad \text{and} \qquad if_x = i(\phi_x + i\psi_x) = -\psi_x + i\phi_x,$$

and that these two complex expressions are equal iff they have the same real part and the same imaginary part; i.e., iff equations (14) hold. Either system characterizes $f = \phi + i\psi$ as being **complex analytic**, and then both its **real part** ϕ and its **imaginary part** ψ are harmonic.† Conversely, as we have seen, in each disk a harmonic function ϕ has a harmonic conjugate ψ for which $f = \phi + i\psi$ is complex analytic. It is customary to denote the point (x, y) by $z = x + iy$. Then, using (14′), it is easy to verify that the functions

$$f(z) = \text{const.}, \qquad f(z) = z, \qquad \text{and} \qquad f(z) = e^z \stackrel{\text{def}}{=} e^x e^{iy}$$

are complex analytic in $\mathbb{R}^2$. For example, in the last case

$$f_x = e^x e^{iy} \qquad \text{and} \qquad f_y = e^x(ie^{iy}) = ie^x e^{iy} = if_x.$$

Similarly, we can establish that a sum, product, quotient, or composite of complex analytic functions is again complex analytic in each domain of definition. (See Problem 4.4C.) However, neither $f(z) = \bar{z} \stackrel{\text{def}}{=} x - iy$ nor $f(z) = |z|^2 = x^2 + y^2$ is complex analytic since in the first case, $f_x = 1$ and $f_y = -i \neq if_x$, and in the second case, $f_x = 2x$ and $f_y = 2y \neq if_x$. (Can you think of any real-valued function that is complex analytic?)

† Although it is not evident, solutions of (14′) in a domain V have C^2 components ϕ and ψ whose harmonicity is then an immediate consequence of (14). See [Ne].

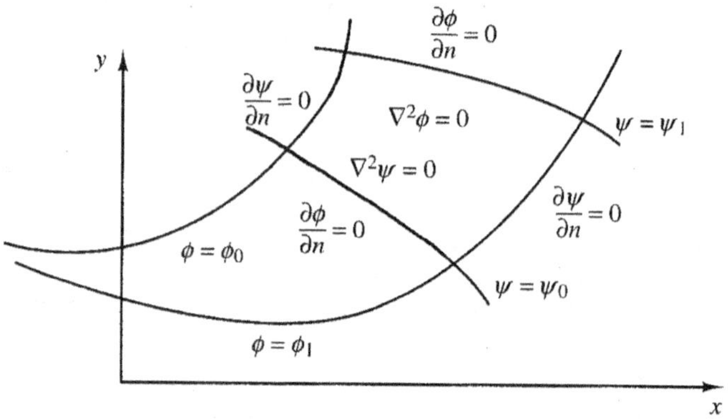

Figure 4.7

Equipotential Curves and Streamlines

If $f = \phi + i\psi$ is complex analytic and *nonconstant* in a domain V, then from (14), we see that

$$\nabla\phi \cdot \nabla\psi = \phi_x\psi_x + \phi_y\psi_y = \psi_y\psi_x + (-\psi_x)\psi_y = 0.$$

But, in general, $\nabla\phi$ is orthogonal to the curves of constant ϕ, and $\nabla\psi$ is orthogonal to those of constant ψ. Since we are in two-dimensional space, it follows that the equipotential curves of constant ϕ are usually orthogonal to those of constant ψ. Exceptional points where both $\nabla\phi$ and $\nabla\psi$ vanish cannot have a limit point in V. (See, e.g., [Ne].) Thus, each nonconstant analytic function provides a harmonic function ϕ with constant values on its equipotential curves along which $\partial\psi/\partial n = 0$, and a harmonic function ψ along whose equipotential curves $\partial\phi/\partial n = 0$, as illustrated in Figure 4.7.

If we interpret ϕ and ψ as the velocity potential and stream function, respectively, of a two-dimensional flow, then $\partial\phi/\partial n = \nabla\phi \cdot \mathbf{n} = \mathbf{v} \cdot \mathbf{n}$ is the flow component normal to a curve in the flow. The condition $\partial\phi/\partial n = 0$ characterizes curves across which no flow occurs, and along each of these **streamlines** ψ is constant. Any streamline may be regarded as an impervious boundary wall in the flow.

Example 6:

$f(z) = z^2 = (x + iy)^2 = x^2 - y^2 + 2ixy$ is complex analytic in $\mathbb{R}^2$ and it can serve as the complex potential of a flow with velocity potential $\phi(x, y) = x^2 - y^2$ and stream function $\psi(x, y) = 2xy$.

Notice that $\psi = 0$ when $x = 0$ or $y = 0$. Consequently, these *lines* may be regarded as walls bounding the flow in, say, the quarter-plane V shown in Figure 4.8. Within V the fluid flows along the streamlines $\psi(x, y) = 2xy =$ const. consisting of equilateral hyperbolas. Along the streamlines, the velocity $\mathbf{v} = \nabla\phi = (2x, -2y)$ is directed as indicated and we have a model for two-dimensional irrotational flow of water around a sharp corner. The origin is called a *stagnation point* of this flow since $\mathbf{v} = \mathbf{0}$ there. Alternatively, we could

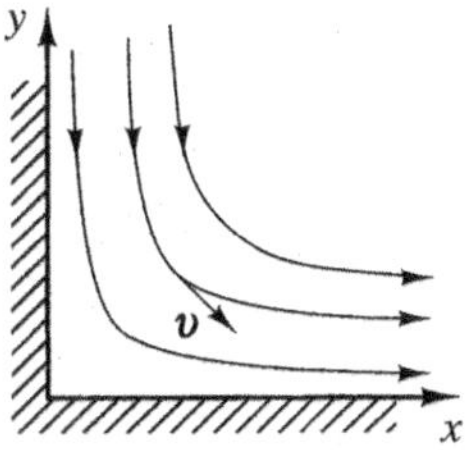

Figure 4.8

make the streamline $\psi = 1$ bound the external region and obtain a model that turns the flow smoothly without stagnation around a curved corner.

If we regard the line $y = 0$ as a boundary wall but permit all values of x, then we have a model for parallel flow directed against a wall as illustrated in Figure 4.9.

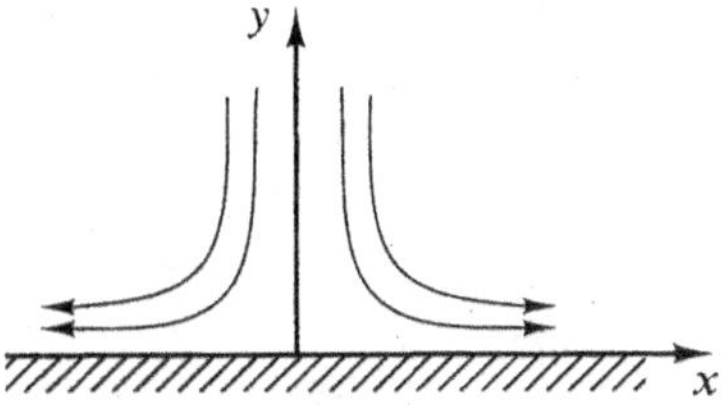

Figure 4.9

We can solve significant boundary value problems in fluid mechanics through the interaction between ϕ and ψ under conformal mappings (to be discussed next). These methods also provide us with solutions to problems in planar elasticity, electrostatics, and magnetostatics.

For example, in electrostatics, the electrical field lines of $\mathbf{E} = -\nabla\phi$ are orthogonal to the equipotential curves of ϕ, so that in two dimensions current flows along the lines $\psi =$ const. See [Ke] and [So].

The following table illustrates some of the interplay between these fields.

Mathematical expression	Heat conduction	Steady-flow hydrodynamics	Electrostatics
$f = \phi + i\psi$ Complex potential	—	—	—
ϕ Harmonic function	Temperature	Velocity potential	Potential
ψ Conjugate harmonic function	—	Stream function	—
$\phi =$ const. } Orthogonal curves	Isothermals	Equipotentials	Equipotentials
$\psi =$ const. } Orthogonal curves	Heat-flux lines	Streamlines	Flux lines
Flux density	$-K\nabla\phi$ (K—Thermal conductivity)	$\nabla\phi = (u, v)$ (Velocity vector)	$-\epsilon\nabla\phi$ (ϵ—Dielectric parameter)
$f' = \frac{1}{2}(f_x - if_y)$ Complex derivative	—	$u - iv$	—

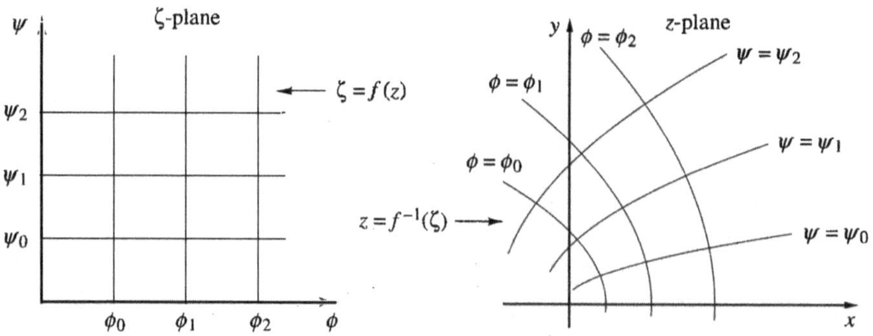

Figure 4.10

Conformal Mapping

Many explicit nonconstant complex analytic functions f are available and each supplies harmonic functions to solve *some* boundary value problems. However, we need considerable skill to find an f that solves a given problem in this manner, and **conformal mapping** can sometimes help. To understand what is involved, let $z = x + iy$ and suppose that $f = \phi + i\psi$ is analytic and *univalent* (one-to-one) in a domain V of the z-plane. Then it can be shown that f maps (transforms) V *onto* a domain $\mathscr{V}$ of the $(\zeta = \phi + i\psi)$-plane and that the inverse function f^{-1} is an analytic function of ζ in $\mathscr{V}$. These mappings are said to be **conformal** since they map curves intersecting at an angle α onto such curves. In particular, horizontal and vertical lines of constant ψ and ϕ, respectively, in the ζ-plane are images of mutually orthogonal equipotential curves in the z-plane, as shown in Figure 4.10. (See [Ne] for further discussion of conformal mappings.)

By studying the image regions in the z-plane of domains in the ζ-plane under various conformal mappings $z = f^{-1}(\zeta)$, we acquire a catalog of solvable problems. The catalog is richer than might be expected from this elementary discussion because the mappings can behave nonconformally at boundary points of domains in which f^{-1} is univalent.

In Figure 4.11 we present a few domains that can be mapped conformally onto the upper-half z-plane $\{y > 0\}$ by elementary functions and indicate the associated boundary correspondences. We also show the positions of four key points under these mappings and use the same direction of cross-hatching in corresponding subregions. In each case, the inverse mapping provides a conjugate pair of harmonic functions ϕ and ψ that solve *some* boundary value problems in the upper half-plane. Moreover, by composition, these domains can be mapped conformally onto one another, and this permits us to solve additional boundary value problems.

Let's illustrate the approach with the following problem from the field of soil mechanics.

Example 7:

A concrete dam holds back a lake of depth h as shown in cross section in Figure 4.12 (p. 172). Because the pressure P at the bottom of the lake exceeds the

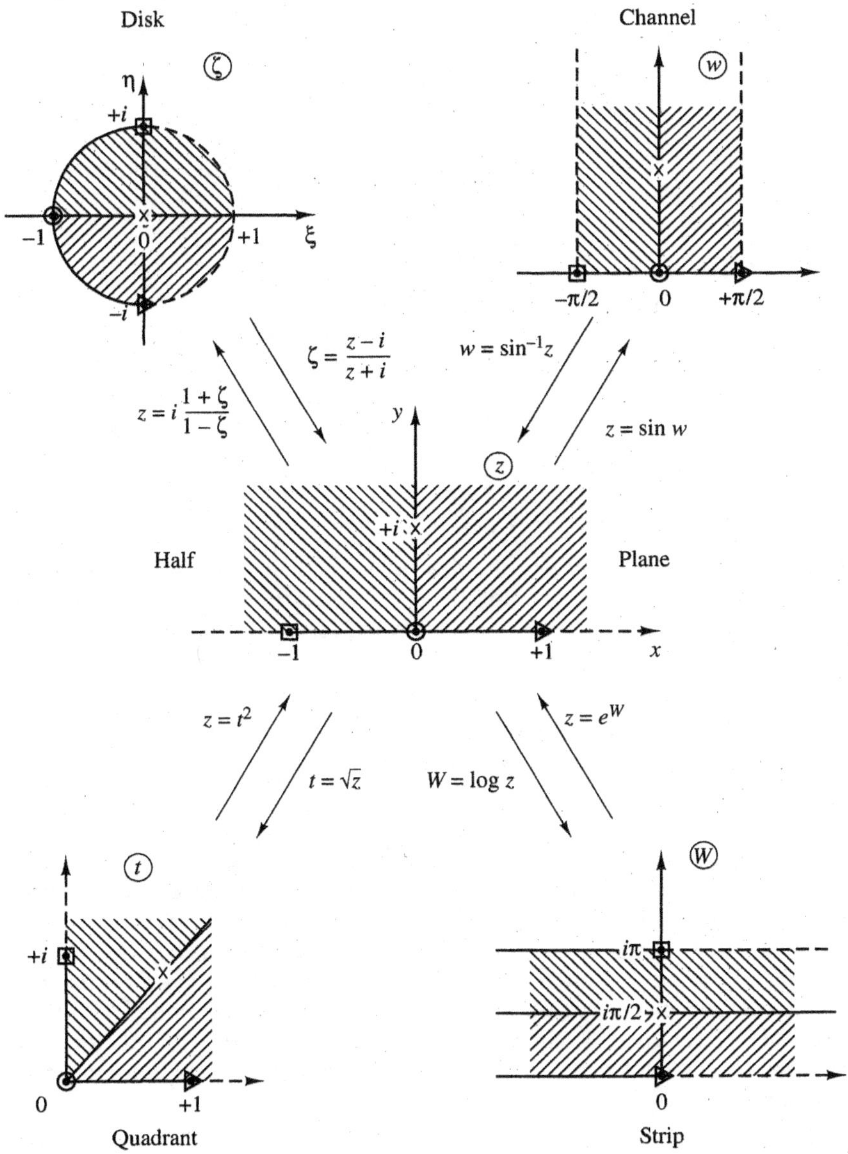

Figure 4.11

atmospheric pressure (of reference value zero) at the toe of the dam, seepage under the dam occurs with discharge velocity **v**. **Darcy's law**‡ postulates that for some constant $k > 0$

$$\mathbf{v} = -k\nabla p \tag{15}$$

‡ The French engineer H. Darcy arrived at this empirical law in 1856 while conducting tests for placement of wells in the city of Dijon.

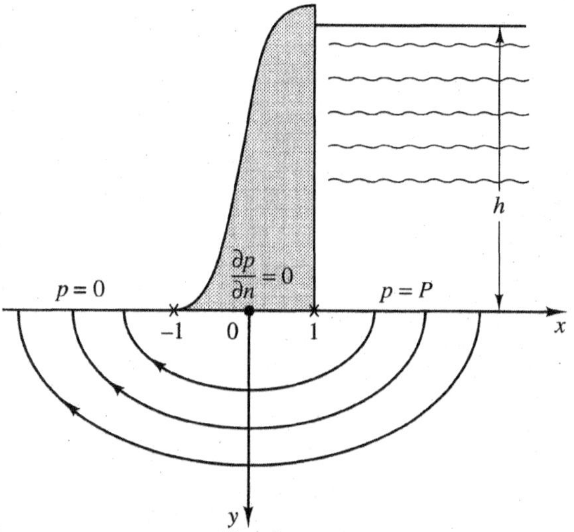

Figure 4.12

where p is the static pressure in the flow. If we choose coordinates as shown, then we see that along the lake-floor level (assumed horizontal) the pressure p satisfies the conditions in the sketch. In particular,

$$\frac{\partial p}{\partial n} = -\frac{1}{k}\mathbf{v}\cdot\mathbf{n} = 0, \qquad -1 < x < 1,$$

because there is no flow through the bottom of the dam.

Finally, assuming the seepage flow to be incompressible, we have

$$\nabla^2 p = -\frac{1}{k}\nabla\cdot\mathbf{v} = 0, \qquad y > 0,$$

so that p is harmonic in the half-space where $y > 0$. If we suppose the dam to be so wide that the flow is essentially two-dimensional, we see that we have a mixed boundary value problem in the upper half-plane. So we look over our half-plane mappings to see if there is one whose real (or imaginary) part behaves suggestively, relative to our requirements. The disk mapping is not too promising, but the channel map has possibilities because its real part ϕ has boundary values $-\pi/2$ and $\pi/2$ for $x < -1$ and $x > 1$, respectively, while $\partial\phi/\partial n = 0$ along the intermediate segment where $\psi = 0$. A little experimenting shows that our problem is solved by taking

$$p(x, y) = \frac{P}{\pi}\phi(x, y) + \frac{P}{2} = \frac{P}{\pi}\mathscr{R}_e \sin^{-1} z + \frac{P}{2} \qquad \textbf{(15')}$$

where $w = \sin^{-1} z$ denotes an appropriate inverse to the mapping $z = \sin w$. In particular, along the bottom of the dam where $z = x$ and $\sin^{-1} z = \sin^{-1} x$ is real-valued, the hydrostatic pressure from seepage is given by

$$p(x, 0) = \frac{P}{\pi}\sin^{-1} x + \frac{P}{2},$$

and the total upward seepage force per unit width of dam is equal to

$$\int_{-1}^{1} p(x, 0)\, dx = \frac{P}{\pi}\int_{-1}^{1} \sin^{-1} x\, dx + \frac{P}{2}\int_{-1}^{1} dx = P,$$

since $\sin^{-1} x$ is an *odd* function on $(-1, 1)$. The weight of the dam must compensate for this uplift, and its distribution in the x-direction should be proportional to $p(x, 0)$, resulting in the familiar profile indicated. The water seeps along curves orthogonal to those of constant pressure following the streamlines shown in Figure 4.12. This flow is irrotational, as explained in Section 4.1.

In the approach just outlined, the conformal mapping function $f = \phi + i\psi$ provides a harmonic solution ϕ or ψ to a boundary value problem in a domain V. In fact, in mapping V onto a domain $\mathscr{V}$, f transfers the harmonic solution h of any boundary value problem in $\mathscr{V}$ to the *harmonic* function $H = h \circ f$, which attempts to solve a corresponding problem in V. Whether it succeeds depends on the boundary behavior of the mapping. Suppose that the mapping f extends continuously to a boundary point z_0 of V. Then H and h take the same boundary values at z_0 and $\zeta_0 = f(z_0)$, respectively. If the mapping is actually conformal at z_0, the normal derivatives at these points are related by

$$\frac{\partial H}{\partial N} = \frac{\partial h}{\partial n} |\nabla \phi|;$$

in particular, homogeneous Neumann conditions are preserved at such points. (See Problem 4.4P.) For example, the solutions h to rectangular and disk problems that were obtained in Section 4.3 supply solutions H to corresponding problems in certain other regions.

Many conformal mappings are presented in [Ko], and the celebrated Riemann mapping theorem asserts that any *proper* subdomain of $\mathbb{R}^2$, which is simply connected (without "holes"), is the conformal image of a disk under *some* univalent analytic mapping. However, Riemann mapping functions of general domains produce boundary correspondence issues of great subtlety. [Du].

Problem Set 4.4

(*Note:* In the following problems, analytic ≡ complex analytic.)

4.4A Use (14′) to verify that each of the following functions is analytic in the $(z = (x, y))$-plane.

1. $f(z) = \text{const.}$ **2.** $f(z) = z = x + iy$

3. $f(z) = z^2 = (x + iy)^2$ **4.** $f(z) = e^{iz} = e^{ix}e^{-y}$

5. $f(z) = z^3 = (x + iy)^3$

4.4B Identify the real part ϕ and the imaginary part ψ of each function in Problem 4.4A.

4.4C If f and $\tilde{f}$ are analytic in a common domain V, show that each combination below is analytic in V because it satisfies (14′):

(a) $s = f + \tilde{f}$ (b) $p = f\tilde{f}$ (c) $q = \tilde{f}/f$ (if $f \neq 0$)

4.4D **1.** Using the combinations in Problem 4.4C and the analyticity of e^z and e^{iz}, explain the analyticity of $\sinh z = (e^z - e^{-z})/2$, $\cosh z = ?$, $\sin z = (e^{iz} - e^{-iz})/2i$, and $\cos z = ?$.

2. Using the combinations in Problem 4.4C finitely often, what is the most general expression that can be formed from the constant analytic functions in conjunction with $f(z) = z$? Where will it be analytic?

4.4E 1. Verify that $\phi(x, y) = 4x^3y - 4xy^3 + x + 1$ is harmonic in $\mathbb{R}^2$.
2. Find a harmonic conjugate ψ to ϕ.
3. Express $f = \phi + i\psi$ as a polynomial "in z."
4. Verify that $\phi(x, y) = \sin x \sinh y$ is harmonic in $\mathbb{R}^2$ and find a harmonic conjugate ψ. Express $f = \phi + i\psi$ in terms of z.

4.4F (*Complex derivative*) When f is complex analytic in a domain V, its complex derivative is

$$f' \stackrel{\text{def}}{=} \tfrac{1}{2}(f_x - if_y) = f_x \qquad \text{(in view of (14'))}.$$

1. If $f(z) = z^2$, $\sin z$, or e^z, verify that $f'(z) = 2z$, $\cos z$, or e^z, respectively.
2. If $f = \phi + i\psi$ is a complex velocity potential for a flow with velocity $(u, v) = \nabla\phi$, show that $f' = u - iv$.

4.4G Show that $\phi(x, y) = \log|z| = \log\sqrt{x^2 + y^2}$ is harmonic in $V = \{|z| > 0\}$ and that in each disk excluding the origin a suitable extension of

$$\psi(x, y) = \arg z \stackrel{\text{def}}{=} \tan^{-1}(y/x) \qquad (x \neq 0)$$

gives one harmonic conjugate to ϕ. Explain why ψ cannot be defined in V.

4.4H Verify that the complex potential $f(z) = Uz$ (for constant $U > 0$) describes a two-dimensional uniform flow above the straight wall $y = 0$. What are the other streamlines and the equipotential curves for this flow?

4.4I 1. Show that for constant $U > 0$, the complex potential

$$f(z) = U\left(z + \frac{1}{z}\right), \qquad z \neq 0$$

describes a flow *outside* the circular wall $\{|z| = 1\}$. Verify that for large $|z|$, the flow is approximately uniform and sketch a few possible streamlines. Does this flow have (stagnation) points where $\mathbf{v} = \mathbf{0}$?
2. What is the effect of allowing $U < 0$ in part 1? of allowing $U = i$? of allowing U to be complex?

4.4J 1. Fluid flows into and out of the 45° corner as shown in the figure. Derive the stream function $\psi(r, \theta) = Ar^4 \sin 4\theta$ and sketch one or two streamlines in the interior of the region. Find a velocity potential for the flow.

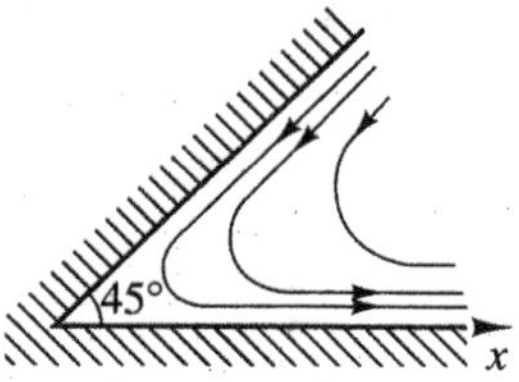

2. Do the same for the 135° corner shown in the figure at the top of the next page.

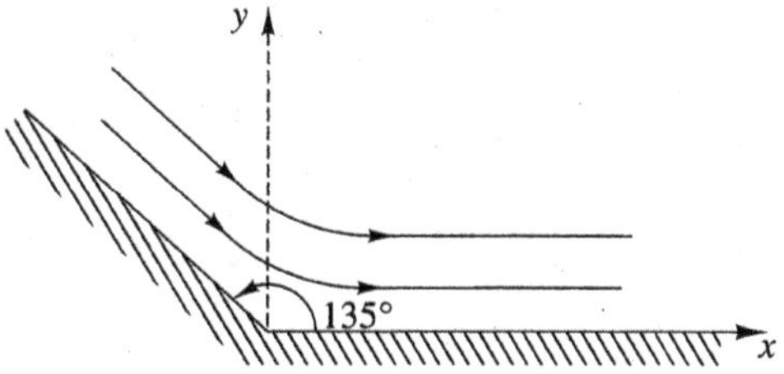

4.4K The mapping used in Example 7 transforms the uniform flow of Problem 4.4H to flow into a closed channel as indicated in the figure. Find the fluid velocity in the channel and sketch some streamlines. Locate stagnation points of the flow.

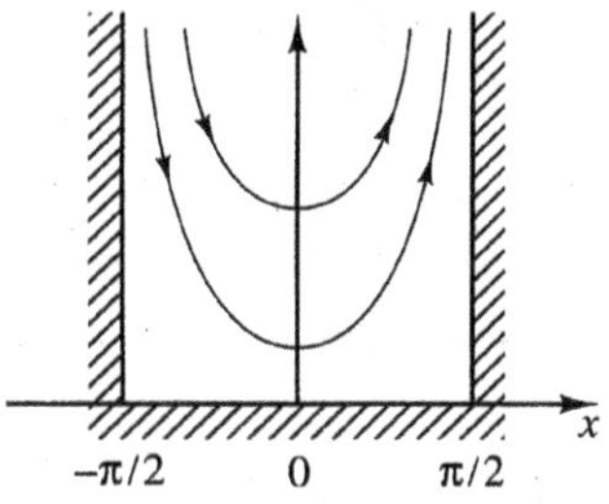

4.4L Explain in words how to use the conformal mappings of Figure 4.11 to find harmonic solutions to the following boundary value problems. Then use the indicated letters to write your solution explicitly.

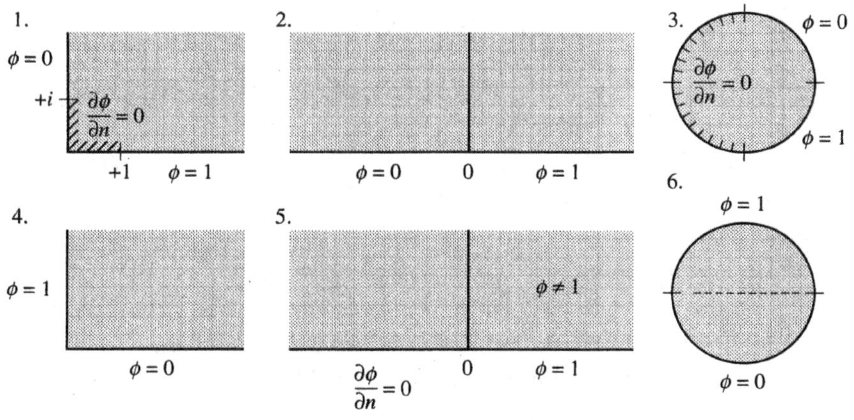

Give one physical interpretation of the problem involving heat conduction and another involving membrane deflection.

4.4M *1.* Verify that for $\alpha \in [0, 2\pi)$, $\varsigma = ze^{i\alpha}$ gives a conformal mapping of $\mathbb{R}^2$ onto itself obtained by rotation about the origin through an angle α.

2. Use such rotations with the results in Problem 4.4L to solve the Dirichlet problems shown in the figures at the top of the next page.

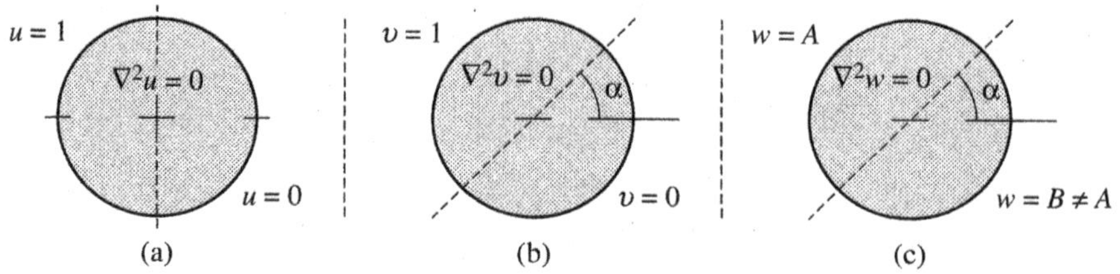

3. Combine the result in part 2(a) with an appropriate mapping to obtain a bounded solution to the problem shown in the following figure.

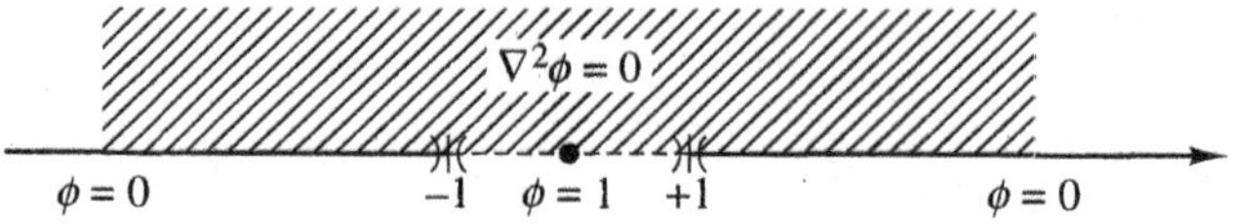

4.4N Explain how the transformation used to solve Problem 4.3L is related to a conformal mapping involving $\zeta = e^z$ expressed in polar coordinates.

4.4O (*Harmonic conjugate*) Let ϕ be harmonic in the disk $\{0 \le x^2 + y^2 < R^2\}$ and define

$$\psi(x, y) = \int_0^x -\phi_y(s, 0)\, ds + \int_0^y \phi_x(x, t)\, dt.$$

1. Differentiate as required to verify that $\psi_y(x, y) = \phi_x(x, y)$ and that

$$\psi_x(x, y) = -\phi_y(x, 0) + \int_0^y \phi_{xx}(x, t)\, dt.$$

2. Use the harmonicity of ϕ to conclude that $\psi_x(x, y) = \phi_y(x, y)$, so that ψ is a harmonic conjugate of ϕ.
3. Explain the origin of the formula defining ψ and why a similar construction is possible in any disk or rectangle.

4.4P (*Composites*) Let $H(z) = h(f(z))$ be defined, where $f = \phi + i\psi$ is analytic in a domain V and h is analytic in $\zeta = \xi + i\eta$.

1. Use (14′) to verify that $H_x = h_\eta(\psi_x - i\phi_x)$ and through a similar expression for H_y, establish that H is analytic in V.
2. Suppose that h is only harmonic. Conclude that H is also harmonic. (*Hint:* In each disk, h has a harmonic conjugate $\tilde{h}$, say.)
3. In part 2 use the chain rule and (14) to verify that

$$|\nabla H|^2 = |\nabla h|^2 |\nabla \phi|^2$$

when f maps V conformally onto a domain $\mathscr{V}$. Explain how this result leads to the normal derivative relation

$$\frac{\partial H}{\partial N} = \frac{\partial h}{\partial n} |\nabla \phi|$$

where **N** and **n** are unit vectors at corresponding points in directions of the exterior normal to $S = \partial V$ and $\mathscr{S} = \partial \mathscr{V}$, respectively.

4.5

Poisson's Equation and Green Functions

In this chapter our goal is to understand the potential equation

$$\nabla \cdot (K\nabla\phi) = -F$$

and how it characterizes steady-state processes, including those involving internal "sources" represented here by F. Thus far, we have solved some source-free problems ($F = 0$) governed by Laplace's equation

$$\nabla^2\phi = 0,$$

so next let's see how to attack these problems when sources *are* present. In particular, we want solutions v for boundary value problems involving **Poisson's equation**

$$\nabla^2 v = F, \qquad \textbf{(16)}$$

where F is prescribed and nonzero. Fortunately, through linear superposition (Section 2.3) we can suppose that *all* boundary conditions are homogeneous and then try the eigenfunction approach that we used in Section 3.4.

In analyzing the mylar roof problem of Examples 1 and 2, we neglected external loading of the membrane — in particular, that of its weight or a possibly nonuniform distribution of snow. To take such distributed loading F into account we must add to our earlier solution u a solution v of the boundary value problem indicated in Figure 4.13 for the rectangular domain

$$R = \{0 < x < \pi, 0 < y < \mathscr{b}\}.$$

In Example 1 we obtained a set of eigenfunctions for the homogeneous boundary conditions at $x = 0$ and $x = \pi$, namely,

$$X_n(x) = \sin nx, \qquad n = 1, 2 \ldots .$$

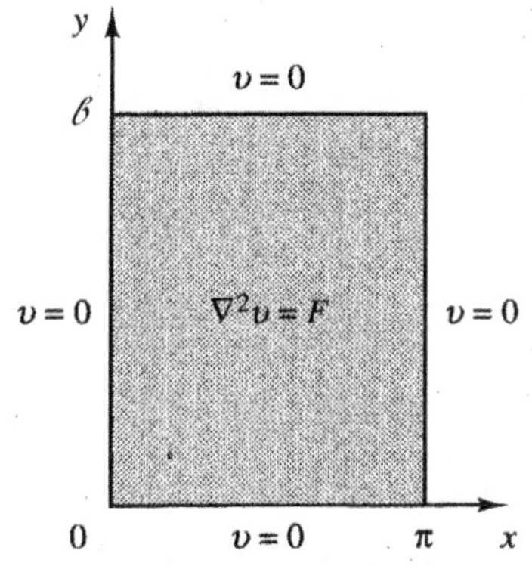

Figure 4.13

Example 8:

Let's first look at the problem when the roof-loading can be described by

$$F(x,y) = \sum_{n=1}^{N} f_n(y)X_n(x)$$

for continuous functions f_n, $n = 1, 2, \ldots, N$. Note that $v_n(x,y) = Y_n(y)X_n(x)$ solves the particular inhomogeneous equation

$$\nabla^2 v_n = Y_n X_n'' + Y_n'' X_n = (Y_n'' - n^2 Y_n)X_n$$
$$= f_n X_n, \text{ say,}$$

provided that Y_n satisfies the equation,

$$Y_n'' - n^2 Y_n = f_n, \qquad \text{where } f_n = f_n(y). \tag{17}$$

Moreover, for v_n to meet the remaining boundary conditions $v(x,0) = v(x,\ell) = 0$, we should make

$$Y_n(0) = Y_n(\ell) = 0. \tag{17'}$$

From Appendix B.1 we know that when f_n is continuous, the integral

$$\tilde{Y}_n(y) = \frac{1}{n}\int_0^y \sinh n(y-t) f_n(t)\, dt \tag{18}$$

solves (17), with $\tilde{Y}_n(0) = 0$, while for arbitrary constants a_n and b_n,

$$a_n \cosh ny + b_n \sinh ny$$

is a general solution to the corresponding homogeneous equation ($f_n = 0$). Thus to have $Y_n(0) = 0$, we consider the combination

$$Y_n(y) = \tilde{Y}_n(y) + b_n \sinh ny, \tag{19}$$

and to have $Y_n(\ell) = 0$, we must take

$$b_n = -\frac{\tilde{Y}_n(\ell)}{\sinh n\ell}, \qquad n = 1, 2, \ldots . \tag{19'}$$

With these choices

$$V(x,y) = \sum_{n=1}^{N} Y_n(y) \sin nx$$

gives us the unique solution to the problem when

$$F(x,y) = \sum_{n=1}^{N} f_n(y) \sin nx,$$

for arbitrary continuous f_n, $n = 1, 2, \ldots, N$.

For example, the sinusoidal loading $F(x,y) = \sin x$, where $f_1(y) = 1$ and all other $f_n(y) = 0$, requires from (18) only the integral

$$\tilde{Y}_1(y) = \int_0^y \sinh(y-t)\, dt = \cosh y - 1.$$

Then from (19), we find that

$$Y_1(y) = \tilde{Y}_1(y) - \frac{\sinh y}{\sinh \ell}\tilde{Y}_n(\ell)$$
$$= (\cosh y - 1) - \frac{\sinh y}{\sinh \ell}(\cosh \ell - 1)$$

or, after some work (see Problem 4.5H), that

$$Y_1(y) = -2\frac{\sinh\left(\frac{y}{2}\right)\sinh\left(\frac{\ell - y}{2}\right)}{\cosh(\ell/2)}. \tag{20}$$

Thus the corresponding deflection of this loaded membrane is given by

$$v_1(x, y) = Y_1(y)\sin x, \qquad (x, y) \in \bar{R}.$$

Observe that $v_1 \le 0$ in this region, so that the deflection v as given by the above solution is directed opposite to the loading F. This is consistent with our earlier analysis, but it must be taken into account properly. In particular, the total deflection of the original roof under this sinusoidal loading is given by $u + v_1$, where u is the solution obtained in Examples 1 or 2.

For more general F, we attempt a formal representation

$$F(x, y) = \sum_{n=1}^{\infty} f_n(y)\sin nx, \qquad 0 < x < \pi, \tag{20a}$$

which requires

$$f_n(y) = \frac{2}{\pi}\int_0^{\pi} F(x, y)\sin nx\,dx, \qquad n = 1, 2, \ldots \tag{20b}$$

and results in the formal series

$$V(x, y) = \sum_{n=1}^{\infty} Y_n(y)\sin nx, \tag{21}$$

meeting the required boundary conditions. But can this formal series solve our problem? To analyze convergence more easily, note that after some work (left to Problem 4.5H), the Y_n of (18) and (19) can be represented as follows:

$$Y_n(y) = -\int_0^{\ell} g_n(y, t) f_n(t)\,dt, \qquad 0 \le y \le \ell \tag{22}$$

where

$$g_n(y, t) = \frac{(\sinh nt)\sinh n(\ell - y)}{n\sinh n\ell}, \qquad t \le y$$
$$= g_n(t, y), \qquad y \le t \tag{23}$$

is the **Green function** for the problem of (17) and (17′). (See Appendix B.2.)

Now, let's estimate the terms of (21). First, recall that when $0 \le t \le y$, then

$0 \le \sinh nt \le \sinh ny$ and since $y \le \ell$, we see that

$$\begin{aligned}|g_n(y,t)| &\le \frac{1}{n}\frac{\sinh ny \sinh n(\ell - y)}{\sinh n\ell}\\ &= \frac{1}{2n}\frac{(e^{ny} - e^{-ny})(e^{n(\ell-y)} - e^{-n(\ell-y)})}{e^{n\ell} - e^{-n\ell}}\\ &\le \frac{1}{2n}\frac{e^{ny}e^{n(\ell-y)}}{e^{n\ell}(1 - e^{-2\ell})} = \frac{B}{n},\end{aligned}$$

for the constant $B = \frac{1}{2}(1 - e^{-2\ell})^{-1}$.

The same estimate is valid when $t \ge y$, and we can use it in (22) to estimate the nth term of (21) as follows:

$$|Y_n(y)\sin nx| \le |Y_n(y)| \le \frac{B}{n}\int_0^\ell 1 \cdot |f_n(t)|\,dt \le \frac{B}{n}\left(\ell \int_0^\ell f_n^2(t)\,dt\right)^{1/2},$$

if we also use the Schwarz inequality for integrals in the last step. (See Section 0.1d.)

Since the "Fourier coefficients" of (20b) satisfy Bessel's inequality (see Section 1.8a) in the form

$$\sum_{n=1}^{N} f_n^2(t) \le \frac{2}{\pi}\int_0^\pi F^2(x,t)\,dx, \qquad N = 1, 2, \ldots,$$

we get a similar inequality for their integrals and, upon passing to the limit as $N \to \infty$, we conclude that

$$\sum_{n=1}^{\infty}\int_0^\ell f_n^2(t)\,dt \le \frac{2}{\pi}\int_0^\ell dt \int_0^\pi F^2(x,t)\,dx.$$

Suppose that F is continuous, so that the last integral is finite and all previous terms are defined. Then we can reproduce the Cauchy inequality arguments of Section 1.8c to conclude that series (21) converges uniformly with a continuous sum on the closed rectangle $\bar{R}$.

It is more delicate to analyze convergence of the termwise differentiated series even when F admits Fourier series representation in the form of (20a). Recall from the discussion in Section 1.8b that to avoid discontinuities in the representation we should make $F(0,y) = F(\pi,y) = 0$. (If $F(x,0) = F(x,\ell) = 0$, we interchange x and y and solve the problem in terms of eigenfunctions on the interval $[0,\ell]$.)

The Green Function

When we replace the dummy variable x in (20b) by s and substitute the resulting integral in (22), the formal series (21) becomes

$$V(x,y) = -\frac{2}{\pi}\sum_{n=1}^{\infty}\left\{\int_0^\pi ds \int_0^\ell dt\, g_n(y,t)\sin ns\, F(s,t)\right\}\sin nx. \qquad \textbf{(24)}$$

If F is continuous so that uniform convergence (as established above) permits interchange of integration and summation, we may express V in the form

$$V(x,y) = -\int_0^{\pi} ds \int_0^{\ell} dt\, G(x,y;s,t)F(s,t). \tag{25}$$

Here, the formal series

$$G(x,y;s,t) = \frac{2}{\pi}\sum_{n=1}^{\infty} g_n(y,t)\sin nx \sin ns \tag{25'}$$

is called the **Green function** for this boundary value problem. Since g_n is known explicitly, G is determined independently of F and, in principle, the solution V at any point (x,y) is obtained through integration of G against F over the rectangle. There are various difficulties associated with this approach, the most significant arising from the fact that G is not finite when $x = s$ and $y = t$. (See Problem 4.5F.) However, for distinct points (x,y) and (s,t) within the rectangle R, G *is* defined and then, as is easily verified, G is **symmetric** in that

$$G(x,y;s,t) = G(s,t;x,y).$$

In this membrane problem $G(x,y;s,t)$ may be interpreted as the deflection at (x,y), due to a unit load at (s,t) under the given boundary conditions. (More accurately, it is the limiting deflection at (x,y) due to distributed loading of total unit force concentrated near (s,t) as the area of concentration shrinks to (s,t).) The symmetry of G is a statement that this deflection is the same when the points (x,y) and (s,t) are interchanged, even when these points are not located symmetrically relative to the boundary!

Example 9:

A very long circular cylinder of unit radius formed from a thin copper sheet is split along a plane through its axis (the z-axis) and the halves are separated slightly. One half is grounded at potential zero, and the other half is maintained at constant potential $\phi_0 \neq 0$ as indicated in Figure 4.14. The resulting electrostatic potential ϕ is harmonic except on the cylindrical surfaces, and we shall suppose that the cylinder is so long that edge effects can be neglected. Then ϕ should be independent of z and we can use polar coordinates as shown. Within the cylindrical cavity, ϕ can be obtained from Poisson's integral. (See

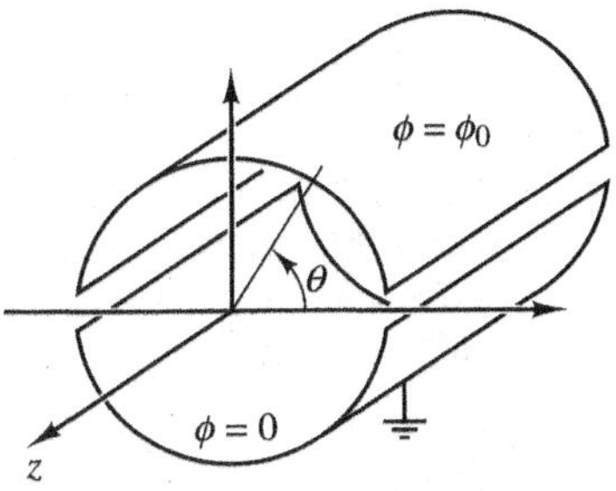

Figure 4.14

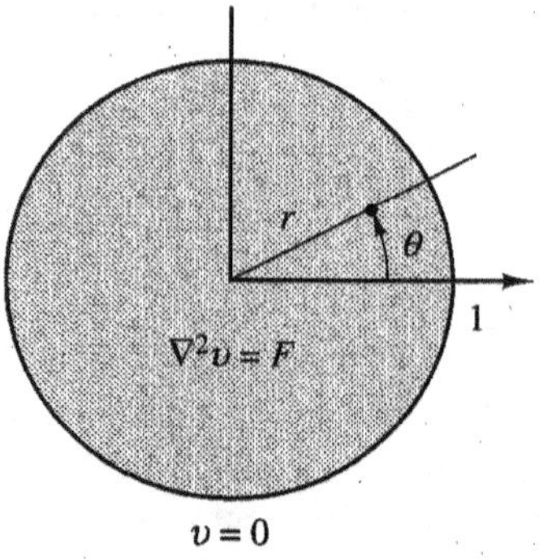

Figure 4.15

Page 160.) However, to permit distributed line charges, we need a solution $v = v(r, \theta)$ of the equation

$$\nabla^2 v = v_{rr} + \frac{1}{r} v_r + \frac{1}{r^2} v_{\theta\theta} = F, \qquad r < 1$$

for suitable F, with $v(1, \theta) = 0$, $-\pi \leq \theta < \pi$ (see Figure 4.15). This can be found through series methods involving the eigenfunctions

$$\Theta_n(\theta) = c_n \cos n\theta + s_n \sin n\theta, \qquad n = 0, 1, 2, \ldots$$

of the homogeneous problem. The arguments are technically complicated (see [Wein]), so we just present the results. If

$$F(r, \theta) = \sum_{n=0}^{\infty} f_n(r) \Theta_n(\theta), \qquad 0 < r < 1, \tag{26}$$

then a formal solution is given by

$$V(r, \theta) = \sum_{n=0}^{\infty} R_n(r) \Theta_n(\theta), \qquad 0 < r < 1, \tag{27}$$

where (see Problem 4.5I)

$$R_0(r) = \log r \int_0^r \rho f_0(\rho)\, d\rho + \int_r^1 (\log \rho) \rho f_0(\rho)\, d\rho,$$

and for $n \geq 1$, **(28)**

$$2nR_n(r) = \int_0^r \rho^n (r^n - r^{-n}) \rho f_n(\rho)\, d\rho + \int_r^1 r^n (\rho^n - \rho^{-n}) \rho f_n(\rho)\, d\rho.$$

When $f_n = 0$, $n > N$, then V so defined is the exact solution to the problem. Otherwise there are the expected difficulties with convergence, differentiation, etc. However, as in the rectangular case, we may use Fourier methods to determine the coefficients c_n and s_n in (26) and after suitable rearrangement, write the solution in the form

$$V(r, \theta) = -\int_0^1 \rho\, d\rho \int_{-\pi}^{\pi} d\phi\, G(r, \theta; \rho, \phi) F(\rho, \phi) \tag{29}$$

for a Green function G defined by a series. But in this case the series for G can be summed when $(r, \theta) \neq (\rho, \phi)$ and we find that

$$G(r, \theta; \rho, \phi) = \frac{1}{4\pi} \log \left[\frac{(1 - r^2)(1 - \rho^2) + D^2}{D^2} \right], \tag{30}$$

when $D = |(r, \theta) - (\rho, \phi)|$ is not zero.

Analysis of G; Fundamental Solutions

Because (30) makes G available in closed form, we can study the properties of this Green function and gain insight about the behavior of more general Green functions. What do we learn?

It is easy to see that G is defined and symmetric when $D \neq 0$ and that for fixed $\rho < 1$: $G(r, \theta; \rho, \phi) \to 0$ as $r \to 1$. Moreover, it is clear that G is positive and approaches $+\infty$ as $(r, \theta) \to (\rho, \phi)$. Physically, G represents the electrostatic potential at (r, θ, z) due to a line charge of unit strength at (ρ, ϕ) in the presence of the fully *grounded* cylinder. In this context, its symmetry is one instance of Maxwell's principle of reciprocity.

To proceed, we note that

$$G(r, \theta; \rho, \phi) + \frac{1}{2\pi} \log D = \frac{1}{4\pi} \log [(1 - r^2)(1 - \rho^2) + D^2], \tag{31}$$

and the right side H, say, is harmonic (in r, θ) for fixed (ρ, ϕ). (See Problem 4.5G.) Thus we can characterize the singularity of G and say that near (ρ, ϕ), G behaves like $-(1/2\pi) \log D$. For this reason,

$$\gamma = -\frac{1}{2\pi} \log D = -\frac{1}{2\pi} \log \sqrt{(x - s)^2 + (y - t)^2},$$

(expressed in standard Cartesian coordinates) is called the **fundamental solution** of Laplace's equation in two dimensions. Observe that it is not domain dependent. With its aid, the Green function G for this Dirichlet problem relative to another bounded domain V of $\mathbb{R}^2$ can be constructed in the form

$$G = \gamma + H$$

provided that for each fixed $(s, t) \in V$ a suitable H can be found. We require that H solve the Dirichlet problem

$$\nabla^2 H = 0 \text{ in } V$$

with $H|_{\partial V} = -\gamma$. Alternatively, if f maps V conformally onto a domain $\tilde{V}$ for which the Green function $\tilde{G}$ is known, then

$$G(x, y; s, t) \overset{\text{def}}{=} \tilde{G}(f(x, y); f(s, t))$$

defines a Green function for V. (Recall the discussion at the end of Section 4.4.)

In any case, once a Green function G has been obtained for a given domain V of $\mathbb{R}^2$, then, in principle, the integral

$$w(x, y) = -\int_V G(x, y; s, t) F(s, t)\, ds\, dt \tag{32}$$

provides a solution to the problem

$$\nabla^2 w = F, \qquad \text{in } V \tag{33}$$

with $w|_{\partial V} = 0$.

This approach is explored in [Ke] and in [G-L].

Problem Set 4.5

4.5A 1. Show how the addition of $s(x, y) = ax + a_0$ for appropriate a and a_0 to u in Example 2 of Section 4.3 yields a solution v to the problem

$$\begin{aligned} &\nabla^2 v = 0 \quad \text{in} \quad \{0 < x < \pi, 0 < y < \ell\} \\ &v(0, y) = 1, \quad v(\pi, y) = 3, \quad 0 < y < \ell \\ &v(x, 0) = s(x, 0), \quad v(x, \ell) = f(x), \quad 0 < x < \pi. \end{aligned}$$

2. What problems could be solved using $s(x, y) = xy$? using $s(x, y) = x^2 - y^2$?

4.5B Obtain a formal solution V to each of the following problems for $\nabla^2 v = F$ in the rectangular domain $\{(x, y): \quad 0 < x < \pi, \quad 0 < y < \ell\}$:

1. $F(x, y) = y \sin x$, with $v = 0$ on the boundary. ($X_n(x) = \sin nx$, $n = 1, 2, \ldots$.)
2. $F(x, y) = x(\pi - x)$, with $v = 0$ on the boundary.
3. $F(x, y) = x^2$, with $v_x(0, y) = v_x(\pi, y) = 0$, $0 < y < \ell$ and $v(x, 0) = v(x, \ell) = 0$, $0 < x < \pi$. ($X_n(x) = \cos nx$, $n = 0, 1, 2, \ldots$.)
4. $F(x, y) = yx^2$ with the boundary conditions of part 3.
5. $F(x, y) = y^2 \sin(3x/2)$, with $v(0, y) = v_x(\pi, y) = 0$, $0 < y < \ell$ and $v(x, 0) = v_y(x, \ell) = 0$, $0 < x < \pi$. ($X_n(x) = \sin(nx/2)$, $n = 1, 3, 5, \ldots$.)

4.5C Obtain a solution to the problem $\nabla^2 v = F$ in the disk $\{(r, \theta): \quad 0 < r < 1\}$ when

1. $F(r, \theta) = \sin\theta$, with $v(1, \theta) = 0$, $-\pi < \theta < \pi$.
2. $F(r, \theta) = r^2(\theta^2 - \pi^2)$, with $v(1, \theta) = 0$, $-\pi < \theta < \pi$.
3. $F(r, \theta) = r^2 \cos\theta$, with $v_r(1, \theta) = 0$, $-\pi < \theta < \pi$.

4.5D Let $u(x, y)$ be the solution to the problem $\nabla^2 u = -1$, $\{|x| < 1, |y| < 1\}$ with $u = 0$, if $|x| = 1$ or $|y| = 1$. Find upper and lower bounds for $u(0, 0)$. (*Hint:* Consider $v(x, y) = u(x, y) + \frac{1}{4}(x^2 + y^2)$.)

4.5E By considering an appropriately deflected square membrane, find the harmonic function in the triangular region bounded by the lines $y = 0$, $x = \pi$, and $y = x$, such that $u(\pi, y) = y$, $0 < y < \pi$, and for $0 < x < \pi$: either (1) $u(x, 0) = u(x, x) = 0$, or (2) $u(x, 0) = (\partial u/\partial n)(x, x) = 0$.

4.5F* 1. Show that if $0 < y < \ell$: $\lim_{n\to\infty} \sinh ny \sinh n(\ell - y)/\sinh n\ell = 1$.

2. Substitute (23) into (25′) and use part 1 to conclude that $G(x, y; x, y)$ behaves like the series

$$\sum_{n=1}^{\infty} \frac{\sin^2 nx}{n} \qquad \text{when} \qquad 0 < x < \pi, \quad 0 < y < \ell.$$

3. Rewrite the last series in the form

$$\sum_{n=1}^{\infty}\frac{1}{2n}-\sum_{n=1}^{\infty}\frac{\cos 2nx}{2n}$$

and use a result from Problem 1.8E to explain why its sum is $+\infty$.

4.5G 1. For fixed (ρ,ϕ) show that $\log D = \log|(r,\theta)-(\rho,\phi)|$ is harmonic as a function of the polar coordinates (r,θ), when $(r,\theta)\neq(\rho,\phi)$. (*Hint:* Use translation.)

2. When $\rho > 0$, verify that the right side of (31) can be expressed as

$$H=\frac{1}{4\pi}\left\{\log\rho^2+\log\left|(r,\theta)-\left(\frac{1}{\rho},\phi\right)\right|\right\}.$$

3. Explain how these arguments establish the harmonicity of H when $r<1$ for fixed $\rho<1$ and ϕ.

4.5H 1. Verify the identity $\sinh(A-B)=\sinh A\cosh B-\cosh A\sinh B$ and use it to show how (23) follows from (18)–(19′).

2. Verify the identity $\cosh 2A = 1+2\sinh^2 A$ and use it together with that in part 1 to obtain (20).

4.5I* 1. Show that in (27), R_n satisfies the equation

$$(rR_n')'-\frac{n^2}{r}R_n=rf_n,\qquad r>0.\quad \text{(See (26).)}$$

2. To obtain formulas (28), use the Green function method of Appendix B.2 with

$$\alpha(r)=1 \text{ and } \beta(r)=-\log r,\qquad \text{when } n=0,$$

and

$$\alpha(r)=\frac{r^n}{2n} \text{ and } \beta(r)=r^n-r^{-n},\qquad \text{when } n=1,2,\ldots .$$

4.6*

Static Equilibrium of Elastic Solids

When we stretch or twist solid bodies with sufficient force, they deform, but some bodies (termed *elastic*) recover their original shape as the applied forces are removed. One such elastic body is the membrane considered in Section 4.2, where we used an energy principle to characterize its static equilibrium state. In this section we extend our energy approach and obtain the equilibrium conditions inside an elastic solid that is slightly deformed from an unstressed state.

We suppose that the undeformed body occupies a closed region $\bar{V}$ of $\mathbb{R}^3$, as shown in Figure 4.16 (p. 186), and that each point $\mathbf{x}=(x_1,x_2,x_3)$ in $\bar{V}$ undergoes a small displacement, $\mathbf{u}=(u_1,u_2,u_3)(\mathbf{x})$. Then, in general, as the body is deformed

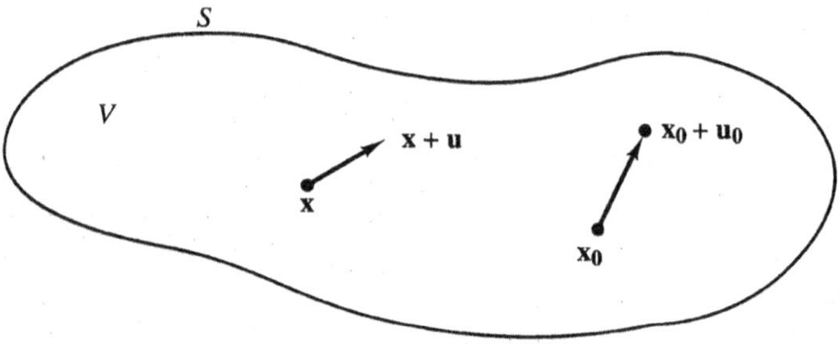

Figure 4.16

it experiences internal *strains*

$$e_{ij} = e_{ji} = \frac{1}{2}\left(\frac{\partial u_i}{\partial x_j} + \frac{\partial u_j}{\partial x_i}\right), \qquad i, j = 1, 2, 3, \tag{34}$$

all of which vanish when $\mathbf{u}$ is a rigid-body motion (see Problem 4.6D). The $e_{ii} = \partial u_i/\partial x_i$, $i = 1, 2, 3$ are called **longitudinal strains** and the other e_{ij} are called **shear strains**.

Suppose the material is homogeneous and isotropic and is acted on by a distributed body force $\mathbf{F} = (F_1, F_2, F_3)(\mathbf{x})$ (per unit volume) such as that of gravity. Then the associated potential energy is given by

$$E(u) = \int_V \left[\frac{\lambda}{2}\theta^2 + \mu \sum_{i,j=1}^{3} e_{ij}^2 - \mathbf{F}\cdot\mathbf{u}\right] dV \tag{35}$$

where $\theta = \nabla \cdot \mathbf{u} = e_{11} + e_{22} + e_{33}$, known as the *dilatation*, measures local expansion or contraction, and λ and μ are the positive material constants of Lamé. In (35), the first terms represent the *strain energy of deformation*. The last term is the work done against the body force $\mathbf{F}$ during the displacement $\mathbf{u}$ that increases the potential energy.

By **Bernoulli's principle**, for given $\mathbf{F}$, the body is in equilibrium when its displacements *minimize* the potential energy as defined above among all those that have the *same* values as $\mathbf{u}$ at the boundary S of V. (We assume that V is a simple domain, that all displacement functions are in $C^2(\bar{V})$, and that each $F_i \in C(\bar{V})$.)

Then (as in Section 4.2) we wish to determine an equilibrium displacement $\mathbf{u}$ such that if $\tilde{\mathbf{u}}$ is *any* displacement that vanishes on S, we have

$$\Delta E \overset{\text{def}}{=} E(\mathbf{u} + \tilde{\mathbf{u}}) - E(\mathbf{u}) \geq 0.$$

To simplify computations, note that when we replace $\mathbf{u}$ by $\mathbf{u} + \tilde{\mathbf{u}}$, then we replace each component u_i by $u_i + \tilde{u}_i$, and so by (34), each e_{ij} by $e_{ij} + \tilde{e}_{ij}$, and finally $\theta = e_{11} + e_{22} + e_{33}$ by $\theta + \tilde{\theta}$, where $\tilde{e}_{ij}$ and $\tilde{\theta}$ refer to $\tilde{\mathbf{u}}$. Then using (35) twice and combining terms, we see that ΔE can be obtained by integrating

$$\frac{\lambda}{2}[(\theta + \tilde{\theta})^2 - \theta^2] = \frac{\lambda}{2}\tilde{\theta}^2 + \lambda\theta\tilde{\theta} \quad (\geq \lambda\theta\tilde{\theta}), \tag{35'}$$

a sum of similar terms obeying corresponding inequalities,

$$\text{and} \quad -\mathbf{F}\cdot[(\mathbf{u} + \tilde{\mathbf{u}}) - \mathbf{u}] = -\mathbf{F}\cdot\tilde{\mathbf{u}}.$$

Consequently,

$$\Delta E \geq \int_V \left[\lambda\theta\tilde{\theta} + 2\mu \sum_{i,j=1}^{3} e_{ij}\tilde{e}_{ij} - \mathbf{F}\cdot\tilde{\mathbf{u}}\right] dV. \tag{36}$$

Now in this integrand, the terms involving $\tilde{u}_i$ only are of the form $\mathbf{T}_i \cdot \nabla\tilde{u}_i - F_i\tilde{u}_i$ for a certain vector function $\mathbf{T}_i$, which is *independent* of $\tilde{\mathbf{u}}$. The last expression may be written

$$\nabla\cdot(\tilde{u}_i\mathbf{T}_i) - (F_i + \nabla\cdot\mathbf{T}_i)\tilde{u}_i,$$

and when the divergence theorem is applied to the integral of the first term the resulting surface integral vanishes (with $\tilde{u}_i$). The remaining integral also vanishes provided that $\mathbf{u}$ is chosen to make

$$F_i + \nabla\cdot\mathbf{T}_i = 0, \tag{37}$$

and with proper identification of $\mathbf{T}_i$ for each $i = 1, 2, 3$, these are the desired equilibrium equations. They can be put in the attractive linear form of Navier (see Problem 4.6A),

$$\mu\nabla^2\mathbf{u} + (\lambda + \mu)\nabla\theta = -\mathbf{F} \tag{37$'$}$$

but they remain a system of second-order equations, coupled through the presence of θ.[†] In special cases, they can be simplified, but their solution with prescribed boundary conditions remains a formidable task. (See [So] and the works of Timoshenko.)

The components of the $\mathbf{T}_i$ are given by

$$\tau_{ij} = 2\mu e_{ij} + \lambda\theta\delta_{ij}, \qquad i, j = 1, 2, 3, \tag{38}$$

and they form the **stress-tensor** $\mathbb{T}$. The τ_{ii} are the **tensile** (or compressive) stresses and the rest are called **shear** stresses. These stresses measure the force per unit area by which the elastic body resists local deformations of the associated types. If the body is inhomogeneous, λ and μ become functions of position, whereas if it is anisotropic, λ and μ should be replaced by appropriate positive-definite matrices. In either case equation (37) is modified accordingly.

In a given material, the relation between stress and strain during controlled deformation can be determined empirically, and the foregoing analysis is limited to materials generating linear stress–strain laws such as (38). Most materials retain linearity only with small deformations, but a few, such as synthetic rubber, exhibit nonlinear behavior throughout their deformation range. (See [Gu] or [Er] for further discussion of nonlinear elasticity.) Finally, stress–strain relations such as (38) do not apply to plastic materials, which can sustain permanent deformation without internal stress.

Although it is quite difficult to obtain nontrivial solutions $\mathbf{u}$ to equations (37$'$) having prescribed boundary values, our approach yields immediately the uniqueness of such solutions under the assumed positivity of λ and μ. (General uniqueness in linear elasticity is treated in [K-P].)

---

† They uncouple in cases of pure shear deformation where $\theta = 0$, and then each component u_i of $\mathbf{u}$ satisfies Poisson's equation.

(4.4) Theorem (*Kirchhoff*): *Let V be a simple domain of $\mathbb{R}^3$ with boundary S. Then there is at most one solution* $\mathbf{u}$ *to the boundary value problem*

$$\mu\nabla^2\mathbf{u} + (\lambda+\mu)\nabla\theta = -\mathbf{F} \qquad \text{in } V,$$

$$\text{with} \qquad \mathbf{u}|_S = \mathbf{f} \qquad \text{prescribed.}$$

Proof: Let $\mathbf{u}$ be one solution, and $\mathbf{u}+\tilde{\mathbf{u}}$ be another. Since both $\mathbf{u}$ and $\mathbf{u}+\tilde{\mathbf{u}}$ minimize E as given by (35), it follows that $\Delta E = E(\mathbf{u}+\tilde{\mathbf{u}}) - E(\mathbf{u}) = 0$. Then *both* sides of the inequality in (36) are zero, which means that $\int_V \lambda\tilde{\theta}^2\,dV = 0$ (look at (35′)); therefore $\tilde{\theta} = 0$ in V since $\lambda > 0$.

But $\tilde{\mathbf{u}}$ satisfies (37′) with $\mu > 0$ and $\mathbf{F} = \mathbf{0}$ (Why?), so that each of its components $\tilde{u}_i$ is *harmonic* in V and vanishes on S. Consequently, $\tilde{u}_i \equiv 0$ in V (Theorem 3.5) so that $\tilde{\mathbf{u}} \equiv \mathbf{0}$ as required. ∎

Problem Set 4.6

4.6A 1. By careful inspection of the integrand of (36), verify that the terms involving $\tilde{u}_1$ only can be expressed as

$$\tau_{11}\frac{\partial\tilde{u}_1}{\partial x_1} + \tau_{12}\frac{\partial\tilde{u}_1}{\partial x_2} + \tau_{13}\frac{\partial\tilde{u}_1}{\partial x_3} - F_1\tilde{u}_1$$

where τ_{1j} is given by (38), $j = 1,2,3$. (*Note:* $\delta_{ij} = 1$ if $i = j$ and is zero otherwise.)

2. Show that when $i = 1$, (37) can be placed in the form (37′). (Similar analysis is possible when $i = 2$ and when $i = 3$.)

4.6B If μ and λ vary with position, then what form do equations (37′) take? (Equation (36) is unchanged.)

4.6C In the analysis given, the effects of distributed external surface forces $\mathbf{f}$ were not of significance. To take them into account, we should add their work $-\int_S \mathbf{f}\cdot\mathbf{u}\,dS$ to the potential energy (35) and now seek $\mathbf{u}$ to make ΔE (as modified) nonnegative among all displacements $\tilde{\mathbf{u}}$.

1. Show that as modified

$$\Delta E \geq \sum_{i=1}^{3}\left\{\int_V \nabla\cdot(\tilde{u}_i\mathbf{T}_i)\,dV - \int_S f_i\tilde{u}_i\,dS - \int_V (F_i + \nabla\cdot T_i)\tilde{u}_i\,dV\right\}$$

2. Apply the divergence theorem to the first integral in part 1 to conclude that the new potential energy is minimized by any $\mathbf{u}$ that satisfies the former equilibrium equations (37) within V, together with the (natural) boundary conditions.

$$\mathbf{T}_i\cdot\mathbf{n}|_S = f_i, \qquad i = 1,2,3.$$

3. Prove that within a rigid-body motion there can be at most one solution $\mathbf{u}$ to the boundary value problem just formulated. (*Hint:* As in the proof of 4.4, reduce this to a consideration of the strains $\tilde{e}_{ij}$ for the difference $\tilde{\mathbf{u}}$ of two solutions. Then see the next problem.)

4.6D* Suppose that all strain components e_{ij} vanish in V. Prove that the associated displacement $\mathbf{u}$ corresponds to that for a rigid-body motion. (*Hint:* Assume that the origin $\mathbf{O} \in V$ and compare $\mathbf{u}(\mathbf{x})$ with $\mathbf{u}(\mathbf{O})$ by means of a line integral in coordinate directions. Suppose that $\mathbf{u}$ is C^3 and integrate by parts as necessary.)

4.6E* (*Torsion of a rectangular column*) A steel column of constant rectangular cross section is fixed at its lower end, while its upper end is twisted (by an external applied torque) into a new position and held fast there. If no other forces are present, then it can be shown that a typical intermediate cross section $\bar{V}$ (in the xy-plane) is deformed longitudinally into a new surface having vertical deflection $z = \alpha u(x, y)$. Here, α is the angle of twist per unit length, and u is chosen to minimize the associated strain–energy function

$$E(u) = \frac{1}{2}\int_V \tau[(u_x - y)^2 + (u_y + x)^2]\, dV$$

over all $u \in C^1(\bar{V})$. $0 < \tau \in C^1(\bar{V})$ permits nonhomogeneity of the bar over its cross sections. (See [So], 109–111.)

1. Assume $\tau \equiv 1$ and use techniques similar to those employed in deriving Theorem 4.1 to show that if $u \in C^2(\bar{V})$ and $v \in C^1(\bar{V})$, then

$$E(u+v) - E(u) \geq -\int_V v\nabla^2 u\, dV + \int_S v\left[\frac{\partial u}{\partial n} - \boldsymbol{\sigma}\cdot\mathbf{n}\right] dS,$$

where $\boldsymbol{\sigma} \stackrel{\text{def}}{=} (y, -x)$ is a vector function on S, and equality holds iff v is constant in V.

2. Conclude that each u as in part 1, which solves the boundary value problem

$$\nabla^2 u = 0, \qquad \text{in } V$$

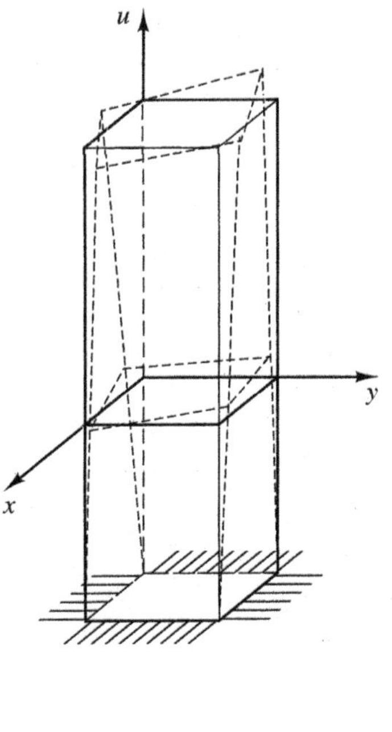

with $(\partial u/\partial n)|_S = \boldsymbol{\sigma}\cdot\mathbf{n}$, will minimize E and provide the torsion function (uniquely within an additive constant).

3. Show that if S is parameterized by its arc length s, then

$$\boldsymbol{\sigma}\cdot\mathbf{n} = \frac{1}{2}\frac{d}{ds}[x^2(s) + y^2(s)],$$

and conclude that for this problem, the Neumann compatibility condition (of Section 4.2)

$$\int_S \frac{\partial u}{\partial n}\,dS = 0$$

is fulfilled. Can you establish this condition directly?

4.* Repeat the preceding analysis when τ is not constant and conclude that part 2 holds if $\nabla\cdot(\tau\nabla u) = y\tau_x - x\tau_y$ in V with $(\partial u/\partial n)|_S = \boldsymbol{\sigma}\cdot\mathbf{n}$ as before.

5. If $V = \{(x,y)\colon\ -1 < x < 1, -1 < y < 1\}$, show that $\boldsymbol{\sigma}\cdot\mathbf{n}$ has values

$$\begin{cases} \pm y, & \text{on } x = \pm 1 \\ \pm x, & \text{on } y = \mp 1 \end{cases}$$

and explain how you would attack the problem in part 2 using $u_0(x,y) = xy$.

6. Suppose V were the disk $\{(x,y)\colon\ 0 \le x^2 + y^2 \le 1\}$. What is $\boldsymbol{\sigma}\cdot\mathbf{n}$? What can you conclude in part 2?

5

Propagation Problems: The Wave Equation

A local disturbance (or signal) propagating naturally within a medium at *finite speed* is referred to as a **wave** because of the form of propagation in water. We create visible waves by dropping a stone into a still pool of water or by snapping one end of a long rope. We produce invisible waves when we generate sounds that propagate in air or electrical signals that transmit along a wire. Actually, we can describe the sound-generating vibrations of a guitar string or drumhead in terms of traveling distortion waves in these media.

Now, wave motion is not just another type of diffusion process, for in diffusion, some effects are transmitted instantaneously throughout the medium. In fact, a wave carries energy whose motion is subject to the laws of dynamics. In this chapter we will extract the equation governing wave motion from Newton's laws as expressed in Hamilton's principle. The resulting wave equation permits both *standing-wave* solutions modeling vibrations in elastic media (Sections 5.1–5.4) and *traveling-wave* solutions modeling propagation (Sections 5.5–5.6).†

Musical instruments, the human voice, and loudspeakers all generate sound by exciting vibrations — rapid oscillatory motions — in some underlying mechanism. To describe this behavior accurately, natural philosophers have always used mathematics. As early as 500 B.C. the Pythagorean school sought numerical relations between the length of a vibrating harp string and the pitch of sound it produces. Then in 1747, to predict the motion of a plucked harp string, d'Alembert produced the first important partial differential equation (the one-dimensional wave equation), together with a method of solution (see Section 5.5). In 1753 Daniel Bernoulli proposed solving this equation by an infinite superposition that used elements of a Fourier series. Then he, Euler, and Lagrange extended these analyses to cover the sounds produced by organ pipes and other wind instruments. They also introduced mathematical models for vibrating membranes (such as drumheads) and other elastic media.

† This wave equation is linear, but some non–linear versions are examined in Section 5.7.

5.1

Membrane Dynamics

Think of the stretched membrane of Section 4.2 as a drumhead — perhaps of an unconventional shape. After it is struck, it is set in motion, which might be specified by its vertical deflection

$$u(\mathbf{x}, t) \qquad \text{at position} \qquad \mathbf{x} \in V \qquad \text{and time } t,$$

where V is a simple planar domain with boundary S as shown in Figure 5.1. $u_t(\mathbf{x}, t)$ is the associated velocity, and we let $F = F(\mathbf{x}, t)$ represent the pressures on the membrane arising from an externally distributed vertical loading or a forcing mechanism. The membrane is assumed to have local (areal) mass density $\rho = \rho(\mathbf{x}, t)$ and distributed tensile resistance function $\tau = \tau(\mathbf{x}, t)$, both positive but permitted the time variations that occur when a drumhead is tightened while it is vibrating.

Then at time t, the membrane has kinetic energy

$$T(u) \stackrel{\text{def}}{=} \frac{1}{2}\int_V \rho u_t^2 \, dV \tag{1}$$

and potential energy

$$P(u) \stackrel{\text{def}}{=} \int_V \left(\tfrac{1}{2}\tau\,|\nabla u|^2 - Fu\right) dV. \tag{2}$$

The first integral term in (2) gives the potential energy of strain against the distributed tension τ, simplified as in the static analysis of Section 4.2. The second term represents the work done by the applied pressure F in displacing the membrane. We assume further that functions appearing in these and subsequent expressions are so smooth that all integrands involved are continuous.

During motion, the membrane may be regarded as a constrained system of particles exchanging kinetic and potential energies. In 1834, Hamilton postulated that such systems actually move along the trajectories that make this exchange efficient in the following sense:

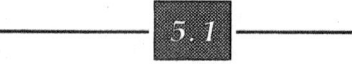

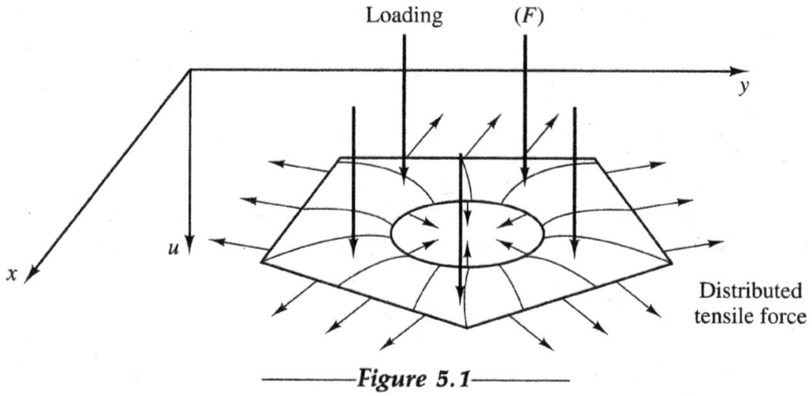

Figure 5.1

(5.1) Hamilton's Principle: *Between times a and b at which positions are specified, a system moves so as to give a* **stationary** *value to its* **action integral**

$$A(u) \stackrel{\text{def}}{=} \int_{a}^{b} [T(u) - P(u)]\, dt. \tag{3}$$

■

We will use this principle to show that during actual motion, u satisfies the *wave equation* (6), which takes the simplified forms (7) and (8). On first reading, you may wish to skip the supporting arguments.

The action integral (3) provides a measure of the energy exchange during some possible motion of the membrane. We seek displacement functions u for which this exchange is optimal in comparison with that of "nearby" motions. Specifically, we want A to have a **stationary value** at u. If we think of the action integral as defining a "surface" lying over the domain of possible displacements, as indicated in Figure 5.2, then A is stationary at u when a very small marble placed on the surface at the point over u will not roll in any "direction" v but remains stationary. Observe that stationarity at u holds if A has either a local maximal value or a local minimal value at u, but it can also hold under other conditions.

Mathematically, we will say that A is **stationary** at u if

$$\lim_{\epsilon \to 0} \frac{A(u + \epsilon v) - A(u)}{\epsilon} = 0, \tag{4}$$

when $u + \epsilon v = u(\mathbf{x}, t) + \epsilon v(\mathbf{x}, t)$, and $v = v(\mathbf{x}, t)$ is any C^2 function that vanishes identically outside some small neighborhood of an arbitrarily chosen fixed point $(\mathbf{x}_1, t_1) \in \mathcal{V} = V \times (a, b)$, as illustrated in Figure 5.3 (p. 194).

Hamilton's principle provides a large class of necessary conditions (one for each test function v) that the actual motion u must satisfy. The task is to show that these conditions, in some sense, determine u. To apply the principle, we form $T(u + \epsilon v)$, and $P(u + \epsilon v)$ from (1) and (2), respectively, and substitute in (3) to

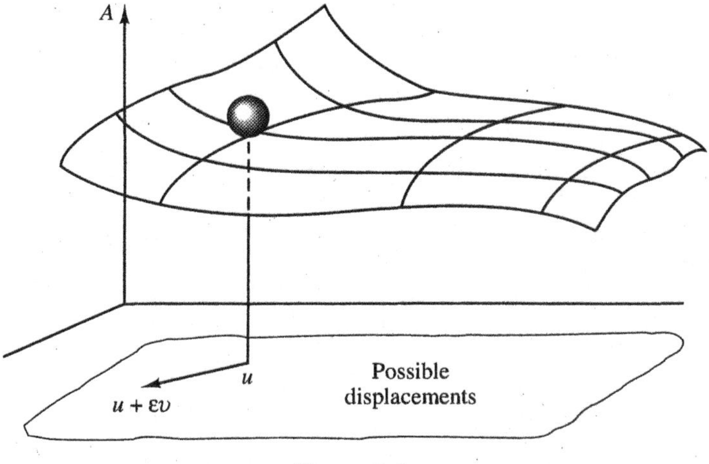

Figure 5.2

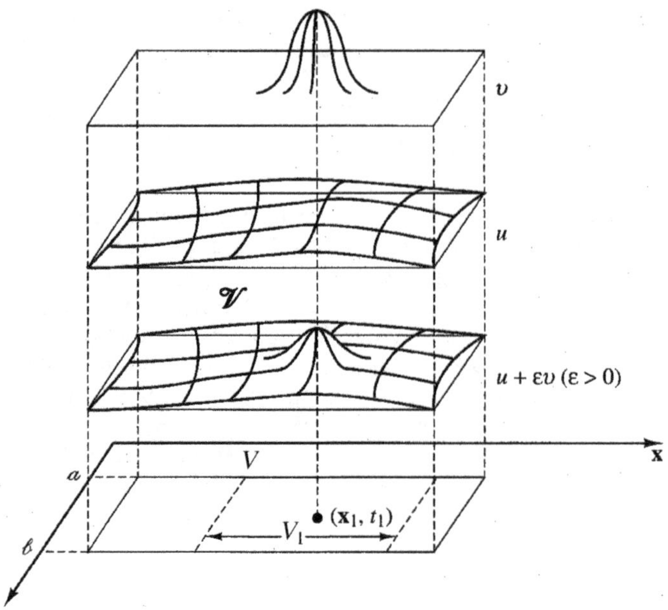

Figure 5.3

obtain

$$A(u+\epsilon v) = \frac{1}{2}\int_a^b \int_V [\rho(u_t + \epsilon v_t)^2 - \tau|\nabla u + \epsilon \nabla v|^2 + 2F(u+\epsilon v)]\, dV\, dt.$$

Then, we subtract $A(u)$ and divide by ϵ ($\neq 0$) to get (after cancellations)

$$\frac{A(u+\epsilon v) - A(u)}{\epsilon} = \frac{\epsilon}{2}\int_a^b \int_V (\rho v_t^2 - \tau|\nabla v|^2)\, dV\, dt$$
$$+ \int_a^b \int_V (\rho u_t v_t - \tau \nabla u \cdot \nabla v + Fv)\, dV\, dt.$$

Hence, in the limit as $\epsilon \to 0$, we conclude from (4) that

$$0 = \int_a^b \int_V (\rho u_t v_t - \tau \nabla u \cdot \nabla v + Fv)\, dV\, dt. \tag{5}$$

Moreover, in (5) we can replace V by a small box $V_1 \subseteq V$ containing $(\mathbf{x}_1, t_1)$, outside of which v and its derivatives vanish identically.

Next, assuming that u is C^2 while ρ and τ are C^1, we use partial integrations on the reduced integral to obtain

$$0 = \int_a^b dt \frac{d}{dt}\int_{V_1} (v\rho u_t)\, dV - \int_a^b dt \int_{V_1} \nabla \cdot (v\tau\nabla u)\, dV$$
$$- \int_a^b \int_{V_1} v[(\rho u_t)_t - \nabla \cdot (\tau \nabla u) - F]\, dV\, dt.$$

Through the fundamental theorem of calculus and the divergence theorem, the first two integral terms become

$$\int_{V_1} (v\rho u_t)\, dV \Big|_{t=a}^{t=b} - \int_a^b dt \int_{S_1=\partial V_1} v\tau \frac{\partial u}{\partial n}\, dS,$$

and each vanishes by our assumptions on v. It follows that

$$0 = \int_a^b \int_{V_1} v[(\rho u_t)_t - \nabla \cdot (\tau \nabla u) - F]\, dV\, dt \tag{5'}$$

for all such functions v. A modification of previous arguments shows that the bracketed term (assumed continuous) cannot be positive at the point $(\mathbf{x}_1, t_1)$, since otherwise an appropriately chosen v would yield a positive integral in (5′). Similarly, it cannot be negative and so must vanish at this point. But since this point is arbitrary, we conclude that inside $V \times (a, b)$, u must satisfy the **wave equation**

$$(\rho u_t)_t = \nabla \cdot (\tau \nabla u) + F. \tag{6}$$

Let's state our results.

(5.2) Theorem: *Let $\mathscr{V} = V \times (a, b)$, where V is a simple domain of $\mathbb{R}^d$. If u_0 makes the action integral*

$$A(u) = \frac{1}{2} \int_a^b \int_V (\rho u_t^2 - \tau |\nabla u|^2 - 2Fu)\, dV\, dt$$

stationary among all those $u \in C^2(V)$ that have the same values as u_0 at $t = a$ and $t = b$, then u_0 will be a solution of (6) *in $\mathscr{V}$.* ■

(5.3) Corollary: *During vertical motion, the loaded stretched membrane of* Section 4.2 *has displacement $u = u(\mathbf{x}, t)$ governed by equation* (6) *in V at each time t.* ■

Now, (6) represents Newton's equation of motion for a membrane element, but our variational derivation avoids having to consider interactive forces from adjacent elements. A nonvariational approach is taken in [Wein].

As usual with such equations, various simplifications are possible. For example, if $\rho_t = 0$, we obtain

$$\rho u_{tt} = \nabla \cdot (\tau \nabla u) + F, \tag{7}$$

which still permits membranes of nonuniform thickness. If both ρ and τ are constant and $F = 0$, we get the **standard wave equation**

$$u_{tt} = c^2 \nabla^2 u, \tag{8}$$

where $c = (\tau/\rho)^{1/2}$ necessarily has the dimensions of speed.

Although the above model is two-dimensional spatially, the derivation is not, and

$$u_{tt} = c^2 u_{xx} \tag{8'}$$

is the appropriate one-dimensional equation for the vertical motion of a uniform elastic string stretched horizontally between two points, such as a cable under tension. This wave equation also governs the longitudinal vibrations of a metal rod or spring. (See Problem 5.1E.) The three-dimensional version is appropriate for describing the compression of air under small perturbations corresponding to sound waves (see Section 6.4).

Problem Set 5.1

5.1A 1. When $\bar{V} = [0, \ell]$ (and $dV = dx$), verify the derivation of the one-dimensional form of (5) from (1) through (4).

2. For small $\delta > 0$,

$$v(x,t) = \begin{cases} \cos^2\left(\dfrac{x - x_1}{2\delta}\right)\cos^2\left(\dfrac{t - t_1}{2\delta}\right), & \begin{array}{l}|x - x_1| \leq \pi\delta \\ |t - t_1| \leq \pi\delta\end{array} \\ 0, & \text{otherwise} \end{cases}$$

is a C^1 function, which is positive near (x_1, t_1).
Explain how such functions could be used to derive the one-dimensional form of (6) from the one-dimensional form of (5).

5.1B 1. Write the two-dimensional form of (8) when $u = u(x, y, t)$.

2. Write this same equation when $u = u(r, \theta, t)$, where r, θ are polar coordinates.

5.1C Verify that the scale changes $\bar{x} = x/\ell$, $\bar{t} = ct/\ell$ transforms (8′) to $\bar{u}_{\bar{t}\bar{t}} = \bar{u}_{\bar{x}\bar{x}}$ where $\bar{u}(\bar{x}, \bar{t}) = u(\ell\bar{x}, \ell\bar{t}/c)$. This shows that when desired, we can assume $c = 1$ in (8′).

5.1D To analyze small transverse planar motions of a chain or cable of length ℓ suspended vertically from one end, note that the tension τ at height x above the *free end* is given by $\tau = \tau(x) = g\rho x$ where ρ is the mass/unit length of the chain and g is the gravitational constant. Explain why the resulting horizontal deflection $u = u(x, t)$ is governed by the equation

$$u_{tt} = g(xu_x)_x.$$

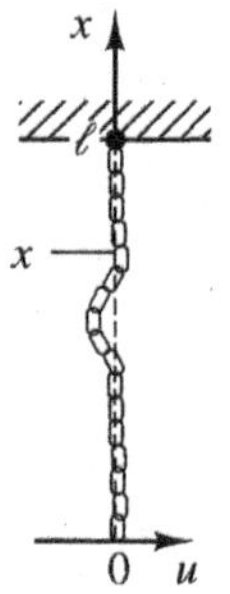

5.1E After a long straight horizontal elastic metal bar of length ℓ is struck at one end, compression waves travel longitudinally *along* the bar. To model this mathematically, suppose that the local longitudinal deflection of a cross section from its equilibrium position is given by a function $u = u(x, t)$, $0 < x < \ell$. Then, nonconstant u (in x) results in local longitudinal strains and proportional stresses. It can be shown that the resulting strain energy at time t is given approximately by

$$P(u) = \frac{1}{2}\int_0^\ell \tau u_x^2 \, dx$$

for an appropriate positive function $\tau = \tau(x, t)$, while the corresponding kinetic energy is $T(u) = \frac{1}{2}\int_0^\ell \rho u_t^2 \, dx$ for a linear density $\rho = \rho(x)$.

1. Apply Hamilton's principle to the relevant action integral and conclude that motions u that occur satisfy a one-dimensional wave equation.

2. Give interpretations for Dirichlet, Neumann, and Robin conditions at the end $x = 0$.

5.1F In considering small vertical dynamical deformations $w = w(x, t)$ of the bar in the previous problem, we can usually neglect the strain energy of stretching in comparison with that of bending, which at time t is approximated by

$$P(w) = \frac{1}{2}\int_0^\ell \mu w_{xx}^2\, dx$$

for an appropriate material stiffness function $\mu = \mu(x, t)$. (See [Tr], Section 6.6.) Such a bar is called a beam.

1. Assume that the associated kinetic energy at time t is given by

$$T(w) = \frac{1}{2}\int_0^\ell \rho w_t^2\, dx,$$

and apply Hamilton's principle to the action integral

$$A(w) = \int_a^b [T(w) - P(w)]\, dt$$

to show that for functions v that vanish outside a neighborhood of a point (x_1, t_1),

$$0 = \int_a^b dt \int_0^\ell dx[\rho w_t v_t - \mu w_{xx} v_{xx}].$$

2. Integrate by parts as required to get

$$0 = \int_a^b dt \int_0^\ell dx[(\rho w_t)_t + (\mu w_{xx})_{xx}]v.$$

3. Conclude that the resulting equation of motion is $(\rho w_t)_t = -(\mu w_{xx})_{xx}$, and discuss some simpler versions.

5.1G* (*Dynamic deformation of elastic solids*) When the elastic body of Section 4.6 is undergoing small deformations over a period of time, perhaps during vibrations, then Hamilton's principle can be employed to obtain the equations of motion. Show that these equations take the form

$$((\rho u_i)_t)_t = F_i + \nabla \cdot \mathbf{T}_i, \qquad i = 1, 2, 3$$

or (as in (37′) of Section 4.6)

$$(\rho \mathbf{u}_t)_t = \mu \nabla^2 \mathbf{u} + (\lambda + \mu)\nabla\theta + \mathbf{F}.$$

5.2

Auxiliary Conditions: Uniqueness

Now let's use uniqueness considerations to help us choose appropriate sets of auxiliary conditions for boundary value problems involving equations (7) or (8). If we wish, we can continue to think about the membrane, but we will use energy arguments that are independent of spatial dimension.

(5.4) Theorem *(Uniqueness): Let $\rho \in C(\bar{V})$ and $\tau \in C^1(\bar{V})$ be positive on V, a simple domain of $\mathbb{R}^d$. Then for given $F = F(\mathbf{x}, t)$, there is at most one C^2 solution $u = u(\mathbf{x}, t) \in C^1(\bar{V})$ to the boundary value problem for the equation*

$$\rho u_{tt} = \nabla \cdot (\tau \nabla u) + F, \text{ in } \mathcal{V} = V \times (0, \infty)$$

with prescribed initial conditions $u|_{t=0}$ and $u_t|_{t=0}$ and boundary conditions in which $u|_S$, or $(\partial u/\partial n)|_S$, or $[(\partial u/\partial n) + hu]|_S$ (for $h > 0$, $h_t = 0$), or a mixture thereof is prescribed on S, the boundary of V at each $t > 0$.

Proof: As usual, let u be the difference of two solutions to the same problem. Then u satisfies the *homogeneous equation*

$$\rho u_{tt} = \nabla \cdot (\tau \nabla u) \tag{9}$$

with zero auxiliary conditions. We shall establish that $u \equiv 0$ by showing that at time $t > 0$ the associated energy integral

$$E(t) = \frac{1}{2}\int_V [\rho u_t^2 + \tau |\nabla u|^2]\, dV \tag{10}$$

vanishes. Upon formal differentiation, (10) yields

$$E'(t) = \int_V [\rho u_t u_{tt} + \tau \nabla u \cdot (\nabla u)_t]\, dV.$$

However, $(\nabla u)_t = \nabla(u_t)$ since, e.g., $(u_x)_t = (u_t)_x$ when u is C^2 in these variables, and we can use a standard vector identity to show that

$$E'(t) = \int_V \nabla \cdot (\tau u_t \nabla u)\, dV + \int_V [\rho u_{tt} - \nabla \cdot (\tau \nabla u)]\, u_t\, dV.$$

The second integral vanishes in view of (9), and the first transforms by the divergence theorem, so that

$$E'(t) = \int_S \tau u_t \frac{\partial u}{\partial n}\, dS.$$

Clearly $E'(t) = 0$ if $u|_S = 0$, or $(\partial u/\partial n)|_S = 0$, or a mixture of these conditions is present. Then $E(t) = \text{const.} = E(0) = 0$, since at $t = 0$, both u and u_t vanish identically. In these cases, it follows that at $t > 0$, the integrand in (10) must vanish, so that both u_t and $|\nabla u|$ vanish in V. But this implies that $u(\mathbf{x}, t)$ is constant $= u(\mathbf{x}, 0) = 0$ and completes the proof when Robin conditions are excluded. (When they are present, we can reach the same conclusion by modifying the expression for E; see Problem 5.2C.) These formal arguments can be made rigorous by approximation from within V. ■

Remarks: Here, the region V is assumed unchanging with time, but we consider uniqueness in other regions of space-time when we discuss characteristics in the next chapter. Observe that the boundary conditions are precisely those considered in Section 3.2 for diffusion problems, but we now require an additional initial condition. On the other hand, this uniqueness argument applies backward as well as forward in time from a reference time ($t = 0$), while that for diffusion does not.

Continuous dependence with respect to auxiliary data is difficult to establish, since the solutions of the wave equation do not satisfy a maximum principle. For example, $u(x,y,t) = x^2 + y^2 + 2t^2$ is a solution of $u_{tt} = \nabla^2 u$ in $\mathbb{R}^2$ for all t, which is larger at times $t \neq 0$ than it is at $t = 0$. This is consistent with physical experience: A drumhead *can* have a larger deflection at a later time than it has at time $t = 0$ or on its boundary.

Problem Set 5.2

5.2A When $V = (0, \ell)$ (and $dV = dx$), write out the details of the proof of the uniqueness theorem (5.4) for $u = u(x,t)$ (omitting the case of Robin conditions).

5.2B *1.* Show that for $t \geq 0$ the argument for uniqueness in the proof of Theorem 5.4 can be retained for solutions of the *damped* wave equation $L(u) \equiv \rho u_{tt} + \delta u_t - \nabla \cdot (\tau \nabla u) = 0$ with the homogeneous Dirichlet condition $u|_S = 0$, provided that $\delta = \delta(\mathbf{x}) > 0$.

2. What happens if other homogeneous boundary conditions are permitted?

5.2C To extend the proof of Theorem 5.4 to admit Robin conditions on S,

1. show that when $[(\partial u/\partial n) + hu]|_S = 0$, for some positive function h with $h_t = 0$, then

$$E'(t) = -H'(t), \qquad \text{where} \qquad 0 \leq H(t) \stackrel{\text{def}}{=} \frac{1}{2}\int_S \tau h u^2 \, dS.$$

2. conclude that $0 \leq E(t) \leq E(t) + H(t) = E(0) + H(0) = 0$ so that $u \equiv 0$ as before.

5.3

Natural Vibrations

There is a well-established connection between the production of sound and vibrating media. To examine simple vibrations of the elastic media whose motions can be described by the wave equation (7), we suppose first that the motions are not being forced ($F = 0$) and then consider those whose positions u have the time-oscillatory product form

$$u(\mathbf{x}, t) = \Phi(\mathbf{x}) \cos \omega t. \tag{11}$$

These motions will be periodic in time with frequency ω (radians/unit time), and they will have the initial velocity $u_t(\mathbf{x}, 0) = 0$, although other initial conditions can be achieved if ωt is replaced by $\omega t + \alpha$.

If we use (11), then

$$\rho u_{tt} - \nabla \cdot (\tau \nabla u) = -[\rho \omega^2 \Phi + \nabla \cdot (\tau \nabla \Phi)] \cos \omega t,$$

and we see that each such u gives a nontrivial solution to (7) (when $F = 0$), provided that Φ is a nontrivial solution of the **Helmholtz equation**

$$\nabla \cdot (\tau \nabla \Phi) + \rho \omega^2 \Phi = 0 \tag{12}$$

in some spatial domain V with boundary S.

If we require $\Phi|_S = 0$ or impose some other homogeneous boundary conditions on Φ, we get an eigenvalue problem for the so-called **natural frequency** ω. Each such ω is associated with a nontrivial solution Φ_ω of (12), and natural (unforced) vibrations can occur at frequency ω in the mode "shape" described by Φ_ω. In this manner the wave equation provides a mathematical model for the natural vibrations of certain elastic media.

Vibrating String

Suppose we want to study the vibrations of a harp string or a guitar string of length ℓ that is tightly stretched between fixed pegs but is otherwise free to move (see Figure 5.4). We can think of it as a one-dimensional membrane, and if its composition and size are uniform, we can assume that both its linear density ρ and its tensile resistance τ are constant. To describe its natural transverse vibrations, we seek product solutions for the one-dimensional wave equation

$$u_{tt} = c^2 u_{xx} \tag{13}$$

in the time-oscillatory form

$$u(x, t) = X(x) \cos \omega t, \tag{13'}$$

with $u(0, t) = u(\ell, t) = 0$. Observe that $u_t(x, 0) = 0$, $0 < x < \ell$.

Substitution in (13) yields the eigenvalue problem

$$X'' + (\omega^2/c^2)X = 0, \qquad (c^2 = \tau/\rho)$$

$$\text{with} \qquad X(0) = X(\ell) = 0,$$

which for appropriate ω has the solutions

$$X(x) = \sin (\omega/c) x.$$

Then $X(\ell) = \sin (\omega/c)\ell = 0$, iff $(\omega\ell/c) = n\pi$, $n = 1, 2, \ldots$, and this gives the *natural frequency*

$$\omega_n = c\frac{n\pi}{\ell} = \frac{n\pi}{\ell}\sqrt{\frac{\tau}{\rho}} = n\omega_1, \tag{14}$$

with its associated *mode shape*

$$X_n(x) = \sin \frac{n\pi x}{\ell}, \qquad n = 1, 2, \ldots .$$

The corresponding solutions

$$u_n(x, t) = \sin \frac{n\pi x}{\ell} \cos \omega_n t \tag{15}$$

describe the natural vibrations of the string in the characteristic sinusoidal modes. Note that $u_n(x, 0) = \sin (n\pi x/\ell)$. Thus if an ideal string is deformed initially in this

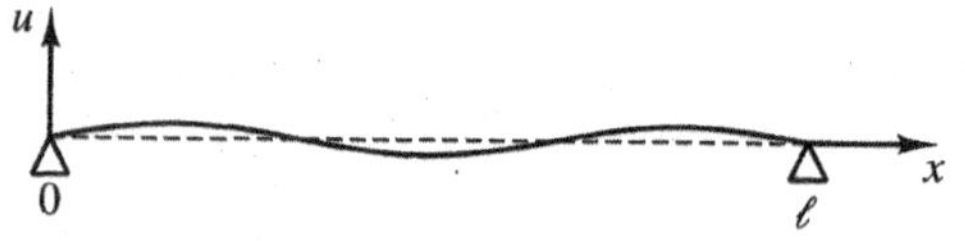

Figure 5.4

"pure" mode and released, it oscillates indefinitely at the corresponding natural frequency ω_n (because of precise interchange of kinetic and potential energies) and provides a pure tone. (An actual string is not perfectly elastic and these vibrations die out as time increases. See Section 5.4.)

Equation (14) shows that the higher natural frequencies are integral multiples of the fundamental frequency ω_1, a fact known to the Pythagorean school. The equation also implies that ω_1 increases as the length ℓ or the density ρ decreases, or as the tension τ increases. Experience confirms this behavior, and in 1636, Mersenne used experimental evidence to predict that ω_1 is proportional to $\sqrt{\tau/\rho}$. [K1].

Now, if the string is plucked, it is released in a shape $u(x,0) = f(x)$, different from a natural mode, and we hear overtones. This observation led Daniell Bernoulli in 1753 to seek a solution in the form of a series of the natural motions; viz.,

$$U(x,t) = \sum_{n=1}^{\infty} s_n \sin\frac{n\pi x}{\ell} \cos n\omega_1 t \tag{16}$$

where the s_n were chosen to make

$$f(x) = U(x,0) = \sum_{n=1}^{\infty} s_n \sin\frac{n\pi x}{\ell}, \qquad 0 < x < \ell.$$

As we now know, this Fourier series representation requires that we select

$$s_n = \frac{2}{\ell}\int_0^{\ell} f(x)\sin\frac{n\pi x}{\ell}\,dx, \qquad n = 1, 2, \ldots$$

and confront the issues of convergence presented in Section 1.5. However, Bernoulli's original approach supplies a physical model for visualizing Fourier series in action and understanding why vibration problems might be solved by the method of superposition.

Example 1:

The triangular shape of a string plucked at its midpoint (shown in Figure 5.5) is described within an amplitude factor by

$$f(x) = \begin{cases} x, & 0 \le x \le \ell' \\ \ell - x, & \ell' \le x \le \ell \end{cases} \qquad \text{where} \qquad \ell' = \ell/2.$$

This initial shape requires

$$s_n = \frac{2}{\ell}\left\{\int_0^{\ell'} x\sin\frac{n\pi x}{\ell}\,dx + \int_{\ell'}^{\ell}(\ell - x)\sin\frac{n\pi x}{\ell}\,dx\right\} \qquad (= 0, n \text{ even})$$

$$= \frac{4}{\ell}\int_0^{\ell'} x\sin\frac{n\pi x}{\ell}\,dx = \frac{4}{n\pi}\int_0^{\ell'}\cos\frac{n\pi x}{\ell}\,dx = \frac{4\ell}{n^2\pi^2}\sin\frac{n\pi}{2}, \qquad n = 1, 3\ldots$$

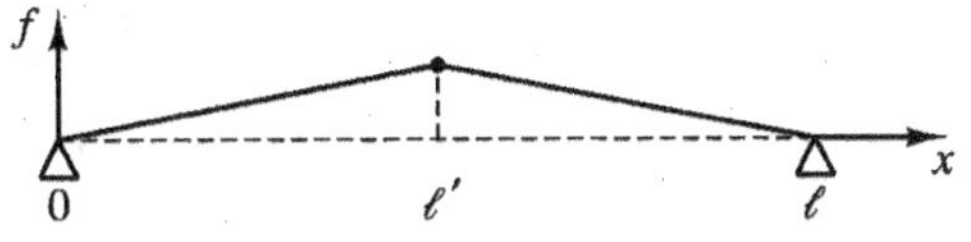

Figure 5.5

and yields the series

$$U(x,t) = \frac{4\ell}{\pi^2}\sum_{n=1}^{\infty}\frac{1}{n^2}\sin\frac{n\pi}{2}\sin\frac{n\pi x}{\ell}\cos n\omega_1 t, \tag{17}$$

which converges uniformly for all x, t. (Why?)

The first term in (17) has the largest amplitude, which explains why the string appears to vibrate predominantly in its fundamental mode at frequency ω_1. This string also produces overtones at the frequencies $3\omega_1, 5\omega_1, 7\omega_1, \ldots$ but not at the frequencies ω_n, $n = 2, 4, \ldots$. Unfortunately, the series in (17) cannot be twice differentiated termwise and retain convergence to guarantee that U represents an actual solution to the wave equation (13). However, each *approximation*

$$U_N(x,t) = \frac{4\ell}{\pi^2}\sum_{n=1}^{N}\frac{1}{n^2}\sin\frac{n\pi}{2}\sin\frac{n\pi x}{\ell}\cos n\omega_1 t$$

does satisfy all requirements exactly, except that for the initial deflection. Moreover,

$$|\, U_N(x,0) - f(x)\,| \le \frac{4\ell}{\pi^2}\sum_{n=N+1}^{\infty}\frac{1}{n^2} \le \frac{4\ell}{\pi^2 N} \qquad (0 \le x \le \ell), \qquad N = 1, 2, \ldots .$$

Example 2:

The torsional oscillations of a circular steel shaft (see Figure 5.6) are also governed by the wave equation $u_{tt} = c^2 u_{xx}$, where $u(x,t)$ is the angle of twist of the cross section at position x, and c^2 is a material constant. (See [Ba].)

When the shaft of length $\ell = \pi$ is fixed at both ends, we essentially have the problem just examined. However, when the ends are free, we must take

$$u_x(0,t) = u_x(\pi,t) = 0,$$

since there can be no torsional strain at these ends. This is the same eigenvalue problem as that for a stretched string of length $\ell = \pi$ with free ends, and the natural modes are $X_n(x) = \cos nx$, $n = 0, 1, 2, \ldots$. For the initial conditions $u_t(x,0) = 0$, $0 < x < \pi$, we consider formal series solutions of the form

$$U(x,t) = \frac{c_0}{2} + \sum_{n=1}^{\infty} c_n \cos nx \cos nct. \tag{18}$$

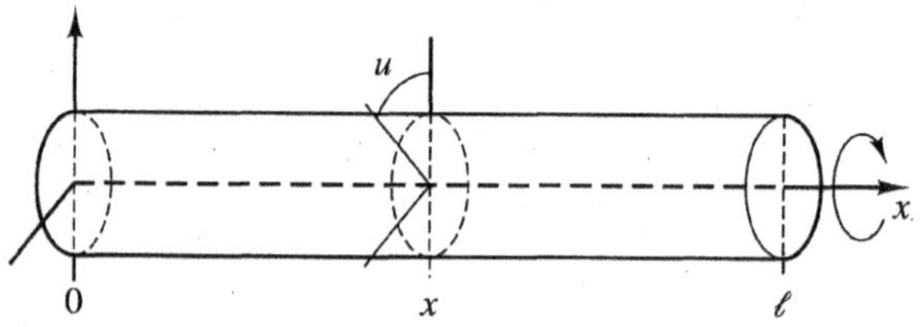

Figure 5.6

To have $U(x,0) = f(x)$, $0 < x < \pi$, we must take

$$c_n = \frac{2}{\pi}\int_0^{\pi} f(x)\cos nx\,dx, \qquad n = 0, 1, 2, \ldots;$$

when f is a polynomial and $n > 0$, then Kronecker's formula (Section 1.3) gives

$$c_n = \frac{2}{\pi}\left[f(x)\frac{\sin nx}{n}\Big\searrow_0 + f'(x)\frac{\cos nx}{n^2} - f''(x)\frac{\sin nx}{n^3}\Big\searrow_0 - f'''(x)\frac{\cos nx}{n^4} \pm \cdots\right]_0^{\pi}. \qquad \textbf{(19)}$$

If, say, $f(x) = (\pi x^2/2) - (x^3/3)$, so that $f'(x) = x(\pi - x)$, then $f'''(x) = -2$ and from (19) we get

$$c_n = \frac{4\cos nx}{\pi n^4}\bigg|_0^{\pi} = \frac{-8}{\pi n^4}, \qquad n = 1, 3,$$

$$= 0, \qquad n = 2, 4, \ldots$$

while $$c_0 = \frac{2}{\pi}\left(\frac{\pi x^3}{6} - \frac{x^4}{12}\right)\bigg|_0^{\pi} = \frac{\pi^3}{6}.$$

Thus (18) becomes

$$U(x,t) = \frac{\pi^3}{12} - \frac{8}{\pi}\sum_{n=1,3,}^{\infty}\frac{1}{n^4}\cos nx\cos nct, \qquad \textbf{(19')}$$

and this series, together with those obtained by differentiating termwise twice, converges uniformly for all x, t. Hence this U gives an actual solution to equation (13) and it is the unique solution to our problem in view of Theorem 5.4.

These examples illustrate why series solutions to the wave equation are not as useful as they were for the heat equation and the potential equation. Unless the prescribed initial conditions admit periodic extensions that are C^2, the series need not be sufficiently differentiable. Irregularities in initial/boundary data are not smoothed as they are for those other equations. This may explain why Daniell Bernoulli encountered so much more resistance than did Fourier some fifty years later in proposing series solutions to partial differential equations. Moreover, already available to Bernoulli's contemporaries was d'Alembert's solution, which we take up in Section 5.6.

Problem Set 5.3

5.3A Obtain a formal series solution U to each of the following problems involving the one-dimensional wave equation $u_{tt} = c^2 u_{xx}$, with $u_t(x,0) \equiv 0$.

1. $u(0,t) = u(\pi,t) = 0$, $\quad t > 0$ and $u(x,0) = x$, $\quad 0 < x < \pi$.
2. $u(0,t) = u(\ell,t) = 0$, $\quad t > 0$ and $u(x,0) = x$, $\quad 0 < x < \ell$.
3. $u_x(0,t) = u_x(\pi,t) = 0$, $\quad t > 0$ and $u(x,0) = x^2$, $\quad 0 < x < \pi$.
4. $u_x(0,t) = u_x(\ell,t) = 0$, $\quad t > 0$ and $u(x,0) = x^2$, $\quad 0 < x < \ell$.
5. $u(0,t) = u_x(\pi,t) = 0$, $\quad t > 0$ and $u(x,0) = x$, $\quad 0 < x < \pi$.

5.3B What changes would be made in the solution to the problem in Example 2 if the initial conditions were $u(x,0) = 0$ and

$$u_t(x,0) = \frac{\pi x^2}{2} - \frac{x^3}{3}, \qquad 0 < x < \pi?$$

5.3C *1.* Obtain a formula analogous to (19) for $s_n = (2/\pi)\int_0^\pi f(x)\sin nx\, dx$, when f is a polynomial.

2. When $\ell = \pi$ in Example 1, use this formula to find a simple nonzero *polynomial* $f(x)$, allowing the series in (16) to be differentiated twice (termwise) and still retain uniform convergence.

5.3D Show that in Example 2, U of (19′) can be written in the form $U(x,t) = \phi(x+ct) + \psi(x-ct)$ for certain (series) functions ϕ, ψ.

5.3E If the string of Example 1 is plucked at $x = p$, then we should take

$$u(x,0) = f(x) = \begin{cases} x, & 0 \le x \le p \\ R(\ell - x), & p \le x \le \ell \end{cases} \qquad \text{where } R = p/(\ell - p).$$

Obtain the formal solution

$$U(x,t) = \frac{2}{\pi^2}\frac{R\ell^2}{p}\sum_{n=1}^{\infty}\frac{1}{n^2}\sin\frac{n\pi p}{\ell}\sin\frac{n\pi x}{\ell}\cos n\omega_1 t,$$

and discuss its physical significance when

$$\frac{p}{\ell} = \frac{1}{7}; \qquad \frac{p}{\ell} = \frac{2}{5}; \qquad \frac{p}{\ell} = \frac{1}{\sqrt{2}}.$$

(In a piano, the point p at which the hammer strikes the string is selected so that $p/\ell \approx 1/7$ thereby minimizing the dissonant contributions of the harmonics for $n = 7, 14, \ldots$ to the resulting sound.)

5.3F For the beam of Problem 5.1F:

1. Verify that $w(x,t) = X(x)\cos\omega t$ satisfies the equation in part 3 of Problem 5.1F when ρ and μ are C^2 functions of x only, provided that X is a solution of the equation $(\mu X'')'' = \rho\omega^2 X$ on $(0,\ell)$.

2. What are the boundary conditions on X that result from

(*a*) the simply-supported end conditions

$$w(0,t) = w(\ell,t) = 0; \qquad w_{xx}(0,t) = w_{xx}(\ell,t) = 0, \qquad t > 0?$$

(*b*) the clamped-end conditions

$$w(0,t) = w(\ell,t) = 0; \qquad w_x(0,t) = w_x(\ell,t) = 0, \qquad t > 0?$$

(*c*) the cantilever conditions

$$w(0,t) = w_x(0,t) = 0; \qquad w_{xx}(\ell,t) = (\mu w_{xx})_x(\ell,t) = 0, \qquad t > 0?$$

3. In the simple case of a uniform beam with $\mu = \rho = 1$, set $\omega = \kappa^2$ and show that when $\kappa \ne 0$, each eigenfunction must take the form

$$X(x) = a\cos\kappa x + b\sin\kappa x + c\cosh\kappa x + s\sinh\kappa x,$$

for constants a, b, c, s, κ, which are to be selected (if possible) to meet the boundary conditions.

4. Show that when $\mu = \rho = 1$, the simply supported conditions in part 2(a) above are satisfied by $X_n(x) = \sin \kappa_n x$, where $\kappa_n = n\pi, n = 1, 2, \ldots$ so that the beam behaves as if it were a stretched string. How could this be used to obtain a corresponding formal solution W to the equation with initial conditions $w(x, 0) = f(x)$, $w_t(x, 0) = 0$, $0 < x < \ell$?
5. Show that when $\mu = \rho = 1$, the clamped-end conditions in part 2(b) above are satisfied by

$$X(x) = c(\cosh \kappa x - \cos \kappa x) + s(\sinh \kappa x - \sin \kappa x)$$

when $\cosh \kappa\ell \cos \kappa\ell = 1$, so that c/s is determined. Verify graphically that this transcendental equation for $\kappa\ell$ has solutions $\kappa_0 = 0 < \kappa_1 < \kappa_2 < \ldots$, where for large n, $\kappa_n\ell \simeq (n + \frac{1}{2})\pi$.

5.4

Forced Vibrations and Damped Vibrations; Resonance

To consider forced motion of the uniform stretched string considered previously, we need solutions v for the nonhomogeneous wave equation

$$L(v) \equiv v_{tt} - c^2 v_{xx} = F \qquad \text{in} \qquad \{0 < x < \ell,\, t > 0\} \tag{20}$$

with the fixed-support conditions

$$v(0, t) = v(\ell, t) = 0, \qquad t > 0$$

and, if desired, the initial conditions

$$v(x, 0) = v_t(x, 0) = 0, \qquad 0 < x < \ell.$$

$F = F(x, t)$ is specified by the particular forcing mechanism employed.

Following the general scheme of Section 2.4, we use eigenfunctions $X_n(x) = \sin(n\pi x/\ell)$ that satisfy both the homogeneous boundary conditions and the equation $X_n'' = -(\omega_n/c)^2 X_n$, for $\omega_n = n\omega_1$, $n = 1, 2, \ldots$, as components of the formal series

$$V(x, t) = \sum_{n=1}^{\infty} T_n(t) X_n(x). \tag{21}$$

Then, formally,

$$L(V) = \sum_{n=1}^{\infty} (T_n'' + \omega_n^2 T_n) X_n = F,$$

provided that

$$F(x, t) = \sum_{n=1}^{\infty} f_n(t) X_n(x) = \sum_{n=1}^{\infty} f_n(t) \sin \frac{n\pi x}{\ell},$$

so that we should take

$$f_n(t) = \frac{2}{\ell} \int_0^{\ell} F(x, t) \sin \frac{n\pi x}{\ell}\, dx, \qquad n = 1, 2, \ldots. \tag{22}$$

The T_n are chosen to satisfy

$$T_n'' + \omega_n^2 T_n = f_n, \qquad \text{with } T_n(0) = T_n'(0) = 0,$$

and when f_n is continuous, this last problem has the solution (from Proposition B.1)

$$T_n(t) = \frac{1}{\omega_n}\int_0^t \sin \omega_n(t-\xi) f_n(\xi)\, d\xi. \tag{23}$$

Example 3:

A passing truck or train sometimes produces room vibrations that elicit sound from an unplayed string of a piano in the room. To model this occurrence in a laboratory, we stretch the string tightly between pegs fixed to a large horizontal table as shown in Figure 5.7. Then we force the table to oscillate vertically with small amplitude a at frequency ω, perhaps by driving an attached motor with an unbalanced flywheel at this frequency. The resulting vertical inertial force per unit mass on the string is approximated by $F(x, t) = A \cos \omega t$ for the amplitude constant $A = a\omega^2$. Assume the string to have length ℓ.

In this case, (22) yields

$$f_n(t) = \left(\frac{2}{\ell}\int_0^\ell \sin\frac{n\pi x}{\ell}\, dx\right) A \cos \omega t = A_n \cos \omega t,$$

where $A_n = 4A/n\pi$ when n is odd, and $A_n = 0$ otherwise.

Therefore (23) gives

$$T_n(t) = \frac{A_n}{\omega_n}\int_0^t \sin \omega_n(t-\xi) \cos \omega\xi\, d\xi$$

$$= \frac{A_n}{2\omega_n}\int_0^t [\sin(\omega_n(t-\xi) + \omega\xi) + \sin(\omega_n(t-\xi) - \omega\xi)]\, d\xi.$$

If $\omega \neq \omega_n$, then

$$T_n(t) = \frac{A_n}{2\omega_n}\left[\frac{\cos(\omega_n(t-\xi)+\omega\xi)}{\omega_n - \omega} + \frac{\cos(\omega_n(t-\xi)-\omega\xi)}{\omega_n + \omega}\right]_0^t$$

or

$$T_n(t) = -A_n \frac{\cos \omega t - \cos \omega_n t}{\omega^2 - \omega_n^2}; \tag{24}$$

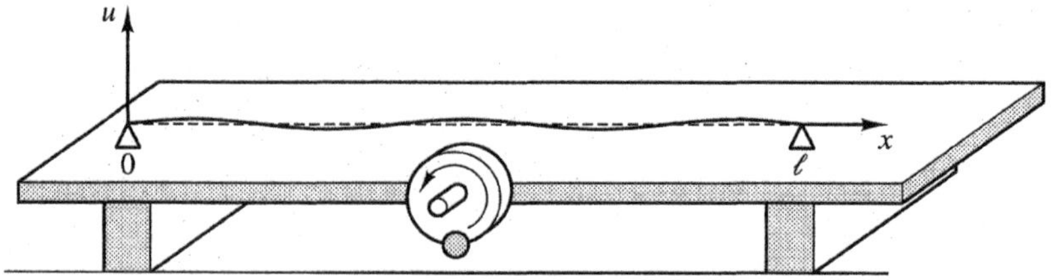

Figure 5.7

while if $\omega = \omega_n$, a simpler computation gives

$$T_n(t) = \frac{A_n t}{2\omega_n} \sin \omega_n t, \tag{24'}$$

which is also the limit of the previous expression as $\omega \to \omega_n$. (See Problem 5.4A.) Upon substituting these expressions in (21), we get

$$V(x,t) = \frac{-4A}{\pi} \sum_{n=1,3,\ldots}^{\infty} \left[\frac{\cos \omega t - \cos \omega_n t}{n(\omega^2 - \omega_n^2)} \right] \sin \frac{n\pi x}{\ell}, \tag{25}$$

assuming that $\omega \neq n\omega_1$, $n = 1, 3, \ldots$.

The series in (25) converges uniformly unless $\omega = \omega_n = n\omega_1$ for some $n = 1, 3, 5, \ldots$. Suppose that $0 < \omega < \omega_1$. Then for fixed $x \in (0, \ell)$ and t, as $\omega \nearrow \omega_1$, the first term approaches the limiting value

$$\frac{2At}{\pi\omega_1} \sin \omega_1 t \sin \frac{\pi x}{\ell},$$

and as t increases, it dominates the rest of the series. This behavior also occurs when ω approaches ω_1 from above and when ω equals ω_1. Hence, forced oscillations at frequency ω_1 induce vibrations of the string in the associated natural mode of increasingly large amplitude. The series behaves similarly as ω approaches or equals each natural frequency $n\omega_1$ for $n = 3, 5, \ldots$ (but not for $n = 2, 4, \ldots$). Consequently, our piano string can be made to resonate, but only at certain frequencies.

This phenomenon, known as **resonance**, also explains why an unbalanced tire on an automobile is only objectionable when driving near the critical speed (usually around 60 mph) corresponding to the first natural frequency of the suspended spring-mass system. (The critical speeds required to excite the higher modes are only reached in racing automobiles.) Were it not for the damping provided by shock absorbers (to be discussed next), the resulting oscillations could be damaging. However, we should not forget that such large amplitudes violate certain linearizing assumptions used in deriving our model.

Damped Vibrations

In actual systems, natural vibrations do not persist indefinitely because various energy-absorbing agencies are present. Unforced motions are reduced, or **damped**, and die out as time increases. With forced vibrations, damping can prevent the occurrence of resonance. Physically, damping of a motion is obtained easily from devices that introduce a velocity-dependent force opposing the acceleration. Such damping may be supplied by a piston in a fluid-filled cylinder (a shock absorber), although other mechanisms can also be imagined.

(a) Unforced damped vibrations: Mathematically, the linearized effects of distributed velocity-dependent damping add to the standard wave equation a term $(-2\delta u_t)$ that opposes the local acceleration as follows.

$$u_{tt} = -2\delta u_t + c^2 u_{xx}, \tag{26}$$

where δ is assumed to be a positive constant. (The factor 2 is used for convenience.)

Observe that when $\delta = 0$, we have the undamped homogeneous equation studied earlier. Also, solutions with given standard boundary conditions and initial u and u_t are unique. (See Problem 5.2B.) When $\ell = \pi$, let's consider the fixed-end boundary conditions

$$u(0,t) = u(\pi,t) = 0 \tag{26'}$$

and require that $u(x,0) = 0$, $0 < x < \pi$.

Again, separation of variables is possible and we obtain elementary product solutions in the form $u_n(x,t) = T_n(t)\sin nx$, where now T_n is chosen to satisfy the more general second-order equation

$$T_n'' + 2\delta T_n' + n^2c^2T_n = 0, \qquad \text{with} \qquad T_n(0) = 0 \qquad \text{and} \qquad T_n'(0) = 1. \tag{27}$$

For $\omega_n = nc$, we consider separately the cases $\omega_n < \delta$, $\omega_n = \delta$, and $\omega_n > \delta$.

From Proposition **B**.1 we see that

$$\left.\begin{aligned} \omega_n < \delta &\Rightarrow T_n(t) = \beta_n^{-1}e^{-\delta t}\sinh\beta_n t \\ \omega_n = \delta &\Rightarrow T_n(t) = e^{-\delta t}t \\ \omega_n > \delta &\Rightarrow T_n(t) = \beta_n^{-1}e^{-\delta t}\sin\beta_n t \end{aligned}\right\} \tag{28}$$

where $\beta_n = |\delta^2 - \omega_n^2|^{1/2}, \qquad n = 1, 2, \ldots .$

Observe that when $\omega_n \neq \delta$, $n = 1, 2, \ldots, N$, finite superposition produces solutions

$$U_N(x,t) = \sum_{n=1}^{N} s_nT_n(t)\sin nx = U_N^{<}(x,t) + U_N^{>}(x,t).$$

Here, $U_N^{<}$ consists of those terms where $\omega_n < \delta$, and the associated motions are nonoscillatory but do die out in time since $\beta_n < \delta$; $U_N^{>}$ consists of those terms where $\omega_n > \delta$, and the associated motions are oscillatory with decreasing amplitude, but at *reduced* frequencies $\beta_n < \omega_n$.

The omitted case where $\omega_n = \delta$, if present, supplies one additional nonoscillatory term which also dies out as $t \to \infty$. (Why?)

We see that a formal series solution

$$U(x,t) = \sum_{n=1}^{\infty} s_nT_n(t)\sin nx \tag{29}$$

may be expressed as $U(x,t) = e^{-\delta t}\tilde{U}(x,t)$ for an appropriate series function $\tilde{U}$. If $\tilde{U}$ is bounded, then U will die out as $t \to +\infty$ as we had anticipated.

(b) Forced damped vibrations: When damping is present, forced vibrations of the same uniform string are governed by the nonhomogeneous equation

$$v_{tt} + 2\delta v_t - c^2v_{xx} = F, \tag{30}$$

which is satisfied by

$$v_n(x,t) = T_n(t)\sin nx,$$

provided that $F(x,t) = f_n(t)\sin nx$ and, as in (27), that

$$\begin{aligned} T_n'' + 2\delta T_n' + \omega_n^2T_n &= f_n, \quad (\omega_n = nc), \\ \text{with} \quad T_n(0) = T_n'(0) &= 0 \quad n = 1, 2, \ldots . \end{aligned} \tag{31}$$

Now, from Proposition B.1, we get the following particular solutions to (31):

$$\left.\begin{aligned}
\omega_n < \delta &\Rightarrow T_n(t) = \frac{e^{-\delta t}}{\beta_n}\int_0^t e^{\delta\xi}\sinh\beta_n(t-\xi)f_n(\xi)\,d\xi \\
\omega_n = \delta &\Rightarrow T_n(t) = e^{-\delta t}\int_0^t e^{\delta\xi}(t-\xi)f_n(\xi)\,d\xi \\
\omega_n > \delta &\Rightarrow T_n(t) = \frac{e^{-\delta t}}{\beta_n}\int_0^t e^{\delta\xi}\sin\beta_n(t-\xi)f_n(\xi)\,d\xi,
\end{aligned}\right\} \tag{32}$$

where $\beta_n = |\delta^2 - \omega_n^2|^{1/2}$, $\quad n = 1, 2, \ldots$.

If we represent the simple forcing term in the series form

$$F(x,t) = A\cos\omega t = \sum_{n=1}^{\infty} f_n(t)\sin nx,$$

then, as in Example 3,

$$f_n(t) = \frac{2}{\pi}\int_0^{\pi} F(x,t)\sin nx\,dx = \frac{4A}{n\pi}\cos\omega t, \qquad n = 1, 3, \ldots, (= 0, n \text{ even})$$

and now we must examine formulas (32) for these f_n. We wish to know whether with damping present ($\delta > 0$), resonance can still occur at some critical frequency ω. The substitution $s = t - \xi$ in these formulas shows that we are concerned with the integrals

$$I(t) = \int_0^t e^{-\delta s}\left\{\begin{matrix}\beta^{-1}\sinh\beta s \\ s \\ \beta^{-1}\sin\beta s\end{matrix}\right\}\cos\omega(t-s)\,ds, \tag{32$'$}$$

for $\beta = \beta_n$, $n = 1, 3, 5, \ldots$. By elementary estimates, for $t > 0$ we have

$$|I(t)| \le \int_0^t e^{-\delta s}\left\{\begin{matrix}\beta^{-1}e^{\beta s} \\ s \\ \beta^{-1}\end{matrix}\right\}ds \le \left\{\begin{matrix}(\beta(\delta-\beta))^{-1} \\ \delta^{-2} \\ (\beta\delta)^{-1}\end{matrix}\right\}$$

since $0 < \beta$ and in the first case, $\beta < \delta$.

Thus in each case $I(t)$ is bounded when $t \ge 0$, and the individual terms in the formal series solution

$$V(x,t) = \sum_{n=1}^{\infty} s_n T_n(t)\sin nx$$

will not become unbounded for any value of ω.

As this problem illustrates, the presence of even a small amount of velocity-dependent damping eliminates the unbounded amplitudes of pure resonance. However, if δ is very small, large amplitudes may still occur, especially when $\omega = \beta_n' = \sqrt{\beta_n^2 - \delta^2}$. Then for t sufficiently large, $T_n(t) \simeq (2A/n\pi\delta\beta_n)\sin\beta_n' t$ (see Problem 5.4I), which shows that the damped system can be forced into approximate oscillations at each reduced frequency β_n' with "large" amplitude $2A/n\pi\delta\beta_n$, $n = 1, 3, 5, \ldots$. This phenomenon is sometimes referred to as "damped resonance," and if its amplitudes are acceptable, it provides an experimentally useful method of maintaining large-scale oscillations with modest inputs of energy.

Problem Set 5.4

5.4A 1. In Example 3, show that when $\omega = \omega_n$, $T_n(t)$ is given by (24′).

2. Verify that as $\omega \to \omega_n$, $T_n(t)$ defined by (24) has the same limit as in part 1. (*Hint:* Use the law of the mean for the function $h(\xi) = \cos \xi t$ (for fixed t) between $\xi = \omega$ and $\xi = \omega_n$.)

3. How often can the series in (25) be differentiated while retaining uniform convergence?

5.4B 1. Show that (25) may be written as $V(x,t) = U(x,t) + A(x)\cos \omega t$, where U is a formal series solution of $L(u) = 0$ for (20) and $A(x)$ is a uniformly convergent series.

2. How could this result be interpreted?

5.4C $F(x,t) = A\sin(\pi x/\ell)\cos \omega t$ gives a distributed forcing term at frequency ω.

1. How might this be realized physically?
2. How should the results of Example 3 be modified?
3. Will resonance still occur as $\omega \to \omega_1$? when $\omega = \omega_n$, $n = 1, 2, \ldots$?

5.4D Repeat 5.4C when

1. $F(x,t) = A\sin(2\pi x/\ell)\cos \omega t$ or 2. $F(x,t) = Ax^2(\ell - x)^2 \cos \omega t$.

5.4E 1. Verify that $u_n(x,t) = T_n(t)\sin nx$ satisfies (26) when T_n satisfies (27).

2. Check that in each case of formula (28) $T_n(0) = 0$ and $T_n'(0) = 1$.

3. Show that each of these $T_n(t)$ dies out exponentially as t increases. (*Hint:* $t \le 2e^{(\delta/2)t}/\delta$ and $0 \le \sinh \beta t \le e^{\beta t}$, for $t \ge 0$.)

5.4F Obtain the solution to the unforced damped problem of (26) and (26′) when $u(x,0) = 0$ and $u_t(x,0) = x(\pi - x)$, $0 < x < \pi$.

5.4G How should the analysis of the unforced damped problem in this section be modified if instead

1. $u_x(0,t) = u_x(\pi,t) = 0$, $t > 0$?
2. $u(x,0) = f(x)$ and $u_t(x,0) = 0$, $0 < x < \pi$?
3. both (1) and (2) hold?

5.4H Obtain the solution to the forced damped problem of this section when

1. $F(x,t) = A\sin x \cos \omega t$, $0 < x < \pi$, and $\delta = c$.
2. $F(x,t) = Ax^2(\pi - x)^2 \cos \omega t$, $0 < x < \pi$, and $\delta = 2c$.

5.4I 1. When $f_n(t) = \cos \omega t$, show by substitution that the equation in (31) has a particular solution of the form

$$T_n(t) = A_n \cos(\omega t + \alpha_n)$$

for a suitable amplitude A_n and phase angle α_n. (*Hint:* Write $\cos \omega t = \cos(\omega t + \alpha_n)\cos \alpha_n + \sin(\omega t + \alpha_n)\sin \alpha_n$.)

2. For large $t > 0$, explain why this particular solution approximates the general solution to the equation when $\delta > 0$.
3. Verify that

$$|A_n| = [(\omega^2 - \omega_n^2)^2 - (2\delta\omega)^2]^{-1/2}$$

so that if $\omega_n > \delta$, $|A_n|$ has its maximal value when

$$\omega = \sqrt{\omega_n^2 - 2\delta^2} = \sqrt{\beta_n^2 - \delta^2} = \beta_n'.$$

What is the maximal value and what is the corresponding α_n in this case?

5.4J 1. What equation for $v = v(x,t)$ makes $u = e^{-\delta t}v$ satisfy (26) ? (30) ?

5.5

Traveling Waves

The operator associated with the standard one-dimensional wave equation (13) factors formally as follows:

$$c^2\frac{\partial^2}{\partial x^2} - \frac{\partial^2}{\partial t^2} = c^2\left(\frac{\partial}{\partial x} - \frac{1}{c}\frac{\partial}{\partial t}\right)\left(\frac{\partial}{\partial x} + \frac{1}{c}\frac{\partial}{\partial t}\right).$$

A related coordinate transformation leads to the general solution of equation (13) in a form that has far-reaching implications about wave equations and their solutions. Indeed, a little experimenting with the chain rule shows that for constant $c > 0$, the transformation

$$\begin{matrix} \xi = x + ct & & x = \dfrac{\xi + \eta}{2} \\ & \Longleftrightarrow & \\ \eta = x - ct & & t = \dfrac{\xi - \eta}{2c} \end{matrix} \tag{33}$$

results in the first-order differential operators

$$\frac{\partial}{\partial \xi} = \frac{1}{2}\left(\frac{\partial}{\partial x} + \frac{1}{c}\frac{\partial}{\partial t}\right); \qquad \frac{\partial}{\partial \eta} = \frac{1}{2}\left(\frac{\partial}{\partial x} - \frac{1}{c}\frac{\partial}{\partial t}\right). \tag{33'}$$

Through (33), a function $u = u(x,t)$ becomes the function $\tilde{u}(\xi,\eta) = u((\xi+\eta)/2, (\xi-\eta)/2c)$, which is C^2 in ξ, η iff u is C^2 in x, t. Then interchange of the mixed partials is permitted and, by (33′),

$$4c^2\frac{\partial}{\partial \eta}\frac{\partial \tilde{u}}{\partial \xi} = c^2\left(\frac{\partial}{\partial x} - \frac{1}{c}\frac{\partial}{\partial t}\right)\left(\frac{\partial u}{\partial x} + \frac{1}{c}\frac{\partial u}{\partial t}\right) = c^2\left(\frac{\partial^2 u}{\partial x^2} - \frac{1}{c^2}\frac{\partial^2 u}{\partial t^2}\right)$$

$$\text{or} \qquad 4c^2\tilde{u}_{\xi\eta} = c^2u_{xx} - u_{tt}, \qquad \text{since } u_{xt} = u_{tx}.$$

Now suppose that u is also a solution of the simple homogeneous wave equation

$$u_{tt} - c^2u_{xx} = 0 \tag{34}$$

in some region $\mathscr{V}$ of x, t space. Then it follows that $\tilde{u}$ is a solution of the simpler equation

$$\tilde{u}_{\xi\eta} = \frac{\partial}{\partial \eta}\frac{\partial \tilde{u}}{\partial \xi} = 0 \qquad (34')$$

in a corresponding region $\tilde{\mathscr{V}}$ of ξ, η space. Let's assume initially that $\mathscr{V} = \mathbb{R}^2$ so that $\tilde{\mathscr{V}} = \mathbb{R}^2$ as well, in view of (33). Then we can integrate equation (34′) with respect to η to get

$$\frac{\partial \tilde{u}}{\partial \xi}(\xi, \eta) = \tilde{\phi}(\xi)$$

for some C^1 function $\tilde{\phi}$; hence we integrate again and find that

$$\tilde{u}(\xi, \eta) = \int \tilde{\phi}(\xi)\, d\xi + \psi(\eta),$$

$$\text{or} \qquad \tilde{u}(\xi, \eta) = \phi(\xi) + \psi(\eta), \qquad (35)$$

for some functions ϕ, ψ, each of only *one* variable. When we retransform to x, t, we see that (34) admits *general* solution in the *functional form*

$$u(x, t) = \phi(x + ct) + \psi(x - ct) \qquad (35')$$

first obtained by d'Alembert in 1746.

Conversely, if ϕ and ψ are any C^2 functions of a *single* variable, then it is straightforward to show that u as defined in (35′) does satisfy equation (34). (See Problem 5.5A.)

For example, if $\phi(\xi) = \xi^2$ and $\psi(\eta) = \cos \eta$, then the combination

$$u(x, t) = (x + ct)^2 + \cos(x - ct)$$

satisfies (34) and in fact, so must either of its terms. (Why?)

According to (35′), at time $t > 0$ the graph of $u(\cdot, t)$ can be found by combining the graph of ϕ translated to the left by ct units with that of ψ translated to the right by ct units, as shown in Figure 5.8.

In particular,

$$u(x, t) = \phi(x + ct)$$

describes a *wave of initial profile* $u(x, 0) = \phi(x)$ *that is traveling to the left with*

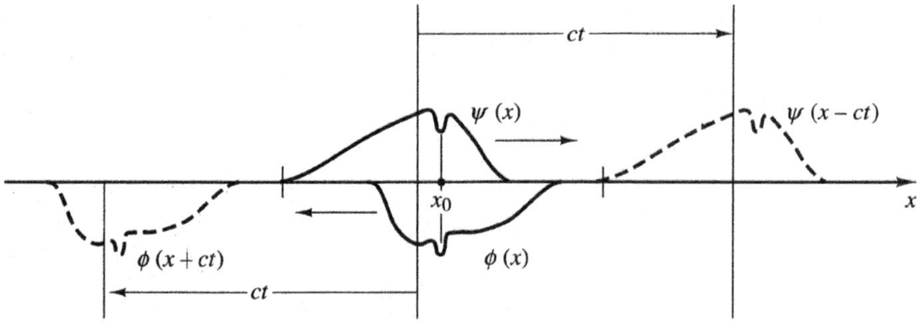

Figure 5.8

speed c. Similarly,

$$u(x,t) = \psi(x - ct)$$

describes a *wave of initial profile* $u(x,0) = \psi(x)$ *that is traveling to the right with speed c*.

The general solution to (34) can be interpreted as the sum of a wave moving to the left and one moving to the right, both with speed c. Actually, each separate *traveling wave* represents the general solution to a simpler first-order equation. For example,

$$u(x,t) = \psi(x - ct) \tag{36}$$

is the general solution to the equation

$$u_t + cu_x = 0 \tag{36'}$$

since then $\tilde{u}_\xi = 0$ so that $\tilde{u} = \tilde{u}(\eta)$ alone.

Characteristics

In the (x, t)-plane, a small "bump" at x_0 in the initial profile of the ϕ-wave "propagates" along the line $x + ct = x_0$, while one for the ψ-wave propagates along the line $x - ct = x_0$. More generally, disturbances in u at a point (x_0, t_0) propagate along the lines through the point with slopes

$$\frac{dx}{dt} = \pm c.$$

Such lines (which correspond to coordinate lines in the ξ, η system) are called the **characteristics of the wave equation** (34). In the ξ, η system, the partial integrations required to produce the solution (35),

$$\tilde{u}(\xi, \eta) = \phi(\xi) + \psi(\eta),$$

are valid in coordinate rectangles (whose sides may have infinite length). (See Figure 5.9.) If (ξ_0, η_0) is a vertex of one such rectangle that contains (ξ, η), then

$$\phi(\xi) - \phi(\xi_0) = \tilde{u}\xi, \eta) - \tilde{u}(\xi_0, \eta) = \tilde{u}(\xi, \eta_0) - \tilde{u}(\xi_0, \eta_0)$$
$$\text{or} \quad \tilde{u}(\xi, \eta) = \tilde{u}(\xi_0, \eta) + \tilde{u}(\xi, \eta_0) - \tilde{u}(\xi_0, \eta_0).$$

Thus the solution within the rectangle is determined by its values on the coordinate lines through (ξ_0, η_0).

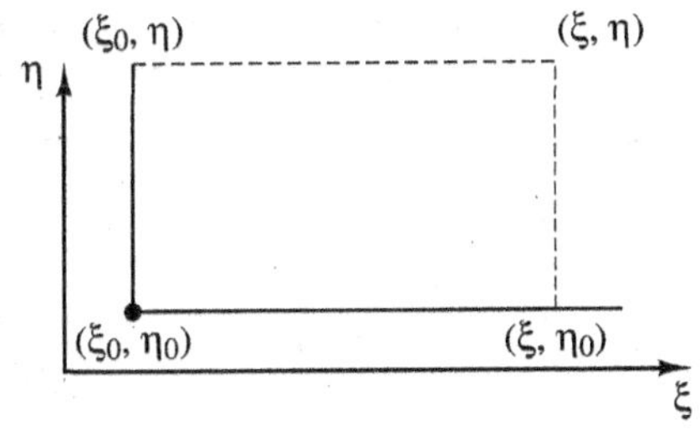

Figure 5.9

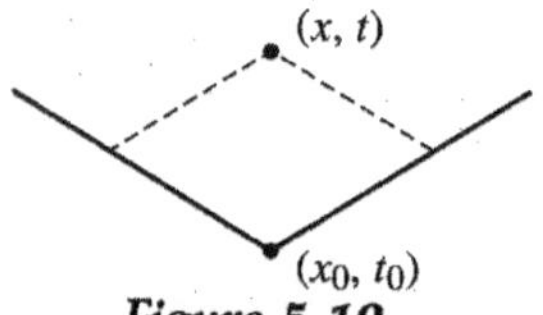

Figure 5.10

In the x, t coordinates, it follows that d'Alembert's solution (35′) is valid in diamond-shaped regions (whose sides may have infinite length) bounded by characteristics. (See Figure 5.10.) Within one such region the solution is determined by its values on the characteristics through a vertex point (x_0, t_0) in the manner indicated above.

Waves in Higher Dimensions

As we have seen, each C^2 function ψ of a single variable provides a *wave profile* for a solution $u(x,t) = \psi(x - ct)$ of the equation $u_{tt} = c^2 u_{xx}$, with $u(x,0) = \psi(x)$, which travels to the right at speed c as time increases.

Similarly, for any constant unit vector $\mathbf{n} = (\lambda, \mu, \nu) \in \mathbb{R}^3$,

$$u(x,y,z,t) = \psi(\lambda x + \mu y + \nu z - ct) = \psi(\mathbf{n} \cdot \mathbf{x} - ct) \tag{37}$$

represents a solution to the standard three-dimensional wave equation

$$u_{tt} = c^2 \nabla^2 u$$

with initial profile

$$u(x,y,z,0) = \psi(\lambda x + \mu y + \nu z) = \psi(\mathbf{n} \cdot \mathbf{x}),$$

which travels in the direction of $\mathbf{n}$ at speed c as time increases. (See Problem 5.5C.) Since in $\mathbb{R}^3$, $\lambda x + \mu y + \nu z$ is constant in each plane normal to $\mathbf{n}$, the direction of travel, these solutions are called **plane waves**. Moreover, suitable integral superposition of these plane waves provides a three-dimensional version of d'Alembert's result. This can be accomplished by means of the Radon transform. ([Jo] and [Wa]). An alternate approach to obtaining such a result is presented in the discussion of Kirchhoff's formula in [C–H].

Problem Set 5.5

5.5A 1. Use the chain rule to verify that u of (35′) satisfies the wave equation (34) when ϕ and ψ are C^2 functions.

2. Show that $\tilde{u}(\xi, \eta) = \psi(\eta)$ satisfies (34′) for an *arbitrary* function ψ, and explain this in relation to part 1.

5.5B Suppose that the values of a solution u of $u_{tt} = c^2 u_{xx}$ are known along a pair of characteristics with equations $x \pm ct = 1$.

1. Show that

$$\tilde{u}(\xi, \eta) = u\left(\frac{\xi + \eta}{2}, \frac{\xi - \eta}{2c}\right) = \phi(\xi) + \psi(\eta)$$

is known along the lines $\xi = 1$, $\eta = 1$.

2. Conclude that $\tilde{u}(\xi,\eta) = \tilde{u}(\xi,1) + \tilde{u}(1,\eta) - \tilde{u}(1,1)$ so that $u(x,t) = \tilde{u}(x+ct, x-ct)$ is determined.

5.5C 1. Verify that $\psi(\lambda x + \mu y - ct)$ satisfies the standard two-dimensional wave equation provided that the constants λ and μ satisfy $\lambda^2 + \mu^2 = 1$.

2. Make a similar computation for $\phi(\lambda x + \mu y + \nu z + ct)$ supposing that $\lambda^2 + \mu^2 + \nu^2 = 1$.

3. Show that $xy \not\equiv \phi(\lambda x + \mu y + t) + \psi(\lambda x + \mu y - t)$ for any functions ϕ, ψ. (*Hint:* Compare representations when $x = 0$ and when $y = 0$.) What does this illustrate?

5.6

d'Alembert's Formula

It is easy to find solutions to the wave equation (34), but in general, we need to determine those that meet given auxiliary conditions. First, let's see how to find ϕ and ψ so that

$$u(x,t) = \phi(x+ct) + \psi(x-ct) \tag{38}$$

has given

$$\textit{initial displacement} \qquad \delta(x) = u(x,0), \tag{38a}$$

and

$$\textit{initial velocity} \qquad v(x) = u_t(x,0). \tag{38b}$$

From the displacement equation (38a) we find that

$$u(x,0) = \delta(x) = \phi(x) + \psi(x)$$

so that $\psi = \delta - \phi$, and from (38),

$$u(x,t) = \delta(x-ct) + [\phi(x+ct) - \phi(x-ct)]. \tag{39}$$

Therefore, the velocity equation (38b) becomes

$$v(x) = u_t(x,0) = -c\delta'(x) + c\phi'(x) - (-c)\phi'(x),$$

which gives the *derivative* relation

$$\phi' = \frac{1}{2}\left(\delta' + \frac{v}{c}\right).$$

Next, we integrate the last equation between the limits of $x - ct$ and $x + ct$ to obtain

$$\phi(x+ct) - \phi(x-ct) = \int_{x-ct}^{x+ct} \phi'(s)\,ds = \frac{\delta(x+ct) - \delta(x-ct)}{2} + \frac{1}{2c}\int_{x-ct}^{x+ct} v(s)\,ds.$$

When we insert this result in (39), we get

$$u(x,t) = \frac{\delta(x+ct) + \delta(x-ct)}{2} + \frac{1}{2c}\int_{x-ct}^{x+ct} v(s)\,ds, \qquad \textit{(d'Alembert's formula)} \tag{40}$$

which gives u simply and explicitly in terms of a desired initial displacement δ and initial velocity v.

Example 4:

The solution u of the wave equation $u_{tt} = c^2 u_{xx}$ with initial displacement $u(x,0) = x^2$ and initial velocity $u_t(x,0) = \cos x$ is obtained from (40) as follows:

$$u(x,t) = \frac{(x+ct)^2 + (x-ct)^2}{2} + \frac{1}{2c}\int_{x-ct}^{x+ct} \cos s \, ds$$

$$= x^2 + c^2t^2 + \frac{1}{2c}[\sin(x+ct) - \sin(x-ct)]$$

$$= x^2 + c^2t^2 + c^{-1}(\sin ct)\cos x.$$

In particular, when $t = \pi/c$, we get $u(x, \pi/c) = x^2 + \pi^2$, and any other values may be found.

d'Alembert's formula has many consequences, some of which will be discussed in the next chapter. For the present, we note that when δ is C^2 and v is C^1, then u as defined by (40) is the unique C^2 solution of (35) with the given initial conditions. This ensures existence, and it is easy to show that u depends continuously on its initial data. (See Problem 5.6E.) Moreover, (40) may be considered as defining a generalized solution to the initial value problem even when δ and v are less smooth. Since boundary conditions have not been implied, uniqueness appears surprising in view of Theorem 5.4. However, for (40) to remain valid with increasing t, the initial conditions must be specified for correspondingly larger values of $|x|$. To have a solution for all $t > 0$, we need to know δ and v for *all* $x \in \mathbb{R}$ so that an associated "string" would be of infinite extent in both directions.

Infinite String

Assume, that we have an elastic string stretched along the x-axis that we deflect initially and then release, so that the initial velocity $v \equiv 0$. Then (40) reduces to

$$u(x,t) = \frac{\delta(x+ct) + \delta(x-ct)}{2}, \tag{41}$$

which may be interpreted as the *average* of the initial deflection after it has moved to the left with speed c and this same deflection after it has moved to the right with speed c.

The resulting traveling waves are illustrated in Figure 5.11 in the simple case where δ is identically zero outside [0, 1], and $v \equiv 0$. Several features are apparent in this figure. First, the solution at the point (x, t) can be affected only by what happens within its (shaded) triangular **domain of dependence**. In particular, when $\delta \equiv 0$ at the base of this triangle, then $u(x,t) \equiv 0$. Second, a point $(x_0, 0)$ where $\delta(x_0) \neq 0$ can affect only the solution for later times t within its (striped) triangular **domain of influence**.† Thus, third, at a particular x, say, $x = 2$, there is a period of time until the initial disturbance δ reaches it, and after some later time, it will again be unaffected.

† The slanted boundaries of both domains are characteristics.

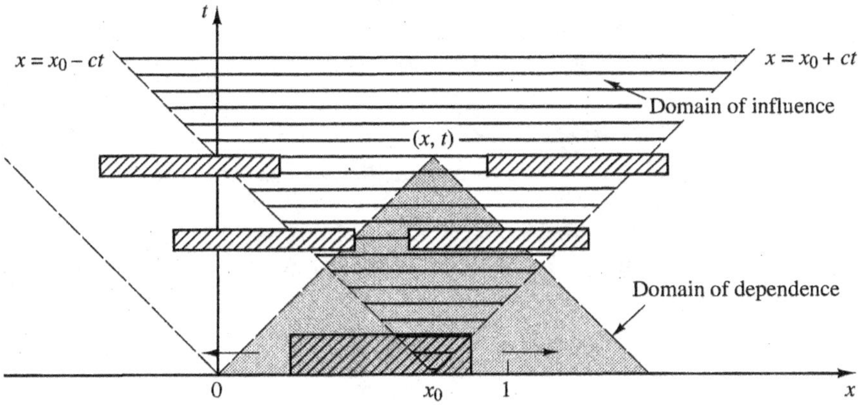

Figure 5.11

Such behavior suggests that the wave equation might be appropriate to describe the transmission of sound, and we will explore this aspect more fully in Section 6.4.

Finite String

Suppose that we wish to have a solution u, given by d'Alembert's formula, that satisfies boundary conditions appropriate to a *finite* string of length $\ell = \pi$. In particular, suppose we want a solution that meets the fixed-end conditions

$$u(0, t) = u(\pi, t) = 0, \qquad \text{for all } t > 0.$$

Such conditions can be realized. For example, if we take $\delta(x) = \sin x$, $x \in \mathbb{R}$, then (41) gives

$$u(x, t) = \frac{\sin(x + ct) + \sin(x - ct)}{2} = \sin x \cos ct$$

and indeed

$$u(0, t) = u(\pi, t) = u(n\pi, t) = 0, \qquad \text{for all } t \text{ when } n = \pm 1, \pm 2, \ldots .$$

Now, let's view this solution as superposed traveling waves. First, we notice that the nodes at the endpoints persist because an infinite succession of translates and their negatives arrive from each direction just at the right time to cancel each other. Next, on the x-interval $[0, \pi]$, as time increases, we watch the initial deflection smoothly dissipate, then become its negative (at time $t = \pi/c$), and after a similar transition, regain its initial values (at time $t = 2\pi/c$). Finally, we see this cycle repeated in each subsequent time interval of duration $2\pi/c$. Graphically, the situation is similar for other initial deflections δ on $[0, \pi]$ extended as *odd* functions of period 2π, as illustrated in Figure 5.12 (inside the back cover). In fact, the traveling initial displacements of a real finite string are *reflected* as they reach the boundaries where they undergo sign reversal and then move toward the opposite ends. As this process is repeated, the string appears to oscillate, and the oscillations may be viewed as "stationary" or standing waves.

We can support our graphical analysis mathematically in the simple case when $v \equiv 0$. Then, to have (41) supply a solution with, say, $u(0,t) = u(\ell,t) = 0$, $t > 0$, we need that $\delta(ct) + \delta(-ct) \equiv 0$ and $\delta(\ell + ct) + \delta(\ell - ct) \equiv 0$. Or, upon replacing ct by x, we want

$$\delta(-x) = -\delta(x) \qquad \text{and} \qquad \delta(\ell + x) = -\delta(\ell - x). \tag{42}$$

The first equation requires that δ be extended as an *odd* function, and then the second requires that $\delta(x + 2\ell) = \delta(\ell + (\ell + x)) = -\delta(-x) = \delta(x)$; i.e., that δ so extended have period 2ℓ, thereby confirming our conjecture.

In particular, if we suppose that on $(0, \ell)$, δ admits a Fourier sine series representation in the form

$$\delta(x) = \sum_{n=1}^{\infty} s_n \sin\frac{n\pi x}{\ell}, \tag{43}$$

then $u(x,t) = [\delta(x + ct) + \delta(x - ct)]/2$ becomes, after rearrangement,

$$u(x,t) = \sum_{n=1}^{\infty} s_n \sin\frac{n\pi x}{\ell} \cos\frac{n\pi ct}{\ell}. \tag{44}$$

This recovers Bernoulli's formal series solution (16) of the same problem without making explicit oscillatory assumptions.

As we observed, the Bernoulli series (17) for the plucked string in Example 1 is continuous but it cannot be twice differentiated termwise with respect to x or t and retain convergence. However, in its traveling-wave form, the solution can be differentiated as often as we please, *except along the characteristics emanating from the points of discontinuity* of δ', viz., the lines with equations

$$x = \frac{n\ell}{2} \pm ct, \qquad n = \pm 1, \pm 3, \pm 5, \ldots,$$

as shown in Figure 5.13.

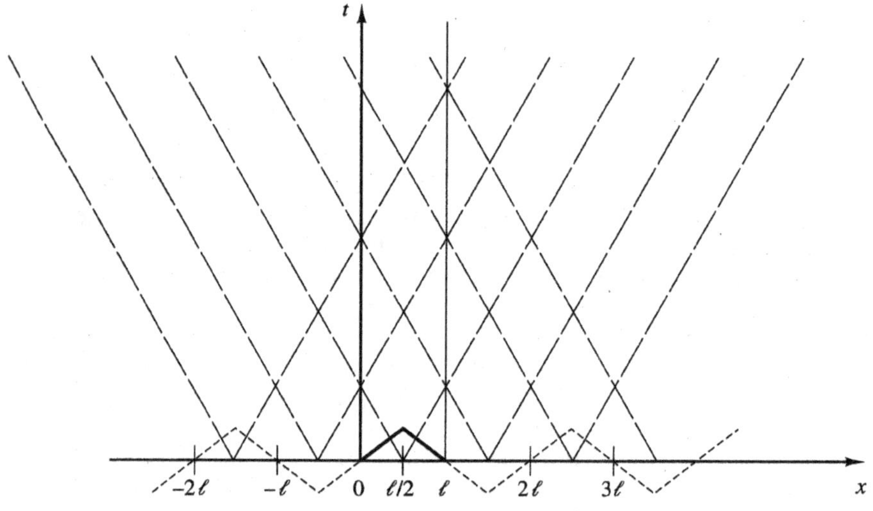

Figure 5.13

These characteristics form common boundaries for adjacent space-time domains inside which series (17) gives a valid solution to the wave equation. But *across* each characteristic boundary, the formal solution exhibits a discontinuity in normal derivative corresponding to a discontinuity of δ'. This example illustrates how discontinuities propagate along characteristics.

d'Alembert's approach extends to the nonhomogeneous wave equation of Section 5.4, but it is more difficult to obtain the results. (See Problem 5.6H.) However, this method does not seem to extend to the damped wave equation.

Example 5 (*Earthquake response*):

During an earthquake, tall buildings are at great risk of being damaged, even demolished. To design buildings that might survive, we must first understand how an ordinary structure responds to having its foundations shaken.

Suppose that a tall building of height ℓ and uniform properties (see Figure 5.14) experiences an earthquake-induced ground motion in a horizontal direction with time history $h = h(t)$. Neglecting bending, we can approximate the associated sway or shearing deflection $u = u(x, t)$ of the center line of the building at height x by regarding the building as an elastic string! Thus u should satisfy the equation

$$u_{tt} = c^2 u_{xx}, \qquad 0 < x < \ell,$$

with $c^2 = \mu/\rho$, where μu_x represents the resistance to shear. We want $u_x(\ell, t) = 0$ (since shear cannot be sustained at the upper free surface of the building) and $u(0, t) = h(t)$ to match the ground motion.

When we use these conditions in (38), we get the solution

$$u(x, t) = \phi(x + ct) + \phi(2\ell + ct - x), \tag{45}$$

where $h(t) = u(0, t) = \phi(ct) + \phi(2\ell + ct)$. In particular, when $c = \ell = 1$, we can show that under the damped oscillatory ground motion $h(t) = e^{-t} \sin \pi t$, the top of the building sways with the motion

$$u(1, t) = -\frac{e^{-t} \sin \pi t}{\cosh 1} = \frac{-h(t)}{\cosh 1}, \qquad t > 0, \tag{46}$$

which is directly out of phase with the ground motion but of reduced amplitude. (See Problem 5.6K.)

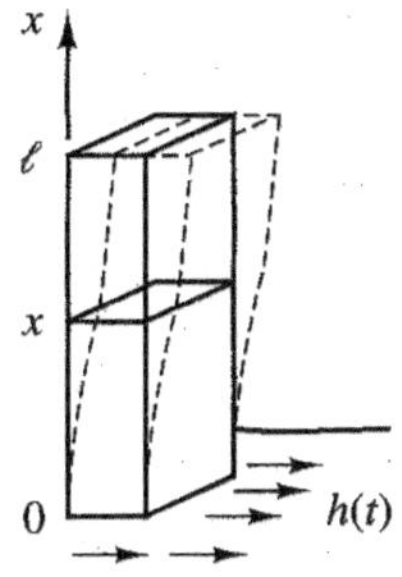

Figure 5.14

Problem Set 5.6

5.6A *1.* Verify that u defined by (40) satisfies (34) when δ is C^2 and v is C^1, with $u(x,0)=\delta(x)$ and $u_t(x,0)=v(x)$.

2. Obtain u from this formula when

(*a*) $\delta(x)=\sin(x^2)$ and $v(x)=e^x$.

(*b*) $\delta(x)=(1-x)/(1+x^2)$ and $v(x)=xe^{x^2}$.

(*c*) $\delta(x)=0$ and $v(x)=\begin{cases}1, & \text{for } |x|\le 1\\ 0, & \text{for } |x|>1.\end{cases}$

(*d*) $\delta(x)=\begin{cases}1, & \text{for } |x|\le 1\\ 0, & \text{for } |x|>1\end{cases}$ and $v(x)=0$.

5.6B Consider the initial value problem for the one-dimensional wave equation (34) with $c=1$.

1. Suppose that $\delta=0=v$ in some interval of length ℓ. Up to what time can one be sure that $u=0$ at the center of the interval?

2. Suppose that $\delta=0=v$ outside the interval $[-1,1]$. Up to what time can one be sure that $u=0$ at (*a*) $x=4$? (*b*) $x=10$? (*c*) $x=-5$?

5.6C *1.* When $v\equiv 0$ for the finite string of length $\ell=\pi$, how would you guess that δ on $(0,\pi)$ be extended so that the solution u given by (41) satisfies $u_x(0,t)=u_x(\pi,t)=0$, $t>0$?

2. Verify your conjecture by formal calculations.

5.6D Repeat the exercise in Problem 5.6C when $u(0,t)=u_x(\pi,t)=0$, $t>0$.

5.6E *1.* From (40), obtain the estimate $|u(x,t)|\le\Delta(\ell)+tV(\ell)$ when $|x\pm ct|\le\ell$, where $\Delta(\ell)=\max_{|s|\le\ell}|\delta(s)|$ and $V(\ell)=\max_{|x|\le\ell}|v(x)|$.

2. Explain how this could be used to establish the continuous dependence of u on its initial data.

5.6F *1.* When $v\not\equiv 0$, show that to have u from (40) satisfy $u(0,t)=u(\ell,t)=0$, $t>0$, then both functions δ, v on $(0,\ell)$ should be extended as odd functions of period 2ℓ.

2. Suppose that δ and v have sine series representations with Fourier coefficients δ_n, v_n, respectively, $n=1,2,\ldots$. Show that, formally, (40) yields a series of the form

$$u(x,t)=\sum_{n=1}^{\infty}(c_n\cos\omega_n t+s_n\sin\omega_n t)\sin\frac{n\pi x}{\ell}.$$

5.6G* *1.* If $v\neq 0$, explain how to extend prescribed functions v and δ on $(0,\ell)$ so that u given by (40) satisfies $u_x(0,t)=u_x(\ell,t)=0$, $t>0$.

2. What type of series representations would be appropriate for such v and δ?

5.6H For the nonhomogeneous wave equation $c^2u_{xx}-u_{tt}=F(x,t)$:

1. Show that (33) transforms this equation to

$$\tilde{u}_{\xi\eta}=\tilde{F}(\xi,\eta)\overset{\text{def}}{=}\frac{1}{4c^2}F\left(\frac{\xi+\eta}{2},\frac{\xi-\eta}{2c}\right).$$

2. Integrate this last equation to obtain

$$\tilde{u}(\xi,\eta) = \phi(\xi) + \psi(\eta) + \int\left(\int \tilde{F}(\xi,\eta)\, d\eta\right) d\xi$$

for appropriate ϕ, ψ.

3. Use this formula when $F(x,t) = x^2 + t$ and retransform variables to get the general solution to the original equation in this special case.

5.6I (*Duhamel's principle*) Suppose that $F \in C^1(V \times \mathbb{R})$, where V is a domain of $\mathbb{R}^d$, and that for each $\tau \in \mathbb{R}$, $U(\mathbf{x},t,\tau)$ solves the initial value problem $U_{tt} = c^2\nabla^2 U$ in $V \times \mathbb{R}$, with $U(\mathbf{x},0,\tau) = 0$ and $U_t(\mathbf{x},0,\tau) = F(\mathbf{x},\tau)$, $\mathbf{x} \in V$. Show that

$$w(\mathbf{x},t) \overset{\text{def}}{=} \int_0^t U(\mathbf{x},t-\tau,\tau)\, d\tau$$

solves the problem $w_{tt} = c^2\nabla^2 w + F$ in $V \times \mathbb{R}$, with $w(\mathbf{x},0) = w_t(\mathbf{x},0) = 0$, $\mathbf{x} \in V$. (*Hint:* Recall Problem 3.4F.)

5.6J 1. When $d = 1$, use (40) to solve Problem 5.6I in the form

$$w(x,t) = \frac{1}{2c}\int_0^t d\tau \int_{x-c(t-\tau)}^{x+c(t-\tau)} F(s,\tau)\, ds,$$

and give a graphical interpretation of this formula.

2. Conclude that the general solution to the nonhomogeneous equation in Problem 5.6H is

$$u(x,t) = \phi(x+ct) + \psi(x-ct) + w(x,t).$$

3. When $F(x,t) = x^2 + t$, compare the result given by this formula to that of Problem 5.6H3.

5.6K* Suppose that the tall building of Example 5 is of homogeneous composition with cross-sectional area $S = S(x)$, linear density $\rho = \rho(x)$, and shear-resistance function $\mu = \mu(x)$. Then, assuming that the building undergoes shearing in one horizontal direction only, the appropriate action integral is

$$A(u) = \frac{1}{2}\int_a^b dt \int_0^\ell [\rho S u_t^2 - \mu S u_x^2]\, dx.$$

1. Explain why this leads to the wave equation

$$(\rho S u_t)_t = (\mu S u_x)_x$$

governing possible motions of the building and how this simplifies when ρ, S, and μ are assumed constant.

2. Show how the shear-free condition $u_x(\ell,t) = 0$ leads from (38) to the solution given in (45). (*Hint:* An additive constant can be absorbed in ϕ.)

3. When $c = \ell = 1$, verify that the ground-motion condition $h(t) = \phi(t) + \phi(t+2)$ can be inverted (formally) to give

$$\phi(t) = \sum_{n=0}^{\infty}(-1)^n h(t+2n), \qquad t > 0$$

provided that the latter series converges and that the resultant $\phi(t) \to 0$ as $t \to +\infty$.

4. Apply the formula in part 3 to $h(t) = e^{-t} \sin \pi t$ to obtain $u(x, t)$ (when $c = \ell = 1$). Simplify your answer as much as possible.
5. Explain how the answer in part 3 changes when $c \neq 1$ or $\ell \neq 1$.
6. To permit the building to sway even though the ground does *not* move, how should ϕ be chosen? Give an example.

5.7*

Simple Waves and Related Nonlinear Equations

Can we use the results of previous sections to model the behavior of real ocean waves? In Section 5.5 we saw that

$$u(x, t) = \psi(x - ct), \qquad c \text{ constant} \tag{47}$$

describes a wave of initial profile ψ that moves to the right with speed c while preserving this profile. When ψ is C^1, u satisfies the simple first-order equation

$$u_t + cu_x = 0. \tag{47'}$$

Although constant-profile water waves can be observed[†], they do not persist. In fact, most ocean waves behave much less regularly, and to model them, we must at least allow points at different wave heights to travel at different speeds.

Simple Nonlinear Waves

If we just replace c in (47) by a C^1 function $p = p(u)$, we obtain the implicit relation

$$u = \psi(x - p(u)\,t) \tag{48}$$

that permits more interesting behavior. Moreover, when ψ is C^1, in general, u, so determined, can be shown to satisfy the *nonlinear wave equation*

$$u_t + p(u)\,u_x = 0. \tag{49}$$

(See Problem 5.7B.) This equation is the nonlinear counterpart of (47′) and it is easily remembered. In Example 7, we will show how to solve this equation with given initial values ψ to recover (48).

Observe that when $p > 0$, (48) describes a wave of initial profile $\psi(x)$ whose parts at elevation u move to the right with speed $p(u)$. In particular, when $p(u) = 1 + u$, then (49) becomes

$$u_t + (1 + u)\,u_x = 0 \tag{50}$$

[†] The earliest recorded account is that of J. Scott Russell (1834), who on horseback, followed a single-humped wave for two miles as it rolled along in a narrow channel.

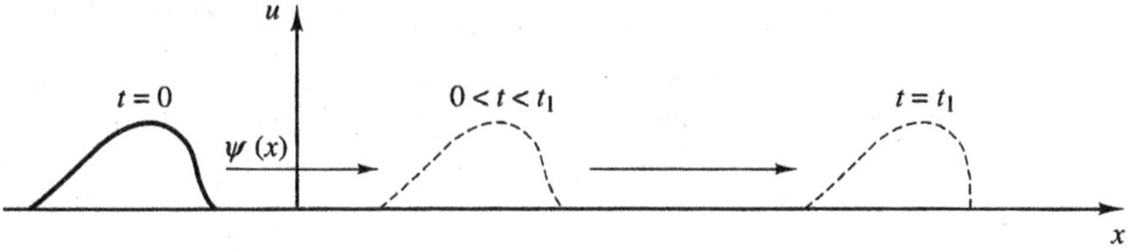

Figure 5.15

and the higher parts of the wave in (48) overtake the lower parts so that we might expect to see the wave steepen and **break**, forming a discontinuity at a critical time t_1 as illustrated in Figure 5.15.

We can watch these waves "break" mathematically when the initial profile has the linear form $\psi(x) = -kx$, where k is a *positive* constant. For then (48) becomes $u = -k(x - (1 + u)t)$, so that

$$u = -\frac{k(x-t)}{1-kt}, \qquad 0 \le t < \frac{1}{k},$$

and the wave slope, $u_x = -\dfrac{k}{1-kt}$, at a given time t

approaches $-\infty$ as t increases to 1/k. On the other hand, when k is negative, the wave appears to flatten out. Both cases are indicated in Figure 5.16.

Thus solutions of equation (50) exhibit some of the behavior of ocean waves. In fact (50) is a simplified version of the equation

$$u_t + (1 + u)u_x = -u_{xxx} \qquad \textit{(KdV equation)} \qquad \mathbf{(50')}$$

derived in 1895 by Korteweg and deVries to model the behavior of "long" water waves in shallow channels. The term on the right permits dispersion and the KdV equation represents one of the more successful attempts to describe water waves mathematically. This equation also allows as solutions certain constant-profile waves (now called *solitons*) modeling those observed by Scott Russell. (See Problem 5.7J.) See also Problems 6.4D, E, and for further discussion, [Sto] is recommended.

Equation (49) has surprisingly many applications and with appropriate choice of p, it provides useful models for traffic flow, glacial creep, and soil erosion, among other phenomena. (See Problems 5.7C–E and [Wh] for examples.)

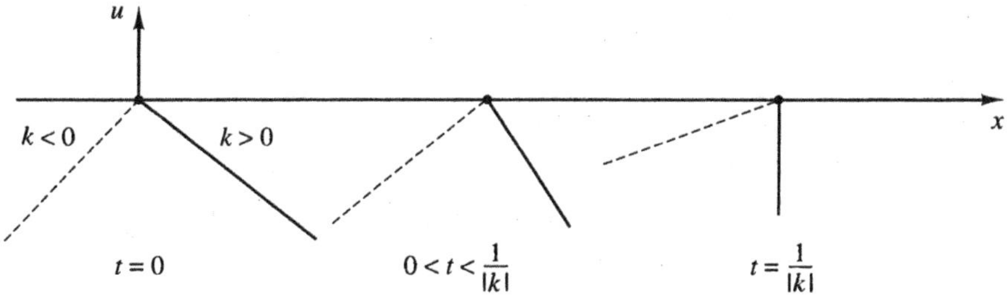

Figure 5.16

Quasilinear First-Order Equations

To gain further insight into Equation (49) and the nature of its solution (48), let's consider the more general **quasilinear equation** in x and t of the form

$$u_t + pu_x = r, \qquad p \neq 0 \tag{51}$$

where now p and r are given C^1 functions of x, t, and u. If we regard (x, t, u) as Cartesian coordinates of a point in $\mathbb{R}^3$, then a solution $u = u(x, t)$ of (51) in a domain D has as its *graph* the set $\mathcal{U} = \{(x, t, u(x, t)) : (x, t) \in D\}$, which is a *surface* in $\mathbb{R}^3$ that lies "over" D. (See Figure 5.17.) Moreover, along a base curve C in this domain parametrized by a C^1 function $x = x(t)$ with slope

$$\frac{dx}{dt} = p, \tag{52a}$$

the solution $u = u(x(t), t)$ has the derivative

$$\frac{du}{dt} = u_x \frac{dx}{dt} + u_t = pu_x + u_t,$$

or, since u satisfies (51),

$$\frac{du}{dt} = r. \tag{52b}$$

Conversely, these **characteristic equations** (52a,b) can be integrated (perhaps numerically) to determine values of u that might satisfy (51) along associated base curves C. Each associated pair (C, u) is described by equations

$$x = x(t) \qquad u = u(t),$$

say, that parametrize a curve $\mathscr{C}$, called a **characteristic** of (51) that lies "over" C, as shown in Figure 5.17.

C itself is said to be **characteristic in D**, or a **base characteristic** for (51), and more than one characteristic $\mathscr{C}$ of (51) may lie over C. However, there is precisely one characteristic $\mathscr{C}_0$ through each point (x_0, t_0, u_0) of definition of p and r, since these functions are C^1 so that system (52a,b) has a *unique* solution with $x(t_0) = x_0$ and $u(t_0) = u_0$. [C–L]. Consequently, most curves that lie over C are *not characteristic*, and many curves Γ lie over planar curves γ that are not base characteristics. In particular, the x-axis or any line parallel to it cannot be a base characteristic for (51). (Why?)

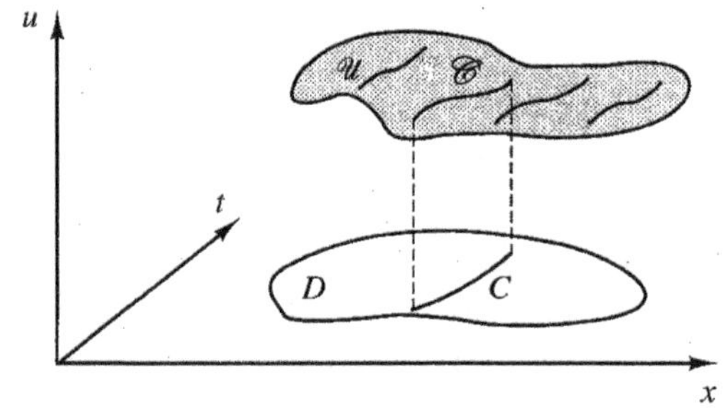

Figure 5.17

Example 6:

Consider the initial value problem

$$u_t + uu_x = -u, \qquad t > 0 \qquad \text{with} \qquad u(x,0) = -x/2.$$

Here $p = u$ and $r = -u$ so that the characteristic equations (52a,b) are just

$$\frac{dx}{dt} = u \qquad \text{and} \qquad \frac{du}{dt} = -u.$$

We can integrate the u equation immediately to find that $u = ce^{-t}$ along a characteristic. Using this, we can integrate the x equation to get

$$x = -ce^{-t} + c_0, \qquad u = ce^{-t},$$

and as t varies we have a family of curves involving arbitrary constants c and c_0.

However, at $t = 0$, $x = -c + c_0$ while $u = c$, so that to satisfy the given initial condition, we must take $c = -\frac{1}{2}(-c + c_0)$ or $c_0 = -c$. Hence

$$x = -c(1 + e^{-t}), \qquad u = ce^{-t}.$$

When we eliminate c, we get

$$u = u(x,t) = \frac{-xe^{-t}}{1 + e^{-t}} = \frac{-x}{1 + e^t}$$

as our solution.

Example 7:

Let's use characteristics to reexamine our original nonlinear equation (49),

$$u_t + p(u)u_x = 0.$$

Here, the characteristic equations (52a,b) are just

$$\frac{dx}{dt} = p(u) \qquad \text{and} \qquad \frac{du}{dt} = 0,$$

so that along a base curve C, u is *constant*, which means that $dx/dt = p(u)$ is also constant. Thus C is a straight line and $\mathscr{C}$ is a parallel line above C, with neither being parallel to the x-axis. To obtain a solution u that has a C^1 initial profile ψ, we suppose that the composite $p \circ \psi$ is defined and observe that $\Gamma_0 = \{(x_0, 0, \psi(x_0)) : x_0 \in \mathbb{R}\}$ is *not* characteristic since it lies above the x-axis, γ_0. Moreover, along the base line C_0 through $(x_0, 0) \in \gamma_0$, we must have

$$u = \psi(x_0) \qquad \text{and} \qquad x = x_0 + p(u)t,$$

and we can regard x_0 as a parameter. If we eliminate x_0, we recover the implicit relation (48),

$$u = \psi(x - p(u)t),$$

which represents a possible solution in some neighborhood of each point $(x_0, 0)$. When $p \circ \psi(x_0)$ *increases* with x_0, this solution is valid for all $t > 0$, since the base lines $x = x_0 + (p \circ \psi)(x_0)t$ for different values of x_0 cannot intersect. However, if $p \circ \psi(x_1) < p \circ \psi(x_0)$ for some $x_1 > x_0$, then the line through

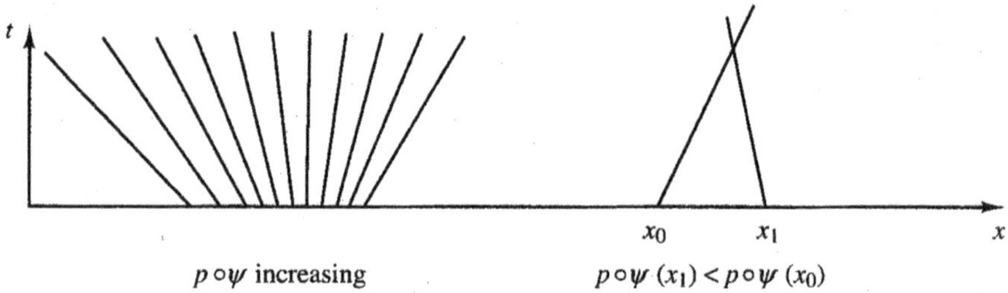

Figure 5.18

$(x_1, 0)$ intersects that through $(x_0, 0)$ at some positive t as shown in Figure 5.18, and solution (48) need not be valid at the point of intersection. This solution and its consequences are further explored in [La].

Characteristic Solution of Initial Value Problem*

In Example 7 we used characteristics to lead us to a solution of equation (49) that has prescribed initial values ψ on the x-axis in the (x, t)-plane. For the more general equation (51), let's seek a solution $u = u(x, t)$ that has given (initial) values u_0 on a curve γ_0 in D parameterized by C^1 functions $x_0 = x_0(s)$ and $t_0 = t_0(s)$, $s \in I \stackrel{\text{def}}{=} (0, 1)$. We shall assume that γ_0 is a *smooth arc*† along which $u_0 = u_0(s)$ is also C^1 and that p and r are defined in a neighborhood of the space-curve

$$\Gamma_0 = \{(x_0(s), t_0(s), u_0(s)) : s \in I\}.$$

Geometrically speaking, Γ_0 lies over γ_0, and we wish to embed Γ_0 in a solution surface $\mathscr{U}$ that lies over a subdomain of D containing γ_0. If γ_0 is characteristic in D this need not be possible, and if it is possible, $\mathscr{U}$ will not be unique. (See Problem 5.7K.)

However, suppose that with the given initial values, γ_0 is *nowhere characteristic* in D, in that at each $s \in I$,

$$x_0' \neq p(x_0, t_0, u_0)\, t_0', \qquad \left(\text{so that } t_0' \neq 0 \Rightarrow \frac{dx_0}{dt_0} \neq p(x_0, t_0, u_0)\right). \tag{53}$$

Then Γ_0 cannot be characteristic. But it is covered by the family of characteristics passing through its points and, at least near Γ_0, this family will form a unique *solution surface* of (51). This construction, known as the *method of characteristics*, is illustrated in Figure 5.19 and can be implemented numerically.

We now have the ingredients for a major result.

(5.5) Proposition: *Under the stated assumptions, if the smooth arc γ_0 satisfies condition* (53), *then there exists a solution u to equation* (51) *in a subdomain D_0 of D containing γ_0, such that $u\big|_{\gamma_0} = u_0$. u is unique within specification of D_0.*

Proof: Existence is established through implicit function theory. (See Problem 5.7L.) Uniqueness follows from the fact that if $\tilde{u}$ is another solution in a domain $\tilde{D}_0$

† γ_0 is *smooth* when its tangent vector $(x_0', t_0') \neq (0, 0)$. γ_0 is an *arc* when its parametrization is one-to-one on I so that γ_0 cannot be closed or cross itself.

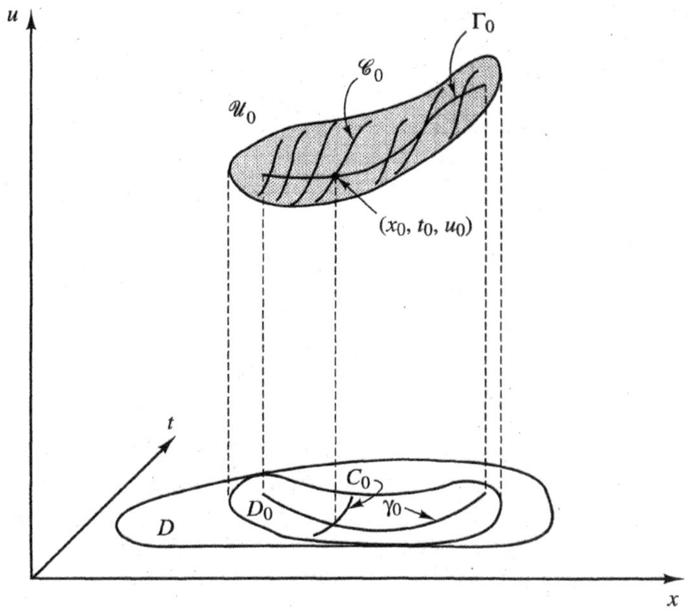

Figure 5.19

and $\tilde{u}$ has the same values as u on γ_0, then through each point of Γ_0 pass unique characteristics — $\mathscr{C}$ for u, $\tilde{\mathscr{C}}$ for $\tilde{u}$ — that also have identical directions at that point. It follows that $\mathscr{C} = \tilde{\mathscr{C}}$, so that u and $\tilde{u}$ agree along a common base C. But since all points $(x, t) \in D_0$ can be reached along base characteristics, we see that u and $\tilde{u}$ agree in a subdomain of $D_0 \cap \tilde{D}_0$ that contains γ_0. ■

Remarks: The decision as to whether γ_0 is nowhere characteristic in D usually depends on which values u_0 are to be specified on γ_0. However, the decision can be made a priori when (as in Example 6) p is a function of x and t only, regardless of whether r depends on u. In this case, equation (51) is said to be *semilinear*. The method can also be applied to the more general quasilinear equation

$$Pu_x + Qu_t = R \tag{54}$$

where P, Q, R are C^1 functions of x, t, and u, provided that P and Q do not vanish simultaneously at some point(s) of interest. Otherwise, we must use more global considerations. (See [C–H, Vol. II] for further discussion.)

Problem Set 5.7

5.7A Suppose that in (48),

$$p(u) = \frac{1}{1+u^2} \qquad \text{and} \qquad \psi(x) = \begin{cases} \sin^2 x, & 0 \le x \le \pi \\ 0, & \text{otherwise.} \end{cases}$$

Give a qualitative description of the behavior of u as t increases.

5.7B Suppose that $u = u(x, y)$ is C^1 and satisfies (48) where ψ and p are C^1 functions of a single variable; show that

$$u_x = \psi'(x - p(u)t)[1 - p'(u)tu_x] = \psi'(\xi)[1 - p'(u)tu_x],$$

say, and obtain a similar equation for u_t to conclude that u is a solution of (49) when $1 + \psi'(\xi)p'(u)t \neq 0$, at relevant ξ, t, u. (Conversely, under this condition, (48) determines u *locally* as a function of x and t by implicit function theory. [Ed].)

5.7C When $p = f'$, equation (49) becomes the *conservation equation* $u_t + f_x = 0$. It describes, for example, the one-dimensional flow of traffic of density u as a function of its flux f. Take

$$f(u) = \begin{cases} u^2(1 - u^2), & 0 \leq u \leq 1 \\ 0, & u < 0 \text{ or } u > 1, \end{cases}$$

and discuss the possibility of traffic pile-up. What happens when $f(u) = u$?

5.7D To describe a flood wave of local height u in water flowing in a rectangular channel at velocity v, use the conservation equation from Problem 5.7C with $f(u) = vu = ku^{3/2}$ (an empirical formula). Conclude that a flood wave moves with speed $p = (3/2)v$. Can these waves break?

5.7E To model erosion of a mountain slope of local height $h = h(x, t)$ above sea level, explain why a functional relation of the form $h_t = -f(h_x)$, where $f > 0$, might be reasonable. Show that $u = h_x$ satisfies the conservation equation of Problem 5.7C, and discuss what happens when breaking occurs. Assume that f is C^1.

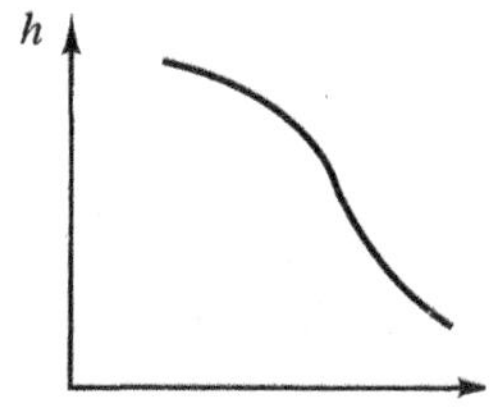

At time t

5.7F Solve the initial value problem $u_t + xu_x = 0$ with **1.** $u(x, 0) = 3x$, or **2.** $u(x, 0) = \psi(x)$, $x \in \mathbb{R}$

5.7G Solve the initial value problem $u_t + xu_x = -tu$ with **1.** $u(x, 0) = 3x$, or **2.** $u(x, 0) = x^2$, $x \in \mathbb{R}$

5.7H Show that the equation $v_t + v_x = v_x^2$ is quasilinear in $u = v_x$, and discuss the behavior of its solutions with $v_x(x, 0) = \psi(x)$ when ψ is C^1.

5.7I* For one-dimensional inviscid isentropic flow, the equations governing the velocity $u = u(x, t)$ and the density $\rho = \rho(x, t)$ are (see Section 6.4),

$$\rho_t + u\rho_x + \rho u_x = 0 \qquad u_t + uu_x + \frac{c^2}{\rho}\rho_x = 0$$

where $c = c(\rho)$ is a known positive function.

1. Show that a solution of the form $u = u(\rho)$ with $u(\rho_0) = 0$ is possible only if $u = \pm\int_{\rho_0}^{\rho}[c(\xi)/\xi]\,d\xi = \pm\ell(\rho)$, say.
2. When $u = \ell$, show that both equations become

$$u_t + a(u)u_x = 0 \qquad \text{where } a(u) = u + c$$

and $c = c(\rho) = c(\rho(u))$ is regarded as a function of u.

3. In this case, conclude that $u = \psi(x - a(u)t)$ should be a wave-type solution with $u(x,0) = \psi(x)$. Discuss the possibility of shock formation (breaking), assuming that $a(u)$ increases with u.
4. Make a similar analysis when $u = -\ell$ in terms of $a(-u)$.

5.7J* 1. In the KdV equation (50′), let $v = 1 + u$ and show that the resulting equation is $v_t + vv_x + v_{xxx} = 0$.
2. To have a soliton solution of this equation in the form $v(x,t) = \psi(x - ct)$, $(c = \text{const.})$ conclude that ψ must satisfy the ordinary equation, $\psi''' + \psi\psi' - c\psi' = 0$.
3. Integrate the last equation *twice* to get

$$3(\psi')^2 + \psi^3 - 3c\psi^2 = a\psi + b$$

for constants a, b. (*Hint:* After one integration, multiply by ψ'.)
4. When $a = b = 0$, verify that $\psi(\xi) = 3c/\cosh^2(\sqrt{c}\,\xi/2)$, $\xi \in \mathbb{R}$ gives a single-humped profile that solves the last equation and can describe a solitary traveling wave. (Other solutions of this equation involve elliptic integrals.)

5.7K Suppose that γ_0 is a smooth arc in the (x,t) plane parametrized by a C^1 function $x = x_0(t)$, with slope

$$\frac{dx_0}{dt} = p\left(x_0(t), t, u_0(t)\right)$$

relative to given C^1 values $u = u_0(t)$.

1. Explain how there could fail to be a solution u to (51) with these initial values u_0 on γ_0. (*Hint:* Consider (52b).)
2. Assume that there is a solution u to the problem in part 1. Why will there be different solutions to the same problem? (*Hint:* Apply Proposition 5.5 to curves γ that cross γ_0.)

5.7L* Refer to Proposition 5.5, its terminology and assumptions, and for $s \in I$, let $\mathscr{C}_s$, the unique characteristic of (51) through the point $\left(x_0(s), t_0(s), u_0(s)\right)$, be parametrized by the functions $x = f(s,t)$, $u = g(s,t)$, say. We can assume that f and g are C^1 [C–L], and by construction, $f_t = p$, $g_t = r$. (Why?)

1. Show that along γ_0: $x_0' = f_s + pt_0'$, so that if (53) holds, then $f_s \neq 0$, and we can use implicit function theory to solve the equation $x = f(s,t)$ in the form $s = \phi(x,t)$, at least in a neighborhood of any given point on γ_0. Moreover ϕ is C^1 in this neighborhood.
2. Use the fact that $x \equiv f(\phi(x,t), t)$ to evaluate ϕ_x and ϕ_t in terms of f_s and $f_t = p$. Then verify that u as defined satisfies (51), so that the graph of $u = g(\phi(x,t), t)$ is a *solution surface element* of (51) that contains a part of Γ_0.
3. Draw pictures illustrating how the surface elements found in part 2 might be assembled into a solution *strip* $\mathscr{U}_0$ that contains Γ_0.

5.7M* Find the solution of the (Euler) equation $tu_t + xu_x = 2u$, for $t > 0$ such that at each t_0 on the parabolic are $\gamma_0 = \{(x_0, t_0) : x_0 = t_0^2,\ t_0 > 0\}$, u has the values $u_0 = 3t_0^3$. (*Hint:* Take $s = t_0$ and divide the equation by t.)

II

Extensions

In Chapters 3–5 we derived the most important second-order partial differential equations modeling physical processes: the heat equation, the potential equation, and the wave equation. In each case we showed how separation methods involving Fourier series components can solve simple boundary value problems for these equations. Other problems do not yield so readily, and in the concluding chapters we confront some of the difficulties.

In Chapter 7, we will see what occurs when separation does not lead to a pure Fourier series but instead produces a series of eigenfunctions of a more general (Sturm–Liouville) eigenvalue problem. Fortunately, many properties of Fourier series—especially orthogonality and its consequences—are also found in the more general case and we can use them to generate formal series solutions to our problems. In particular, using series of Bessel functions or Legendre functions, we will find formal solutions to boundary value problems for standard equations in cylindrical or spherical regions. Of course, we would like to have actual solutions to these problems, so in Chapter 9 we will examine Sturm–Liouville problems in more depth in order to understand the behavior of our series.

In some problems involving spatially unbounded regions, there are "too many" eigenfunctions to permit solution by series superposition. Instead we try **integral** superposition, and when the region is "rectangular" (for example, a quarter-plane or half-space), this superposition usually takes the form of an **integral transform**. In Chapter 8 we will treat various integral transforms—including the cosine, sine, Fourier, and Laplace transforms—together with applications. These transforms involve infinite integrals that are not as manageable as infinite series, primarily because the improper integral $\int_0^\infty f(x)\,dx$ can converge even though $f(x) \not\to 0$ as $x \to +\infty$.

But first, in Chapter 6, we will look at more general partial differential equations and learn how the local character of their solutions is related to the global behavior of the processes they describe.

6

General Second-Order Linear Equations; Systems

In two variables x, y, the most general second-order linear partial differential equation for $u = u(x, y)$ can be expressed as

$$au_{xx} + 2bu_{xy} + cu_{yy} = du_x + eu_y + gu + f = \mathscr{L}(u, u_x, u_y), \text{ say,} \tag{1}$$

where the coefficient *functions*, $a, b, \ldots, f, g$, are given in some planar domain V in which a solution $u = u(x, y)$ is desired. (a, b, and c are assumed continuous and not simultaneously zero.)

In previous chapters we examined three special cases of equation (1) that arise in modeling physical phenomena: the potential equation, where $b = 0$ and $ac > 0$; the heat equation, where $b = c = 0$ and $ae > 0$, and the wave equation, where $b = 0$ and $ac < 0$. We focused most of our attention on equations with constant coefficients, although the original derivations permitted such terms as $(\tau u_x)_x = \tau u_{xx} + \tau_x u_x$, for nonconstant τ.

In this chapter we discover that *locally* the solutions to equation (1) behave qualitatively as if they were solutions of one of the three standard types. In particular, each equation exhibits local domains of influence or dependence of one of these types as indicated by the existence of special curves (characteristics) that allow discontinuities in its solutions. Then we use this insight to discuss the qualitative behavior of phenomena in electrodynamics and fluid mechanics that are governed by second-order linear equations. We conclude the chapter with a brief discussion of quasilinear systems and their applications.

6.1

The Cauchy Problem

Let's begin our analysis of equation (1) by considering the behavior of its solutions in the neighborhood of a given point. The solution of an *ordinary* second-order linear differential equation is usually determined locally by its value at a point, together with that of its derivative at the same point (an exception arises only when the coefficient of the second-derivative term vanishes at the point). Is there an analogous result for partial differential equations? Perhaps. d'Alembert's formula of Section 5.6 expresses the solution $u = u(x, t)$ of the wave equation $u_{xx} - u_{tt} = 0$ in terms of its values on a segment γ of the line $t = 0$, together with those of u_t on this segment, and we note that u_t is a *derivative of u normal* to γ. For equation (1), we replace the segment by a *smooth curve* γ with an interior point (x_0, y_0), and on γ we specify values for both u and u_n, its derivative *normal to* γ. This constitutes a set of **Cauchy data** for u on γ, and the associated **Cauchy problem** consists of finding a solution u to equation (1) in a neighborhood V_0 of (x_0, y_0) that has these data on $\gamma_0 = \gamma \cap V_0$. The Cauchy problem is seen to be a kind of local boundary value problem for equation (1) relative to γ, but we consider γ as located within a larger domain V, as shown in Figure 6.1.

Suppose that γ is parametrized by C^1 functions $x(s)$, $y(s)$ with derivatives $\dot{x}(s)$, $\dot{y}(s)$, where s is an arc length along γ so that $\dot{x}^2 + \dot{y}^2 = 1$. When u is prescribed on γ, then $u(x(s), y(s))$ is specified, and through the chain rule, its derivative with respect to s is given by

$$u_s = u_x\dot{x} + u_y\dot{y}.$$

Here we assume that u is C^1 in V and evaluate its derivatives on γ. Then on γ, in the direction $\mathbf{n} = (-\dot{y}, \dot{x})$ indicated in Figure 6.1, u has the normal derivative

$$u_n = \nabla u \cdot \mathbf{n} = -u_x\dot{y} + u_y\dot{x}.$$

Thus, if Cauchy data for u is prescribed on γ, the left sides of these equations are

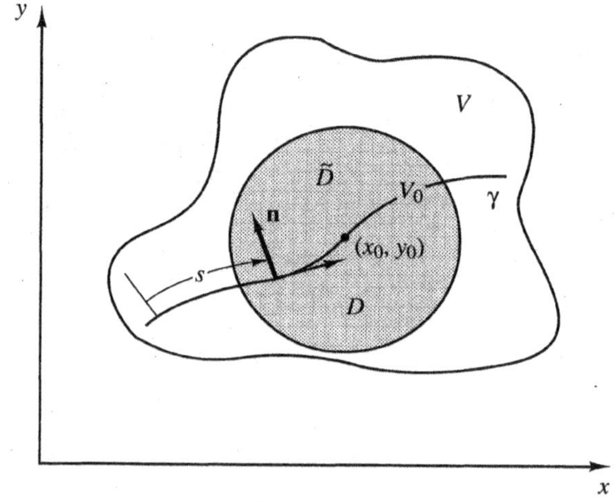

Figure 6.1

known and the equations can be solved for u_x and u_y on γ since the relevant determinant is just $\dot{x}^2+\dot{y}^2$. Consequently, $u_x(x(s),y(s))$ and $u_y(x(s),y(s))$ are specified, and assuming that u is C^2 in V, we can calculate their s-derivatives as follows:

$$\begin{aligned} u_{xs} &= u_{xx}\dot{x}+u_{xy}\dot{y} \\ u_{ys} &= u_{xy}\dot{x}+u_{yy}\dot{y} \qquad (\text{since } u_{yx}=u_{xy}). \end{aligned} \tag{2}$$

When u also satisfies equation (1), then along γ we have three equations in the *three* unknowns u_{xx}, u_{xy}, and u_{yy}. These equations have unique solutions provided that the determinant of their coefficients is nonvanishing; i.e., if

$$\begin{vmatrix} a & 2b & c \\ \dot{x} & \dot{y} & 0 \\ 0 & \dot{x} & \dot{y} \end{vmatrix} = a\dot{y}^2 - 2b\dot{x}\dot{y} + c\dot{x}^2 \neq 0. \tag{3}$$

We will extend these arguments and consider higher derivatives presently, but first let's see what we can learn using what we have thus far.

Characteristics; Weak Discontinuities

Since γ is C^1, we may take the neighborhood V_0 of (x_0,y_0) so small that γ divides V_0 into two subdomains D and $\tilde{D}$ as shown in Figure 6.1.

Suppose that v is C^1 in V_0 and satisfies equation (1) both in D and in $\tilde{D}$. Must v be a solution to equation (1) in V_0? We know that $u=v|_D$ and $\tilde{u}=v|_{\tilde{D}}$ have the same Cauchy data on $\gamma_0=\gamma\cap V_0$. Assume provisionally that each second derivative of u and of $\tilde{u}$ has a continuous extension to γ_0. Then, when condition (3) holds, the preceding argument shows that the respective extensions of each second derivative must agree on γ_0, and thus v is a C^2 solution of equation (1) in V_0.

However, when along γ condition (3) does *not* hold but instead

$$a\dot{y}^2 - 2b\dot{x}\dot{y} + c\dot{x}^2 = 0, \tag{4}$$

γ is said to be a **characteristic** (curve) for equation (1). Then it is possible for a C^1 function v to have second normal derivatives with jump discontinuities across γ even though v is an actual solution to (1) on each side of γ. Such **weak discontinuities**, if present at a point (x_0,y_0), can "propagate" only along these characteristic curves. Moreover, as we saw in Section 5.6, these characteristics can also carry discontinuities in the *first* derivatives of u. Finally, since characteristics permit such discontinuities, they provide the natural one-sided boundaries for domains of influence and dependence. Unless they are present, the solution at one point affects — and is affected by — the solution at any other point in the same domain.

Along γ, the slope of the tangent line at any point where $\dot{x}\neq 0$ is given by $y'=dy/dx=\dot{y}/\dot{x}$. If γ is a characteristic, then dividing the terms of (4) by $\dot{x}^2$, we see that y' must satisfy the quadratic equation

$$ay'^2 - 2by' + c = 0. \tag{4'}$$

Now, what does (4′) tell us about our standard equations? First, consider the Laplace equation $u_{xx}+u_{yy}=0$, where $a=c=1$, $b=0$, and (4′) reduces to $y'^2+1=0$. Since this is not possible when y' is real valued, we conclude that Laplace's equation does *not* have characteristics and therefore its solutions *cannot* have weak discontinuities.

Second, note that the heat equation $u_{xx} - u_y = 0$ has $a = 1$ and $b = c = 0$, so that (4′) becomes $y'^2 = 0$ and permits only the lines $y = \text{const.}$ as characteristics. Discontinuities of u_{yy} are possible at any line $y = y_0$, but they cannot propagate into the regions $\{y > y_0\}$ or $\{y < y_0\}$.

Finally, look at the wave equation $u_{xx} - u_{yy} = 0$, where $a = 1$, $b = 0$, $c = -1$, and (4′) becomes $y'^2 - 1 = 0$ or $y' = \pm 1$. This equation has as characteristics the lines $y \pm x = \text{const.}$, whose effects were discussed in Section 5.6. In particular, there is one pair of characteristic lines through each point (x_0, y_0).

For the general equation (1), condition (4) permits real characteristics only when the discriminant $\delta \stackrel{\text{def}}{=} b^2 - ac \geq 0$. Then, if $a \neq 0$, we solve (4′) for y' to obtain the ordinary differential equations

$$y' = \frac{dy}{dx} = \frac{b + \sqrt{b^2 - ac}}{a} \tag{5+}$$

$$y' = \frac{dy}{dx} = \frac{b - \sqrt{b^2 - ac}}{a} \tag{5−}$$

whose respective integral curves $\xi(x, y) = \text{const.}$ and $\eta(x, y) = \text{const.}$, say, are characteristics of equation (1). If $\delta(x_0, y_0) > 0$, then two characteristics pass through (x_0, y_0) with distinct slopes.

In Problems 6.1C-E, we explore extensions of these concepts to first-order linear systems. In fact they also apply to so-called quasilinear systems (see Section 6.5) and to equations in higher dimensions. For example, the two-dimensional wave equation $u_{tt} = c^2\nabla^2 u$ has as characteristics in $\mathbb{R}^3$ conical surfaces that separate the present's dependence on the past from its influence on the future. (See [Ga] or [C-H].)

Local Existence and Uniqueness

There remains the important question of whether given Cauchy data on a *non-characteristic* curve γ does determine locally a corresponding unique solution to equation (1). This question is quite deep and we will only indicate the nature of some of its answers.

First, suppose that the coefficient functions of equation (1) are infinitely differentiable (C^∞), together with the data functions $u|_\gamma$ and $u_n|_\gamma$, and that we seek a C^∞ solution u to equation (1). Then we may differentiate equation (1) as often as desired with respect to x or y and evaluate it at (x_0, y_0). In this way, if (3) holds, we can "solve" for all of the values of the partial derivatives of u at (x_0, y_0) in a manner similar to that used in obtaining u_{xx}, u_{xy}, and u_{yy} along γ. If such knowledge is sufficient to determine u locally, then it is clear that the given Cauchy data permits at most one solution to this Cauchy problem.

This is the case when u is *real analytic*, in that for some $r > 0$, u admits an absolutely and uniformly convergent power series representation in the form

$$u(x, y) = \sum_{m,n=0}^{\infty} u_{mn} \frac{(x - x_0)^m}{m!} \frac{(y - y_0)^n}{n!}$$

when $|x - x_0| + |y - y_0| \leq r$. Here, $u_{mn} = \dfrac{\partial^{m+n} u}{\partial x^m \partial y^n}(x_0, y_0)$, $m, n = 0, 1, 2, \ldots$.

For existence of u, the difficulty lies in showing by a priori estimates that the coefficients u_{mn} are small enough to guarantee convergence of this series. The necessary analysis was initiated in 1842 by Cauchy and completed by Sonia Kowalewsky in 1875; it yields the following celebrated result.

(6.1) Theorem: *If each of the coefficients $a, \ldots, f, g$ of equation* (1) *is real-analytic near (x_0, y_0), a point on an analytic*† *curve γ where condition* (3) *holds, then equation* (1) *has a unique real-analytic solution $u = u(x, y)$ near (x_0, y_0) with given real-analytic Cauchy data $u|_\gamma$* and *$u_n|_\gamma$.*

Proof: See [Ga]. ■

(6.2) Remarks: With appropriate modifications, this result holds for first-order linear equations and systems thereof and so for higher-order linear equations. (See Problem 6.1E.) It remains valid for certain quasilinear equations; for example, it holds in equation (1) if the coefficient functions depend analytically on u, u_x, and u_y. Considerable efforts have been made to relax the various requirements of real analyticity with much success. However, unified treatment of the resulting problems is complicated because the solution to the Cauchy problem need *not* depend continuously on its Cauchy data (Problem 6.1F), so that standard approximation techniques cannot be applied.

Instead, the problems are subdivided into the three types suggested by the classification scheme in the next section. For each category special techniques have evolved to attack the Cauchy problem and associated boundary value problems. (See, for example, [G-T], [Fr], and [Wh].)

Many of these results are made possible by enlarging the class of functions we accept as solutions. We have already made the distinction between actual and formal solutions to a given problem, the latter being perhaps insufficiently differentiable. With the use of distributional methods introduced by Sobolev (c 1933) and extended by L. Schwartz, the sense in which these formal solutions do "solve" the equation can be made precise. (However, the sense in which these new objects can be said to have requisite boundary values is a very delicate matter—especially when the boundary of the domain is irregular). (See [Ho] and [L-M].) The hope that such efforts might completely eliminate analyticity requirements ended in 1957 when H. Lewy discovered a simple linear system, $\mathbb{D}\mathbf{u}_x + \mathbb{E}\mathbf{u}_y = \mathbf{F}$, which has local solutions for real-analytic $\mathbf{F}$ but does not have even C^1 solutions for certain nonanalytic $\mathbf{F} \in C^\infty$. (See Problem 6.5M.) However, we also have Carleman's proof from 1939 that nonanalytic solutions to such systems are uniquely determined by their Cauchy data on any analytic noncharacteristic curve. See the discussion of Holmgren's uniqueness result in [C-H].

---

† A planar curve is *analytic* when it can be parametrized by a pair of real-analytic functions on a common interval.

Problem Set 6.1

6.1A Explain carefully why equations (1) and (2) along γ lead to equation (3).

6.1B Obtain integral curves for equations (5+) and (5−) when

1. $a = 1,\quad b = 3,\quad c = 3$
2. $a = 4,\quad b = 2,\quad c = 1$
3. $a = 1,\quad b = x + y,\quad c = 2xy + x^2$
4. $a = \cos x,\quad b = 1,\quad c = \cos x$
5. $a = 1,\quad b = \sinh x,\quad c = -1$

6.1C Suppose we consider, instead of equation (1), the first-order linear equation $du_x + eu_y = f + gu$ for continuous coefficients d, e, f, g in some domain $\mathscr{D} \subseteq \mathbb{R}^2$, in which $d^2 + e^2 \neq 0$.

1. How should the Cauchy problem for this equation be formulated along a curve γ in $\mathscr{D}$?
2. Show that if $u(x(s), y(s))$ is known along γ, then so are u_x and u_y when $d\dot{y} - e\dot{x} \neq 0$.
3. How should the characteristics be identified, and what should be considered "weak" discontinuities for this equation?

6.1D 1. Show that in terms of new unknown variables $v = u_x$ and $w = u_y$, equation (1) can be replaced by the following first-order linear system in (u, v, w):

$$u_x = v, \qquad u_y = w, \qquad av_x + 2bv_y + cw_y = f + dv + ew + gu.$$

2. For this system to be equivalent to equation (1), v and w must be compatible. What equation expresses this compatibility?

6.1E A first-order linear system in x, y [such as that in the preceding example for the *vector* $\mathbf{u} = (u, v, w)$] can be written in the matrix form $\mathbb{D}\mathbf{u}_x + \mathbb{E}\mathbf{u}_y = \mathbf{F} + \mathbb{G}\mathbf{u}$ for appropriate *matrices* $\mathbb{D}$, $\mathbb{E}$, $\mathbf{F}$, $\mathbb{G}$.

1. What are the corresponding matrices for the system in Problem 6.1D?
2. Comparing with Problem 6.1C, what is the Cauchy problem for such a system?
3.* What should be the equation for its characteristics?

6.1F The Cauchy problem for Laplace's equation is improperly posed (see Section 3.5). To establish this assertion in the half-plane $V = \{(x, y) : x > 0\} \subseteq \mathbb{R}^2$, consider the following example of Hadamard (1917).

1. Show that $u_m(x, y) = (1/m^2) \sinh mx \sin my$ satisfies $\nabla^2 u = 0$ in V, with $u_m(0, y) = 0$. Also
2. $(\partial u_m/\partial x)\,(0, y) = (1/m) \sin my \to 0$, as $m \to \infty$, but
3. $u_m(1, \pi/2) \nrightarrow 0$ as $m \to \infty$.

6.2

Classification

Now let's use the characteristic condition (4) to classify equation (1),

$$au_{xx} + 2bu_{xy} + cu_{yy} = \mathscr{L}(u, u_x, u_y),$$

in a domain $\mathscr{D}$ of $\mathbb{R}^2$. Since (4) is quadratic in $\dot{y}/\dot{x}$ (or in $\dot{x}/\dot{y}$), we have the usual trichotomy associated with the sign of $b^2 - ac$. It is customary to identify each case by the name of the conic section with equation $ax^2 + 2bxy + cy^2 = 1$, when a, b, and c are held constant.

Assume provisionally that the coefficient a does not vanish in $\mathscr{D}$.

Elliptic case: $b^2 - ac < 0$ in $\mathscr{D}$. Then real characteristics are not present and equation (1) is said to be **elliptic** in $\mathscr{D}$. Its solutions cannot exhibit weak discontinuities or restricted domains of influence/dependence. The standard elliptic equation is that of Laplace, $u_{xx} + u_{yy} = 0$, for which $b^2 - ac = -1$ everywhere. Moreover, equation (1) can be replaced locally by the simpler equation

$$\rho\tilde{u}_{\xi\xi} + \tilde{u}_{\eta\eta} = \tilde{\mathscr{L}}(\tilde{u}, \tilde{u}_\xi, \tilde{u}_\eta) \tag{6a}$$

for an appropriate function $\tilde{\mathscr{L}}$ (see equation (9)) and positive function ρ, under the transformation $u(x, y) = \tilde{u}(\xi, \eta)$, where

$$\frac{dy}{dx} = \frac{b}{a} \quad \text{has integral curves } \xi(x, y) = \text{const.}; \qquad \text{and } \eta = x. \tag{6b}$$

In fact, $\rho = (ac - b^2)\xi_y^2/a$, and when ρ is constant, a simple scale change in which ξ is replaced by $\xi/\sqrt{\rho}$ reduces the left side of (6a) to the Laplace operator. (See Problem 6.2F.)

Hyperbolic case: $b^2 - ac > 0$ in $\mathscr{D}$. Then through each point in $\mathscr{D}$ pass two characteristics with distinct slopes, namely, the appropriate integral curves of equations (5+) and (5−). Equation (1) is said to be **hyperbolic** in $\mathscr{D}$ and its solutions exhibit domains of influence/dependence bounded by characteristics, across which weak discontinuities can occur. Moreover, such discontinuities, if present, propagate only along characteristics. These facts have been developed into an efficient scheme for approximating solutions known as the **method of characteristics**. (See [St], Ch. 9.) The standard hyperbolic equation is $\sigma^2 u_{xx} - u_{yy} = 0$, for which $b^2 - ac = \sigma^2 > 0$. As we saw in Section 5.5, when σ is constant this can be reduced to $\tilde{u}_{\xi\eta} = 0$ by the transformation to characteristic coordinates $\xi = x - \sigma y$, $\eta = x + \sigma y$. Similarly, equation (1) can be replaced locally by the simpler equation

$$\tilde{u}_{\xi\eta} = \tilde{\mathscr{L}}(\tilde{u}, \tilde{u}_\xi, \tilde{u}_\eta) \tag{7}$$

through transformation to the characteristic coordinates ξ, η obtained from the integral curves of (5+) and (5−). (See Problem 6.2G.)

Parabolic case: $b^2 - ac \equiv 0$ in $\mathscr{D}$. Then the curves of (5+) and (5−) coincide and there is only one characteristic through each point in $\mathscr{D}$. In this case equation

(1) is said to be **parabolic** in $\mathscr{D}$ and its solutions can exhibit weak discontinuities across characteristics, which bound domains of influence/dependence.

The standard parabolic equation is $u_{xx} = u_y$, and locally, equation (1) can be replaced by the simpler equation

$$\tilde{u}_{\eta\eta} = \tilde{\mathscr{L}}(\tilde{u}, \tilde{u}_\xi, \tilde{u}_\eta) \tag{8a}$$

by means of the transformation $u(x, y) \equiv \tilde{u}(\xi, \eta)$ where

$$\frac{dy}{dx} = \frac{b}{a} \quad \text{has integral curves } \xi(x, y) = \text{const.}; \qquad \text{and } \eta = x. \tag{8b}$$

(See Problem 6.2F.)

(6.3) Remarks: In the normal forms (6)–(8) above, $\tilde{\mathscr{L}}$ is given by

$$\tilde{\mathscr{L}}(\tilde{u}, \tilde{u}_\xi, \tilde{u}_\eta) = D\tilde{u}_\xi + E\tilde{u}_\eta + G\tilde{u} + F, \tag{9}$$

where D, E, F, and G may depend on ξ, η. The underlying transformations are not unique, and in specific instances other choices may be preferable.

We have assumed that a, b, and c are continuous with values a_0, b_0, and c_0 at $(x_0, y_0) \in \mathscr{D}$; then $\delta = b^2 - ac$ is also continuous with value δ_0 at this point. When $\delta_0 \neq 0$, then near this point δ has the sign of δ_0 and local ellipticity or hyperbolicity of equation (1) can be inferred. However, when $\delta_0 = 0$, we can deduce only that there is one characteristic *direction* at (x_0, y_0); i.e., one direction for a tangent line to a characteristic curve through this point. (See Example 1.)

In obtaining (5±) we assumed that $a \neq 0$, which would follow near (x_0, y_0) if $a_0 \neq 0$. If $a_0 = 0$ but $c_0 \neq 0$, then we can simply interchange x and y (which does not affect the value of $b_0^2 - a_0 c_0$). If $a_0 = c_0 = 0$, then $b_0 \neq 0$ and equation (1) is hyperbolic near (x_0, y_0) with characteristics through this point that have horizontal and vertical slopes. Indeed, when $a \equiv c \equiv 0$, in $\mathscr{D}$, then equation (1) is already of the normalized form (7) and has only the lines $x = \text{const.}$, $y = \text{const.}$ as characteristics. ■

It is possible to make similar classifications of second-order linear equations in higher dimensions relative to the existence of characteristic hypersurfaces. For example, Laplace's equation, $\nabla^2 u = 0$, is elliptic and the standard wave equation, $u_{tt} = \nabla^2 u$, is hyperbolic, while the heat equation, $u_t = \nabla^2 u$, is parabolic. (See [Ga] or [C-H].)

Example 1:

The equation

$$u_{xx} + 4u_{xy} + xu_{yy} + 3(\sin x)u_y = 0 \tag{10}$$

has coefficients $a = 1, b = 2, c = x$, so that

$$b^2 - ac = 4 - x.$$

Thus in the half-plane where $x < 4$, this equation is hyperbolic; in the complementary half-plane where $x > 4$, the equation is elliptic.

In particular, near the point (3, 0), equations (5 ±) provide characteristics through (3, 0) with slopes

$$y' = 2 \pm \sqrt{4-x};$$

$$\text{namely,} \qquad \begin{cases} y_1(x) = 2x - \frac{2}{3}\left[(4-x)^{3/2} + 8\right] \\ y_2(x) = 2x + \frac{2}{3}\left[(4-x)^{3/2} - 10\right] \end{cases}, \tag{11}$$

whose graphs for $x < 4$ are shown in Figure 6.2. More generally, we see from equations (11) that for $x < 4$, characteristic coordinates are given by

$$\xi = y - 2x + \tfrac{2}{3}(4-x)^{3/2}; \qquad \eta = y - 2x - \tfrac{2}{3}(4-x)^{3/2},$$

and these can be used to normalize (10). (See Problem 6.2C.)

Strictly speaking, we cannot classify equation (10) at (4, 0). (Why?) However, if we perform the same computations as above at (4, 0), then we obtain for $x < 4$ the functions

$$\begin{aligned} y_1(x) &= 2x - \tfrac{2}{3}(4-x)^{3/2} - 8 \\ y_2(x) &= 2x + \tfrac{2}{3}(4-x)^{3/2} - 8 \end{aligned}, \tag{12}$$

which have the same limiting slope 2 as $x \nearrow 4$, also shown in Figure 6.2.

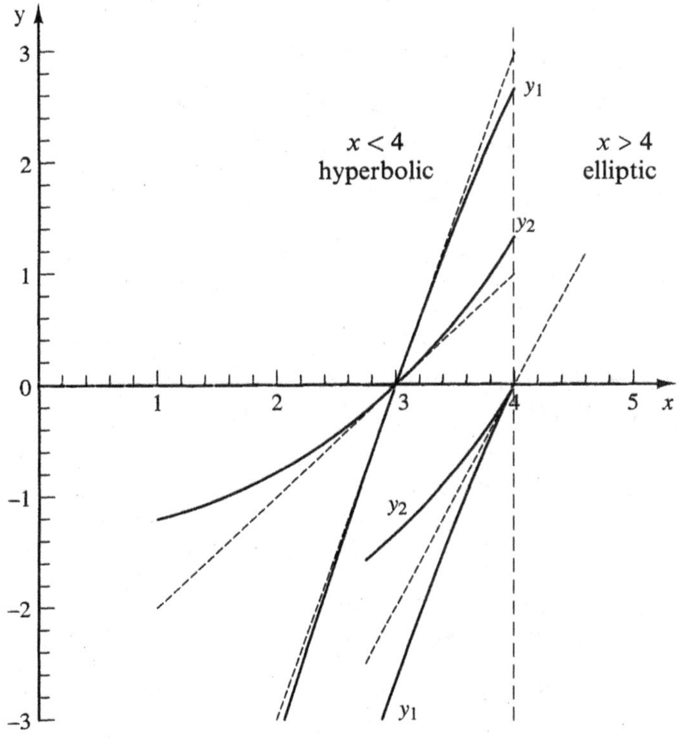

Figure 6.2

In this sense we might say that there is "one" characteristic direction at $(4, 0)$; however, each associated characteristic is one-sided in extent.

Such complications cannot arise when a, b, and c are constants.

Example 2:

$$u_{xx} + 2u_{xy} + u_{yy} + 3(\sin x)\,u_y - 5yu = 0 \qquad \text{has} \qquad b^2 - ac = 1 - 1 = 0, \tag{13}$$

and the equation is parabolic everywhere. Now, equations (5+) and (5−) do give a single characteristic through each point, namely, the straight line with slope $y' = 1/1 = 1$. Hence, such characteristics have the equation $\xi(x, y) \equiv y - x = \text{const}$.

According to (8b), the transformation

$$\xi = y - x, \qquad \eta = x; \qquad \text{or} \qquad x = \eta, \qquad y = \xi + \eta$$

results in a reduced equation for

$$\tilde{u}(\xi, \eta) \stackrel{\text{def}}{=} u(\eta, \xi + \eta);$$

upon substitution, we find that

$$\tilde{u}_{\eta\eta} + (3\sin\eta)\tilde{u}_\xi - 5(\xi + \eta)\tilde{u} = 0, \tag{14}$$

and the resemblance to the heat equation is apparent.

Problem Set 6.2

6.2A Classify each of the following equations in the indicated regions.

1. $4u_{xx} + 6u_{xy} + 5u_{yy} = 6yu_x + 7.$ $(\mathbb{R}^2)$.
2. $2u_{xx} + 6u_{xy} - 5u_{yy} + 6xu_y + 4u = 3.$ $(\mathbb{R}^2)$.
3. $2xyu_{xx} - 2u_{xy} = 3(\sin y)\,u.$ $(xy > 0)$.
4. $(\cos x)\,u_{xx} + 2u_{xy} = -(\cos x)\,u_{yy}.$ $(x > 0)$.
5. $y^2u_{xx} + 2xu_{xy} = -x^2u_{yy}.$ $(|y| < 1)$; $(|y| > 1)$.

6.2B For Example 2, verify that the given transformation results in the reduced equation (14). (*Hint:* $\tilde{u}_{\eta\eta} = u_{xx} + 2u_{xy} + u_{yy}$.)

6.2C *1.* In Example 1, when $x < 4$ show that in terms of characteristic coordinates $\xi, \eta : x = \alpha$ and $y = 2\alpha + [(\xi + \eta)/2]$ $\quad (\xi > \eta)$ where $\alpha(\xi, \eta) = 4 - (3(\xi - \eta)/4)^{2/3}$ (so that $\alpha_\xi = -\alpha_\eta$).

*2.** Verify that for $\tilde{u}(\xi, \eta) = u(\alpha, 2\alpha + [(\xi + \eta)/2])$,

$$\tilde{u}_{\xi\eta} = (u_{xx} + 4u_{xy} + xu_{yy})\,\alpha_\xi\alpha_\eta + (u_x + 2u_y)\alpha_{\xi\eta}.$$

3. How does (10) simplify in these coordinates?

6.2D *1.* Verify that the equation $u_{xx} - 2u_{xy} + 2u_{yy} \equiv 1$ is elliptic in $\mathbb{R}^2$.

2. Show that (6b) yields $\xi = x + y$, $\eta = x$ and $\tilde{u}(\xi, \eta) = u(\eta, \xi - \eta)$.

3. What is (6a) in this case? (*Hint:* Calculate $\tilde{u}_{\xi\xi} + \tilde{u}_{\eta\eta}$.)

6.2E *1.* Tricomi's equation, $u_{xx} - xu_{yy} = 0$, is important in transonic gas dynamics. Give as much information as possible about the solutions of this equation when (a) $x > 0$ and (b) $x < 0$.

2. What happens when $x = 0$?

6.2F* *1.* Suppose that a family of smooth curves is represented by $\xi(x, y) = \text{const}$. Then, in general, the slope of the tangent line to each such curve satisfies the differential equation $dy/dx = -\xi_x/\xi_y$, when $\xi_y \neq 0$.

2. Thus if also as in (6b) or (8b) $dy/dx = b/a$ and $\eta = x$, show that $u(x, y) = \tilde{u}(\xi, \eta) = \tilde{u}(\xi(x, y), x)$ has derivatives

$$u_x = \tilde{u}_\xi \xi_x + \tilde{u}_\eta \qquad \text{and} \qquad u_y = \tilde{u}_\xi \xi_y.$$

3. In the same manner, find u_{xx}, u_{xy}, and u_{yy} and conclude that

$$au_{xx} + 2bu_{xy} + cu_{yy} = \alpha\tilde{u}_{\xi\xi} + a\tilde{u}_{\eta\eta} + (\text{terms linear in } \tilde{u}_\xi, \tilde{u}_\eta),$$

where $\alpha \overset{\text{def}}{=} a\xi_x^2 + 2b\xi_x\xi_y + c\xi_y^2$.

4. Show that in part 3, $\alpha = [(ac - b^2)/a]\xi_y^2$ (since $a\xi_x = -b\xi_y$). Why does this establish (8a)? (6a)?

5. In the elliptic case with constant a, b, c,where $a > 0$, make a scale change to reduce (6a) to a case where $\rho = 1$.

6.2G* *1.* When $\xi = \xi(x, y)$ and $\eta = \eta(x, y)$ define a C^2 coordinate transformation, verify that for $u(x, y) \overset{\text{def}}{=} \tilde{u}(\xi(x, y), \eta(x, y))$, we have

$$u_x = \tilde{u}_\xi \xi_x + \tilde{u}_\eta \eta_x \qquad \text{and} \qquad u_y = \tilde{u}_\xi \xi_y + \tilde{u}_\eta \eta_y.$$

2. Obtain corresponding expressions for u_{xx}, u_{xy}, and u_{yy}, and show that

$$au_{xx} + 2bu_{xy} + cu_{yy} = \alpha\tilde{u}_{\xi\xi} + 2\beta\tilde{u}_{\xi\eta} + \gamma\tilde{u}_{\eta\eta} + (\text{terms linear in } \tilde{u}_\xi, \tilde{u}_\eta),$$

where α is as in Problem 6.2F, with similar expressions for β, γ.

3. Conclude that when $b^2 > ac$ (the hyperbolic case), where ξ and η give the respective integral curves of (5+) and (5−), then $\alpha = \gamma \equiv 0$. This analysis establishes (7) and provides another approach to classification that is often used. (However, it does not explain the real significance of characteristics.)

6.3

Applications to Electrodynamics

In applications, second-order equations of the form (1) sometimes arise when unknowns are eliminated from a system of first-order equations. In this section we consider two important cases where the first-order system is itself linear.

(a) Transmission line equations: Both the telegraph and the telephone convince us that electrical signals can be transmitted effectively over long distances along conducting wires that are either suspended above the earth or buried—possibly submerged— within it. A modern transmission cable usually

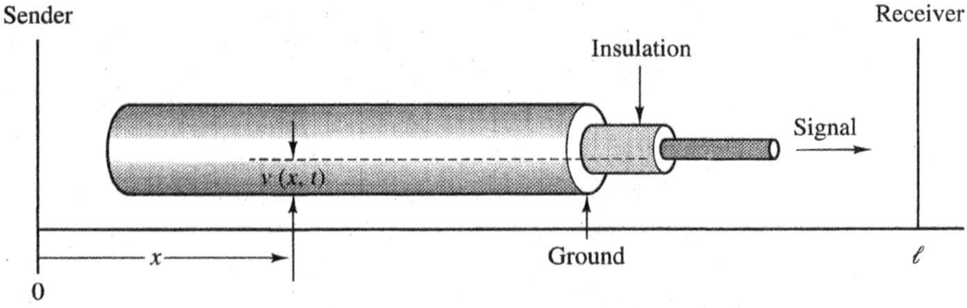

Figure 6.3

consists of a plastic or rubber insulating sleeve separating an inner "sending" core, such as copper, from a grounded outer conductor (see Figure 6.3).

The signal at time t consists of a voltage difference $v = v(x, t)$ between core and ground at distance x from the sending end where $v(0, t)$ is controlled electromechanically. Effective transmission over distance ℓ is achieved when a time history of measurements of v at ℓ can be used to infer what $v(0, t)$ must have been. The flow of electrical charge along the wire is a current $i = i(x, t)$, and experiments show that at time t, voltage is lost along the wire because of the resistance to the passage of current (Ohm's law) and the inductive effects of the time rate of change of current.

In a limiting form, this relationship can be expressed as follows:

$$-v_x = Ri + Li_t, \tag{15a}$$

where in suitable units R represents the resistance and L the inductance of a unit length of cable. Due to its construction, a unit length of cable acts as if it were a capacitor of capacitance C (to ground), and imperfect insulation permits current leakage to ground with conductance G. The resulting current loss at time t takes the limiting form

$$-i_x = Gv + Cv_t. \tag{15b}$$

In a cable of uniform properties, we can reasonably assume that the cable parameters R, L, G, and C are nonnegative constants, and under these conditions, the validity of equations (15a and 15b) has been established through extensive laboratory testing. (See [Fle].)

Next, suppose that both v and i are C^2 functions of x and t. Then upon differentiating (15a) with respect to x and substituting from (15b) as required, we can eliminate i as follows:

$$v_{xx} = -Ri_x - Li_{xt} = R(Gv + Cv_t) + L(Gv_t + Cv_{tt})$$

$$\text{or} \quad v_{xx} - LCv_{tt} - (RC + LG)v_t - RGv = 0. \tag{16}$$

Let's look at two special cases of this equation.

(1) When LC is zero, equation (16) is parabolic, and in low-frequency telegraphy the terms in equations (15a and 15b) involving L and G are usually negligible in comparison to those involving R and C, respectively. Then equation (16) reduces to the **telegraph equation**

$$v_t = (RC)^{-1}v_{xx}$$

and, as Lord Kelvin pointed out in 1855, under these conditions signal transmission is a process of diffusion, with diffusivity constant $k = (RC)^{-1}$. In particular, from (8′) in Section 3.1, we see that the simple oscillatory signal $v(0, t) = \cos \omega t$ generates the line voltage

$$v(x, t) = e^{-ax} \cos(\omega t - ax), \qquad \text{for} \qquad a = \sqrt{\omega/2k}$$

that exhibits both amplitude loss and phase shift with distance of transmission, and both effects are frequency dependent! Although such cables allow us to distinguish a dot from a dash in Morse code, they cannot transmit the complexity of human speech effectively over long distances.

(2) However, suppose we reexamine (16). When $LC \neq 0$, we see that this equation is *hyperbolic* with the same space-time characteristics as the wave equation

$$u_{tt} = c^2 u_{xx}, \qquad \text{where } c^2 = (LC)^{-1}.$$

We cannot solve (16) in the d'Alembert form, but under the additional substitutions $\alpha = R/L$, $\beta = G/C$, it becomes

$$v_{tt} - c^2 v_{xx} + (\alpha + \beta) v_t + \alpha\beta v = 0; \tag{17}$$

then, if $v(x, t) = e^{-(\alpha+\beta)t/2} u(x, t)$, we obtain the following equation for u:

$$u_{tt} - c^2 u_{xx} = \left(\frac{\alpha - \beta}{2}\right)^2 u, \tag{17′}$$

which *does* reduce to the wave equation when $\alpha = \beta$. (See Problem 6.3A.) If we could manufacture a transmission line for which $R/L = G/C$, then (16) would have the general solution

$$v(x, t) = e^{-\alpha t}[\phi(x + ct) + \psi(x - ct)], \qquad \text{where } \alpha = \frac{R}{L} = \frac{G}{C}, \tag{17″}$$

and the cable would transmit signals of any complexity with speed $c = (LC)^{-1/2}$ as waves without distortion in shape but with diminished amplitude. Such "distortionless" telephone cables were proposed by Heaviside and a modification of their design is employed today. In particular, following Pupin (1900), induction is deliberately introduced to improve transmission even though some older telegraphers regarded the resulting voltage loss as anathema! (See [Fle].)

(b) Maxwell's equations: One of the most important first-order linear systems is that of Maxwell (1864), which governs electromagnetic phenomena. We will not derive these equations, since to do so we would need substantial knowledge of both electricity and magnetism in addition to Stokes' law of vector fields. (See [Lo].) However, in some simple situations we will show that Maxwell's system is related to familiar second-order equations and then draw significant conclusions about the nature of electrodynamics.

If we suppose that a vector current is flowing in a homogeneous isotropic medium in $\mathbb{R}^3$ in which the charge density is ρ, then at time t and Cartesian position $\mathbf{x} = (x, y, z)$, Maxwell's equations relate the resulting electrical and magnetic vector-field strengths $\mathbf{E}$ and $\mathbf{H}$, respectively. In the mixed system of units where the ratio of electromagnetic to electrostatic units is denoted by c, **Maxwell's equations** can be stated as follows:[†]

$$\begin{aligned} &\text{(a)}\quad \nabla\cdot(\epsilon\mathbf{E})=\rho &&\text{(b)}\quad \nabla\cdot(\mu\mathbf{H})=0\\ &\text{(c)}\quad \nabla\times\mathbf{E}=-\frac{1}{c}(\mu\mathbf{H})_t &&\text{(d)}\quad \nabla\times\mathbf{H}=\frac{1}{c}\sigma\mathbf{E}+\frac{1}{c}(\epsilon\mathbf{E})_t. \end{aligned} \tag{18}$$

Equations (18a) and (18b) apply to the fields separately and express respectively Gauss's law relating the electrical field to the production of charge and that of Poisson affirming the fact that magnetic poles are not found in isolation. The remaining equations embody the experimental discoveries of Oersted, Ampére, Faraday, and Lenz as to how these fields interact, augmented by Maxwell's crucial observation that to avoid inconsistency with (18a) the dynamic term $(1/c)(\epsilon\mathbf{E})_t$ must be added in (18d). These equations also incorporate certain constitutive laws, each involving a nonnegative parameter μ, ϵ, or σ that measures respectively the magnetic permeability, dielectric strength, or conductivity of the medium. For our immediate purposes, each parameter can be assumed constant[‡] and in free space, both μ and ϵ are assigned the reference value 1. (See [Lo].)

For a C^1 vector field $\mathbf{f}=(u,v,w)$ in $\mathbb{R}^3$, $\nabla\times\mathbf{f}$ ($\equiv$ curl $\mathbf{f}$) is the vector field with components $(w_y-v_z, u_z-w_x, v_x-u_y)$. Then, if $\mathbf{f}$ is spatially C^2 it can be shown that

$$\begin{aligned} &\text{(a)}\quad \nabla\cdot(\nabla\times\mathbf{f})=0\\ &\text{(b)}\quad \nabla\times(\nabla\times\mathbf{f})=\nabla(\nabla\cdot\mathbf{f})-\nabla^2\mathbf{f}. \end{aligned} \tag{19}$$

(See Problem 6.3B.)

Now, let's specialize to a nonconducting medium, such as air or glass, for which $\sigma=0$ and assume that $\mathbf{E}$ and $\mathbf{H}$ have C^2 components. Then using (19b), (18b), (18d), and (18c) successively, we find that

$$\nabla^2\mathbf{H}=\nabla(\nabla\cdot\mathbf{H})-\nabla\times(\nabla\times\mathbf{H})=-\nabla\times\left(\frac{\epsilon}{c}\mathbf{E}_t\right)=-\frac{\epsilon}{c}(\nabla\times\mathbf{E})_t,$$

$$\text{or,}\qquad \nabla^2\mathbf{H}=-\frac{\epsilon}{c}\left(-\frac{\mu}{c}\mathbf{H}_t\right)_t=\frac{\epsilon\mu}{c^2}\mathbf{H}_{tt}.$$

Therefore each component of the magnetic field-strength $\mathbf{H}$ satisfies the three-dimensional wave equation!

It is shown in Problems 6.3C, D that $\sigma=0\Rightarrow\rho_t=0$ and that when the residual charge $\rho\equiv0$, each component of $\mathbf{E}$ satisfies this *same* wave equation. Hence, in such media, both electricity and magnetism manifest waves propagating with speed $c/\sqrt{\epsilon\mu}$. Experiments reveal that in free space (where $\epsilon=\mu=1$) c has the same value as the velocity of light; thus, these waves travel through free space with light speed. Conversely, with Maxwell, we might conjecture that light is electromagnetic in origin and wavelike in behavior.

Observe that when all quantities are independent of time, then from (18c), $\nabla\times\mathbf{E}\equiv\mathbf{0}$. It is easily verified that this equation holds when $\mathbf{E}=-\nabla\phi$ for some

---

[†] In these equations, a factor of 4π is absorbed into both ρ and σ, and the ratio c can also be eliminated by redefinition.

[‡] In fact, μ, ϵ, and σ are functions, which for anisotropic materials such as crystals, should be regarded as square matrices.

C^2 function ϕ (called an electrostatic potential). Conversely, having $\nabla \times \mathbf{E} \equiv \mathbf{0}$ ensures that $\mathbf{E} = -\nabla\phi$ for some function ϕ in each ball and in certain other domains. (See Problem 6.3B.) When $\mathbf{E} = -\nabla\phi$, (18a) becomes

$$\nabla \cdot (\epsilon \nabla \phi) = -\rho,$$

which is just our potential equation from Section 4.1, and this suggests a mechanical-electrothermal analogy. When a variable dielectric ϵ is C^1, this equation takes the form

$$\epsilon \nabla^2 \phi + \nabla\epsilon \cdot \nabla\phi = -\rho;$$

when ϵ is constant, it becomes Poisson's equation,

$$\nabla^2 \phi = -\frac{\rho}{\epsilon}.$$

Other special cases of Maxwell's equations suggest themselves. After suitable elimination, most of them can be recognized in relation to those emphasized in this book, and solutions can be obtained in sufficiently simple cases. (See Problems 6.3D,E.) Because of its importance, Maxwell's system (18) has been examined extensively, and many of its properties have been discovered, including the auxiliary conditions appropriate to various applications. (See [Mu] for a mathematical account.)

Problem Set 6.3

6.3A *1.* Verify that (17) follows from (16).

2. Show that (17′) follows from (17) when $v = ue^{-(\alpha+\beta)t/2}$.

6.3B *1.* For $\mathbf{f} = (u, v, w)$, verify equations (19a) and (19b), and if ϕ is C^2, show that $\nabla \times (\nabla\phi) = 0$.

2. Suppose that for a C^1 vector field $\mathbf{f} = (u, v, w)$, we have curl $\mathbf{f} \equiv \mathbf{0}$ in a sphere V centered at the origin. Then inside V, we can *define* a function ϕ with $\nabla\phi = \mathbf{f}$ by an integral of its desired partial derivatives along coordinate lines; viz.,

$$\phi(x,y,z) \overset{\text{def}}{=} \int_0^x u(\xi,0,0)\,d\xi + \int_0^y v(x,\eta,0)\,d\eta + \int_0^z w(x,y,\zeta)\,d\zeta.$$

(*i*) Verify that $\phi_z(x,y,z) = w(x,y,z)$.

(*ii*) Verify that $\phi_y(x,y,z) = v(x,y,0) + \int_0^z w_y(x,y,\zeta)\,d\zeta$ and use $w_y(x,y,\zeta) = v_\zeta(x,y,\zeta)$ (from where?) to eliminate the integral to get $\phi_y = v$.

(*iii*)* Use similar means to show that $\phi_x = u$.

6.3C *1.* Apply (19a) to (18d) with σ, ϵ, μ constant and substitute as necessary to get

$$\rho_t + a\rho = 0, \qquad \text{for } a = \sigma/\epsilon.$$

2. Conclude that $\rho = \rho_0 e^{-at}$, where $\rho_0 = \rho|_{t=0}$, so that in a conductor an initial charge ρ_0 dies out exponentially in time.

6.3D When $\rho = 0$ (and σ, ϵ, μ are constants) show that each component of $\mathbf{E}$ in (18) satisfies a *damped* wave equation. Obtain a similar equation for $\mathbf{H}$. What happens to these equations when $\rho \neq 0$?; when $\sigma = 0$?

6.3E When all quantities in equation (18) are independent of time, show that

$$\nabla \cdot (\sigma \mathbf{E}) = 0, \qquad \text{while } \nabla \cdot (\epsilon \mathbf{E}) = \rho.$$

What can you conclude?

6.3F The (Navier–Cauchy) equations for the dynamic deformation of a three-dimensional isotropic elastic body may be written as follows:

$$(\rho \mathbf{u}_t)_t = (\lambda + \mu)\nabla \theta + \mu \nabla^2 \mathbf{u} + \mathbf{F}$$

where the dilatation $\theta = \nabla \cdot \mathbf{u}$, $\rho = \rho(x, y, z, t)$ is the mass density, λ and μ are the Lamé constants, $\mathbf{F}$ is the distributed external force, and $\mathbf{u} = \mathbf{u}(x, y, z, t) = (u, v, w)$ is the (vector) displacement. (See Problem 5.1G.)

1. When ρ is constant, $\mathbf{F} = \mathbf{0}$, and $\mathbf{u}_y = \mathbf{u}_z = \mathbf{0}$, show that u and v each satisfy one-dimensional wave equations — but with different speeds of propagation, giving rise to the possibility of both longitudinal and transverse waves. (See Problem 5.1E.)
2.* More generally, when ρ is constant, show that $\theta = \nabla \cdot \mathbf{u}$ satisfies a three-dimensional wave equation, as does the vector $\nabla \times \mathbf{u}$. (*Hint:* Use equations (19).) This indicates how an earthquake can produce seismic waves of different types and speeds in an elastic earth. Moreover, since only transverse waves are observed in free space, it advances a theoretical argument against the presence of an elastic "aether" to support electromagnetic waves.

6.4

Applications to Fluid Mechanics

Matter in gaseous or liquid state is called a **fluid**, and a fluid in motion is called a **flow**. In particular, both water and air are fluids, and their flow determines not only our technology but also our lives. For this reason we want to use the equations governing a flow to make inferences about its behavior. If we regard the flow as a continuum, then its thermodynamic-Eulerian state at time t and position $\mathbf{x} = (x, y, z)$ relative to a *fixed* rectangular coordinate system is specified by its velocity vector $\mathbf{v} = (u, v, w)$, its static pressure p, its mass density ρ, and its absolute temperature T.

We need equations relating these variables, and to simplify their derivation, let's assume that our flow is inviscid (frictionless) and neglect the effects of gravity and chemical reactions. Then consider a fixed control volume V (such as a ball) with boundary S, where $\bar{V} = V \cup S$ is located within the fluid, as shown in Figure 6.4.

(a) The continuity equation: In the absence of internal sources, conservation of mass relative to V requires that the mass change within V be accounted for by mass-flux across S. Thus,

$$\frac{\partial}{\partial t}\int_V \rho \, dV = -\int_S \rho \mathbf{v} \cdot \mathbf{n} \, dS,$$

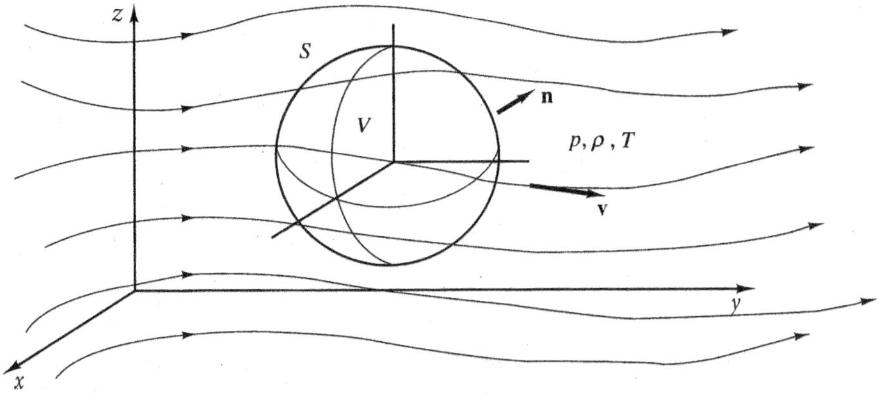

Figure 6. 4

where $\mathbf{n} = (n_1, n_2, n_3)$ denotes the outward-directed unit vector normal to S, so that $-\mathbf{v}\cdot\mathbf{n}$ gives the normal flow velocity *into* V.

Upon transforming the surface integral through the divergence theorem (0.2) we see that

$$\int_V [\rho_t + \nabla\cdot(\rho\mathbf{v})]\, dV = 0,$$

for each fixed control volume, V. If we assume that the integrand is continuous, we obtain our first important result:

$$\rho_t + \nabla\cdot(\rho\mathbf{v}) = 0. \qquad \textit{(Continuity equation)} \quad \mathbf{(20)}$$

The continuity equation may be expressed in the form

$$\frac{d\rho}{dt} \overset{\text{def}}{=} \rho_t + \mathbf{v}\cdot\nabla\rho = -\rho(\nabla\cdot\mathbf{v}), \qquad \mathbf{(20')}$$

where $d\rho/dt$ is the total time rate of change of ρ along a particle path.† In this regard, ρ can experience both a convective effect $(\mathbf{v}\cdot\nabla)\rho = u\rho_x + v\rho_y + w\rho_z$ as a result of the flow, and a substantive effect, ρ_t.

The flow is said to be **incompressible** when $d\rho/dt = 0$, and from (20′) we see that this occurs iff the velocity field $\mathbf{v}$ is divergence-free. If, additionally, $\mathbf{v} = \nabla\phi$, for some *velocity potential* ϕ, then (20′) reduces to the Laplace equation $\nabla^2\phi = \nabla\cdot(\nabla\phi) = 0$.

Note that when $\mathbf{v} = \nabla\phi$, then the *vorticity* vector $\boldsymbol{\omega} \overset{\text{def}}{=} \nabla\times\mathbf{v} = \mathbf{0}$. Conversely, a vorticity-free flow is said to be **irrotational**, and if we suppose $\mathbf{v}$ to have C^2 components, then locally $\mathbf{v} = \nabla\phi$. (See Problem 6.3B.)

(b) Euler's equations of motion: Next, we must find equations that express Newton's second law of dynamics. It will be enough to understand the origin of the equation that treats acceleration in the z-direction. The associated momentum change within a control volume V is related to that gained by flux

† The derivative d/dt, sometimes denoted D/Dt, is called variously the Stokes' derivative, or the material derivative, or the particle derivative.

through $S = \partial V$ as follows:

$$\frac{\partial}{\partial t}\int_V \rho w \, dV = -\int_S \rho w(\mathbf{v}\cdot\mathbf{n})\, dS - \int_S pn_3 \, dS, \tag{21}$$

where the last term is the z-component of the net external force exerted on V by the static pressure p distributed over S. (In an inviscid fluid there are no other surface forces, and body forces have been neglected provisionally.)

Observe that $\int_S pn_3 \, dS = \int_S \mathbf{f}\cdot\mathbf{n}\, dS$, if $\mathbf{f} = (0,0,p)$ so that $\nabla\cdot\mathbf{f} = p_z$ only. Therefore, when we apply the divergence theorem to (21), we find that

$$\int_V [(\rho w)_t + \nabla\cdot(\rho w \mathbf{v}) + p_z]\, dV = 0,$$

for each control volume V, and by the usual argument it follows that

$$(\rho w)_t + \nabla\cdot(\rho w\mathbf{v}) + p_z = 0,$$

$$\text{or,} \qquad \rho w_t + (\rho_t + \nabla\cdot\rho\mathbf{v})\, w + \rho\mathbf{v}\cdot\nabla w = -p_z.$$

Hence, recalling (20) and (20′), we conclude that

$$\frac{dw}{dt} = w_t + (\mathbf{v}\cdot\nabla)\, w = -\frac{1}{\rho}\, p_z.$$

This result constitutes the z-component of **Euler's equations of motion** which take the vector form

$$\frac{d\mathbf{v}}{dt} \overset{\text{def}}{=} \mathbf{v}_t + (\mathbf{v}\cdot\nabla)\mathbf{v} = \frac{-1}{\rho}\nabla p. \qquad (\textit{Euler's equations}) \tag{22}$$

The left side of (22) represents the acceleration of the mass element located at $(\mathbf{x}, t)$. If we replace the right side by $-(\nabla p/\rho) + \mathbf{F}$, we can account for the contribution of a body force $\mathbf{F}$ (per unit mass) such as that of gravity.

In simple hydrodynamics, ρ is assumed to be constant. Then, if the flow is *irrotational* in that $\nabla\times\mathbf{v} = \mathbf{0}$ and *steady* so that $p_t = 0$ and $\mathbf{v}_t = \mathbf{0}$, it obeys **Bernoulli's equation** of energy conservation

$$\frac{1}{2}|\mathbf{v}|^2 + \frac{p}{\rho} = \text{const.} \tag{22′}$$

However, for gases and real liquids, we cannot suppose ρ to be constant and (22′) must be modified. (See Problem 6.4C.) A gas is termed **ideal** when in thermal equilibrium its state variables are related by the equation

$$p = \rho R T \tag{23}$$

for some gas constant R. If a flow process for an ideal gas is **isothermal** in that $T = T_0$ remains constant, it follows that $p = \rho R T_0$ is a function of ρ alone, and with (20) and (22), we obtain a system of four scalar equations in the four variables u, v, w, and ρ. For any fluid, similar systems govern those processes where p can be expressed as a function of ρ. Then on physical grounds we expect that

$$c^2(\rho) \overset{\text{def}}{=} \frac{dp}{d\rho} \geq 0, \tag{23′}$$

so that $\nabla p = c^2 \nabla \rho$ and (22) takes the form

$$\mathbf{v}_t + (\mathbf{v} \cdot \nabla)\mathbf{v} = -c^2 \frac{\nabla \rho}{\rho}. \tag{24}$$

There are several useful linearizations of the system obtained by combining (24) with (20). Let's look at two cases.

Example 3 *Acoustic propagation (small disturbances)*:

Most sounds that reach our ears without deafening us take the form of small disturbances traveling through otherwise undisturbed air. Let's see what our equations tell us about such acoustic propagation.

Suppose that the density of an undisturbed fluid is a constant ρ_0 and that disturbances are so small that

$$\frac{\rho}{\rho_0} = 1 + \sigma \qquad \text{with} \qquad |\sigma| \ll 1, \qquad \text{and} \qquad |\mathbf{v}| \ll 1.$$

(σ is called the **condensation.**) Substituting $\rho = \rho_0(1+\sigma)$ in (20) and (24), we obtain after simplification

$$\sigma_t + \nabla \cdot ((1+\sigma)\mathbf{v}) = 0$$

and, since $\nabla \rho = \rho_0 \nabla \sigma$,

$$(1+\sigma)[\mathbf{v}_t + (\mathbf{v} \cdot \nabla)\mathbf{v}] = -c^2 \nabla \sigma.$$

These equations are also nonlinear, but if we assume that we can neglect terms that involve products of σ, u, v, w, or their derivatives in comparison with terms without such products, we obtain the system

$$\sigma_t + \nabla \cdot \mathbf{v} = 0,$$
$$\mathbf{v}_t = -c_0^2 \nabla \sigma,$$

where we have approximated $c^2 = p'(\rho)$ by $c_0^2 = p'(\rho_0)$.† Finally, we can eliminate $\mathbf{v}$ to get

$$\sigma_{tt} = -\nabla \cdot (\mathbf{v}_t) = c_0^2 \nabla \cdot (\nabla \sigma) = c_0^2 \nabla^2 \sigma, \tag{25}$$

which shows that with small disturbances, the condensation is governed by the three-dimensional *wave equation* of Chapter 5. Hence, such small disturbances travel as condensation waves through the medium with speed c_0. The associated pressure changes are experienced as sound and we conclude that c_0 is essentially the speed of sound propagation in the *undisturbed fluid.* (We actually "hear" not p but p_t; however, with small disturbances, $p_t \approx \rho_0 c_0^2 \sigma_t$ also satisfies (25) and is propagated accordingly.) In particular, if we consider a sound source at the origin in three-dimensional space that produces radially symmetric (small) disturbances, then $\sigma = \sigma(r,t)$ should satisfy (25) in the form

$$\sigma_{tt} = c_0^2\left(\sigma_{rr} + \frac{2}{r}\sigma_r\right) = \frac{c_0^2}{r}(r\sigma)_{rr}.$$

† In effect, it is these assumptions that define a "small" disturbance.

(See Problem 4.1D.)

This may be converted to the equation

$$(r\sigma)_{tt} = c_0^2 (r\sigma)_{rr}, \tag{25'}$$

which, by d'Alembert's result of Section 5.5, has as its only *outward-bound* solution

$$r\sigma(r, t) = \psi(r - c_0 t)$$

for appropriate ψ. Hence $\sigma(r, t) = (1/r)\psi(r - c_0 t)$, which shows that small-sound waves travel outward from a source at speed c_0 without distortion in shape but with amplitude that diminishes with distance. This is fortunate, for otherwise we would hear many more sounds than we might wish.

(6.4) Remarks: Both Newton and Euler supposed that in air, sound propagates isothermally, which gave the erroneous result that $c_0^2 = p'(\rho_0) = RT_0$. In 1816 Laplace observed correctly that the process is isentropic[†] so that $c_0^2 = \gamma RT_0$ where for air $\gamma \approx 1.4$. For gases, but not liquids, the speed of sound is directly related (through γ) to molecular composition.

The sounds made by an organ pipe are caused by exciting natural longitudinal vibrations (standing waves) in the mass of air within the pipe, as shown in Figure 6.5. The cross-sectional deflections $u = u(x, t)$ that describe these vibrations are governed as above by the simple wave equation $u_{tt} = c_0^2 u_{xx}$, where c_0 is the speed of sound in the undisturbed air. (See also Problem 5.1E.) If the pipe is of length ℓ and open at both ends as shown, we must have $u_x(0, t) = u_x(\ell, t) = 0, t > 0$ to give vanishing longitudinal strain, and the problem is essentially that for the shaft in Example 2 of Section 5.3. If we neglect other end-effects, the natural frequencies are $\omega_n = n\pi c_0/\ell$, with modes $X_n(x) = \cos(n\pi x/\ell)$, $n = 1, 2, \ldots$ .The fundamental frequency $\omega_1 = \pi c_0/\ell$ is the one that is usually heard, and the pipe is tuned by proper selection of ℓ. However, c_0^2 is proportional to the absolute temperature T_0, a fact of concern to organ builders. Other wind instruments, such as clarinets or trumpets, are tuned to compensate for room conditions by hand-adjustment of their effective lengths ℓ just prior to performance. Variation in the character of sound from

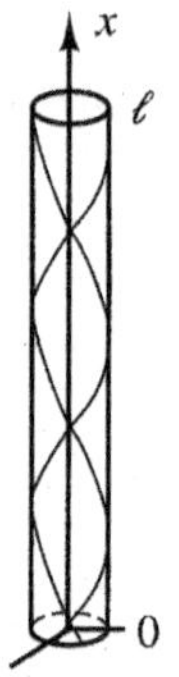

Figure 6.5

[†] See the discussion of energy considerations that follows Example 4.

such instruments is produced by permitting or suppressing certain vibrations at the higher natural frequencies.

Example 4 *Steady flow (small perturbations)*:

If we suppose that in each control volume the conditions are not changing with time, then we get, from (20) and (24), the *steady-state equations*

$$\nabla \cdot (\rho \mathbf{v}) = 0 \qquad \text{or} \qquad \nabla \cdot \mathbf{v} = -\mathbf{v} \cdot (\nabla \rho / \rho)$$
$$\text{and} \qquad (\mathbf{v} \cdot \nabla)\mathbf{v} = -c^2(\nabla \rho / \rho),$$

from which $\nabla \rho / \rho$ may be eliminated to give the single scalar equation

$$\mathbf{v} \cdot [(\mathbf{v} \cdot \nabla)\mathbf{v}] = c^2 \nabla \cdot \mathbf{v}. \tag{26}$$

The last equation can be used to describe steady flow of air over a small model in the test chamber of a wind tunnel (see Figure 6.6) in which the tunnel conditions (fan speed, temperature, etc.) are static. In this case, the velocity $\mathbf{v}$ is essentially a perturbation of the undisturbed one-dimensional flow with velocity $(U, 0, 0)$ that would occur if the model were not present.

Thus we have

$$u = \tilde{u} + U, \qquad v = \tilde{v}, \qquad \text{and} \qquad w = \tilde{w},$$

so that

$$\nabla \cdot \mathbf{v} = \frac{\partial \tilde{u}}{\partial x} + \frac{\partial \tilde{v}}{\partial y} + \frac{\partial \tilde{w}}{\partial z},$$

where $\tilde{u}, \tilde{v}, \tilde{w}$ are supposed small in relation to U, the "free-stream" velocity. It is also consistent to take $c \approx C$, the corresponding free-stream speed of sound.

Assuming that we can neglect terms with products of $\tilde{u}, \tilde{v}, \tilde{w}$, and their derivatives in comparison to those terms without such products,† we obtain

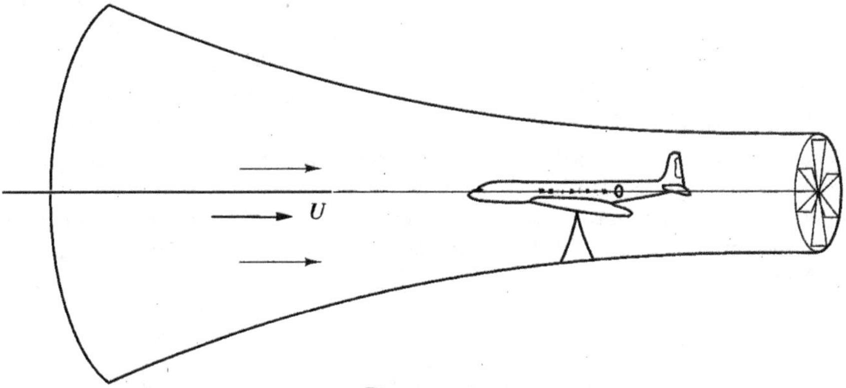

Figure 6.6

† This assumption may be regarded as defining a "small" perturbation in steady flow.

from (26)

$$U^2 \frac{\partial \tilde{u}}{\partial x} = C^2 \left(\frac{\partial \tilde{u}}{\partial x} + \frac{\partial \tilde{v}}{\partial y} + \frac{\partial \tilde{w}}{\partial z} \right),$$

which gives one equation in three unknowns. If we suppose that the perturbation velocity, $\tilde{\mathbf{v}} = (\tilde{u}, \tilde{v}, \tilde{w}) = \nabla \phi$, for some C^2 function ϕ (which is locally equivalent to supposing that the flow is "vortex free" or irrotational as in Problem 6.3B), then we see that ϕ satisfies the equation

$$(1 - M^2)\phi_{xx} + \phi_{yy} + \phi_{zz} = 0, \tag{26$'$}$$

where $M \stackrel{\text{def}}{=} U/C$ is the free-stream **Mach number**.

Now, if conditions are independent of z, then (26′) reduces to the equation

$$(1 - M^2)\phi_{xx} + \phi_{yy} = 0, \tag{27}$$

whose classification depends upon M:

1. When $M > 1$, then $U > C$, the free-stream speed of sound, so that the flow is **supersonic** and equation (27) is hyperbolic with a pair of characteristics of constant slopes $y' = \pm \kappa = \pm |M^2 - 1|^{-1/2}$ emanating from each point (x, y). The flow is essentially traveling wavelike, and the general solution is given by a sum of terms of the form

$$\phi(x, y) = f(x \pm \kappa y).$$

Weak discontinuities in ϕ are propagated along characteristics which define restricted domains of dependence/influence. When we fly supersonically, we leave behind the (small) disturbance we create, since it propagates only with the speed of sound.

2. When $M < 1$, then $U < C$ so that the flow is **subsonic** and equation (27) is elliptic. There are no characteristics, and the scale change $\bar{x} = \kappa x$ reduces (27) to Laplace's equation in $\bar{x}, y$. In this case, there are no restricted domains of dependence/influence, and theoretically, each small disturbance affects the entire flow.

3. When $M = 1$, then $U = C$, and the equation is parabolic, with a single characteristic $x = x_0$ through each point. However, near $M = 1$, a careful analysis shows that a nonlinear term neglected in deriving (27) is actually of equal importance to those retained.† Thus the appropriate **transonic** equation is inherently nonlinear and it will not be considered further here. (See [L-R].)

The foregoing classifications, observations, and terminology also apply to equation (26′). An actual flow problem can involve one region in which the flow is subsonic and another in which it is supersonic, but an interface sustains a strong discontinuity in normal momentum as well as in the thermodynamic state variables (pressure and entropy). Such an interface is called a **shock wave** and is

---

† In each of the preceding examples, linearized equations were obtained by assuming that certain terms were "small" in comparison with others, but it is difficult to predict circumstances under which this is permissible. To select parameters that describe a flow regime in which the relevant suppositions *are* valid requires an artful combination of scaling and similarity analyses guided by experimental findings. (See [L-S] and [Os].)

an idealization of what is responsible for the sonic booms associated with supersonic aircraft. (See also Section 6.5.)

(c) Energy considerations: In the derivation of (24) and subsequent results, we supposed that p was related to ρ in such a manner that $c^2 = dp/d\rho$ is defined. This is a simplification of the actual situation, and to the momentum equation in the form (22)

$$\mathbf{v}_t + (\mathbf{v} \cdot \nabla)\mathbf{v} = -\frac{\nabla p}{\rho}, \tag{28}$$

we should add an equation of energy conservation incorporating the first law of thermodynamics. It is shown in Problem 6.4A that for inviscid flow of a non-heat-conducting gas, this equation can be expressed as follows:

$$\frac{de}{dt} \overset{\text{def}}{=} e_t + \mathbf{v} \cdot \nabla e = q - \frac{p}{\rho} \nabla \cdot \mathbf{v} \tag{29}$$

where ρe is the local mean random kinetic energy in the gas and q represents the local heat per unit mass per unit time added externally. (When $q \equiv 0$, the process is termed **adiabatic**.)

The resulting system is still incomplete. To progress, we use the second law of thermodynamics to introduce the **entropy** s per unit mass in the differential form

$$T\,ds = de + p\,d\left(\frac{1}{\rho}\right); \tag{30}$$

i.e., we postulate that upon dividing by T, the right side of this expression becomes an exact differential in the same way that upon dividing by x^2, $x\,dy - y\,dx$ becomes the exact differential of y/x. When (30) is combined with (29) and (20′), we find that

$$T\frac{ds}{dt} = \frac{de}{dt} - \frac{p}{\rho^2}\frac{d\rho}{dt} = q.$$

Thus the flow is adiabatic iff $ds/dt = 0$, and then the entropy s is constant *along the path* of each particle in the flow. Similarly, from (28), (29), and (30), we can derive **Crocco's equation,**

$$T\nabla s = \mathbf{v}_t + \nabla\left[e + \frac{p}{\rho} + \frac{|\mathbf{v}|^2}{2}\right] - \mathbf{v} \times (\nabla \times \mathbf{v}). \tag{31}$$

In (31), the bracketed term is constant for many steady flows (recall Bernoulli's equation (22′)). Then if the flow is also irrotational ($\nabla \times \mathbf{v} \equiv \mathbf{0}$), it is necessarily **isentropic** in that s is constant, and therefore it is adiabatic. In general, an adiabatic process in which work is done reversibly on a non-heat-conducting gas *is* isentropic, and the transmission of a small sound is considered to be such a process. (For further discussion, see [L-R].)

For an ideal gas obeying (23), it can be shown that $de = c_v\,dT$, where for our purposes c_v is a positive gas constant. (See Problem 6.4F.) Then for an isentropic process, (30) gives

$$c_v\,dT = -p\,d\left(\frac{1}{\rho}\right) = \frac{p}{\rho^2}\,d\rho, \tag{32}$$

and when this result is used with (23) in the differential form

$$\frac{dp}{p} = \frac{d\rho}{\rho} + \frac{dT}{T},$$

it is seen that

$$\frac{dp}{p} = \frac{d\rho}{\rho}\left(1 + \frac{R}{c_v}\right) = \gamma\frac{d\rho}{\rho}, \text{ say,} \qquad \text{where} \qquad \gamma = 1 + \frac{R}{c_v},$$

so that

$$c^2 = \frac{dp}{d\rho} = \gamma p/\rho = \gamma RT,$$

as claimed in an earlier remark. It follows that (see Problem 6.4G)

$$\frac{dc}{d\rho} = \frac{\gamma - 1}{2}\frac{c}{\rho}, \tag{33}$$

and a final integration gives the isentropic relations

$$p = A_0\rho^\gamma, \qquad c^2 = A_0\gamma\rho^{\gamma - 1}, \tag{33'}$$

where A_0 is a flow-dependent constant for the ideal gas.

Problem Set 6.4

6.4A For the control volume V used in deriving (21), conservation of energy may be expressed by the requirement that

$$\frac{\partial}{\partial t}\int_V E\,dV - \int_V \rho q\,dV = -\int_S p\mathbf{v}\cdot\mathbf{n}\,dS - \int_S E\mathbf{v}\cdot\mathbf{n}\,dS,$$

where E represents the local total energy of the flow, ρq represents the rate of heat being added, and the first term on the right represents the work done by the pressure acting normal to the surface S. (We are neglecting the work done by body forces and the effects of heat conduction.)

γ. What is represented by the last term on the right?

2. Use the divergence theorem and the standard appeal to continuity to obtain the energy equation $E_t + \nabla\cdot(E\mathbf{v}) = \rho q - \nabla\cdot(p\mathbf{v})$.

3. Express $E = \rho(|\mathbf{v}|^2/2) + \rho e$ (where the first term is the local kinetic energy of the flow and the second represents that for the mean random molecular motions). Use (20') and (22) as required to obtain (29).

6.4B In an organ pipe of length ℓ that is closed at the lower end ($x = 0$) and open at the other end ($x = \ell$), the propagation of sound is governed by the wave equation $\sigma_{tt} = c_0^2\sigma_{xx}$, $0 < x < \ell$ with the boundary conditions $\sigma(0, t) = \sigma_x(\ell, t) = 0$. Show that natural sound oscillations can occur only at frequencies $\omega_n = n\pi c_0/2\ell$, $n = 1, 3, 5, \ldots$ and find the corresponding modes X_n. (The absence of other frequencies accounts for the hollow sound of a clarinet in its low register.)

6.4C (*Bernoulli's equation*)

1. When body forces $\mathbf{F}$ per unit mass (such as that of gravity) are acting on a fluid or gas, show that the momentum equation (22) should be $\mathbf{v}_t + (\mathbf{v}\cdot\nabla)\mathbf{v} = \mathbf{F} - (\nabla p/\rho)$.

2. Verify that $(\mathbf{v}\cdot\nabla)\mathbf{v} = (1/2)\nabla(|\mathbf{v}|^2) - \mathbf{v}\times(\nabla\times\mathbf{v})$. (See (19b).)
3. Conclude that when ρ is constant and $\mathbf{F}$ is conservative ($\mathbf{F} = -\nabla\Omega$), then

$$\nabla\left(\frac{|\mathbf{v}|^2}{2} + \frac{p}{\rho} + \Omega\right) = -\mathbf{v}_t + \mathbf{v}\times(\nabla\times\mathbf{v}).$$

When the flow is also irrotational prove that it satisfies *Bernoulli's equation* in the form $(|\mathbf{v}|^2/2) + (p/\rho) + \Omega = \phi_t + \text{const.}$ for an appropriate velocity potential ϕ. (See Problem 6.3B.)
4. If in part 3 the flow is also steady, show that $(|\mathbf{v}|^2/2) + (p/\rho) + \Omega = \text{const.}$, or if (23) is valid, $(|\mathbf{v}|^2/2) + RT + \Omega = \text{const.}$ How does this explain the cooling effect of a fan?
5. Take the curl of the first equation in part 3, and use the fact that $\nabla\times\nabla\phi \equiv 0$ to obtain the **vorticity equation** $d\boldsymbol{\omega}/dt = (\boldsymbol{\omega}\cdot\nabla)\mathbf{v}$, where $\boldsymbol{\omega} \overset{\text{def}}{=} \nabla\times\mathbf{v}$ and $d/dt = (\partial/\partial t) + (\mathbf{v}\cdot\nabla)$.
6. Within a two-dimensional flow, where $\mathbf{v} = (u, v, 0)$, show that $d\boldsymbol{\omega}/dt = \mathbf{0}$.

6.4D (*Water waves in a channel: Linearized analysis*)

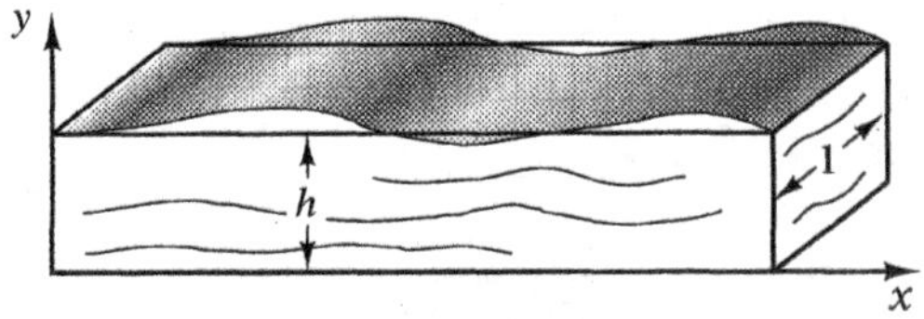

The irrotational horizontal flow of water of mean depth h in a long straight channel of unit width may be described by a velocity potential $\phi = \phi(x, y, t)$ that satisfies Laplace's equation in x and y. ϕ must also satisfy the following boundary conditions: $\phi_y(x, 0, t) = 0$, to indicate that there is no vertical velocity component at the floor of the channel where $y = 0$; and $\phi_y(x, h, t) \approx -(1/g)\phi_{tt}(x, h, t)$, a consequence of Bernoulli's equation (of Problem 6.4C) applied at the mean surface where $y = h$. g is the gravitational acceleration. (The effects of surface tension are neglected.)

1. Show that both the equation and the floor condition are satisfied by $\phi(x, y, t) = T(t)\cosh\kappa y\, e^{i\kappa x}$ for any real κ and any function $T = T(t)$.
2. Conclude that to meet the surface condition, T should satisfy the equation $T'' + \omega^2 T = 0$, where $\omega^2 = g\kappa\tanh\kappa h$. This means that for each real κ the flow admits sinusoidal-shaped waves with potential $\phi(x, y, t) = \cosh\kappa y\cos(\kappa x - \omega t)$, which travel with wave-speed $c = \omega/\kappa = \sqrt{(g\tanh\kappa h)/\kappa}$ and wave length $2\pi/\kappa$. (κ is called the wave number.)
3. If $\kappa h \ll 1$ (which is the case with tidal waves), show that $c \approx \sqrt{gh}$. If $\kappa h \gg 1$, show that $c \approx \sqrt{g/\kappa}$ and explain the dispersive behavior of most ocean waves.

6.4E If the restoring force on the water waves of the previous problem is caused by surface tension of strength τ_0 instead of gravity, then it can be shown that the resulting surface condition is $\phi_{tt} \approx \delta_0\phi_{xxy}$ (at $y = h$), where $\delta_0 = \tau_0/\sigma_0$ and σ_0 is the weight-density of water.

1. Show that pure surface-tension waves can occur at frequency ω and wave number κ, if $\omega^2 = \delta_0 \kappa^3 \tanh \kappa h$.
2. If κ is very large, conclude that the wave velocity $c = \omega/\kappa \approx \sqrt{\delta_0 \kappa}$.
3. If both surface tension and gravity are acting, then the surface condition is $\phi_{tt} \approx \delta_0 \phi_{xxy} - g\phi_y$, $(y = h)$. In this case, what relation between ω and κ governs wave dispersion?

6.4F In thermodynamics, it is usual to suppose that there are at most two independent state variables. In (30), assume that both s and e are C^2 functions of T and v, where $v = 1/\rho$.

1. Show that

$$\frac{\partial s}{\partial T} = \frac{1}{T}\frac{\partial e}{\partial T} \quad \text{and} \quad \frac{\partial s}{\partial v} = \frac{1}{T}\left(\frac{\partial e}{\partial v} + p\right)$$

2. When (23) holds,

$$\frac{\partial s}{\partial v} = \frac{1}{T}\frac{\partial e}{\partial v} + \frac{R}{v}.$$

(Why?) By cross-differentiating, eliminate s and conclude that $\partial e/\partial v = 0$; i.e., $e = e(T)$ so that $de = c_v\, dT$, where $c_v = e'(T)$.

6.4G If $c^2 = \gamma p/\rho$, explain how (33) is obtained and how (33′), the final isentropic relations involving A_0, follow.

6.5*

Linear and Quasilinear Systems of First-Order Equations

In the last two sections we encountered phenomena described by systems of first-order equations that could be converted into second-order equations—perhaps after linearization. It is natural to ask for a method that deals directly with first-order systems, especially since (as in Problem 6.1D) linear equations of arbitrary order can always be replaced by an equivalent first-order system. In this section we uncover a few facts about systems of the special **quasilinear** form

$$\mathbf{u}_t + \mathbb{A}\mathbf{u}_x = \mathbf{F} \tag{34}$$

in a domain D of $\mathbb{R}^2$, where $\mathbf{u}$ is an unknown m-dimensional vector function in D, $\mathbf{F}$ is a given m-dimensional vector function of x, t, and $\mathbf{u}$, and $\mathbb{A}$ is a given $m \times m$ matrix function of x, t, and $\mathbf{u}$, all with C^1 components. We use the information to gain insight into the nature of transmission line equations and those of gas dynamics.

System (34) is **linear** when $\mathbf{F} = \mathbb{B}\mathbf{u} + \mathbf{f}$, and the components of the matrix $\mathbb{A}$, the matrix $\mathbb{B}$, and the vector $\mathbf{f}$ are C^1 functions of x and t only.

For example, the transmission line equations (15a and 15b) provide the following system of form (34) in $\mathbf{u} = (i, v) = (\mathrm{u}_1, \mathrm{u}_2)$:

$$\mathbf{u}_t + \mathbb{A}\mathbf{u}_x = \mathbf{F} \overset{\text{def}}{=} \begin{pmatrix} -R\ell^2 u_1 \\ -Gc^2 u_2 \end{pmatrix}, \tag{35}$$

$$\text{where} \quad \mathbb{A} = \begin{bmatrix} 0 & \ell^2 \\ c^2 & 0 \end{bmatrix}, \quad \text{and} \quad \begin{matrix} \ell = L^{-1/2} \\ c = C^{-1/2} \end{matrix}. \qquad \textbf{(35')}$$

We see that this system is linear; we shall take up its solution shortly.

Normal Systems

In (34), when $\mathbb{A}$ is a *diagonal* matrix Λ whose only possible nonzero entries λ_j, $j = 1, 2, \ldots, m$ are arranged along the principal diagonal, then the system takes the **normal** form

$$\frac{\partial u_j}{\partial t} + \lambda_j \frac{\partial u_j}{\partial x} = f_j \qquad j = 1, 2, \ldots, m, \qquad \textbf{(36)}$$

where u_j and f_j are the component functions of $\mathbf{u}$ and $\mathbf{F}$, respectively. Our first goal is to transform system (34) where possible into one of this special form in new unknowns. To do this we try using linear algebraic methods to convert the matrix $\mathbb{A}$ to a diagonal matrix Λ whose diagonal elements λ_j are **eigenvalues** of $\mathbb{A}$; that is, each λ_j is a real or complex number λ for which

$$p(\lambda) = \det[\mathbb{A} - \lambda \mathbb{I}] = 0 \qquad \textbf{(37)}$$

(where $\mathbb{I}$ is the $m \times m$ identity matrix), or equivalently, for which the equation

$$\mathbb{A}\mathbf{V} = \lambda \mathbf{V} \qquad \textbf{(37')}$$

has a nonzero eigenvector solution $\mathbf{V}$. From (37) it follows that $\mathbb{A}$ has eigenvalues, and at most m different eigenvalues, since p is an mth-order polynomial. When all λ_j are real and distinct, the associated nonzero solutions $\mathbf{V}_j$ of (37′) are linearly independent so that the $m \times m$ matrix $\mathbb{V}$, having these $\mathbf{V}_j$ as *columns*, is invertible (see [Ner]), and by (37′), $\mathbb{A}\mathbb{V} = \mathbb{V}\Lambda$.

Hyperbolic Systems

For the transmission line problem above, using (37) we easily find that in (35′), the 2×2 matrix of constants $\mathbb{A}$ has the distinct pair of *real* eigenvalues $\lambda_1 = \ell c$, $\lambda_2 = -\ell c$. It is possible that the general matrix $\mathbb{A}$ has no real eigenvalues (see Problem 6.5B) and in this case system (34) is said to be **elliptic.**† At the other extreme is a matrix $\mathbb{A} = \mathbb{A}(x, t, \mathbf{u})$ that has m *distinct real* eigenvalues $\lambda_j = \lambda_j(x, t, \mathbf{u})$, $j = 1, 2, \ldots, m$, at each "point" $(x, t, \mathbf{u})$ of interest. Then we say that system (34) is **hyperbolic,**† and henceforth we will consider only hyperbolic systems. Observe that in general this characterization depends on values of $\mathbf{u}$ and therefore on the unknown solution $\mathbf{u}$ of (34) at a point $(x, t, \mathbf{u})$. When $\mathbf{u}$ is a C^1 solution that makes system (34) hyperbolic in D, it can be assumed that the λ_j and the elements of $\mathbb{V}$ and $\mathbb{V}^{-1}$ are C^1 in D. (See [C-L].) When $\mathbb{A}=\mathbb{A}(x,t)$, it is straightforward to differentiate $\mathbf{w} \overset{\text{def}}{=} \mathbb{V}^{-1}\mathbf{u}$ and substitute as required to verify that the components of $\mathbf{w}$ satisfy the so-called **normal system**

$$\frac{\partial w_j}{\partial t} + \lambda_j \frac{\partial w_j}{\partial x} = g_j, \qquad j = 1, 2, \ldots, m, \qquad \textbf{(38)}$$

† When second-order linear equations are converted to first-order systems, this terminology is consistent with that introduced in Section 6.2.

where each $g_j = g_j(x, t, \mathbf{w})$ is a function that can be determined explicitly from $\mathbb{A}$, $\mathbb{V}$, and $\mathbf{F}$. If system (34) is also linear, this process can be carried through without prior knowledge of a solution $\mathbf{u}$. Then the normal system will also be linear and each of its solutions $\mathbf{w}$ gives a corresponding solution $\mathbf{u} = \mathbb{V}\mathbf{w}$ of (34). (See Problem 6.5G.)

Characteristics

Each equation (38) recalls the nonlinear equation (51) introduced in Section 5.7, and it is natural to examine associated characteristics. In D, assuming that a solution $\mathbf{u}$ of (34) is at hand or equivalently that a solution $\mathbf{w}$ of (38) is available, consider

$$\text{a curve } C_j \text{ with slope} \qquad \frac{dx}{dt} = \lambda_j \tag{39}$$

$$\text{along which} \qquad \frac{d}{dt} w_j(x(t), t) = g_j, \qquad j = 1, 2, \ldots, m. \tag{39$'$}$$

When system (34) is hyperbolic, then through each point $(x, t) \in D$ pass m such (base) **characteristic curves** with distinct slopes, none parallel to the x-axis. If t represents time, it follows from (39) that $\lambda_j = \lambda_j(x, t)$ is the local velocity in the x-direction of a point tracing the path $x = x_j(t)$ of the characteristic C_j.

Characteristics play a significant role in the behavior of solutions to a given system (34). To understand why, let $\mathbf{u}$ be continuous in D and satisfy (34) on either side of a smooth arc γ_0 that divides D as in Figure 6.1. If $\mathbf{u}_x$ and $\mathbf{u}_t$ extend continuously to an interior point of γ_0 from either side, it can be shown that $\mathbf{u}$ is C^1 near this point *unless* γ_0 *is characteristic at the point*. In the latter case, $\mathbf{u}$ can sustain a discontinuity in its normal derivative to γ_0 that "propagates" along γ_0 or some other characteristic. (See Problem 6.5L.) In particular, unless characteristics for $\mathbf{u}$ are present in D, such **weak discontinuities** in $\mathbf{u}$ cannot occur.

Method of Characteristics

When system (34) is *linear* and *hyperbolic*, its characteristic curves C_j can be found by integrating system (39) (perhaps numerically). These in turn can be used to recover the w_j from system (39′) (which is linear in $\mathbf{w}$) and a set of prescribed initial values on a *noncharacteristic* curve. By this **method of characteristics** we can find the solution $\mathbf{u}$ to a hyperbolic linear system that has given **initial data**

$$\mathbf{u}(x, 0) = \psi(x) \tag{40}$$

on the *noncharacteristic* curve $\gamma_0 = \{(x_0, 0), x_0 \in \mathbb{R}\}$ (the x-axis).

Example 5 (*Transmission line equations*):

Let us resume our analysis of the transmission line system of (35) where the eigenvalues are $\lambda_1 = \ell c$ and $\lambda_2 = -\ell c$ and the associated eigenvectors are $\mathbf{V}_1 = \binom{1}{c/\ell}$ and $\mathbf{V}_2 = \binom{1}{-c/\ell}$, respectively—results that are easily verified.

Thus if we set $\mathbf{u} = \mathbb{V}\mathbf{w}$; i.e.,

$$\begin{pmatrix} u_1 \\ u_2 \end{pmatrix} = \begin{bmatrix} 1 & 1 \\ c/\ell & -c/\ell \end{bmatrix} \begin{pmatrix} w_1 \\ w_2 \end{pmatrix} = \begin{pmatrix} w_1 + w_2 \\ (w_1 - w_2)\, c/\ell \end{pmatrix}, \tag{41}$$

it follows that

$$\begin{aligned} w_1 &= \tfrac{1}{2}(u_1 + u_2\ell/c) \\ w_2 &= \tfrac{1}{2}(u_1 - u_2\ell/c). \end{aligned} \tag{41'}$$

Therefore, the normal system (38) becomes

$$\begin{aligned} \frac{\partial w_1}{\partial t} + \ell c \frac{\partial w_1}{\partial x} &= \mathscr{a} w_1 + \mathscr{b} w_2 \\ \frac{\partial w_2}{\partial t} - \ell c \frac{\partial w_2}{\partial x} &= \mathscr{a} w_2 + \mathscr{b} w_1 \end{aligned} \tag{42}$$

where $\quad 2\mathscr{a} \stackrel{\text{def}}{=} -R\ell^2 - Gc^2 \quad$ and $\quad 2\mathscr{b} \stackrel{\text{def}}{=} -R\ell^2 + Gc^2.$ **(42′)**

To obtain these results, we substitute equations (41′) in the left sides of equations (42), use (35) to eliminate derivatives, and finally, set $\mathbf{u} = \mathbb{V}\mathbf{w}$. (See Problem 6.5E.)

In this case, the characteristic equations (39) reduce to $dx/dt = \pm \ell c$, and we see that through each point (x_0, t_0) in the (x, t)-plane pass characteristic straight lines C_+ and C_-, say, with equations

$$x(t) = x_0 \pm \ell c (t - t_0). \tag{43}$$

Moreover,

$$\begin{aligned} \text{along } C_+ &: \frac{dw_1}{dt} = \mathscr{a} w_1 + \mathscr{b} w_2 \\ \text{and along } C_- &: \frac{dw_2}{dt} = \mathscr{a} w_2 + \mathscr{b} w_1. \end{aligned} \tag{44}$$

These equations suggest a numerical procedure for finding a solution $\mathbf{w}$ to system (42) for $t > 0$ that has given *initial values* $\mathbf{w}(x, 0) = \boldsymbol{\phi}(x)$ on the *non-characteristic* base curve $\gamma_0 = \{(x, 0): \; x \in \mathbb{R}\}$ (the x-axis).

Indeed, as Figure 6.7 indicates, this specification of $\mathbf{w}$ on γ_0 permits the linear approximations

$$\begin{aligned} w_1(x, t_1) &\approx [\phi_1 + t_1(\mathscr{a}\phi_1 + \mathscr{b}\phi_2)](x - \ell c t_1) \\ w_2(x, t_1) &\approx [\phi_2 + t_1(\mathscr{a}\phi_2 + \mathscr{b}\phi_1)](x + \ell c t_1) \end{aligned} \tag{45}$$

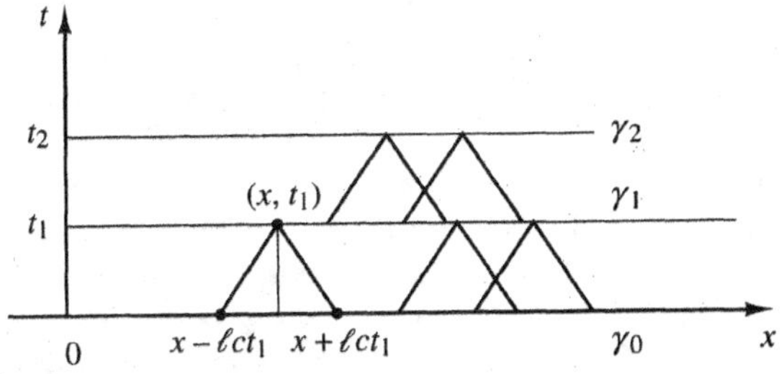

Figure 6.7

for small $t_1 > 0$. Then we can use *these* values of $\mathbf{w}$ on $\gamma_1 = \{(x, t_1):\ x \in \mathbb{R}\}$ to approximate $\mathbf{w}$ on $\gamma_2 = \{(x, t_2):\ x \in \mathbb{R}\}$ for small $t_2 - t_1 > 0$, etc. These results can be refined using standard predictor-corrector methods of numerical analysis (see [B-F]), and they provide an approximate solution $\mathbf{u}$ of the transmission line system (35) for $t > 0$ that has prescribed initial data

$$\mathbf{u}(x, 0) = \boldsymbol{\psi}(x) \stackrel{\text{def}}{=} \nabla\phi(x). \tag{46}$$

When the transmission line is *distortionless* in that $R/L = G/C$, or equivalently, that $\ell = 0$ in (42′), equations (42) uncouple and they can be integrated to yield an explicit solution to this Cauchy problem. (See Problem 6.5F.)

When system (34) is nonlinear, its characteristics and any associated normal system (38) can only be determined from a *known* solution $\mathbf{u}$. Consequently, it is far more difficult to use this approach effectively, as we shall see next.

*One-Dimensional Inviscid Isentropic Flow**

To model the inviscid flow of a gas within a long straight tube we can use the one-dimensional form of equations (20) and (22) from the previous section, namely,

$$\begin{aligned} &\rho_t + u\rho_x + \rho u_x = 0 && (\textit{Continuity}) \\ &u_t + uu_x + \frac{c^2}{\rho}\rho_x = 0. && (\textit{Momentum}) \end{aligned} \tag{47}$$

Here, $\rho = \rho(x, t)$ and $u = u(x, t)$ are respectively the density and velocity of the flow, and we have introduced the isentropic assumption that the static pressure $p = p(\rho)$ so that $c(\rho) \stackrel{\text{def}}{=} \sqrt{dp/d\rho}$ is positive and has the dimensions of speed.

For $\mathbf{u} = (\rho, u)$, these equations form the quasilinear system

$$\mathbf{u}_t + \mathbb{A}\mathbf{u}_x = \mathbf{0} \tag{48}$$

$$\text{with} \qquad \mathbb{A} = \begin{bmatrix} u & \rho \\ c^2/\rho & u \end{bmatrix}. \tag{48′}$$

It is easy to verify that the eigenvalues of $\mathbb{A}$ are just $\lambda_+ = u + c$ and $\lambda_- = u - c$. (See Problem 6.5H.) Hence the system is hyperbolic, and through each point (x, t) passes a pair of characteristics C_+, C_-, say, with slopes

$$\frac{dx}{dt} = u \pm c. \tag{49}$$

If $u > 0$, then a small disturbance moves along C_+ *with the flow* at local velocity $u + c$, and it may be identified with sound that would propagate at speed c through an otherwise undisturbed gas; C_- would carry this same small sound *against the flow* at local velocity $u - c$.

Now, normal equations (38) associated with (48) and (49) are not tractable (Try to obtain them!), so to make progress, we look for new functions that do not change along one of the characteristics. For the general quasilinear system (34), this search might not be fruitful. Even in the present case, there are no obvious candidates for functions v such that, say, $v_t + (u + c)v_x = 0$. But in 1860,

Bernhard Riemann (1824–1866) observed that in terms of the modified "density"†

$$\ell = \ell(\rho) = \int_{\rho_0}^{\rho} \frac{c(\xi)}{\xi}\, d\xi, \qquad \text{for which} \qquad \ell_t = \frac{c}{\rho}\rho_t \qquad \text{and} \qquad \ell_x = \frac{c}{\rho}\rho_x,$$

equations (47) take the more symmetrical form

$$\begin{aligned} \ell_t + u\ell_x + cu_x &= 0 \\ u_t + uu_x + c\ell_x &= 0. \end{aligned} \tag{50}$$

Then it is easy to add and subtract the new equations to obtain the equivalent system

$$\begin{aligned} v_t + (u+c)v_x &= 0 \qquad & v &= \frac{(\ell+u)}{2} \\ & \text{where} & & \\ w_t + (u-c)w_x &= 0 \qquad & w &= \frac{(\ell-u)}{2}. \end{aligned} \tag{51}$$

v and w are referred to as **Riemann invariants** of system (47) since

$$\begin{aligned} &\text{along} \quad C_+\text{: } v \text{ is constant} \\ &\text{and along} \quad C_-\text{: } w \text{ is constant.} \end{aligned} \tag{51$'$}$$

Moreover, from v and w we can recover

$$u = v - w \qquad \text{and} \qquad \ell = v + w, \tag{52}$$

and, in principle, c or ρ, so that a numerical attack modeled after that of Example 5 is feasible.

To simplify matters, assume that the gas is ideal. Then we can introduce (33) and conclude that for the gas constant $\gamma > 1$,

$$\frac{d\ell}{d\rho} = \frac{c}{\rho} = \frac{2}{\gamma - 1}\frac{dc}{d\rho},$$

so that $c = c_0 + [(\gamma - 1)/2]\ell$, when $c_0 = c(\rho_0)$ or $\ell_0 = 0$.

Using this result and (52), we get

$$u \pm c = (v - w) \pm \left[c_0 + \frac{\gamma - 1}{2}(v + w)\right], \tag{53}$$

which allows us to calculate the local sound velocities in terms of the Riemann invariants and a reference sound-speed c_0.

Let's look at some basic flows that satisfy equations (47).

Constant-state flow: Equations (47) are clearly satisfied in any flow region D_0 in the (x, t)-plane where the state of the system is constant, i.e., where $\rho = \rho_0$ (so $c = c_0$) and $u = u_0$. In particular, they are satisfied in a region D_0 in which the ideal gas is at rest ($u_0 = 0$). (See Figure 6.8.) In D_0 *both* families of characteristics are parallel lines of slopes $u_0 \pm c_0$, and both v and w will maintain

† See also Problem 5.7I.

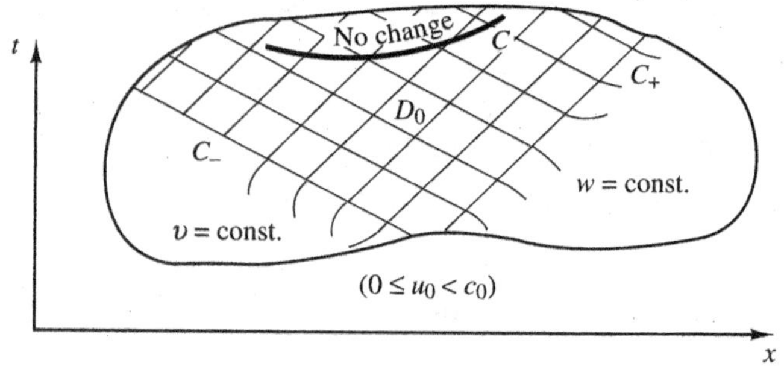

Figure 6.8

their constant values along respective characteristics into any domain adjacent to D_0 that shares a boundary curve C crossed by curves of *both* families. Thus by (53), this region of constant state cannot end in some larger solution domain of (47) with a boundary curve C unless C is crossed by only *one* family of characteristics—i.e., unless C is itself a characteristic *straight line*.

Let us state this surprising result in a more useful form.

(6.5) Proposition: *In a solution domain of* (47), *a constant-state flow can change only into one in which* $w =$ const. *across a* C_+ *characteristic line or into one in which* $v =$ const. *across a* C_- *characteristic line.*

Proof: It is enough to look at the first case where the separating curve C above is a C_+ characteristic straight line, and w *is* constant on the family of C_- characteristics crossing $C = C_+$. But those curves cover regions adjacent to C. ■

Simple waves; shocks: A flow in which $w =$ const. is called a **forward simple wave**; one in which $v =$ const. is called a **backward simple wave**. Either can permit subregions of constant state, but our proposition may be restated to say that a *constant-state flow can change isentropically only into a simple wave*.

To better appreciate this terminology, let's look at a *forward wave* in an ideal gas where in some region the flow is at rest ($u_0 = 0$) and the speed of sound is c_0. Then from (51) we have $w =$ const. $= w_0 = \ell_0/2 = 0$, so that $u = \ell = v$ satisfies the equation

$$u_t + (u + c)u_x = 0;$$

or, using (53) with $v = u$, $w = 0$,

$$u_t + \left(c_0 + \frac{\gamma + 1}{2}u\right)u_x = 0.$$

Now, the last equation is a special case of the simple nonlinear wave equation (49) in Section 5.7. In Example 7 of that section, we used characteristics to obtain the implicit solution

$$u = \psi\left[x - \left(c_0 + \frac{\gamma + 1}{2}u\right)t\right] = \psi(\xi), \text{ say,}$$

representing a wave of initial profile $u(x,0) = \psi(x)$ that moves to the right with local velocity $c_0 + [(\gamma+1)/2]u$. If ψ is C^1, then at time $t > 0$, the particle velocity profile has slope (see Problem 5.7B)

$$u_x = \frac{\psi'(\xi)}{1 + \dfrac{\gamma+1}{2}\psi'(\xi)\,t}, \qquad \text{where defined,} \tag{54}$$

and in particular, this holds if $\psi' > 0$. However, if $\psi' < 0$, the velocity profile can steepen and break as t increases to some critical time t_1 resulting in a discontinuity called a **shock** in gas dynamics.†

In any *forward wave*, the particle *velocity u increases with density* (or pressure) since

$$\frac{du}{d\rho} = \frac{d\ell}{d\rho} = \frac{c}{\rho} > 0 \qquad \left(\text{and } \frac{dp}{d\rho} = c^2 > 0\right). \tag{55}$$

If $\psi' < 0$ as above, the forward wave is **compressive** because parts of greater density (velocity) travel faster and overtake those of lesser density, and, as we have just seen, *compressive waves can generate shocks*. On the other hand, shocks will not arise when $\psi' > 0$; then the forward wave is **expansive** since (54) gives density/velocity profiles that tend to flatten as time increases. A forward compressive wave is produced ahead of an idealized piston accelerated *to the right* in a gas-filled tube (as shown in Figure 6.9), and a forward expansive wave is left behind.

Backward waves are produced by moving the same piston to the left, and, as this model suggests, backward waves are just forward waves moving in the opposite direction.

We have seen two kinds of flow that satisfy equations (47), and learned why a solution in a region of constant state can change only into a simple wave in an adjoining region. We have also seen that simple compressive waves can develop into shocks that are inherently nonisentropic. Of course, simple waves can also change to more complicated isentropic flows, but their analysis is outside the scope of this text. (See [L-R] and [C-F].)

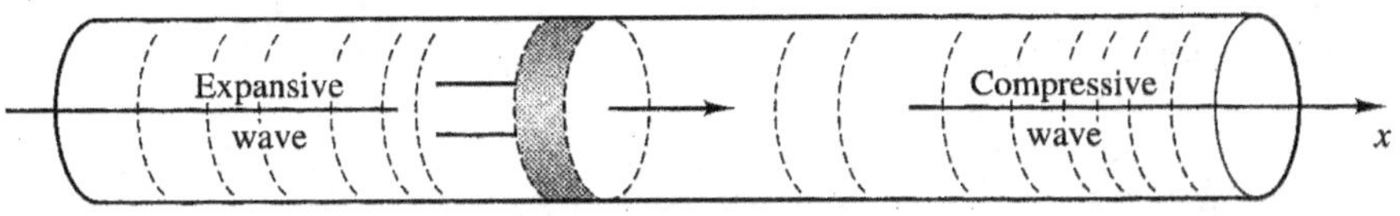

Figure 6.9

Problem Set 6.5

6.5A *1.* Show that $\lambda = \pm \ell c$ are the only eigenvalues for the matrix $\mathbb{A}$ of (35′).

---

† A shock involves *strong* discontinuities in the flow parameters ρ and u that require a corresponding change in *entropy*. We cannot use our isentropic equations (47) to model flow *across* shocks, but we can use them to describe flow on either side of a shock region. The shock itself moves at local velocities *different* from those of sound.

2. For $\lambda_1 = \ell c$, show that $\mathbf{V}_1 = \left|\begin{smallmatrix} 1 \\ \ell/c \end{smallmatrix}\right|$ is one eigenvector of $\mathbb{A}$ and that $\tilde{\mathbf{V}}_1 = \left|\begin{smallmatrix} c \\ \ell \end{smallmatrix}\right|$ is another.

6.5B Write the Cauchy–Riemann equations of Section 4.4 in the form $u_x = v_t$, $u_t = -v_x$ and show that the system in $\mathbf{u} = (u, v)$ is linear and **elliptic**.

6.5C 1. Show that for $a, b \in \mathbb{R}$, the system

$$u_t + a v_x = 0$$

$$v_t + b u_x = 0$$

is elliptic when $ab < 0$ and hyperbolic when $ab > 0$.

2. Show that this classification agrees with that of the second-order equation satisfied by u or v assumed to be C^2.

6.5D 1. Show that for $\mathbf{u} = (u, v)$ the system

$$u_t + u_x + v = 0$$

$$v_t - x^2 v_x + u = 0$$

is linear and hyperbolic with eigenvalues $\lambda_1 = 1$ and $\lambda_2 = -x^2$.

2. Identify the characteristics.

3.* Find the corresponding normal system (38), and obtain as much information as possible about w_1 and w_2.

6.5E 1. Use equations (41) and (41′) to convert the transmission line system (35 and 35′) into the normal system (42). (*Hint:* Follow the outlined procedure.)

2. When $a = b = 0$, explain how to find $\mathbf{w}$ explicitly. Could these conditions be realized for a real transmission cable?

6.5F For a distortionless transmission line, $b = 0$ in system (42). Explain how to solve the initial value problem explicitly in this case.

6.5G 1. Show that when system (34) is linear and hyperbolic, it can be converted to a *linear* normal system (38) in $\mathbf{w} = \mathbb{V}^{-1}\mathbf{u}$, provided that the elements of $\mathbb{A} = \mathbb{A}(x,t)$ are C^1. (*Hint:* $\mathbb{A}\mathbb{V} = \mathbb{V}\Lambda$.)

2. If the original system is also homogeneous in that $\mathbf{F} = \mathbb{B}\mathbf{u}$, must the normal system be homogeneous as well when $\mathbb{B} = \mathbb{B}(x,t)$?

6.5H 1. Show that the matrix $\mathbb{A}$ in (48′) has the real eigenvalues $u \pm c$ at each point.

2. For $\lambda_- = u - c$, verify that $\mathbf{V}^- = \binom{\rho}{-c}$ is an eigenvector and find a corresponding eigenvector $\mathbf{V}^+$ for $\lambda_+ = u + c$.

3.* Try to obtain a normal system using the matrix

$$\mathbb{V} = \begin{pmatrix} \rho & \rho \\ c & -c \end{pmatrix}$$

of unknown functions.

6.5I* 1. Show that in two dimensions x and y, the isentropic equations for *steady*, *irrotational* flow can be written in the form

$$u_y - v_x = 0$$
$$(u^2 - c^2)u_x + uv(u_y + v_x) + (v^2 - c^2)v_y = 0.$$

(*Hint:* Eliminate $\nabla p = c^2 \nabla \rho$ from equations (20) and (22).)

2. Assuming that $v^2 \neq c^2$, write this system in the quasilinear form

$$\mathbf{u}_y + \mathbb{A}\mathbf{u}_x = \mathbf{0}$$

where

$$\mathbf{u} = (u, v) \qquad \text{and} \qquad \mathbb{A} = \begin{bmatrix} 0 & -1 \\ \dfrac{u^2 - c^2}{v^2 - c^2} & \dfrac{2uv}{v^2 - c^2} \end{bmatrix}.$$

3. Show that the new system is elliptic when $|\mathbf{u}|^2 = u^2 + v^2 < c^2$ (subsonic flow) and hyperbolic when $|\mathbf{u}|^2 > c^2$ (supersonic flow).
4. In the hyperbolic case, find the eigenvalues.

6.5J The general nonlinear first-order equation in x and t has the functional form

$$F(x, t, u, u_x, u_t) = 0$$

for a function F defined on some domain of $\mathbb{R}^5$.

1. If F is C^1 and u is a C^2 solution of this equation, show that $\mathbf{v} = (u_x, u_t)$ will satisfy a quasilinear system of the form

$$\mathbb{A}\mathbf{v}_x + \mathbb{B}\mathbf{v}_t = \mathbf{G}.$$

2. What condition permits inversion of $\mathbb{B}$? When will the resulting system be hyperbolic?

6.5K 1. For the forward wave analyzed at the end of this section, where $w = 0$ and $c = c_0 + [(\gamma - 1)/2]u$, obtain an explicit equation of the form $\rho = \rho(u)$. (*Hint:* Use (55).)

2. Make an analysis of a backward simple wave using equations (51) with $v = 0$. Identify backward compressive waves and backward expansive waves. (*Hint:* In this case, $du/d\rho < 0$. [Why?])

6.5L If $\mathbf{u} = \mathbf{u}(x, t)$ is a solution in a domain D of a given quasilinear system (34), let C be a curve in D parameterized by a C^1 function $x = x(t)$.

1. Along C show that

$$\left(\mathbb{A} - \frac{dx}{dt}\mathbb{I}\right)\mathbf{u}_x = \mathbf{F} - \frac{d\mathbf{u}}{dt}$$

is known.

2. Conclude that at each t, we can solve for $\mathbf{u}_x$ on C *unless* $dx/dt = \lambda$ where $\lambda = \lambda(x, t, \mathbf{u})$ is an eigenvalue of $\mathbb{A}$ at the point $(x, t, \mathbf{u}(x(t), t))$.
3. Use this to explain why a solution $\mathbf{u}$ to (34) may exhibit a *weak discontinuity* but only *across* a characteristic C. (*Hint:* Refer to the discussion in the text.)

6.5M (*A Lewy example*) Consider the system

$$u_x - xv_y = f$$

$$v_x + xu_y = 0,$$

where f is C^∞ in a ball $B_R(\mathbf{0})$ in $\mathbb{R}^2$.

1. Show that at the origin the line $x = 0$ is not characteristic for this system, but the line $y = 0$ is.
2. Suppose that (u, v) is a C^2 solution of the system. What second-order equations are satisfied by u and v? Is the line $x = 0$ noncharacteristic for these equations?

(It is known that this system has *no* C^1 solutions for certain choices of f. See [Co].)

6.5N* 1. With the substitutions (y, x) for (x, t) and $\mathbf{u} = (u_x, u_y)$, transform Tricomi's equation from 6.2E into a system of the form (35) with $\mathbf{F} = \mathbf{0}$.

2. When $x > 0$, find the relevant eigenvalues and identify the characteristics.

3.* Put the system in normal form (38).

7

Series Methods

7.1

Introduction to Sturm–Liouville Problems

In Chapter 5 we examined the vibrations of a string of constant density ρ and tension τ. For gut strings, which are not perfectly uniform, we need to consider the free vibrations of a nonuniform string of length ℓ governed by equation (7) of Section 5.1,

$$\rho u_{tt} = (\tau u_x)_x, \qquad 0 < x < \ell, \tag{1}$$

where $\rho = \rho(x)$ and $\tau = \tau(x)$ are given positive functions on $(0, \ell)$.

The search for time-oscillatory solutions of form $u(x, t) = X(x) \cos \omega t$ leads to the modal equation

$$(\tau X')' = -\lambda \rho X, \qquad \text{for } \lambda = \omega^2 > 0, \tag{1'}$$

which should have nontrivial solutions for certain natural frequencies ω that depend on the boundary conditions present. For example, to have the fixed-end conditions $u(0, t) = u(\ell, t) = 0$, we must take

$$X(0) = X(\ell) = 0.$$

To permit the end at ℓ to be free we need

$$X(0) = X'(\ell) = 0;$$

and to allow there an elastic restraint of the Robin type $u_x(\ell, t) = Bu(\ell, t)$, we should take

$$X(0) = 0 \qquad \text{and} \qquad X'(\ell) = BX(\ell).$$

Similarly, the temperature distribution in a nonhomogeneous rod of length ℓ and insulated sides is governed by the equation

$$\nu u_t = (K u_x)_x, \qquad 0 < x < \ell, \tag{2}$$

$$\text{with} \qquad \nu = \nu(x) \qquad \text{and} \qquad K = K(x).$$

Separation of variables in the form

$$u(x,t) = X(x)e^{-\lambda t}, \qquad \lambda > 0,$$

leads to the X-equation

$$(KX')' = -\lambda \nu X, \qquad 0 < x < \ell, \tag{2'}$$

which is the *same* as (1′) if $K = \tau$ and $\nu = \rho$, and this suggests another thermomechanical analogy.

Even uniform conditions permit complications: In polar coordinates, the radial equation for free vibrations of a homogeneous circular membrane of unit radius is (see Section 7.4)

$$(rR')' - \frac{m^2}{r}R = -\lambda r R, \qquad 0 < r < 1, \qquad m = 0, 1, 2, \ldots \tag{3}$$

and $1/r$ blows up as r approaches zero. Here again, physically realistic homogeneous boundary conditions for $u = u(r, \theta, t)$ suggest boundary requirements for $R = R(r)$. For example, R might be bounded with $R(1) = 0$, while $rR'(r) \to 0$ as $r \searrow 0$.

We hope that each of the above problems has enough (eigen)values λ and associated (eigen)functions X (or R) to match a given f via series superposition and to attack related nonhomogeneous problems such as those for forced vibrations. However, because the coefficient functions in (1′) are rather general and those in (3) are not well behaved, our hopes might not be realized. Even when they are, the answers may not be simple. To proceed, let's set forth a category of eigenvalue problems that encompasses most of those above in cases of physical significance.

Sturm–Liouville Problems

Consider solutions $y = y(x)$ of the equation

$$(\tau y')' - qy = -\lambda \rho y, \qquad a < x < b \tag{4}$$

where τ, q, and ρ are given continuous functions on the *fixed* open interval (a, b) in which τ is C^1 and $\rho\tau$ is C^2. We assume that τ and ρ are *positive* functions with finite endpoint limits from within (a, b).† For technical reasons, the endpoint behavior of the coefficient functions influences the type of boundary conditions that we can impose on y and still have a tractable problem.

(7.0) Definition: *Each finite endpoint e at which $(\rho\tau)(e) > 0$, and q, τ', and $(\rho\tau)''$ have finite limits, is said to be* ***regular****. It is assigned a* ***regular boundary***

† If $v \in C(a, b)$, then a limit of v from within (a, b) at a or at b will be denoted $v(a)$ or $v(b)$, respectively.

***condition** of the form*

$$B_{e}(y) \stackrel{\text{def}}{=} E'y'(e) - Ey(e) = 0 \tag{4'}$$

*for e-dependent real constants E' and E not both zero. An endpoint $\tilde{e}$ with $\tau(\tilde{e}) = 0$ is said to be **singular**,† and y is required to meet related boundedness conditions.* ■

For a given interval (a, b) under these assumptions, equation (4), together with an assignment of regular boundary conditions (4′) or singular conditions, constitutes a **Sturm–Liouville (S–L) eigenvalue problem**. The problem is **regular** when *both* endpoints are regular; otherwise it is **singular**, and for the present, we will suppose that $q \geq 0$ in singular problems. (See Section 9.1.)

Eigenfunctions and Eigenvalues; Normalization

A *nontrivial* bounded solution y of (4) with bounded derivative for which $B_{e}(y) = 0$ at each **regular** endpoint e is called an **eigenfunction** to the **eigenvalue** λ of the problem. We consider only *real-valued* eigenfunctions since the eigenvalues must be real (see Remarks 7.1′). An eigenfunction y for which $\int_a^b \rho y^2\,dx = 1$ is said to be **normalized**.‡ An eigenfunction Y can be replaced by the normalized eigenfunction $y = Y/a$ to the *same* eigenvalue λ through use of the normalizing constant

$$a = \| Y \| \stackrel{\text{def}}{=} \left(\int_a^b \rho Y^2\,dx \right)^{1/2}, \tag{5}$$

called the ***ρ*-norm** of Y. Note that y and $-y$ are the *only* normalized eigenfunctions to λ. (See Problem 7.1E.)

ρ-Orthogonality

At this point we cannot be certain that a given S–L problem has any eigenfunctions or eigenvalues, but if it does, we have elements for a generalized Fourier series because of the following.

(7.1) Proposition: *Eigenfunctions y and $\tilde{y}$ to **distinct** eigenvalues λ and $\tilde{\lambda}$ respectively of the **same** S–L problem are **ρ-orthogonal** in that*

$$\int_a^b \rho y \tilde{y}\,dx = 0. \tag{6}$$

Proof: On each interval $[a', b'] \subseteq (a, b)$, we have from (4) and its counterpart for $\tilde{y}$ that

$$(\tilde{\lambda} - \lambda) \int_{a'}^{b'} \rho y \tilde{y}\,dx = \int_{a'}^{b'} (-\lambda \rho y)\tilde{y}\,dx - \int_{a'}^{b'} (-\tilde{\lambda} \rho \tilde{y}) y\,dx$$

† More generally, any nonregular endpoint is said to be singular, but in this text we consider only nonregular endpoints where τ is zero.

‡ Henceforth we will suppress dummy variables of integration whenever possible.

$$= \int_{a'}^{b'} [(\tau y')'\tilde{y} - (\tau \tilde{y}')'y]\,dx$$

$$= \tau[y'\tilde{y} - \tilde{y}'y]\Big|_{a'}^{b'},$$

by partial integration and cancellations. Now, as $a' \searrow a$ and $b' \nearrow b$, the last expression approaches zero! This occurs because τ vanishes at a singular endpoint while at each regular endpoint e, $B_e(y) = B_e(\tilde{y}) = 0$, for the *same* constants E' and E in (4'). Hence,

$$\text{if } E' = 0, \qquad \text{then} \qquad E \neq 0 \qquad \text{so that} \qquad y(e) = \tilde{y}(e) = 0;$$

$$\text{if } E' \neq 0, \qquad \text{then} \qquad (\tau y'\tilde{y})(e) = \frac{E}{E'}(\tau y\tilde{y})(e) = (\tau\tilde{y}'y)(e).$$

Therefore in the limit, as $a' \searrow a$ and $b' \nearrow b$,

$$(\tilde{\lambda} - \lambda)\int_a^b \rho y\tilde{y}\,dx = 0,$$

but $\tilde{\lambda} - \lambda \neq 0$, so that (6) holds. ■

(7.1') Remarks: 1. By examining this proof, we see that ρ-orthogonality occurs because, in a limiting sense,

$$(\tau y'\tilde{y})(e) = (\tau\tilde{y}'y)(e), \qquad \text{at each endpoint } e \tag{6'}$$

and our boundary requirements were chosen in each case to guarantee this. If $\tau(a) = \tau(b)$, then ρ-orthogonality also follows when we impose the **periodic conditions**

$$y(a) = y(b); \qquad y'(a) = y'(b) \tag{6''}$$

on each eigenfunction. There are certain other "feedback" conditions that suffice. (See Problem 7.2C.)

2. We can adapt the proof to show why an S–L problem has only real eigenvalues. Indeed, if λ and y are complex-valued solutions of an S–L eigenvalue problem, then their complex conjugates $\tilde{\lambda}$ and $\tilde{y}$, say, are also solutions of the problem. (See Problem 7.1G.) Hence the last equation of the proof still holds, but now

$$\int_a^b \rho y\tilde{y}\,dx = \int_a^b \rho\,|y|^2\,dx > 0 \text{ (why?) so that } \lambda (= \tilde{\lambda}) \text{ is real.}$$

S–L Series

Because of their ρ-orthogonality, eigenfunctions $y_1, y_2, \ldots, y_N$ to *distinct* eigenvalues $\lambda_1, \lambda_2, \ldots, \lambda_N$, respectively are **linearly independent** on (a, b). More generally, if each y_n is *normalized*, and

$$f(x) = \sum_{n=1}^{N} c_n y_n(x), \qquad a < x < b \tag{7}$$

for some constants $c_1, c_2, \ldots, c_N$, then we may use ρ-orthogonality as in Section 1.2

to conclude that for each $m \leq N$:

$$\int_a^b \rho f y_m \, dx = \sum_{n=1}^{N} c_n \int_a^b \rho y_n y_m \, dx = c_m \int_a^b \rho y_m^2 \, dx.$$

Thus for each $n \leq N$, the coefficients in (7) are given by the formula

$$c_n = \frac{\int_a^b \rho f y_n \, dx}{\int_a^b \rho y_n^2 \, dx} = \int_a^b \rho f y_n \, dx. \tag{7'}$$

This suggests the possibility of representing a given function f on (a, b) by the formal S–L series

$$F(x) = \sum_{n=1}^{\infty} c_n y_n(x), \qquad a < x < b,$$

where the y_n are normalized eigenfunctions to distinct eigenvalues λ_n, and the c_n are given by $(7')$ for $n = 1, 2, \ldots$.

Equivalently, we can consider representing f by the series

$$F(x) = \sum_{n=1}^{\infty} C_n Y_n(x), \qquad a < x < b, \tag{8}$$

where Y_n is any eigenfunction to λ_n that has finite norm $\| Y_n \|$, $n = 1, 2, \ldots$. From $(7')$, the associated coefficients are seen to be as follows:

$$C_n = \frac{1}{\| Y_n \|^2} \int_a^b \rho f Y_n \, dx, \qquad n = 1, 2, \ldots . \tag{8'}$$

Example 1:

Consider the S–L problem for the equation

$$y'' = -\lambda y, \qquad 0 < x < \pi$$

with boundary conditions $y(0) = y(\pi) = 0$.

Here $\tau(x) = \rho(x) = 1$ and $q = 0$, so that the problem is *regular*. It has *eigenfunctions* $\quad Y_n(x) = \sin nx \quad$ to *eigenvalues* $\quad \lambda_n = n^2, \quad n = 1, 2, \ldots$.

To get normalized eigenfunctions, we divide each Y_n by its norm

$$\| Y_n \| = \left(\int_0^\pi \sin^2 nx \, dx \right)^{1/2} = \sqrt{\pi/2}, \qquad n = 1, 2, \ldots .$$

Then we may consider representing a given integrable f on $(0, \pi)$ either by

$$\sum_{n=1}^{\infty} c_n \sqrt{\frac{2}{\pi}} \sin nx, \qquad \text{with} \qquad c_n = \sqrt{\frac{2}{\pi}} \int_0^\pi f(x) \sin nx \, dx, \qquad n = 1, 2, \ldots$$

or by

$$\sum_{n=1}^{\infty} C_n \sin nx, \qquad \text{with} \qquad C_n = \frac{2}{\pi} \int_0^\pi f(x) \sin nx \, dx, \qquad n = 1, 2, \ldots .$$

With the identification $C_n = \sqrt{2/\pi}\, c_n$, $n = 1, 2, \ldots$, these series are identical, and we recognize the second as the Fourier sine series generated by f (see Section 1.4).

When we recall the difficulty in relating a given f to its Fourier series (Section 1.8), we appreciate the issues associated with general series representation as in (8). To the extent that such representation can be attained, we have a means of solving associated boundary value problems.

Example 2:

For $n = 1, 2, \ldots$, assume that y_n is a normalized eigenfunction to the eigenvalue $\lambda_n > 0$ for the S–L problem

$$(\tau y')' = -\lambda \rho y, \qquad 0 < x < \ell,$$

with ℓ finite and $y(0) = y(\ell) = 0$.

If this problem is regular, the series

$$U(x, t) = \sum_{n=1}^{\infty} c_n\, y_n(x) \cos \sqrt{\lambda_n} t$$

provides a formal solution to equation (1) that describes transverse vibrations of a nonuniform stretched string of length ℓ released from rest with initial displacement

$$U(x, 0) = \sum_{n=1}^{\infty} c_n\, y_n(x), \qquad 0 < x < \ell.$$

Each y_n represents a modal shape in which this string can vibrate at the "natural" frequency $\omega_n = \sqrt{\lambda_n}$, and it is physically plausible that a few such y_n and λ_n exist. However, it is less evident that enough y_n exist to represent a prescribed $f(x) = U(x, 0)$ in this manner.

Regular problems are simpler to investigate, and for them we have the following major results.

(7.2) Theorem: *A regular* S–L *problem on* (a, b) *has normalized eigenfunctions* y_n *to eigenvalues* λ_n, $n = 1, 2, 3, \ldots$, *such that*

(1) $\lambda_1 < \lambda_2 < \cdots < \lambda_n$ *and* $\lambda_n \nearrow +\infty$ *as* $n \to +\infty$. **(9)**

(2) *each* $f \in \hat{C}^1[a, b]$ *that vanishes at the same endpoint(s) as the eigenfunctions can be represented by the absolutely and uniformly convergent series*

$$f(x) = \sum_{n=1}^{\infty} c_n\, y_n(x), \qquad a \le x \le b. \tag{10}$$

Hence,

$$\int_a^b \rho f^2\, dx = \sum_{n=1}^{\infty} c_n^2. \tag{10'}$$

Proof: See Sections 9.2 and 9.3. ■

(7.3) Remarks: Regularity demands that the interval (a, b) be bounded. In (9), λ_1 is finite but it may be negative. The endpoint conditions imposed on f in part 2 are essential for uniform convergence in (10). If, for example, the S–L problem requires that its eigenfunctions vanish at a while their derivative vanishes at b, then (10) holds for those $f \in \hat{C}^1[a, b]$ with $f(a) = 0$. (Were $f(a) \neq 0$, it can be shown that (10) holds pointwise, but *not uniformly*, on the half-open interval $(a, b]$). For this problem, if f is only sectionally smooth on $[a, b]$ (as in Section 1.5), then at each point $x \in (a, b)$, the series in (10) converges with sum $[f(x+) + f(x-)]/2$. (10′) is an immediate consequence of uniform convergence in (10) (see Problem 7.1F), but it holds for any f for which the integral is finite.

Singular S–L problems arise naturally in important applications. Their analysis involves improper integrals on possibly unbounded intervals, and it is more difficult to characterize admissible eigenfunctions y and representable functions f for such problems. For example, in equation (3), $\tau(r) = r$ on $(0, 1)$. Therefore, although the endpoint $r = 1$ where $\tau(1) = 1$ is regular and requires only a regular boundary condition (4′), the endpoint $r = 0$ is singular since $\tau(0) = 0$. Near $r = 0$, it is not clear what behavior should be imposed on eigenfunctions R, their derivatives R', or functions f to obtain results comparable to those in Theorem 7.2. Observe that on a finite interval $(1, b)$ corresponding problems for (3) must be regular.

Several applications resulting in singular problems are taken up in Sections 7.4–7.6, and supporting theory is developed in Chapter 9.

Multidimensional S–L Problems

For $d > 1$, a d-dimensional version of equation (1′) is as follows:

$$\nabla \cdot (\tau \nabla \psi) = -\lambda \rho \psi \qquad \text{in } V, \tag{11}$$

where V is a simple domain of $\mathbb{R}^d$ with boundary $S = \partial V$. To avoid complications we consider only conditions that produce *regular* S–L problems in V. Accordingly, we suppose τ and ρ to be continuous *positive* functions on $\bar{V}$, with $\tau \in C^1(V)$. These conditions are surely fulfilled when $\tau(\mathbf{x}) \equiv \rho(\mathbf{x}) \equiv 1$, and then (11) reduces to the **Helmholtz equation**

$$\nabla^2 \psi = -\lambda \psi.$$

To complete the specification of an S–L problem involving equation (11), we need a regular boundary condition of the form

$$B(\psi) \overset{\text{def}}{=} \left(g \frac{\partial \psi}{\partial n} - h\psi \right) \bigg|_S = 0 \tag{11′}$$

where g and h are continuous functions on S that do not vanish simultaneously at any point. The simplest examples of such a condition are

$$\begin{aligned} &\text{(i) the Dirichlet condition} \qquad \psi|_S = 0 \\ &\text{(ii) the Neumann condition} \qquad \left.\frac{\partial \psi}{\partial n}\right|_S = 0. \end{aligned} \tag{12}$$

Under either condition in (12), if $\psi \in C^1(\bar{V})$ is an eigenfunction to λ in (11) and ψ is *normalized* in that $\int_V \rho\psi^2\, dV = 1$, then it can be shown that

$$\lambda = \int_V \tau\,|\nabla\psi|^2\, dV \geq 0. \tag{12'}$$

(See Problem 7.1H.) It is difficult to determine eigenfunctions or eigenvalues explicitly except in a few simple cases (see Section 7.3), but equation (12′) affords valuable comparisons between the eigenvalues in related problems.

For a regular problem as described above, a multidimensional version of Theorem 7.2 exists (see Section 9.6), but it does not guarantee strict inequalities in (9). Indeed, a significant feature of multidimensional S–L problems is that a single eigenvalue can admit several linearly independent eigenfunctions. Such **degenerate** eigenvalues occur in the problem of Example 4 in Section 7.3.

Problem Set 7.1

7.1A Classify each of the following problems as regular or singular; if singular, explain why. In each case, identify τ, q, and ρ, and write $\|y\|^2$ explicitly.

1. $(xy')' - y = -e^x\lambda y$ on $(1, 2)$ with $y(1) = y'(2) = 0$.
2. $(xy')' - y = -e^x\lambda y$ on $(0, 1)$ with $y(1) = 0$, and y, y' bounded.
3. $[(\sin x)y']' - x^2y = -\lambda y$ on $(0, \pi)$ with y, y' bounded.
4. $[(\sin x)y']' - x^2y = -\lambda y$ on $(\pi/4, 3\pi/4)$ with $y(\pi/4) = y(3\pi/4) = 0$.

7.1B Heat conduction in a long thin circular rod of length ℓ and uniform properties is governed by the equation

$$\nu u_t = (Ku_x)_x - qu, \qquad 0 < x < \ell, \qquad t > 0,$$

where ν and K are positive constants and q is a function that, when positive, allows for heat lost by radiation at the lateral surface to the surroundings (supposed) at reference temperature zero. Suppose $\nu = K = 1$, $q =$ constant, and the ends of the rod are insulated so that $u_x(0, t) = u_x(\ell, t) = 0$, $t > 0$.

1. If an initial distribution $u(x, 0) = f(x)$ is prescribed, show that the eigenvalues of the associated S–L problem are given by

$$\lambda_n = q + \left(\frac{n\pi}{\ell}\right)^2, \qquad n = 0, 1, 2, \ldots .$$

2. What is $\lim_{n\to\infty} \lambda_n/n^2$?
3. Give a corresponding normalized eigenfunction y_n to λ_n, $n = 0, 1, 2, \ldots$.
4. Must there be negative eigenvalues when $q < 0$? Could 0 be an eigenvalue?

7.1C Show that the general second-order linear equation

$$y'' + p(x)y' + q(x)y = f(x)$$

with continuous coefficients p, q, f can be put in the self-adjoint form

$$(Py')' + (Pq)y = Pf$$

by means of a suitable positive function P.

7.1D 1. Verify that $(x^2y')' = -\lambda y$ is an equation of Euler type and use the transformation $x = e^t$ to find its general solution on $(0, 1]$.

2. Conclude that its only nontrivial solutions with $y(1) = 0$ are constant multiples of

$$y(x) = \frac{1}{\sqrt{x}} \sin\left[\sqrt{\lambda - (\tfrac{1}{4})}\log x\right], \qquad (\lambda > \tfrac{1}{4}),$$

none of which is bounded near $x = 0$.

3. Formulate a singular S–L problem that has no eigenfunction.

*7.1E** Suppose that y_1 and y_2 are linearly independent eigenfunctions to the *same* eigenvalue λ of an S–L problem for (4) involving a *regular* boundary condition $B_a(y) \stackrel{\text{def}}{=} A'y'(a) - Ay(a) = 0$.

1. Use the general theory of second-order linear equations to show that for this λ, *every* solution u of (4) is a linear combination of y_1 and y_2.
2. It is also possible to construct a solution v of (4) with $B_a(v) \neq 0$. (See Appendix B.3.) Explain the resulting contradiction and why it guarantees that if y is normalized, then $-y$ is the only other normalized eigenfunction. (This conclusion also holds if ℓ is a regular endpoint of the S–L problem or even when both endpoints are singular.)

7.1F Show how uniform convergence of the series in (10) guarantees that the c_n are given by (7′) and hence, that (10′) holds.

7.1G Suppose that λ and y are complex-valued solutions of an S–L eigenvalue problem involving (4). Show that their complex conjugates $\tilde{\lambda}$ and $\tilde{y}$, say, are solutions of the same problem. (Remember that τ, ρ, q, and any boundary-constants *are* real-valued.)

7.1H Suppose that ψ is a solution of

$$\nabla^2\psi = -\lambda\psi$$

in a simple domain V of $\mathbb{R}^d$, with $\int_V \psi^2\, dV = 1$, that meets either of the boundary conditions in (12).

1. Use the divergence theorem to show that the eigenvalue λ satisfies $\lambda = \int_V |\nabla\psi|^2\, dV \geq 0$.
2. If instead of (12), ψ meets the mixed conditions of (11′), then what relation between g and h will ensure $\lambda \geq 0$?

7.2

Modified Fourier Series

In Section 2.4 we saw that linear boundary value problems involving boundary conditions of a simple strip type on $[0, \pi]$ can frequently be separated by using trigonometric functions of the form $\sin nx$, $\cos nx$, or possibly $\sin(nx/2)$, for integral values of n. When this separation is achieved, a solution involving Fourier series representation of boundary functions is feasible. However, for the same problems Robin conditions lead to complications, which we now illustrate.

Example 3:

Ultralight human-powered aircraft now exist that have made flights of distances exceeding sixty miles, but some models lack maneuverability due to the absence of control surfaces. A proposed rectangular aileron (wing-control surface) model is fashioned from a homogeneous membrane whose leading edge and fuselage edge are fixed as shown in Figure 7.1a. The outboard side is a flexible spar so designed that along the trailing edge, the normal slope of the membrane is proportional to the deflection there. The response of the aircraft to aileron deflection depends on the shape of the deflected surface. If we suppose the panel to be of unit length and width b as shown, then its vertical deflection is described by $u = u(x, z)$, for $0 < x < 1$ and $0 < z < b$.

According to the analysis in Section 4.2, for small aileron deflection, u satisfies Laplace's equation

$$u_{xx} + u_{zz} = 0 \qquad \text{in} \qquad \{0 < x < 1, 0 < z < b\}$$

subject to the boundary conditions

$$u(0, z) = 0, \qquad u_x(1, z) = Bu(1, z), \qquad 0 < z < b$$
$$u(x, 0) = 0, \qquad u(x, b) = f(x), \qquad 0 < x < 1,$$

for some nonzero design constant B and a compatible spar-deflection function f. In particular, both the *sign* and magnitude of B can be prescribed.

Separation of variables in the form

$$u(x, z) = y(x)Z(z)$$

leads to the following (regular) eigenvalue problem for the modal functions y:

$$y'' = -\lambda y \qquad \text{on } (0,1),$$
$$\text{with} \qquad y(0) = 0; \qquad y'(1) = By(1). \tag{13}$$

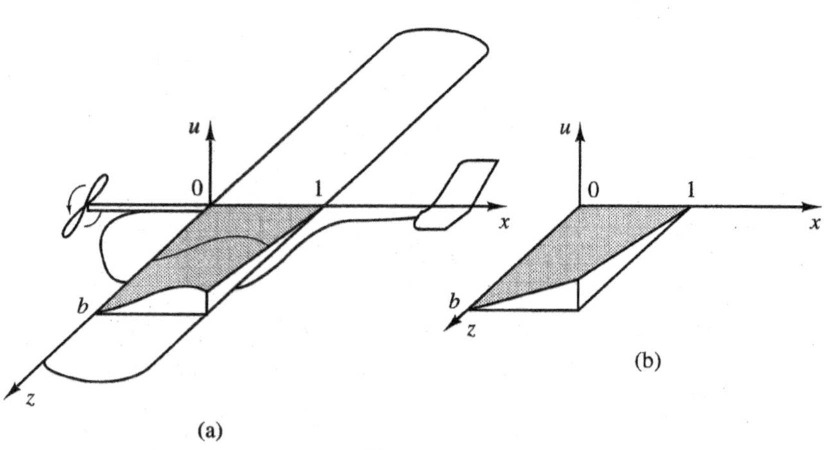

Figure 7.1

Since $\rho(x) = 1$, an eigenfunction y for this problem is normalized when $\int_0^1 y^2\,dx = 1$, and then y is related to its eigenvalue λ as follows:

$$\lambda = \int_0^1 (\lambda y)y\,dx = -\int_0^1 y''y\,dx = \int_0^1 (y')^2\,dx - y'y\Big|_0^1$$

so that

$$\lambda = \int_0^1 y'^2\,dx - By^2(1) \geq -By^2(1). \qquad \textbf{(13')}$$

In particular, $\lambda > 0$ if $B < 0$. In this case, we can set $\lambda = \omega^2$ and see that for $y(0) = 0$, an eigenfunction y must be of the form

$$y(x) = \sin \omega x \qquad (\text{so } y'(x) = \omega \cos \omega x)$$

with ω chosen to meet the trailing-edge condition (13)

$$y'(1) = \omega \cos \omega = B \sin \omega = By(1) \qquad \textbf{(14)}$$

or, since $B < 0$, $\qquad \tan \omega = \omega/B.$

As Figure 7.2 indicates graphically, for each $B < 0$, this transcendental equation has an infinite set of positive solutions $\omega_1 < \omega_2 < \cdots$. Here, for large n, $\omega_n \approx (n - \frac{1}{2})\pi$, so that $\lambda = \lambda_n = \omega_n^2 \approx n^2\pi^2$, but ω_n is *not* an integral multiple of ω_1 for any $n > 1$.

To each λ_n there is an eigenfunction

$$y_n(x) = \beta_n \sin \omega_n x,$$

which is normalized when we choose the constant β_n so that

$$1 = \beta_n^2 \int_0^1 \sin^2 \omega_n x\,dx = \frac{\beta_n^2}{2}\left(1 - \frac{\sin \omega_n \cos \omega_n}{\omega_n}\right)$$

$$= \frac{\beta_n^2}{2}(1 - B^{-1}\cos^2 \omega_n), \qquad \text{by (14).}$$

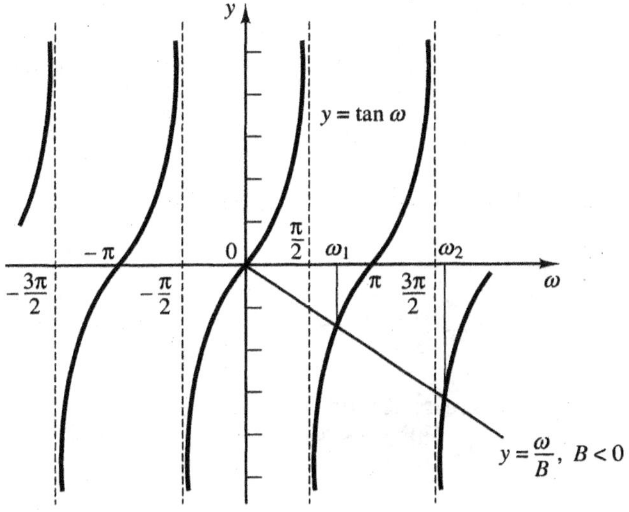

Figure 7.2

Hence,

$$\beta_n^2 = 2[1 - B^{-1}\cos^2\omega_n]^{-1}, \qquad n = 1, 2, \ldots .$$

As a result of Theorem 7.2, we see that each

$$f \in \mathscr{S}_0 \stackrel{\text{def}}{=} \{f \in \hat{C}^1[0,1] : f(0) = 0\}$$

has the absolutely and uniformly convergent trigonometric series representation

$$f(x) = \sum_{n=1}^{\infty} C_n \sin \omega_n x,$$

$$\text{where} \qquad C_n = \beta_n^2 \int_0^1 \sin(\omega_n x) f(x)\,dx, \qquad n = 1, 2, \ldots .^{\dagger} \tag{14'}$$

Clearly, $u_n = y_n Z$ satisfies Laplace's equation, when

$$0 = y_n''Z + y_n Z'' = (-\lambda_n Z + Z'')y_n,$$

or when $Z = Z_n$ satisfies the equation

$$Z_n'' - \omega_n^2 Z_n = 0.$$

Since we need $Z_n(0) = 0$, we can take $Z_n(z) = \sinh \omega_n z$ and

$$u_n(x, z) = \sin \omega_n x \sinh \omega_n z, \qquad n = 1, 2, \ldots .$$

These choices lead us to the formal solution

$$U(x, z) = \sum_{n=1}^{\infty} b_n \sin \omega_n x \sinh \omega_n z, \tag{15}$$

for which we wish to have

$$U(x, b) = f(x) = \sum_{n=1}^{\infty} b_n \sinh \omega_n b \sin \omega_n x, \qquad 0 < x < 1.$$

Here, as above, if $f \in \mathscr{S}_0$, we should take

$$b_n \sinh \omega_n b = C_n = \beta_n^2 \int_0^1 \sin(\omega_n x) f(x)\,dx, \qquad n = 1, 2, \ldots$$

and conclude that

$$U(x, z) = \sum_{n=1}^{\infty} C_n \sin \omega_n x \frac{\sinh \omega_n z}{\sinh \omega_n b}, \tag{15'}$$

where the series converges absolutely and uniformly for $0 \le x \le 1$ and $0 \le z \le b$ (since then, $0 \le \sinh \omega_n z \le \sinh \omega_n b$).

---

† This is not a pure Fourier series because ω_n is not an integral multiple of ω_1. It is in this sense that the series is "modified"; the official name for such series is **nonharmonic**.

For $B < 0$, the control deflections are as illustrated in Figure 7.1a. However, we can also envision those of Figure 7.1b where $B > 0$. Then there may be negative eigenvalues, so we must consider additional eigenfunctions $y = y(x)$ associated with $\lambda = -\omega^2 < 0$. To have $y(0) = 0$, we can take these eigenfunctions to be of the form $y(x) = \sinh \omega x$, where now we select ω to make

$$\tanh \omega = \frac{\omega}{B},$$

if this is possible. But as Figure 7.3 reveals, this equation has precisely one positive solution ω_0 if $B > 1$ and *none* if $B \leq 1$. Thus, if $0 < B \leq 1$, our analysis is as before.

However, when $B > 1$, there is one new eigenfunction $y_0(x) = \beta_0 \sinh \omega_0 x$ (for an appropriate normalizing constant β_0) corresponding to the eigenvalue $\lambda_0 = -\omega_0^2$. The augmented set of eigenfunctions $y_0, y_1, \ldots$ provides representation of functions in $\mathscr{S}_0$, and we obtain formal solutions to our control problem in the form

$$U(x, z) = b_0 \sinh \omega_0 x \sin \omega_0 z + \sum_{n=1}^{\infty} b_n \sin \omega_n x \sinh \omega_n z. \tag{16}$$

Note that in this case, uniqueness of solution is *not* assured by Corollary 3.2.

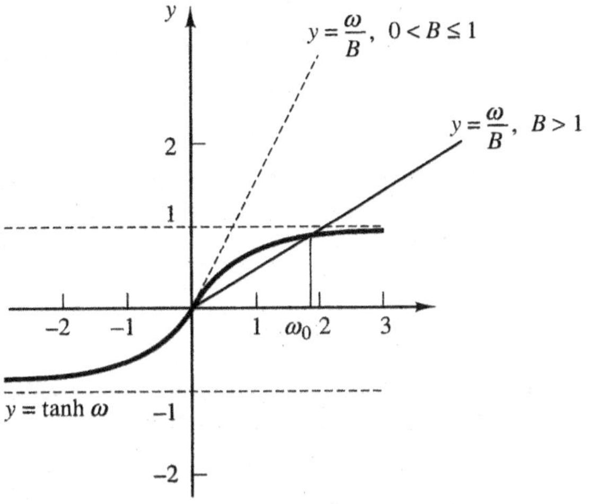

Figure 7.3

Problem Set 7.2

7.2A The end at $x = 0$ of a long thin iron rod of length $\ell = 1$ and insulated sides is kept at reference temperature zero, while the end at $x = 1$ experiences the cooling determined by $u_x(1, t) = Bu(1, t)$. (Recall Section 3.3)

1. If the rod is initially at temperature

$$u(x, 0) = f(x) = x, \qquad 0 < x < 1,$$

show that the required eigenfunctions are those found in Example 3 and find the formal series solution U. Does it converge uniformly when $t \geq 0$? (Assume that B is negative.)

2. Explain the physical difference in the problem when $B < 0$ and when $B > 0$. Which of these cases is the more natural? Describe an experiment to realize the other case.

7.2B Suppose that the tall building in Example 5 of Section 5.6 is placed on an elastic foundation to reduce the effects of horizontal ground motion from an earthquake.

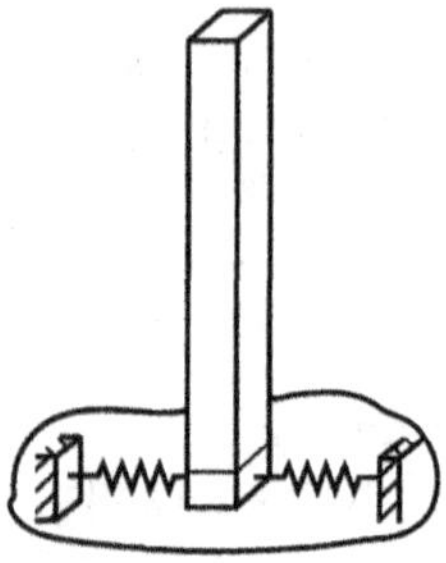

1. In the absence of ground motion explain why a string model for the horizontal sway $u = u(x, t)$ might satisfy the wave equation $u_{tt} = c^2 u_{xx}$, $0 < x < \ell$, $t > 0$ with $\mu u_x(0, t) = ku(0, t)$ and $u_x(\ell, t) = 0$, $t > 0$, and with some initial displacement and velocity. Here k is a positive elastic stiffness constant of the foundational material while μ, ρ, and $c^2 = \mu/\rho$ are as before. Is it reasonable that

$$u(0, t) = \frac{\mu}{k} u_x(0, t) \to 0 \quad \text{as} \quad k \to \infty?$$

2. When $c^2 = 1$ and $k\ell = 2\mu$, find a formal series solution to the boundary value problem in part 1 with the initial conditions

$$u(x, 0) = 0, \qquad u_t(x, 0) = 1.$$

7.2C (*Feedback conditions*) Consider the heat conduction problem of 7.1B where $\nu = K = 1$ and $q = 0$, $\ell = 1$, but the insulated-end conditions are replaced by mutual feedback conditions in the form

$$u_x(0, t) = -Au(1, t); \qquad u_x(1, t) = Au(0, t), \qquad t > 0$$

for some constant $A \neq 0$.

1. Show that $u(x, t) = e^{-\lambda t} y(x)$ is a solution, provided that y is an eigenfunction to an eigenvalue λ for the problem

$$y'' = -\lambda y$$

$$\text{with} \quad y'(0) = -Ay(1) \quad \text{and} \quad y'(1) = Ay(0).$$

Prove that Proposition 7.1 is valid for these eigenfunctions.

2. Show that for this problem, 0 is not an eigenvalue unless $A = -2$.
3. Show that when $A = 1$, then 0 and -1 are not eigenvalues, but there are negative eigenvalues $\lambda = -\omega^2$ whenever $\sinh \omega = 2\omega/(\omega^2 - 1)$ (so $\cosh \omega = (\omega^2 + 1)/(\omega^2 - 1)$). Verify graphically that this transcendental equation has precisely two nonzero solutions $\omega_0 \approx \pm 1.54$, so that the problem has one negative eigenvalue $\lambda_0 = -\omega_0^2$. What is a corresponding eigenfunction y_0? Is there another eigenfunction to λ_0 linearly independent of y_0?

4. In part 3, show graphically that the problem has an infinite number of positive eigenvalues $\lambda_n = \omega_n^2 > 1$, where $\sin \omega_n = -2\omega_n/(\omega_n^2+1)$, and explain why associated eigenfunctions are given by
$$y_n(x) = \omega_n \cos \omega_n x - \sin \omega_n (x-1), \qquad n = 1, 2, \ldots .$$
Verify that $\lim_{n\to\infty}(\lambda_n/n^2) = \pi^2$.
5. For the differential equation $y'' = -\lambda y$, obtain the integral $y'^2(x) = -\lambda y^2(x) + c$, where $c = y'^2(0) + \lambda y^2(0)$. Then show that
$$\lambda \int_0^1 y^2\,dx = \int_0^1 y'^2\,dx - (y'y)\Big|_0^1$$
$$= -\lambda \int_0^1 y^2\,dx + y^2(1) + \lambda y^2(0) - 2y(1)y(0),$$
under the boundary conditions of part 3, and conclude that
$$\int_0^1 y_n^2\,dx = \frac{(\lambda_n - 1)^3}{2(\lambda_n+1)^2}, \qquad n = 1, 2, \ldots .$$
(*Hint:* $y_n(1) = y_n(0) = \omega_n \cos \omega_n$, $n = 1, 2, \ldots$.)
6. Explain how you could use the preceding information to obtain a formal solution U to the boundary value problem under consideration with the initial condition $u(x,0) = f(x)$, $0 < x < 1$. Show that unless $\int_0^1 y_0 f\,dx = 0$, $\lim_{t\to\infty} |U(x,t)| = +\infty$.
7. When $A = -1$, show that there are no negative eigenvalues, so that we should expect $\lim_{t\to\infty} U(x,t) = 0$ in part 6. Explain this fact based on the difference between the physical problem when $A = +1$ and when $A = -1$.

7.2D* For the regular S–L problem
$$y'' = -\lambda y \qquad \text{on } (0,\ell) \qquad (\ell \text{ finite})$$
$$\text{with} \qquad y'(0) = Ay(0) \qquad \text{and} \qquad y'(\ell) = By(\ell),$$

1. Establish that $\lambda_0 = -\omega_0^2$ is an eigenvalue
$$\text{iff} \qquad \tanh \omega_0 \ell = \frac{(A-B)\omega_0}{AB - \omega_0^2}$$
and show graphically why there can be at most one such λ_0. (Consider cases where AB is positive, negative, and zero.)
2. Verify that $\lambda = \omega^2$ is a positive eigenvalue
$$\text{iff} \qquad \tan \omega\ell = \frac{(A-B)\omega}{AB + \omega^2}$$
and show graphically why there are eigenvalues $\lambda_n = \omega_n^2$, $n = 1, 2, \ldots$ with $\omega_1 < \omega_2 < \cdots$, and $\omega_n \ell \approx n\pi$ for large n.
3.* For large n, set $\omega_n \ell = n\pi + \delta_n$ where $|\delta_n| < \pi/2$, so that
$$\tan \delta_n = \tan \omega_n \ell = \frac{(A-B)\omega_n}{AB + \omega_n^2}$$
is small.

Conclude that

$$|\delta_n| \le |\tan \delta_n| \le \frac{2|A-B|}{\omega_n} \le \frac{4|A-B|\ell)}{n\pi},$$

when n is so large that $|AB| < 2\omega_n^2$.

7.2E Combine the results of Example 3 and the last problem to conclude that a regular problem for $y'' = -\lambda y$ on $(0, \ell)$ (ℓ finite) has at most one eigenvalue $\lambda_0 \le 0$. Show that it has positive eigenvalues

$$\lambda_n = \omega_n^2,\ n = 1, 2, \ldots$$

where for large n: $$\omega_n \approx \left(n - \frac{\sigma(2-\sigma)}{2}\right)\frac{\pi}{\ell},$$

if σ counts the number of endpoints (0, 1, or 2) at which the eigenfunctions vanish, and *we index the* ω_n *properly*.

7.3

Multiple Fourier Series

Separation of variables in problems involving more than two rectangular coordinates (x, y) may be feasible by terms involving products of the form

$$\cos mx \cos ny \qquad m, n = 0, 1, 2, \ldots,$$

or similar products in which either cosine is replaced by sine. When such separation is achieved, solution of the problem usually entails corresponding multiple Fourier series representation of boundary functions $f(x, y)$. For example, we might wish to have

$$f(x, y) = \sum_{m,n=1}^{\infty} c_{mn} \sin mx \sin ny \qquad \text{in the square} \quad S = \{0 < x < \pi, 0 < y < \pi\}, \tag{17}$$

for some constants c_{mn}, $m, n = 1, 2, \ldots$, with absolute and uniform convergence of the series. Then it is straightforward to invoke the separate orthogonality of the trigonometric functions involved to conclude that

$$c_{mn} = \left(\frac{2}{\pi}\right)^2 \int_0^{\pi} dx \int_0^{\pi} dy f(x, y) \sin mx \sin ny, \qquad m, n = 1, 2, \ldots \tag{17'}$$

Conversely, if $f \in C^1(\bar{S})$ with $f|_{\partial S} = 0$, and we use (17′) to define coefficients c_{mn}, then we can call on the one-dimensional convergence results of Section 1.5 to infer that (17) holds with absolute and uniform convergence. Independent scale changes can be made in each coordinate direction x and y with corresponding results for a rectangle. Moreover, we can obtain corresponding results when we replace sines by cosines and make the usual Fourier series adjustments when m or n is zero.

Example 4 (*Vibrations of rectangular membranes*):

Let's consider a membrane of uniform properties stretched tightly over a horizontal frame in the shape of the rectangle $V = \{0 < x < a, 0 < y < b\}$. (See Figure 7.4.) When the resulting "drumhead" is struck, it vibrates transversely and produces sound.

According to the small-deflection analysis in Section 5.1, at time t, the vertical deflection $u(x, y, t)$ of the membrane at $(x, y) \in V$ obeys the simple wave equation

$$\nabla^2 u = \frac{1}{c^2} u_{tt}, \qquad \text{for constant } c^2, \tag{18}$$
$$\text{with} \qquad u = 0 \qquad \text{on } \partial V.$$

Through the scale changes $\bar{x} = x/c$, $\bar{y} = y/c$, we obtain the same equation with $c^2 = 1$ for $\bar{u}(\bar{x}, \bar{y}, t) = u(x, y, t)$, and we shall simply assume that $c^2 = 1$ in (18).

Vibrations of the new membrane at frequency ω can occur in the form

$$u(x, y, t) = \psi(x, y) \cos \omega t$$

for mode shapes ψ that satisfy the Helmholtz equation

$$\nabla^2 \psi = \psi_{xx} + \psi_{yy} = -\omega^2 \psi, \qquad \text{in } V, \tag{19}$$
$$\text{with} \qquad \psi|_{\partial V} = 0.$$

This is an eigenvalue problem for $\lambda = \omega^2$, and we seek those λ providing nontrivial solutions ψ in V.

It is natural to attempt separation in the form

$$\psi(x, y) = X(x) Y(y),$$
$$\text{with} \qquad X(0) = X(a) = Y(0) = Y(b) = 0.$$

This is feasible when $X''Y + XY'' = -\lambda XY$, or when

$$-\frac{X''}{X} = \left(\frac{Y''}{Y} + \lambda\right) = \alpha, \qquad \text{a constant}, \tag{19'}$$

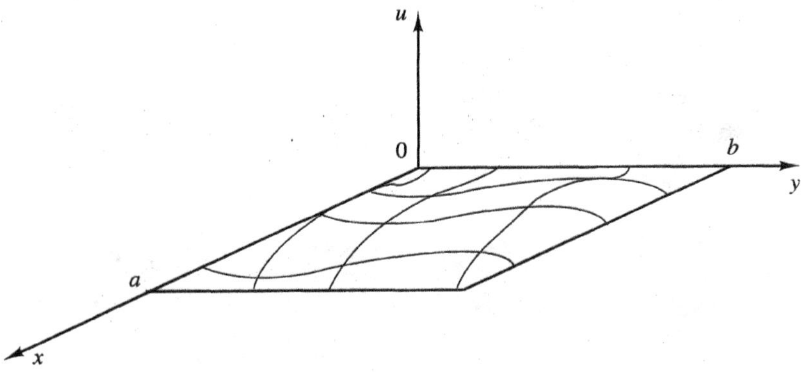

Figure 7.4

and by the familiar arguments we see that we must take

$$\alpha = \left(\frac{m\pi}{a}\right)^2 \text{ and } X_m(x) = \sin\frac{m\pi x}{a}, \quad \text{for } m = 1, 2, \ldots .$$

The associated Y_m must solve the eigenvalue problem

$$Y_m'' = -\left(\lambda - \frac{m^2\pi^2}{a^2}\right)Y_m, \qquad \text{with } Y_m(0) = Y_m(b) = 0.$$

For each m, this problem has the nontrivial solution

$$Y_{mn}(y) = \sin\frac{n\pi y}{b},$$

provided that

$$\lambda - \frac{m^2\pi^2}{a^2} = \frac{n^2\pi^2}{b^2}, \qquad n = 1, 2, \ldots .$$

Thus the only admissible eigenvalues λ are of the form

$$\lambda_{mn} = \omega_{mn}^2 = \pi^2\left(\frac{m^2}{a^2} + \frac{n^2}{b^2}\right), \qquad m, n = 1, 2, \ldots \tag{20}$$

and for each λ_{mn}, an associated eigenfunction is given by

$$\psi_{mn}(x, y) = \sin\frac{m\pi x}{a}\sin\frac{n\pi y}{b}. \tag{21}$$

The rectangular membrane of side lengths a and b can vibrate indefinitely at each natural frequency $\omega_{mn} = \sqrt{\lambda_{mn}}$, in mode shape(s) ψ_{mn}, with motions described by the standing wave

$$u_{mn}(x, y, t) = \psi_{mn}(x, y)\cos\omega_{mn}t. \tag{22}$$

In Figure 7.5 we indicate how a square drumhead of side π behaves when

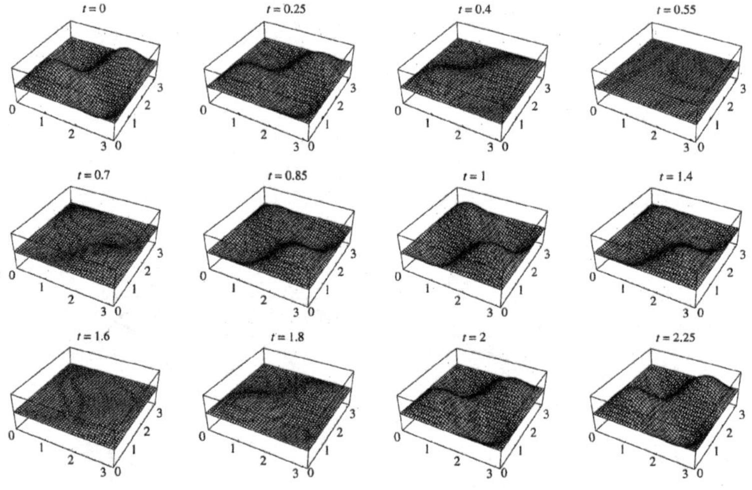

Figure 7.5

$m = n = 2$, over a time cycle of motion. If $a = b$, then in (20) $\lambda_{mn} = \lambda_{nm}$, so that, for example, ψ_{23} and ψ_{32} will be linearly independent mode shapes for the *same* frequency ω_{23}. Such degenerate behavior, predictable from the symmetry of a square membrane, can also occur in nonsquare cases. However, if b^2/a^2 is *irrational*, there can only be one mode shape to a given frequency. (See Problem 7.3C.)

To conclude, suppose our membrane is released from rest with initial deflection

$$u(x, y, 0) = f(x, y), \qquad (x, y) \in V.$$

Then the resulting motion is described by the formal series

$$U(x, y, t) = \sum_{m,n=1}^{\infty} c_{mn}\psi_{mn}(x, y) \cos \omega_{mn}t, \tag{23}$$

where the summation is performed in order of ascending values of $m + n$. As in our introductory analysis, the coefficients c_{mn} in (23) are given by

$$c_{mn} = \frac{4}{ab}\int_0^a \sin\frac{m\pi x}{a}\,dx \int_0^b f(x, y) \sin\frac{n\pi y}{b}\,dy, \qquad m, n = 1, 2, \ldots . \tag{24}$$

For example, if $f(x, y) = x(a - x)y(b - y)$ in V, then

$$c_{mn} = \begin{cases} \dfrac{8^2a^2b^2}{\pi^6 m^3n^3}, & m, n = 1, 3, 5, \ldots \\ 0, & \text{otherwise} \end{cases}$$

and (23) becomes

$$U(x, y, t) = \frac{8^2a^2b^2}{\pi^6} \sum_{m,n=1,3,}^{\infty} \frac{1}{m^3n^3} \sin\frac{m\pi x}{a} \sin\frac{n\pi y}{b} \cos \omega_{mn}t.$$

Since

$$\sum_{m,n=1,3,}^{\infty} \frac{1}{m^3n^3} \le \sum_{m=1}^{\infty} \frac{1}{m^3} \sum_{n=1}^{\infty} \frac{1}{n^3} < +\infty,$$

then by the M-test, this series for U converges uniformly, together with those of its first partial derivatives. U is C^1 everywhere and meets the initial condition and the boundary conditions, but it is difficult to show that U satisfies the wave equation (18). (Recall that we encountered similar difficulty with series solutions for vibrating string problems in Section 5.3.) Convergence difficulties can be alleviated by making f sufficiently well-behaved at the boundaries. (See Problem 7.3D.)

Problem Set 7.3

7.3A Find a Fourier series representation of the form

$$f(x, y) = \sum_{m,n+1=1}^{\infty} c_{mn} \sin mx \cos ny$$

in the square $\{0 < x < \pi, 0 < y < \pi\}$ for

1. $f(x, y) = y \sin x.$
2. $f(x, y) = xy.$ (The case $n = 0$ requires care.)
3. $f(x, y) = \begin{cases} x, & x \le y \\ y, & y \le x. \end{cases}$

7.3B Find eigenvalues $\lambda = \omega^2$ and orthogonal eigenfunctions of

$$\nabla^2\psi + \omega^2\psi = 0$$

for the rectangle $\{0 < x < 1, 0 < y < 2\}$, where

1. $\psi(0, y) = \psi_x(1, y) = 0, \quad 0 < y < 2$
   $\psi(x, 0) = \psi(x, 2) = 0, \quad 0 < x < 1$
2. $\psi(0, y) = \psi(1, y) = 0, \quad 0 < y < 2$
   $\psi_y(x, 0) = \psi_y(x, 2) = 0, \quad 0 < x < 1$
3. $\psi_x(0, y) = \psi(1, y) = 0, \quad 0 < y < 2$
   $\psi(x, 0) = \psi_y(x, 2) = 0, \quad 0 < x < 1$
4. $\psi(0, y) = \psi(1, y) = 0, \quad 0 < y < 2$
   $\psi_y(x, 0) = \psi_y(x, 2) + \psi(x, 2) = 0, \quad 0 < x < 1$

7.3C (*Other eigenvalues for a rectangle*) Find the eigenvalues ν and eigenfunctions $u = u(x, y)$ for the problem

$$\nabla^2 u = -\nu u \quad \text{in} \quad \{0 < x < a, 0 < y < b\}$$

with $u_x(0, y) = u_x(a, y) = 0$, $0 < y < b$, when (1) $u_y(x, 0) = u_y(x, b) = 0$, $0 < x < a$; or with $\nu = \mu$, when (2) $u(x, 0) = u_y(x, b) = 0$, $0 < x < a$.

How do these eigenvalues ν and μ compare with each other and with the membrane eigenvalues λ? Discuss degeneracy of eigenvalues when $b^2 = \sqrt{2}a^2$.

7.3D 1. Suppose that the series

$$f(x, y) = \sum_{m,n=1}^{\infty} c_{mn} \sin\frac{m\pi x}{a} \sin\frac{n\pi y}{b}$$

converges uniformly for $0 \le x \le a$, $0 \le y \le b$. Establish that

$$c_{mn} = \frac{4}{ab}\int_0^a dx \int_0^b dy\, f(x, y) \sin\frac{m\pi x}{a} \sin\frac{n\pi y}{b}, \qquad m, n = 1, 2, \ldots .$$

2. Calculate these c_{mn} when $a = 1$, $b = 2$ for $f(x, y) = x^2(1 - x)y(2 - y)$ and show that the resulting series does converge uniformly. Is its sum f? Explain.
3. Repeat part 2 when $f(x, y) = xy$. Can the series still converge uniformly? (*Hint:* Consider appropriate periodic extensions of f in the coordinate directions.)

4. Give an example of a nontrivial function f for which the associated series for U given by (23) will provide an actual solution for the motion of this membrane with initial deflection f (and initial velocity 0).

7.3E To solve $\nabla^2 u = u_{tt}$ in $\{0 < x < \pi, 0 < y < 2\pi, t > 0\}$ with

$$u_x(0,y,t) = u_x(\pi,y,t) = 0, \qquad 0 < y < 2\pi,\ t > 0,$$
$$u(x,0,t) = u(x,2\pi,t) = 0, \qquad 0 < x < \pi,\ t > 0,$$
and $u(x,y,0) = f(x,y)$, $\quad u_t(x,y,0) = 0, \quad 0 < x < \pi, \quad 0 < y < 2\pi,$

explain why we should consider the formal series

$$U(x,y,t) = \sum_{m=0,n=1}^{\infty} c_{mn} \cos mx \sin\frac{ny}{2} \cos\sqrt{\lambda_{mn}}\,t.$$

Find λ_{mn} and c_{mn}. (The case $m = 0$ requires care.)

7.3F To solve Laplace's equation $\nabla^2 u = 0$ in the box

$$V = \{0 < x < a, 0 < y < b, 0 < z < \ell\} \subseteq \mathbb{R}^3$$

with $u(x,y,\ell) = f(x,y), 0 < x < a, 0 < y < b$, and $u = 0$ on the other faces of the box, explain why we should consider the formal series

$$U(x,y,z) = \sum_{m,n=1}^{\infty} c_{mn} \sin\frac{m\pi x}{a} \sin\frac{n\pi y}{b} \sinh\sqrt{\lambda_{mn}}\,z.$$

Find λ_{mn} and c_{mn} and give a physical interpretation of this problem.

7.3G A square metal plate (of uniform composition and thickness) with insulated faces and given initial temperature is suddenly dropped in the ocean. To describe its internal heat conduction, why would it suffice to solve $\nabla^2 u = u_t$ in $\{0 < x < \pi,\ 0 < y < \pi,\ t > 0\}$, with $u(x,y,0) = f(x,y)$, $0 < x < \pi, 0 < y < \pi$, and $u = 0$ on the edges?

Obtain a formal series solution U by first finding elementary product solutions that separate t. Express the coefficients of this series in terms of f.

7.3H (*Damped vibrations of a membrane*) The vertical deflection $w = w(x,y,t)$ of a horizontal drumhead whose motion is opposed by air resistance is described approximately by the *damped wave equation* $w_{tt} = c^2\nabla^2 w - 2\delta w_t$, for a damping constant $\delta > 0$.

1. Assume that $w(x,y,t) = u(x,y)T(t)$, and conclude that u should satisfy the equation $\nabla^2 u = -\lambda u$ for some constant λ that makes $T'' + 2\delta T' + \lambda c^2 T = 0$.
2. For the square membrane $V = \{0 < x, y < \pi\}$ find the eigenvalues λ and associated products if $w|_{\partial V} = 0$ and $w(x,y,0) = 0$. (*Hint:* See Section 5.4.)
3. Obtain a formal series solution W to the problem in part 2 if $w_t(x,y,0) = f(x,y)$ is prescribed and continuous.

4. Carry out the computations in part 3 when $\delta = 1 = c^2$ and $f(x, y) = xy$, and discuss briefly the convergence of the series.

7.31 Find the eigenvalues λ and associated eigenfunctions for the Dirichlet problem

$$\nabla^2 u + \lambda u = 0 \qquad \text{in } V$$

with $u|_{\partial V} = 0$, where V is the box $\{0 < x < a, 0 < y < b, 0 < z < \ell\} \subseteq \mathbb{R}^3$.

7.4

Bessel Functions and Related Series

With differential equations involving the Laplacian in polar coordinates, separation methods frequently lead to S–L problems for some form of Bessel's equation. For instance, in the previous section we considered transverse vibrations of a uniform rectangular membrane. These can be described by solutions u of the two-dimensional standardized wave equation

$$\nabla^2 u = u_{tt} \tag{25}$$

that vanish on the boundary of the rectangle. To examine corresponding vibrations of a uniform circular membrane or drumhead of, say, unit radius, we assume that it is stretched horizontally over the disk $V = \{r < 1\}$ represented in polar coordinates (r, θ). (See Figure 7.6.)

Vibration at natural frequency ω of the form $u(r, \theta, t) = \psi(r, \theta)\cos \omega t$ can occur in mode shapes ψ that satisfy the Helmholtz equation

$$\nabla^2\psi = \psi_{rr} + \frac{1}{r}\psi_r + \frac{1}{r^2}\psi_{\theta\theta} = -\omega^2\psi, \qquad \text{in } V. \tag{26}$$

Separation of this equation in the product form $\psi(r, \theta) = R(r)\Theta(\theta)$ is possible if the functions R and Θ make

$$\frac{R''}{R} + \frac{1}{r}\frac{R'}{R} + \frac{1}{r^2}\frac{\Theta''}{\Theta} = -\omega^2$$

$$\text{or} \qquad r^2\frac{R''}{R} + r\frac{R'}{R} + \omega^2 r^2 = -\frac{\Theta''}{\Theta} = \text{const.} \tag{26'}$$

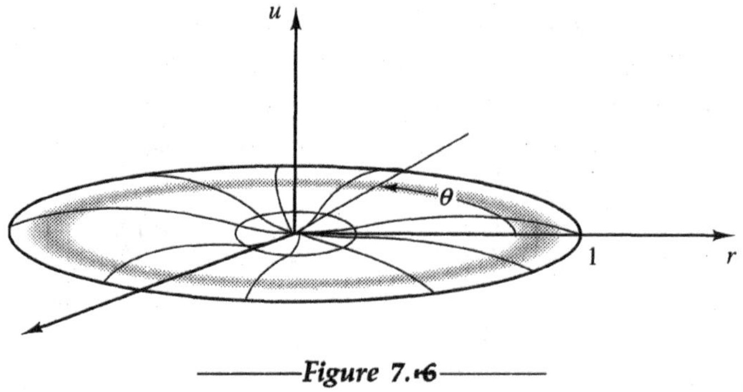

Figure 7.6

For reasonable solutions in a full disk, we need for Θ to have period 2π. Separation is then feasible when

$$\frac{\Theta''}{\Theta} = \text{const.} = -m^2, \qquad \text{for } m = 0, 1, 2, \ldots ,$$

provided that for each m, $R = R_m$ is a solution of the equation

$$\frac{R''}{R} + \frac{1}{r}\frac{R'}{R} - \frac{m^2}{r^2} = -\omega^2, \quad 0 < r < 1.$$

At the boundary of the membrane we require $R(1) = 0$. Since the equation for R may be written

$$(rR')' - \frac{m^2}{r}R = -\omega^2 rR, \tag{27}$$

we see that we have an S–L problem for eigenvalues $\lambda = \omega^2$ on (0,1) with $\tau(r) = \rho(r) = r$ and $q(r) = -m^2/r$. The problem is *singular* because $\tau(0) = 0$, but the endpoint $r = 1$ is regular.

Example 5 (*Axisymmetric vibrations of a circular drumhead*):

If the circular drumhead in Figure 7.6 is struck exactly at its center, we expect the ensuing vibrations to be independent of θ. For natural vibrations at frequency ω, radial mode shapes are given by functions $R = R(r)$, satisfying equation (27) with $m = 0$ or equivalently by $R(r) = y(\omega r)$, where $y = y(x)$ is a solution of **Bessel's equation of order zero,**

$$xy'' + y' = -xy, \qquad 0 \le x \le \omega, \tag{28}$$

with $y(\omega) = R(1) = 0$. From symmetry, we expect that $y'(0) = (1/\omega)R'(0) = 0$. Equation (28) is not elementary, but it has an analytic solution

$$y(x) = \sum_{k=0}^{\infty} a_k \frac{x^k}{k!},$$

provided that appropriate coefficients

$$a_k = y^{(k)}(0), \quad k = 0, 1, 2, \ldots$$

can be determined. (See Appendix B.1.) Observe that necessarily $a_1 = y'(0) = 0$, since the remaining terms in (28) vanish when $x = 0$.

If we differentiate equation (28) k times using the Leibniz formula as required, we find that for $k = 1, 2, \ldots$

$$xy^{(k+2)} + (k+1)y^{(k+1)} = -xy^{(k)} - ky^{(k-1)}.$$

Upon evaluating at $x = 0$, we obtain the coefficient **recursion formula**

$$(k+1)a_{k+1} = -ka_{k-1}, \qquad k = 1, 2, \ldots .$$

Hence $0 = a_1 = a_3 = \cdots = a_k$, k odd, and if $a_0 = 1$, then $a_2 = -1/2$, $a_4 = -3a_2/4 = (1 \cdot 3)/(2 \cdot 4)$, $a_6 = -(1 \cdot 3 \cdot 5)/(2 \cdot 4 \cdot 6)$, etc.

Therefore, if convergent, the series

$$J_0(x) = y(x) = 1 - \frac{x^2}{2 \cdot 2!} + \frac{1 \cdot 3}{2 \cdot 4} \frac{x^4}{4!} - + \cdots$$
$$= \sum_{j=0}^{\infty} \frac{(-1)^j (x/2)^{2j}}{(j!)^2} \tag{29}$$

will provide a solution analytic near $x = 0$. This series converges for all $x \in \mathbb{R}$ (by the ratio test) and it can be shown that within a constant factor, J_0 is the only nontrivial bounded solution of (28) on $(0, \omega)$. (See Problem 7.4I.)

The natural frequencies ω are those for which $J_0(\omega) = 0$, and although it is not immediately evident that such ω exist, we note that series (29) resembles that for $\cos x$. It is easy to show that $|J_0(x)| \leq 1$ and $|J_0'(x)| \leq 1$ because the associated energy function $E = y'^2 + y^2$ for $y = J_0$ *decreases* from its value of 1 at $x = 0$. Indeed, from equation (28), we see that $E' = 2(y'' + y)y' = -2y'^2/x$, which is negative when $x > 0$. Moreover, on comparing equations satisfied by $J_0(x)$ and $\cos x$, it can be shown that J_0 vanishes at least once in every positive interval of length π; see Theorem B.3. In fact, for large values of x, $J_0(x)$ behaves somewhat like $(\cos x)/\sqrt{x}$ and the positive zeros of J_0 can be arranged in an unbounded sequence ω_n, $n = 1, 2, \ldots$ with $0 < \omega_1 < \omega_2 < \cdots < \omega_n < \omega_{n+1} < \cdots$ where $\omega_1 \simeq 2.405$, $\omega_2 \simeq 5.520$, $\omega_3 \simeq 8.654$, $\omega_4 \simeq 11.792$, and $\omega_n \simeq n\pi - (\pi/4)$, $n \geq 5$. (See Figure 7.7.)

It follows that associated with each ω_n is a natural mode shape $R_n(r) = J_0(\omega_n r)$, and a drumhead initially deflected radially in this shape will vibrate axisymmetrically in the form

$$u_n(r, t) = J_0(\omega_n r) \cos \omega_n t.$$

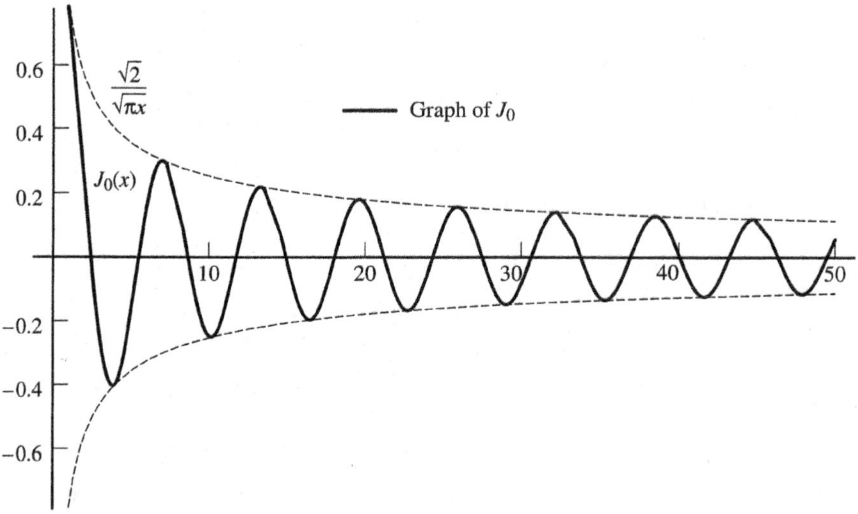

Figure 7.7

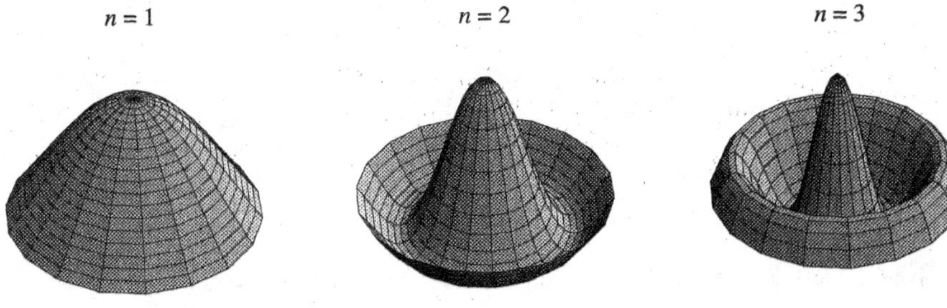

Fig. 7.8

Figure 7.8 shows initial deflection shapes for $n = 1, 2$, and 3, and as we see the number of ripples increases with n.

Now suppose that the initial radial deflection of our drumhead is prescribed by

$$u(r, 0) = f(r), \qquad 0 \le r \le 1,$$

where $f(1) = 0$. Following the Fourier series model, we attempt a solution in the form

$$U(r, t) = \sum_{n=1}^{\infty} C_n J_0(\omega_n r) \cos \omega_n t$$

where we desire that the coefficients C_n ensure convergence and make

$$U(r, 0) = f(r) = \sum_{n=1}^{\infty} C_n J_0(\omega_n r), \quad 0 \le r \le 1. \tag{30}$$

For each n, $Y_n(r) = J_0(\omega_n r)$ is an eigenfunction to the eigenvalue $\lambda_n = \omega_n^2$ of the S–L problem of equation (27) for $m = 0$. Therefore, Proposition 7.1 applies with $\rho(r) = r$ and, according to (8′), if the series in (30) converges uniformly, we should take

$$C_n = \frac{1}{j_n} \int_0^1 r f(r) J_0(\omega_n r)\, dr, \tag{31}$$

with

$$j_n = \int_0^1 r J_0^2(\omega_n r)\, dr, \quad n = 1, 2, \ldots .$$

Formulas (31) show that such representation of f is unique if it is possible, but it is quite difficult to prove that a given f does admit Bessel series representation in the form (30). (See Proposition 9.6.) When this has been established, we can regard the finite sums

$$U_N(r, t) = \sum_{n=1}^{N} C_n J_0(\omega_n r) \cos \omega_n t$$

as approximate solutions to our problem. However, there remain the usual questions about convergence, differentiability, etc. of the formal solution U.

Calculation of the integrals in (31) is facilitated by use of special properties of Bessel functions. (See Problems 7.4 E–G.)

Example 6 (*General vibrations of a circular drumhead*):

If our circular drumhead is struck off-center, then the resulting vibrations will not be axisymmetric, and in equation (27) we must also allow that $m = 1, 2, 3, \ldots$. Consequently, we need nontrivial solutions $y(x) = R(x/\omega)$ of **Bessel's equation of order** m:

$$x^2 y'' + xy' - m^2 y = -x^2 y \quad \text{on } [0, \omega] \tag{32}$$

with $y(\omega) = 0$. A solution of this equation is given by the series

$$J_m(x) = \sum_{j=0}^{\infty} \frac{(-1)^j (x/2)^{2j+m}}{j!(j+m)!}, \tag{32'}$$

which converges for all x. It can also be established that for each m, there is an unbounded sequence ω_{mn}, $n = 1, 2, \ldots$, for which $J_m(\omega_{mn}) = 0$, where

$$0 < \omega_{m1} < \omega_{m2} < \cdots < \omega_{mn} < \cdots$$

(see Problems 7.4F, H). Comparative graphs of J_0, J_1, and J_2 are shown in Figure 7.9.

Each ω_{mn} (including those ω_{0n} found previously) represents a natural frequency of the membrane, and vibrations occur in associated mode shapes

$$R_{mn}(r)\begin{cases}\cos m\theta \\ \sin m\theta\end{cases} = J_m(\omega_{mn} r)\begin{cases}\cos m\theta \\ \sin m\theta\end{cases}, \qquad \begin{matrix} m = 0, 1, \ldots \\ n = 1, 2, \ldots, \end{matrix}$$

two of which are illustrated in Figure 7.10.

Now we must consider double series of the form

$$U(r, \theta, t) = \sum_{m+1, n=1}^{\infty} (c_{mn} \cos m\theta + s_{mn} \sin m\theta) J_m(\omega_{mn} r) \cos \omega_{mn} t \tag{33}$$

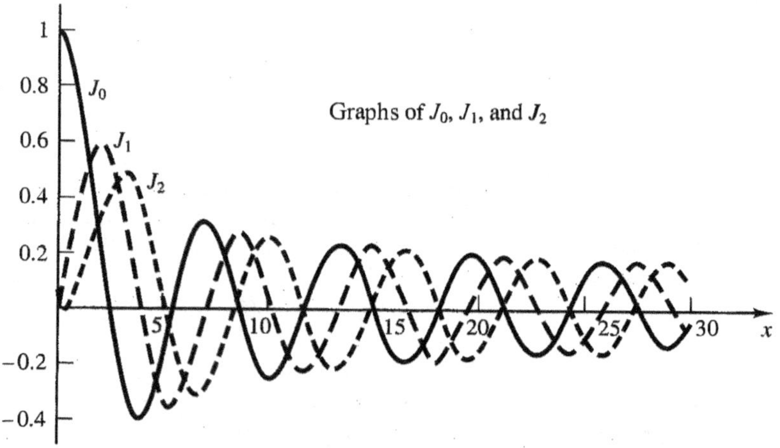

Figure 7.9

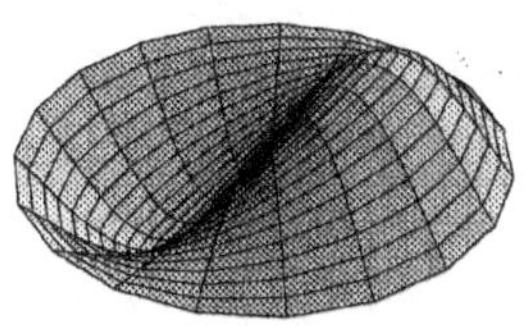

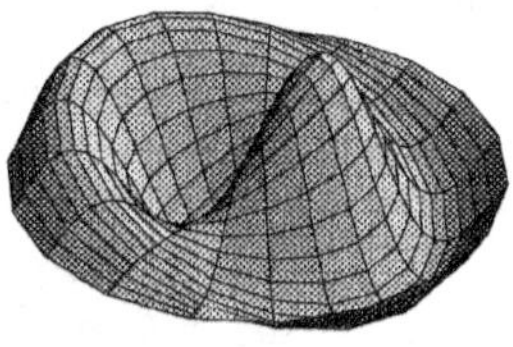

Figure 7.10

in order to match a prescribed initial deflection

$$U(r, \theta, 0) = f(r, \theta), \qquad 0 \le r < 1, \qquad |\theta| \le \pi.$$

For compatibility, we should require that $f(1, \theta) = 0$, $|\theta| \le \pi$ and that $f(r, -\pi) = f(r, \pi)$, $0 \le r \le 1$, but the resulting technical difficulties remain formidable. (See [Wein].)

When equation (25) is replaced by the standardized heat equation

$$\nabla^2 u = u_t, \qquad \text{in } V \times (0, \infty),$$

then separation is possible in the form

$$u(r, \theta, t) = R(r)\,\Theta(\theta)e^{-\lambda t}$$

provided that $\Theta''/\Theta = -m^2$, $m = 0, 1, 2, \ldots$, and $R = R(r)$ satisfies (27) with ω^2 replaced by λ. In this manner, Bessel functions feature in the solution of problems in heat conduction in either a thin disk with insulated surfaces or a very long circular shaft. Similarly, these functions generate harmonic functions u of the cylindrical coordinates (r, θ, z) in $\mathbb{R}^3$. On the other hand, similar problems for the hollow cylinder $\{(r, \theta, z)\colon 1 < r < \ell\}$ lead to S–L problems for Bessel's equation (27) that are regular for finite ℓ, so that Theorem 7.2 applies. (See Problem 7.4J.)

Behavior of $J_m(x)$ for Large $x > 0$*

To characterize $J_m(\omega_{mn} r)$, $0 < r < 1$, for large values of ω_{mn}, we should first consider the behavior of $J_m(x)$ for large x. Now, it is straightforward to verify that $y(x) = \sqrt{x}\,J_m(x)$ satisfies the equation

$$y'' + y = \frac{M}{x^2}\,y, \qquad x > 0, \qquad \text{where } M = m^2 - \frac{1}{4} \tag{33'}$$

and by energy estimates to conclude that for $x \ge a > 0$, $|y(x)|$ is bounded by Y, say. (See Problem 7.4L.)

Therefore, using the first formula of Proposition B.1, we can "solve" the last equation in the form

$$\begin{aligned} y(x) &= A\cos(x - \alpha) + M\int_a^x \sin(x - t)\,\frac{y(t)}{t^2}\,dt \\ &= A\cos(x - \alpha) + M\,\mathscr{I}_m\left\{e^{ix}\int_a^x e^{-it}\,\frac{y(t)}{t^2}\,dt\right\}, \end{aligned}$$

for appropriate constants A and α.[†] Since $|y|$ is bounded and $\int_a^\infty t^{-2}dt < +\infty$, the *improper integral,*

$$C + iS \overset{\text{def}}{=} \int_a^\infty e^{-it}\frac{y(t)}{t^2}\,dt$$

exists, and hence for *new constants* $A \geq 0$ and α,

$$y_1(x) \overset{\text{def}}{=} y(x) - A\cos(x-\alpha) = -M\mathscr{I}_m\left\{e^{ix}\int_x^\infty e^{-it}\frac{y(t)}{t^2}\,dt\right\}.$$

Moreover, for $x \geq a$ the difference function $y_1(x)$ admits the easy estimate

$$|y_1(x)| \leq |M| \int_x^\infty \frac{|y(t)|}{t^2}\,dt \leq \frac{|M|\,Y}{x} = \frac{A_1}{x}, \quad \text{say.}$$

Therefore, for this y_1

$$J_m(x) = \frac{y(x)}{\sqrt{x}} = \frac{A}{\sqrt{x}}\cos(x-\alpha) + \frac{y_1(x)}{\sqrt{x}}, \qquad x \geq a, \tag{34}$$

and, assuming $A \neq 0$, it follows that for large x, J_m oscillates approximately sinusoidally with period 2π but dies out "like" $1/\sqrt{x}$. In particular, for large n, $\omega_{mn} \approx n\pi + (\pi/2) + \alpha_m$ where the phase α_m is independent of n.

The above approximation may be improved if in the integral defining $y_1(x)$, we replace $y(t)$ by

$$y(t) = A\cos(t-\alpha) + y_1(t) = A\,\frac{e^{i(t-\alpha)} + e^{-i(t-\alpha)}}{2} + y_1(t).$$

Then the first term can be integrated immediately, and the remaining terms can be estimated. It follows that

$$y(x) = A\left[\cos(x-\alpha) - \frac{M}{2x}\sin(x-\alpha)\right] + y_2(x), \tag{34'}$$

where $|y_2(x)| \leq A_2/x^2$ for an explicitly available constant $A_2 > 0$. (See Problem 7.4M.) Consequently, for $x \geq a$

$$J_m(x) = \frac{A}{\sqrt{x}}\left[\cos(x-\alpha) - \frac{M}{2x}\sin(x-\alpha)\right] + \frac{y_2(x)}{\sqrt{x}},$$

and successively better results can be obtained as desired. Despite such information it is surprisingly difficult to obtain precise values for the constants A and α. However, in Section 8.6 we prove that $A = \sqrt{2/\pi}$ and that within a multiple of 2π, $\alpha = \alpha_m = (m\pi/2) + (\pi/4)$. (See Theorem 8.10.)

Problem Set 7.4

7.4A 1. Verify that equation (25) has a solution of the time-oscillatory form $u(r,\theta,t) = \psi(r,\theta)\cos\omega t$, if ψ satisfies (26).

2. Show that separation of (26) is feasible if (26′) holds.

[†] For any real constants c and s, we have

$$c\cos x + s\sin x = A\cos(x-\alpha), \qquad x \in \mathbb{R}$$

where $A = \sqrt{c^2+s^2}$ and $\alpha \in [0,\pi]$ with $\cos\alpha = c/A$.

7.4B Show that $R = R(r)$ satisfies (27) for $\omega \neq 0$ iff $y(x) \stackrel{\text{def}}{=} R(x/\omega)$ satisfies (32).

7.4C A flat steel disk of radius 2 feet with insulated faces is initially at a uniform temperature of 20°. Then at time $t = 0$, its edge temperature is lowered to a reference temperature of 0°.

1. Assuming that its thermal diffusivity $k = 3$ in compatible units, show that separation of the heat equation $u_t = k\nabla^2 u$ in the axisymmetric form $u(r, \theta, t) = T(t)\,R(r)$ leads to (27) with $m = 0$.
2. Obtain a formal series involving J_0 for the subsequent temperature distribution.

7.4D A chain of length ℓ suspended freely from one end undergoes transverse planar motions $u = u(x, t)$ governed by the equation

$$u_{tt} = g(xu_x)_x$$

(see Problem 5.1D) where x is measured from the *free* end and g is the gravitational acceleration.

1. Show that oscillations of frequency ω in the form $u(x, t) = X(x)\cos \omega t$ are possible if

$$(xX')' = -\frac{\omega^2}{g}X = -\lambda X, \qquad \text{say,} \qquad \text{where } X(\ell) = 0.$$

2. Let $X(x) = y(2\sqrt{\lambda x})$, and show that $y = y(s)$ satisfies (28), $sy'' + y' = -sy$.
3. Conclude that vibration can occur in modes $X(x) = J_0(2\omega\sqrt{x/g})$ at natural frequencies ω for which $J_0(2\omega\sqrt{\ell/g}) = 0$. (This problem, analyzed by D. Bernoulli in 1738 and by Euler in 1746, led to the first examination of a series for a Bessel function. Both mathematicians predicted the existence of an infinite set of natural frequencies, and Euler found approximate values for the first three. See [Wat].)

7.4E 1. Show that $y(t) = J_0(t)$ satisfies the equation

$$\frac{d}{dt}[(t^2y'^2) + (t^2y^2)] = 2ty^2.$$

2. Conclude that for $\omega > 0$,

$$\int_0^1 xJ_0^2(\omega x)\,dx = \frac{1}{\omega^2}\int_0^\omega tJ_0^2(t)\,dt = \frac{1}{2}\,[J_0^2(\omega) + J_0'^2(\omega)].$$

3. What value does this give for j_n in (31)?
4. Use (32) and generalize the result in part 2 to conclude that

$$\int_0^1 xJ_m^2(\omega x)\,dx = \frac{1}{2}\left[\left(1 - \frac{m^2}{\omega^2}\right)J_m^2(\omega) + J_m'^2(\omega)\right].$$

7.4F 1. Differentiate (32′) and combine appropriately to obtain the identity

$$[x^{-m}J_m(x)]' = -x^{-m}J_{m+1}(x), \qquad x > 0,$$

which shows that between successive positive zeros of J_m (so of $x^{-m}J_m(x)$) there must be a zero of J_{m+1} (by Rolle's theorem).

2. Similarly, for $m = 1, 2, \ldots$, show that $[x^m J_m(x)]' = x^m J_{m-1}(x)$, and conclude that the positive zeros of J_{m+1} *interlace* those of J_m. (Fig. 7.9)

7.4G 1. Use (32′) to verify the recursion formula

$$xJ_{m+1}(x) = 2mJ_m(x) - xJ_{m-1}(x), \qquad m \geq 1.$$

2. Combine this result with that of Problem 7.4F1 to show that

$$2J_m'(x) = J_{m-1}(x) - J_{m+1}(x), \qquad m = 1, 2, \ldots .$$

7.4H* The series (32′) for J_m can be obtained directly by seeking an analytic solution to (32) in the form $y(x) = \sum_{k=0}^{\infty}(a_k/k!)x^k$, with coefficients $a_k = y^{(k)}(0)$, $k = 0, 1, 2, \ldots$.

1. Use equation (32) to show that $m \neq 0 \Rightarrow a_0 = 0$.
2. Differentiate (32) and evaluate at $x = 0$ to show that

$$m \neq 1 \Rightarrow a_1 = 0.$$

3. Differentiate (32) $k \geq 2$ times and evaluate at $x = 0$ to get the following recurrence formula for the coefficients a_k:

$$k(k-1)a_k + ka_k - m^2 a_k = -k(k-1)a_{k-2}$$

$$\text{or} \qquad (k^2 - m^2)\,a_k = -k(k-1)a_{k-2}.$$

4. Therefore if $k = m$, we can have $a_m \neq 0$, although all $a_k = 0$ for $k < m$. Consequently $a_{m+1} = a_{m+3} = a_{m+5} = \cdots = 0$, while for

$$k = m+2: \quad a_{m+2} = -\frac{(m+2)(m+1)}{2 \cdot 2m+2}\,a_m = -\frac{m+2}{4}\,a_m$$

$$k = m+4: \quad a_{m+4} = -\frac{(m+4)(m+3)}{4 \cdot (2m+4)}\,a_{m+2} = \frac{(m+4)(m+3)}{8 \cdot 4}\,a_m$$

$$k = m+6: \quad a_{m+6} = -\frac{(m+6)(m+5)}{6 \cdot (2m+6)}\,a_{m+4} = -\frac{(m+6)(m+5)(m+4)}{12 \cdot 8 \cdot 4}\,a_m$$

5. Verify that if $a_m = 1/2^m$, then this procedure leads to the series (32′) for J_m. Observe that within choice of a_m, this is the only solution to (32) that is analytic in a neighborhood of $x = 0$.

7.4I Use reduction of order (as in Theorem B.2) to argue that a solution y_m of (32) linearly independent of J_m must be of the form vJ_m, with $v(x) = \int [xJ_m^2(x)]^{-1}\,dx$, and so $y_m(x)$ is unbounded near $x = 0$.

7.4J* When properly normalized, the function y_m from the previous problem is denoted Y_m, so that the general solution of (32) on an interval $(a, b) \subseteq (0, +\infty)$ is given by $y(x) = aJ_m(x) + bY_m(x)$, where a and b are arbitrary constants.

1. Use this fact to analyze the axially symmetric free vibrations of a uniform annular membrane $\{1 < r < 2\}$, governed by (25), where the more general solution to (28) must be permitted.

2. How are the natural frequencies ω characterized? (Note that our comparison and asymptotic results for J_m also apply to y.)

7.4K (*The zeros of* J_0)

1. Show that $|x| \le 1 \Rightarrow J_0(x) > 0$.

$$\left(\textit{Hint}: \text{ For } |x| \le 1, \left| \sum_{j=1}^{\infty} \frac{(-1)^j (x/2)^{2j}}{(j!)^2} \right| \le \sum_{j=1}^{\infty} 4^{-j}. \right)$$

2. For $x > 1$, verify that $y(x) = \sqrt{x}\, J_0(x)$ satisfies the equation

$$y'' + \left(1 + \frac{1}{4x^2}\right) y = 0.$$

3. Note that if $x > 1$, then $1 < 1 + (1/4x^2) < \frac{5}{4}$, and use Theorem B.3 to show that the distance d between successive positive zeros of y (or of J_0) is in the interval $[2\pi/\sqrt{5}, \pi]$.
4. Make sharper estimates on $[a, +\infty)$ and show that as $a \to \infty$, the distance between successive zeros in this interval approaches π.

7.4L (*The zeros of* J_m). For $m = 1, 2, \ldots$

1. verify that $y(x) = \sqrt{x}\, J_m(x)$ satisfies the equation (33')

$$y'' + \left(1 - \frac{M}{x^2}\right) y = 0, \qquad x > 0, \qquad \text{where } M = m^2 - \frac{1}{4}.$$

2. obtain as much information as possible about the positive zeros of J_m for $m \ge 1$. (*Hint*: See Problems 7.4K and 7.4F.)
3. show that the energy function $E = y'^2 + y^2$ satisfies the differential inequality

$$E' = \frac{2Myy'}{x^2} \le \frac{ME}{x^2}, \qquad x > 0.$$

4. Conclude that $E(x) \le e^M E(1) = M_1^2$, say, for $x > 1$, so that $|J_m(x)| \le M_1/\sqrt{x}$, when $x > 1$.

7.4M (*Behavior of* $J_m(x)$ *for large* $x > 0$) In the analysis at the end of this section,

1. verify that if $y(x) = \sqrt{x}\, J_m(x)$, $x > 0$, then

$$y(x) = A \cos(x - \alpha) + y_1(x), \qquad \text{where } |y_1(x)| \le \frac{A_1}{x}.$$

2. use the result in the *integral* defining y_1 to get (34') where

$$|y_2(x)| \le \left| \frac{MA}{2} e^{i\alpha} \int_x^{\infty} \frac{e^{-2it}}{t^2}\, dt + M \int_x^{\infty} e^{-it} \frac{y_1(t)}{t^2}\, dt \right|$$

Integrate the first term by parts and estimate the second to establish that $|y_2(x)| \le A_2/x^2$.

3. indicate how you would further improve these results.

7.5

Legendre Polynomials and Related Series

In the standard spherical coordinates (r, ϕ, θ) shown in Figure 7.11, the Laplacian of a C^2 function $u = u(r, \phi)$ that is independent of θ takes the form

$$\nabla^2 u = u_{rr} + \frac{2}{r} u_r + \frac{1}{r^2 \sin \phi} (\sin \phi \; u_\phi)_\phi, \qquad r > 0. \tag{35}$$

If we introduce $s = \cos \phi \in [-1, 1]$, then the Laplacian of $\psi(r, s) \stackrel{\text{def}}{=} u(r, \cos^{-1} s)$ becomes

$$\nabla^2 \psi = \psi_{rr} + \frac{2}{r} \psi_r + \frac{[(1 - s^2)\psi_s]_s}{r^2}, \qquad r > 0. \tag{35'}$$

(See Problem 7.6A.) For example, separation of Laplace's equation $\nabla^2 \psi = 0$ in "conical" coordinates (r, s) entails finding products $R(r)y(s)$ for which

$$r^2 \frac{R''}{R} + 2r \frac{R'}{R} = -\frac{[(1 - s^2)y']'}{y} = \lambda, \qquad \text{a constant.} \tag{36}$$

Now,

$$[(1 - s^2)y']' = -\lambda y, \qquad -1 < s < 1 \tag{37}$$

is **Legendre's equation of order zero**, and we want values of λ for which it has bounded solutions y on $(-1, 1)$. (This is an S–L problem on $(-1, 1)$ with $\tau(s) = 1 - s^2$, $q(s) = 0$, and $\rho(s) = 1$. Since $\tau(\pm 1) = 0$, both ends are singular and the eigenfunctions are not required to meet any regular boundary conditions.)

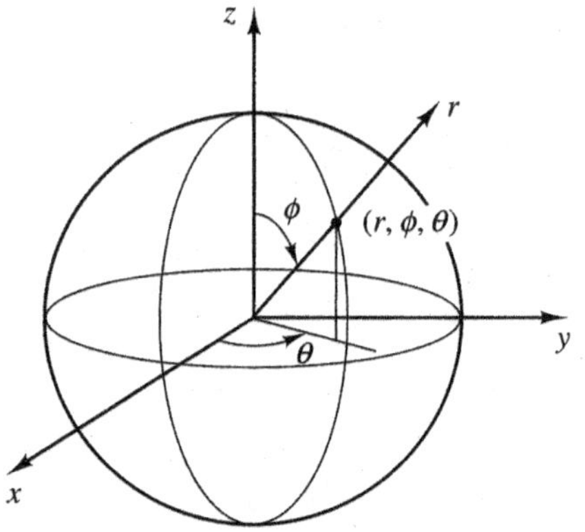

Figure 7.11

Polynomial Solutions

By inspection, $y_0(s) = 1$ is an eigenfunction to the eigenvalue $\lambda_0 = 0$, and $y_1(s) = s$ is an eigenfunction to the eigenvalue $\lambda_1 = 2$. Although s^2 is not a solution for any value of λ, $y_2(s) = s^2 - \frac{1}{3}$ is an eigenfunction (to $\lambda_2 = ?$) and it is natural to consider *polynomials* as possible solutions to (37).

A polynomial of the form

$$y(s) = \sum_{k=0}^{n} a_k \frac{s^k}{k!}$$

will be a solution to (37) provided that its coefficients

$$a_k = y^{(k)}(0), \qquad k = 0, 1, 2, \ldots, n$$

can be determined appropriately. If we differentiate (37) k times, using the Leibniz formula on the product $(1-s^2)y'$, we obtain the equation

$$(1-s^2)y^{(k+2)} + (k+1)(-2s)y^{(k+1)} - (k+1)ky^{(k)} = -\lambda y^{(k)}. \qquad \textbf{(37')}$$

Evaluation at $s = 0$ yields the **recursion formula**

$$a_{k+2} = [k(k+1) - \lambda]a_k, \qquad k = 0, 1, 2, \ldots$$

for the coefficients. (See Appendix B.1.)

In particular, if $\lambda = \lambda_n = n(n+1)$, $n = 0, 1, 2, \ldots$, then $a_{n+2} = 0$, and the recursion formula generates a corresponding polynomial solution $y(s) = P_n(s)$ with the *degree and parity* of n. In this case, $a_n = y^{(n)}(1) \neq 0$, so that $y(1) \neq 0$,† and if a_n is selected to make $P_n(1) = 1$, the resulting P_n is called the **Legendre polynomial of degree *n***. For example,

$$P_0(s) = 1, \qquad P_1(s) = s, \qquad P_2(s) = \frac{3s^2 - 1}{2}, \qquad \text{and} \qquad P_3(s) = \frac{5s^3 - 3s}{2}.$$

We present graphs of the first six Legendre polynomials in Figure 7.12 (p. 302).

For general n, we have **Rodrigues' formula**

$$P_n(s) = \frac{1}{n!\,2^n} \frac{d^n}{ds^n}(s^2 - 1)^n, \qquad n = 0, 1, 2, \ldots \qquad \textbf{(38)}$$

from which we get the normalizing integral evaluation

$$\int_{-1}^{1} P_n^2(s)\,ds = \frac{2}{2n+1}, \qquad n = 0, 1, 2, \ldots \qquad \textbf{(39)}$$

and the recurrence relation

$$(n+1)P_{n+1}(s) = (2n+1)sP_n(s) - nP_{n-1}(s), \qquad n = 1, 2, \ldots . \qquad \textbf{(39')}$$

(See Problems 7.5B–D.)

We have actually found *all* the solutions to our singular S–L problem. To see this, suppose that y is an arbitrary eigenfunction to some eigenvalue λ for the problem. If $\lambda = \lambda_n = n(n+1)$ for an integer $n \geq 0$, we can use reduction of order (Appendix B.1) to show that y is proportional to P_n, since otherwise y is unbounded near $x = 1$. Moreover, unless $\lambda = \lambda_n$ for some integer $n \geq 0$, we can use the orthogonality of y to P_0, to P_1, to P_2 (so, to s^2), to P_3 (so, to s^3), etc. to conclude that y is zero. (Recall Theorem 1.17.) A different argument will be given in Lemma 9.7.

---

† Because in (37′), $y(1) = 0 \Rightarrow y'(1) = 0 \Rightarrow y''(1) = 0 \Rightarrow \cdots \Rightarrow y^{(n)}(1) = 0$.

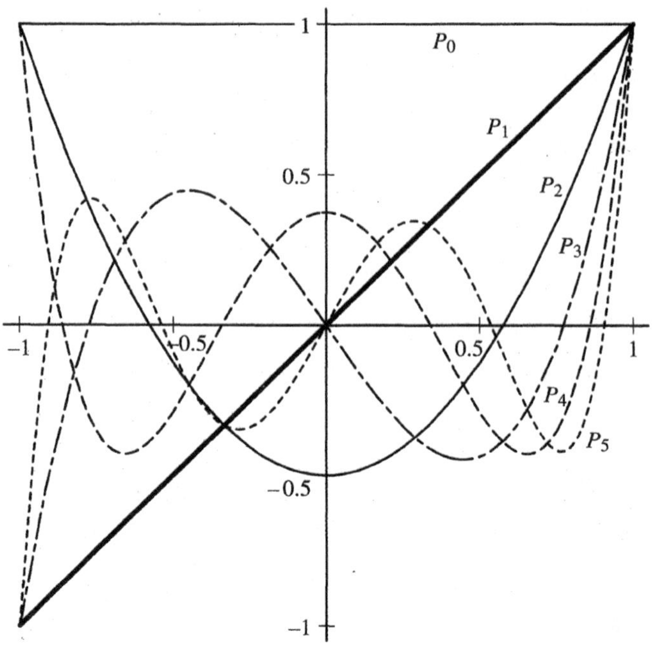

Figure 7.12

A Spherical Dirichlet Problem

Now that we have compatible solutions $y = P_n$, $\lambda = n(n+1)$ of equation (37), let's return to our original problem of finding product solutions $R(r)y(s)$ to Laplace's equation and consider the radial part of the separation equation (36). We see that for each $n = 0, 1, 2, \ldots$, we need a radial function $R = R_n$ that satisfies the equation

$$r^2R'' + 2rR' - n(n+1)R = 0, \qquad \textbf{(40)}$$

and it is easy to verify that $R_n(r) = r^n$ satisfies this Euler equation. Thus, product solutions of Laplace's equation in these coordinates are as follows:

$$\psi_n(r, s) = r^n P_n(s), \qquad n = 0, 1, 2, \ldots \qquad \textbf{(41)}$$

A formal solution ψ to a Dirichlet problem for the unit sphere $V = \{r < 1\}$ where $\psi|_{\partial V} = f(s)$ is given by the series

$$\Psi(r, s) = \sum_{n=0}^{\infty} C_n r^n P_n(s) \qquad \textbf{(41')}$$

provided that coefficients C_n can be found to give f the **Legendre series representation**

$$f(s) = \Psi(1, s) = \sum_{n=0}^{\infty} C_n P_n(s), \qquad -1 < s < 1. \qquad \textbf{(42)}$$

For distinct n (and so distinct $\lambda_n = n(n+1)$), the P_n are 1-orthogonal on $(-1, 1)$ by Proposition 7.1, and in view of (39), we should take

$$C_n = \frac{2n+1}{2} \int_{-1}^{1} f(s)\, P_n(s)\, ds, \qquad n = 0, 1, 2. \tag{42'}$$

The question of which f admit Legendre series representation on $(-1, 1)$ cannot be answered until we obtain suitable bounds for the Legendre polynomials. Fortunately, it can be proved that for $n = 0, 1, 2, \ldots$

$$|P_n(s)| \le 1, \qquad |P_n'(s)| \le n^2, \qquad \text{and} \qquad |P_n''(s)| \le n^4, \qquad -1 \le s \le 1. \tag{43}$$

Consequently, if (42) holds with absolute and uniform convergence, then Ψ defined in (41′) provides the unique solution to our Dirichlet problem. (See Problems 7.5E and F.)

Example 7 (*A heart pump model*):

A simplified model for the left ventricle of a human heart is provided by a sphere of time-dependent radius $R = R(t)$ with a circular aortic opening of *constant* area A, as shown in Figure 7.13. During contraction we suppose that the opening remains fixed while the center of the sphere moves directly toward the center of the opening and the radius $R(t)$ decreases accordingly. As a result, some of the blood filling the ventricle cavity is ejected through the opening with mean speed $U = U(t)$ into the attached cylindrical aorta. This occurs sufficiently rapidly that viscosity effects are negligible and we can assume that the flow is irrotational, incompressible, and symmetric with respect to the aortal axis. Consequently (see Section 6.4), at each time t, the flow velocity can be derived

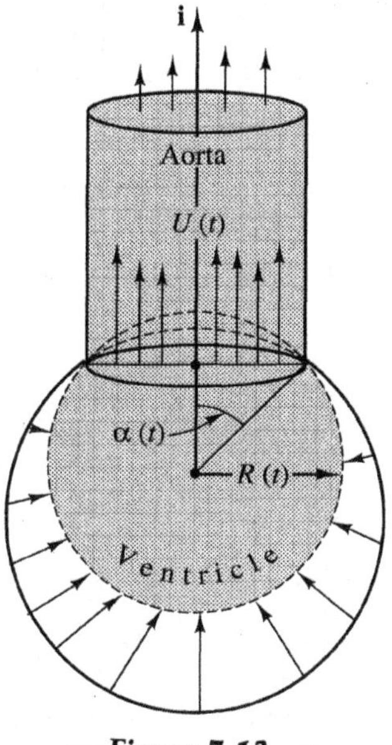

Figure 7.13

from a potential ψ, which is independent of θ in spherical coordinates (r, ϕ, θ).

Thus, within the full spherical cavity V at time t:

$$\nabla^2\psi = 0, \qquad 0 \le r \le R(t) \tag{44}$$

where at the surface S the normal velocity

$$\frac{\partial\psi}{\partial n} = \frac{\partial\psi}{\partial r} = f(s), \qquad s = \cos\phi \in (-1, 1) \tag{44'}$$

is known. We have a Neumann problem for ψ, which by compatibility has a solution only if (see Section 4.2)

$$\int_S \frac{\partial\psi}{\partial n}\, dS = 2\pi \int_{-1}^{1} f(s)\, ds = 0.$$

Our introductory analysis suggests that we might find a formal solution Ψ to our Neumann problem in the form

$$\Psi(r, s) = \sum_{n=0}^{\infty} c_n \left(\frac{r}{R}\right)^n P_n(s)$$

provided that the coefficients c_n give

$$f(s) = \left.\frac{\partial\Psi}{\partial r}\right|_{r=R} = \sum_{n=1}^{\infty} \frac{nc_n}{R} P_n(s), \qquad -1 < s < 1. \tag{45}$$

From (42) and (42′) it follows that we should take $c_0 = 0$ and

$$c_n = \frac{R}{n}\left(\frac{2n+1}{2}\right)\int_{-1}^{1} f(s)\, P_n(s)\, ds, \qquad n = 1, 2, \ldots \tag{46}$$

for an appropriate normal-velocity function f. If f is a polynomial, the series is finite. More generally, when $f \in \hat{C}^1[-1, 1]$, the representation holds, and the presence of geometric terms $(r/R)^n$ in the Ψ series, together with appropriate bounds for Legendre polynomials and their derivatives, guarantee that Ψ is a solution to (44) when $r < R$. But, since the aortal radius $R(t) \sin\alpha(t)$ (say) is constant, the normal velocity function is not continuous. In fact,

$$f(s) = \begin{cases} Us, & a \le s \le 1 \\ \dot{R}\left(1 - \dfrac{s}{a}\right), & -1 \le s \le a, \end{cases} \tag{47}$$

which is discontinuous at the aortal junction where $a = \cos\alpha(t)$. However, it can be shown using Abel summation (of Section 1.9), that $\lim_{r \nearrow R}(\partial\Psi/\partial r)\,(r, s) = f(s)$ except at $s = a$ where this limit is $Ua/2$. Moreover, Ψ itself *is* continuous when $r \le R$ and $|s| < 1$. (See Problem 7.5G and [Pe] for further discussion.)

Problem Set 7.5

7.5A *1.* The surface of a solid iron ball $\{r < 2\}$ is kept at the temperature $\psi(2, s) = s^2$, $|s| \le 1$, where conical coordinates (r, s) are used. Find the steady-state temperature $\psi(r, s)$ within the ball assuming that $\nabla^2\psi = 0$ there. (*Hint*: A series is not required.)

2. Suppose that in part 1 $\psi(2, s) = e^s$, $|s| \le 1$. Write a formal series for $\psi(r, s)$, $r < 2$. Can the coefficients be calculated exactly?

7.5B 1. To establish (38) (*Rodrigues' formula*) let $D = d/dx$, set

$$R_n(x) = D^n(x+1)^n(x-1)^n = (x+1)^n D^n(x-1)^n + nD(x+1)^n D^{n-1}(x-1)^n + \cdots ,$$

and show that when $x = 1$, $R_n(1) = 2^n D^n(x-1)^n|_{x=1} = 2^n n!$.

2. Show that $\int_{-1}^{1} x^k R_n(x)\,dx = 0$ if $k < n$. (*Hint*: $R_n = D^n v$ where $v = (x^2-1)^n$.) Now integrate by parts k times.

3. Express $R_n = \sum_{k=0}^{n} c_k P_k$ and use the orthogonality of the P_j (however normalized) to conclude that $c_j = 0$ for $j < n$. Thus $R_n = c_n P_n$ can serve as the nth eigenfunction as can $P_n(x) = (1/2^n n!)\,R_n(x)$ which has value 1 at $x = 1$.

7.5C 1. Use the recurrence relation of Bonnet (1852)

$$(n+1)\,P_{n+1}(x) = (2n+1)\,xP_n(x) - nP_{n-1}(x), \qquad n = 1, 2, \ldots$$

to obtain P_3, P_4, P_5, P_6, and P_7.

2. Derive this relation by verifying that

$$P_{n+1}(x) - P_{n-1}(x) = \frac{2n+1}{2^n n!}\,D^{n-1}v(x)$$

while
$$P_{n+1}(x) - xP_n(x) = \frac{n}{2^n n!}\,D^{n-1}v(x),$$

where $v(x) = (x^2-1)^n$. (*Hint*: Use (38).)

7.5D 1. Prove that $P'_{n+1} - P'_{n-1} = (2n+1)\,P_n$, $n = 1, 2, \ldots$, so that

$$(2n+1)\int_x^1 P_n(t)\,dt = P_{n-1}(x) - P_{n+1}(x).$$

(Christoffel, 1856.)

2. Conclude that $(2n+1)\int_{-1}^{1} P_n^2\,dt = \int_{-1}^{1} P_n P'_{n+1}\,dt = (P_n P_{n+1})(t)|_{-1}^{1} = 2$. (*Hint*: A Legendre polynomial is orthogonal to any polynomial of lesser degree. [Why?])

3. Show that

$$C_{n+1} \overset{\text{def}}{=} \frac{2n+3}{2}\int_{-1}^{1} e^s P_{n+1}(s)\,ds = (2n+3)\left(\frac{C_{n-1}}{2n-1} - C_n\right), \quad n = 1, 2, \ldots .$$

7.5*E (*Bounds for Legendre polynomials* P_n) To obtain the bounds in (43):

1. From the equation $(1-x^2)\,P_n'' - 2xP_n' = -n(n+1)\,P_n$, and the normalization $P_n(1) = 1$, deduce that if $x \in [0, 1]$, then

$$n(n+1)\,[P_n^2(x) - 1] = -(1-x^2)(P_n'(x))^2 - 2\int_x^1 s(P_n'(s))^2\,ds \le 0, \ n = 0, 1, 2, \ldots .$$

(*Hint*: Multiply by P_n' and integrate over $[x, 1]$.)

2. Since $P_n^2(-x) = P_n^2(x)$, conclude from part 1 that $|x| \le 1 \Rightarrow |P_n(x)| \le 1$, $n = 0, 1, \ldots$.

3. Use part 2 with the result of Problem 7.5D1 in the form $P'_{n+2} = P_n' + (2n+3)\,P_{n+1}$ to prove that $|x| \le 1 \Rightarrow |P_n'(x)| \le n^2$, $n = 0, 1, 2, \ldots$. (*Hint*: Use mathematical induction.)

4. Prove that $|x| \leq 1 \Rightarrow |P_n''(x)| \leq n^4$, $n = 0, 1, 2, \ldots$.
5. Explain how the bounds in parts 2, 3, and 4 permit differentiation of the function

$$\Psi(r, s) = \sum_{n=0}^{\infty} C_n r^n P_n(s), \qquad r \leq r_0 < 1, \qquad |s| \leq 1$$

twice with respect to r or s, if $|C_n| \leq nC$, $n = 1, 2, \ldots$.

7.5F* (*Convergence of Legendre series*)

1. If $f \in \hat{C}^1[-1, 1]$, use the first result of Problem 7.5D to show that in (42′) $|C_n| \leq 2\sqrt{2}\,\|f'\|/\sqrt{2n-1}$, $n = 1, 2, \ldots$. (*Hint*: The Schwarz inequality.)
2. If $f' \in \hat{C}^1[-1, 1]$, show that $|C_n| \leq 4\sqrt{2}\,\|f''\|/(2n-3)^{3/2}$, $n = 2, 3, \ldots$, and use (43) to establish that the Legendre series for f converges absolutely and uniformly on $[-1, 1]$ with sum f.
3. If the Legendre series in (42) converges uniformly on $[a, b] \subseteq [-1, 1]$, use Abel's test to conclude that the function $\Psi(r, s)$ in (41′) is continuous on $[0, 1] \times [a, b]$.

7.5G For the heart pump model in Example 7:

1. During contraction, show that a point on the surface of the ventricle has velocity

$$\dot{R}\mathbf{n} - (\dot{R}a)\mathbf{i}$$

where $\mathbf{n}$ is the outward unit normal at time t, $\mathbf{i}$ is the unit vector in the aortic flow direction, and $a = \cos\alpha(t)$. Show that having $(R\sin\alpha)^2 = R^2(1-a^2)$ constant gives (47).

2. Conclude that since $\int_{-1}^{1} f(s)\,ds = 0$ then $\dot{R} = [a(a-1)/(a+1)]U$, which relates the geometry to the mean aortal speed. (*Hint*: Use Neumann compatibility.)
3. Let $V = V(t)$ denote both the interior of the sphere at time t and its volume. The total momentum in the direction of $\mathbf{i}$ is the blood-density times the integral

$$I = \int_V \nabla\Psi \cdot \mathbf{i}\, dV = \int_S (s\Psi)|_{r=R}\, dS$$

where $S = \partial V$. (Why?) Show that

$$I = 2\pi R^2 \sum_{n=0}^{\infty} c_n \int_{-1}^{1} sP_n(s)\,ds = \tfrac{4}{3}\pi R^2 c_1 = \tfrac{3}{2}V \int_{-1}^{1} f(s)s\,ds.$$

4. Use part 2 and (47) to obtain that $4I = (1-a)(a^2 + 4 + 3a)VU$.

7.5H (*The Gram–Schmidt process*) In the linear space $C[a, b]$, define $(f, g) = \int_a^b \rho f g\,dx$ for a given continuous positive weight function ρ.

1. Suppose that y_1 and f_2 are linearly independent functions in this space and $(y_1, y_1) = 1$. Show that for some constants c_1 and $c_2 \neq 0$, $y_2 = c_1 y_1 + c_2 f_2$ is ρ-orthogonal to y_1 with $(y_2, y_2) = 1$.
2. If f_3 is linearly independent of y_1 and f_2, find $y_3 = c_1 y_1 + c_2 y_2 + c_3 f_3$ that

is ρ-orthogonal to y_1 and y_2 with $(y_3, y_3) = 1$, and explain how this process could be continued.

3. Apply this process to the (linearly independent) monomials $f_n(x) = x^{n-1}$, for $n = 1, 2, 3$, and 4 on $[-1, 1]$ with $\rho(x) = 1$, and verify that the y_n so obtained are (essentially) the Legendre polynomials.

7.6 Spherical Problems

In the last section, we examined the Laplacian in "conical" coordinates (r, ϕ) (or (r, s) where $s = \cos\phi$). In the standard spherical coordinates, (r, ϕ, θ) of Figure 7.14, the Laplacian of a C^2 function u is for $r > 0$:

$$\nabla^2 u = u_{rr} + \frac{2}{r} u_r + \frac{1}{r^2 \sin\phi} (\sin\phi\, u_\phi)_\phi + \frac{1}{r^2 \sin^2\phi} u_{\theta\theta}, \tag{48}$$

and in the companion system (r, s, θ) the Laplacian of $\psi(r, s, \theta) = u(r, \cos^{-1} s, \theta)$ is

$$\nabla^2 \psi = \psi_{rr} + \frac{2}{r}\psi_r + \frac{[(1-s^2)\psi_s]_s}{r^2} + \frac{1}{r^2(1-s^2)}\psi_{\theta\theta}. \tag{49}$$

(See Problem 7.6A.)

In the latter coordinates, let's find solutions to Laplace's equation $\nabla^2\psi = 0$ in the form of products $R(r)y(s)\Theta(\theta)$. If we multiply (49) by r^2, we see that separation is possible when

$$r^2\frac{R''}{R} + 2r\frac{R'}{R} = -\frac{[(1-s^2)y']'}{y} - \frac{1}{1-s^2}\frac{\Theta''}{\Theta} = \lambda, \tag{49$'$}$$

for some constant λ. For solution in a sphere we expect that Θ has period 2π and,

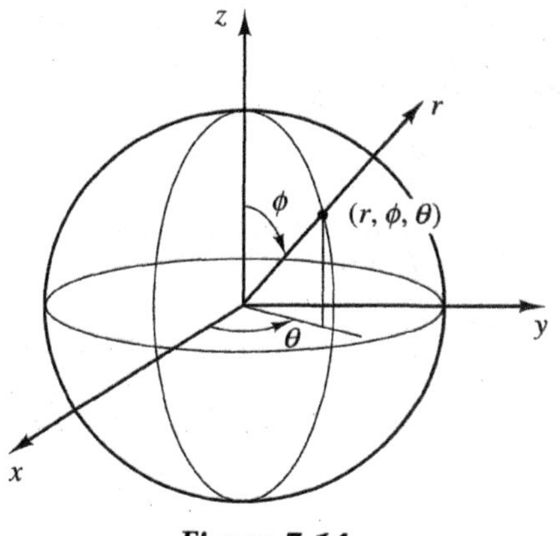

Figure 7.14

as in Section 7.4, we take

$$\frac{\Theta''}{\Theta} = \text{const.} = -m^2, \qquad m = 0, 1, 2, \ldots .$$

For each m, it follows that y and λ are related through the requirement

$$[(1-s^2)y']' - \frac{m^2}{1-s^2}y = -\lambda y, \qquad -1 < s < 1, \tag{50}$$

which is **Legendre's equation of order** m. The corresponding S–L problem has singular endpoints at both ends.

When $m = 0$, we found as solutions to this problem the Legendre polynomials P_n for $\lambda = \lambda_n = n(n+1)$, $n = 0, 1, 2, \ldots$. Moreover, from (37′) we see that $w = P_n^{(m)}$, (the mth derivative of P_n) satisfies the equation

$$(1-s^2)w'' - 2(m+1)sw' = -[\lambda_n - m(m+1)]w \tag{51}$$

and consequently, it can be shown that

$$y(s) \stackrel{\text{def}}{=} (1-s^2)^{m/2} w(s), \qquad -1 < s < 1,$$

is a solution of (50) for $\lambda = \lambda_n = n(n+1)$.[†] (See Problem 7.6B.) Therefore, for each integer $n \geq m$, we have an eigenvalue $\lambda_{mn} = n(n+1)$ to the **associated Legendre function**

$$P_n^m(s) \stackrel{\text{def}}{=} (1-s^2)^{m/2} P_n^{(m)}(s), \qquad 0 \leq m \leq n. \tag{52}$$

As before, the corresponding radial equation (40) is satisfied by $R_{mn}(r) = r^n$, and it follows that for $0 \leq m \leq n$, each product

$$\psi(r, s, \theta) = r^n P_n^m(s) \begin{cases} \cos m\theta \\ \sin m\theta \end{cases} \tag{53}$$

is a solution of Laplace's equation for $r \geq 0$. These products produce harmonic functions in standard spherical coordinates (r, ϕ, θ) of the form

$$u_{mn}(r, \phi, \theta) = r^n (\sin\phi)^m P_n^{(m)}(\cos\phi)\Theta_{mn}(\theta), \qquad 0 \leq m \leq n, \tag{54}$$

where for coefficients c_{mn}, s_{mn},

$$\Theta_{mn}(\theta) = c_{mn}\cos m\theta + s_{mn}\sin m\theta. \tag{54'}$$

Dirichlet Problem for the Sphere

Now suppose that we need a solution $u = u(r, \phi, \theta)$ of Laplace's equation $\nabla^2 u = 0$ in the unit sphere $V = \{r < 1\}$ that is continuous in V and that has prescribed surface values $u(1, \phi, \theta) = f(\phi, \theta)$. (See Figure 7.15.)

It is natural to consider the formal series

$$U(r, \phi, \theta) = \sum_{0 \leq m \leq n}^{\infty} r^n P_n^m(\cos\phi)(c_{mn}\cos m\theta + s_{mn}\sin m\theta), \tag{55}$$

where the coefficients c_{mn} and s_{mn} are to be found if possible to represent f as the

---

[†] When $\lambda = n(n+1)$, (50) is known as the **associated Legendre equation** of order m.

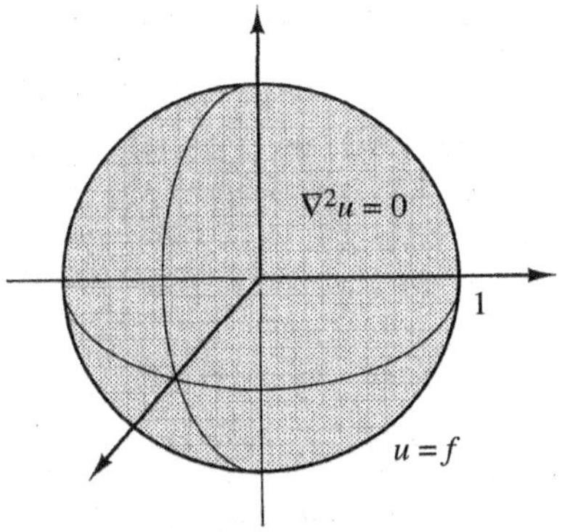

Figure 7.15

Fourier–Legendre series

$$f(\phi, \theta) = U(1, \phi, \theta) = \sum_{0 \le m \le n}^{\infty} P_n^m(\cos \phi)(c_{mn} \cos m\theta + s_{mn} \sin m\theta), \qquad \textbf{(55')}$$

By exploiting the separate properties of Fourier series and Legendre series we can show that the terms of series (55′) provide a set of orthogonal functions on the surface of the unit sphere, so that the coefficients for a given f can be determined. (See Problem 7.6G.) However, we must still consider questions of convergence, differentiability, etc. The resulting technical complexity is somewhat forbidding, and we refer the interested reader to [Wein] or to more general treatises such as [S] for details. We remark only that the solution can be put in the Poisson integral form

$$U(r, \phi, \theta) = \frac{1 - r^2}{4\pi} \int_0^{\pi} d\tilde{\phi} \int_{-\pi}^{\pi} d\tilde{\theta} \frac{\sin \tilde{\phi} f(\tilde{\phi}, \tilde{\theta})}{|(r, \phi, \theta) - (1, \tilde{\phi}, \tilde{\theta})|^3}, \qquad \textbf{(56)}$$

where the denominator of the integrand indicates the cube of the *Euclidean* distance between the points with the designated spherical coordinates.

Corresponding results for the exterior of the sphere are considered in Problems 7.6E, F.

The three-dimensional Laplace equation is of great importance in many fields, including celestial mechanics, elasticity, magnetostatics, and hydrodynamics. Spherical coordinates supply one of several known special systems in which solutions can be obtained by separation methods. (See [C–H, I].)

Spherical Harmonic Functions

When expressed in the rectangular coordinates

$$x = r \sin \phi \cos \theta, \qquad y = r \sin \phi \sin \theta, \qquad z = r \cos \phi$$

each harmonic function u_{mn} in (54) can be shown to be a **homogeneous polynomial of degree** n. That is, it is a linear combination of terms of the form

$$x^i y^j z^k, \qquad \text{where} \qquad i + j + k = n.$$

Homogeneous polynomials $u = u(x, y, z)$ that satisfy Laplace's equation

$$u_{xx} + u_{yy} + u_{zz} = 0, \qquad \text{in } \mathbb{R}^3$$

are called (solid) **spherical harmonic functions**. For example, $x + y$ and $x^2 + y^2 - 2z^2 + 3xz$ are spherical harmonic functions, while $x + yz$ is not, even though it is a harmonic polynomial. It can be proved that every harmonic polynomial is a linear combination of those given through (54). In fact, for each $n = 0, 1, 2, \ldots$, the $2n + 1$ functions

$$r^n Y_n^m(\phi, \theta), \qquad |m| \le n \tag{57}$$

form a basis for the spherical harmonic polynomials of degree n, where

$$Y_n^m(\phi, \theta) \overset{\text{def}}{=} \begin{cases} P_n^m(\cos\phi)\cos m\theta, & 0 \le m \le n \\ P_n^{|m|}(\cos\phi)\sin|m|\theta, & -n \le m < 0. \end{cases} \tag{57'}$$

The functions Y_n^m defined in (57′) are called **surface harmonics.** $Y_n^0(\phi, \theta) = P_n(\cos\phi)$ is called a **zonal** (surface) **harmonic** because it vanishes on n latitudes positioned symmetrically with respect to $\phi = \pi/2$, and, when $n \ne 0$, each Y_n^n is called a **sectoral** (surface) **harmonic** because it vanishes on great circles through the poles.

Example 8 (*Structure of the hydrogen atom*):

In the Rutherford–Bohr model, an atom of hydrogen consists of an electron in orbital motion about a proton of far greater mass which remains essentially at rest at the origin. (See Figure 7.16.) Then for a stationary state, the quantum-mechanical† wave function $\psi = \psi(\mathbf{x})$ should satisfy the reduced Schrödinger equation which in Hartree units may be expressed (see [Wa])

$$\nabla^2\psi - 2V\psi = -2E\psi, \qquad 0 < |\mathbf{x}| < +\infty. \tag{58}$$

Here, $V = V(|\mathbf{x}|)$ is the electrical potential of the central field of the proton in which the electron moves, and E is the energy level of the state. We wish to determine the values of E (not necessarily positive) that permit nontrivial solutions of (58) that remain bounded as $|\mathbf{x}| \to \infty$.

In terms of the spherical coordinates (r, s, θ), (58) has product solutions

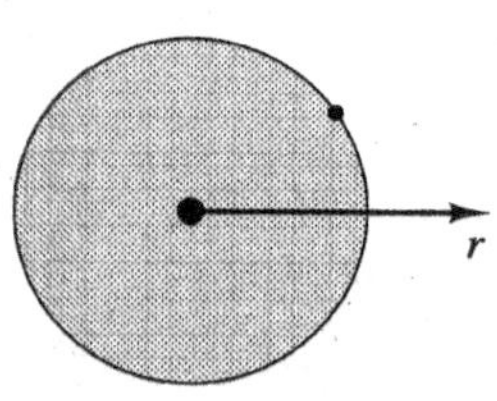

Figure 7.16

† See also the discussion of the uncertainty principle in Section 8.3.

$R(r)y(s)\Theta(\theta)$, provided that for some constant λ

$$r^2\frac{R''}{R}+2r\frac{R'}{R}+2r^2(E-V)(r)=-\frac{[(1-s^2)y')]'}{y}-\frac{1}{1-s^2}\frac{\Theta''}{\Theta}=\lambda.$$

Except for the left side, this equation is identical to (49′). Therefore, if we again take $\Theta''/\Theta=-m^2$, $m=0,1,2,\ldots$, we will obtain (50) and its nontrivial solutions $P_n^m(s)$, for $\lambda=\lambda_n=n(n+1)$, $n\geq m$. From (57′), we see that equation (58) has solutions of the form

$$\psi(r,\phi,\theta)=R(r)\,Y_n^m(\phi,\theta),\qquad |m|\leq n, \tag{58′}$$

provided that R and E satisfy the radial equation

$$r^2R''+2rR'-\lambda_nR+2r^2(E-V(r))R=0,$$

or equivalently, that $v(r)\stackrel{\text{def}}{=}rR(r)$ and E satisfy the simpler equation

$$v''-\frac{n(n+1)}{r^2}v+2(E-V(r))v=0,\qquad 0<r<+\infty. \tag{59}$$

This equation is not elementary. To proceed, let's first specify V.

For electrical neutrality of the hydrogen atom, its electron and proton must have opposing charges of the same magnitude. If we assume an inverse-square law of attraction between them (as in classical electrostatics), then the associated Coulomb potential is given by $V(r)=-q^2/r$ for an appropriately chosen charge-constant, q. With this choice equation (59) becomes

$$v''-\frac{n(n+1)}{r^2}v=\left(-2E-2\frac{q^2}{r}\right)v, \tag{60}$$

which has bounded solutions of the form

$$v(r)=e^{-\mu r}r^{n+1},\qquad \text{for appropriate } \mu>0 \text{ and } E<0.$$

Indeed, then $v''(r)=e^{-\mu r}[n(n+1)r^{n-1}-2\mu nr^n+\mu^2r^{n+1}]$ or

$$v''-\frac{n(n+1)}{r^2}v=\left(\mu^2-\frac{2(n+1)\mu}{r}\right)v, \tag{60′}$$

and upon comparison with (60), we see that we should take $\mu=q^2/(n+1)$, $n=0,1,2,\ldots$, and $2E=-\mu^2$. With more work, we can show that for each integer $N>n$, equation (60) has one solution of the form $v(r)=e^{-\mu r}p_N(r)$ where p_N is a polynomial of degree N, if $2E=-\mu^2=-q^4/N^2$. (When $E<0$, this equation has no other solutions that are bounded near infinity.) See Problem 7.6L.

Thus, each *negative* energy level $E_N=-q^4/2N^2$, $N=1,2,\ldots$ provides a possible stationary state for the hydrogen atom according to our assumptions. Consequently, $E\geq -q^4/2$, and this fact can be used to explain mathematically why the electron always has sufficient kinetic energy to prevent collapse of the atom. At each energy level E_N the atom is stable, but it can exhibit this stability in N^2 independent eigenstates corresponding to the admissible values of m and

n with $|m| \leq n < N$.† What is more surprising is that *every* positive value of E provides nontrivial bounded solutions of (59). (See Problem 7.6M.) However, in the associated states, the electron apparently has too much kinetic energy to maintain a stable orbit.

The above model is predicated on the assumptions that the nucleus remains fixed and that V has the Coulomb expression $V(r) = -q^2/r$, $r > 0$. Other forms of V may be hypothesized and then (59) must be reexamined. Moreover, a complete solution to the problem should permit slight shifts in the position of the nucleus although, as in classical mechanics, the center of mass of the combined electron-proton system does remain fixed. (See [Wa].)

Problem Set 7.6

7.6A Observe that the standard spherical coordinate transformation expressed in the form

$$\begin{cases} x = \rho \cos \theta \\ y = \rho \sin \theta \end{cases}, \quad -\pi \leq \theta < \pi; \qquad \begin{cases} z = r \cos \phi \\ \rho = r \sin \phi \end{cases}, \quad 0 \leq \phi \leq \pi,$$

appears as a *pair of successive polar coordinate transformations.*

1. Use the result of Problem 2.2G *twice* to verify that in spherical coordinates, the Laplacian becomes

$$\nabla^2 = \left(\frac{\partial^2}{\partial x^2} + \frac{\partial^2}{\partial y^2}\right) + \frac{\partial^2}{\partial z^2} = \left(\frac{\partial^2}{\partial z^2} + \frac{\partial^2}{\partial \rho^2}\right) + \frac{1}{\rho}\frac{1}{\partial \rho} + \frac{1}{\rho^2}\frac{\partial^2}{\partial \theta^2}$$

$$= \frac{\partial^2}{\partial r^2} + \frac{2}{r}\frac{\partial}{\partial r} + \frac{1}{r^2 \sin \phi}\frac{\partial}{\partial \phi}\left(\sin \phi \frac{\partial}{\partial \phi}\right) + \frac{1}{r^2 \sin^2 \phi}\frac{\partial^2}{\partial \theta^2}.$$

(*Hint*: $\partial/\partial \rho = \sin \phi(\partial/\partial r) + ((\cos \phi)/r)(\partial/\partial \phi)$.)

2. Conclude that when $s = \cos \phi$, then $\partial/\partial s = (1/\sin \phi)(\partial/\partial \phi)$ and $\nabla^2 \psi$ is as given in (49).

7.6B Show that if w satisfies (51) then $y(s) = (1 - s^2)^{m/2} w(s)$ is a solution of (50) when $\lambda = n(n+1)$. (Write $y = vw$ and show that $(1 - s^2)v' = -msv$.)

7.6C Find the solution $u = u(r, \phi, \theta)$ to the Dirichlet problem $\nabla^2 u = 0$ in the sphere $\{r < 2\}$, with

1. $u(2, \phi, \theta) = 3 \sin \phi \cos \phi \sin \theta, \quad 0 \leq \phi \leq \pi, \quad |\theta| \leq \pi$

or

2. $u(2, \phi, \theta) = 5 \sin^2 \phi \cos 2\theta, \quad 0 \leq \phi \leq \pi, \quad |\theta| \leq \pi.$

(*Hint*: Use (54) with P_2.)

7.6D When $\lambda = n(n+1)$, a radial equation corresponding to (50) is (40), viz., $r^2 R'' + 2rR' - n(n+1)R = 0$.

---

† There are $2n + 1$ values of m with $|m| \leq n$, and therefore

$$\sum_{n=0}^{N-1} (2n+1) = \sum_{n=0}^{N-1} [(n+1)^2 - n^2] = N^2$$

different states with $|m| \leq n < N$.

1. Verify that $R(r) = r^n$ is one solution.
2. Show that $R(r) = r^{-(n+1)}$ is a linearly independent solution when $r > 0$.
3. Conclude that for $0 \le m \le n$, each nontrivial function of the form

$$\tilde{u}_{mn}(r, \phi, \theta) = r^{-(n+1)} P_n^m(\cos \phi)\, \Theta_{mn}(\theta)$$

is harmonic when $r > 0$, where Θ_{mn} is defined in (54′). Observe that *none* of these solutions remains bounded as $r \searrow 0$.

7.6E Use the result of the previous problem to find a solution $u = u(r, \phi, \theta)$ to the Dirichlet problem $\nabla^2 u = 0$ in $\{r > 2\}$ where $u(\mathbf{x}) \to 0$ as $r = |\mathbf{x}| \to \infty$ and

1. $u(2, \phi, \theta) = \sin^2 \phi \cos 2\theta$,

or

2. $u(2, \phi, \theta) = 3 \sin 2\phi \cos \theta - \sin^2 \phi \cos 2\theta$.

(*Hint*: Refer to Problem 7.6C.)

7.6F The surface of an iron ball of radius 3 is maintained at the prescribed temperature $u(3, \phi, \theta) = f(\phi, \theta), \quad 0 \le \phi \le \pi, \quad |\theta| \le \pi$.

1. Write a formal series $U = U(r, \phi, \theta)$ representing the steady-state temperature within the ball.
2. Write a formal series $U = U(r, \phi, \theta)$ representing the steady-state temperature in the *exterior* of the ball. (*Hint*: Refer to Problem 7.6E.) Try to find the coefficients.

7.6G (*Orthogonality*)

1. Show that if $0 \le m \le n$ and $0 \le m' \le n'$, then

$$I(m, m'; n, n') \stackrel{\text{def}}{=} \int_{-1}^{1} P_n^m(s)\, P_{n'}^{m'}(s)\, ds \int_{-\pi}^{\pi} \Theta_{mn}(\theta)\, \Theta_{m'n'}(\theta)\, d\theta$$

$$= 0, \text{ unless } m' = m \text{ and } n' = n$$

when (52) and (54′) are used to define the terms. (*Hint*: When $m' \neq m$, consider the second integral, and when $m' = m$, invoke a general proposition about eigenfunctions of S–L problems in regard to the *first* integral.)

2. Suppose that the series in (55′) converges absolutely and uniformly with sum f. Explain why

$$s_{mn} = \frac{1}{\pi I_{nm}} \int_0^{\pi} d\phi \int_{-\pi}^{\pi} d\theta \sin \phi\, P_n^m(\cos \phi) \sin m\theta\, f(\phi, \theta)$$

where $\quad I_{nm} = \int_{-1}^{1} [P_n^m(s)]^2\, ds, \qquad 0 \le m \le n.$

Obtain corresponding expression(s) for c_{mn} where the case $m = 0$ is considered separately.

3.* Use (52) to show that

$$I_{nm} = \frac{(n+m)!}{(n-m)!} \int_{-1}^{1} P_n^2(s)\, ds = \frac{(n+m)!}{(n-m)!} \frac{2}{2n+1}.$$

(*Hint*: Refer to Problems 7.5B, D.)

7.6H 1. If in (55′) $f(\phi, \theta)$ is an *even* function of θ for each ϕ, then show that U given by (55) should have $U_\theta(r, \phi, 0) = U_\theta(r, \phi, \pi) = 0$.

2. Explain how this observation might yield the steady-state temperature distribution in a *hemispherical* iron ball whose flat surface is *insulated* and whose spherical surface is maintained at prescribed temperature f.

7.6I Explain how to obtain a formal solution to the following boundary value problem:

$$\nabla^2 u = 0 \quad \text{in} \quad \left\{0 < r < 2, 0 < \theta < \frac{\pi}{2}, 0 < \phi < \pi\right\} \quad \text{with} \quad u(2, \phi, \theta) = f(\phi, \theta)$$

and

1. $u(r, \phi, 0) = u(r, \phi, \pi/2) = 0$

or

2. $u_\theta(r, \phi, 0) = u_\theta(r, \phi, \pi/2) = 0.$

7.6J 1. Use the Legendre polynomials $P_1(s) = s$ and $P_2(s) = (3s^2 - 1)/2$ in (57′) to find three linearly independent spherical harmonic polynomials of degree 1 and five of degree 2.

2. Explain why (57′) provides $(2n + 1)$ linearly independent spherical harmonic polynomials with $|m| \leq n$. (*Hint*: Consider orthogonality in θ.)

3. Show that as in Problem 7.6G, unless $m' = m$ and $n' = n$,

$$\int_0^\pi \sin\phi \, d\phi \int_{-\pi}^{\pi} Y_n^m(\phi, \theta)\, Y_{n'}^{m'}(\phi, \theta)\, d\theta = 0.$$

7.6K $P(x, y, z) = \sum_{i+j+k=n} c_{ijk} x^i y^j z^k$ represents an arbitrary polynomial in x, y, and z that is homogeneous of degree n.

1. Show that the coefficient $c_{ijk} = \dfrac{1}{i!j!k!} \dfrac{\partial^i}{\partial x^i} \dfrac{\partial^j}{\partial y^j} \dfrac{\partial^k}{\partial z^k} P.$

2. When P is also harmonic explain why each c_{ijk} in part 1 can be represented as a sum of the $2n + 1$ coefficients where $i = 0$ or 1 only. (*Hint*: The equation $P_{xx} = -P_{yy} - P_{zz}$ can be differentiated as often as desired with respect to x or y or z.)

3. Conclude that there can be no more than $2n + 1$ linearly independent spherical harmonic polynomials of degree n.

7.6L (*The hydrogen atom*)

1. Verify that in spherical coordinates (58) separates as indicated, provided that (59) holds for $v = Rr$.

2. Show that when $v(r) = e^{-\mu r} r^{n+1}$ is substituted in (60), then (60′) follows.

3.* Show that $v(r) = e^{-\mu r}(a_0 r^{n+1} + a_1 r^{n+2} + \cdots)$ is a formal solution of (60) when $\mu^2 = -2E$, provided that the coefficients a_k satisfy the recursion relation

$$a_k = -2a_{k-1}\left[\frac{q^2 - (n+k)\mu}{k(2n+k+1)}\right], \; k = 1, 2, \ldots ,$$

which terminates (when $k = N - n$) iff $\mu = q^2/N$ for some integer $N = n + 1, n + 2, \ldots$.†

4. Conclude that each energy level $E_N = -q^4/2N^2$, $N = 1, 2, \ldots$, provides

† The resulting polynomials are essentially derivatives of those of Laguerre. (See (34) in Section 9.2.)

an admissible solution to (59) for $0 \le n < N$, and so a solution to (58) when $V(r) = -q^2/r$.

5.* If the recursion relation in part 3 does not terminate (and $\mu > 0$), then, for large k, show why it can be approximated by $a_k = 2a_{k-1}(\mu/k)$, which generates the corresponding terms of the power series for $e^{2\mu r}$. Thus the associated solutions v should grow exponentially as $r \to +\infty$, and this behavior eliminates their consideration. Can you provide a rigorous verification of this assertion?

7.6M* To prove that (60) has an admissible solution for each positive value of E, let $\mu = i\sqrt{E} = i\nu$ in the recursion relation of Problem 7.6L3. For large k, the approximation in Problem 7.6L5 is valid, which indicates that there should be a well-defined *complex-valued* solution to (60) for $0 \le r < +\infty$. We wish to prove that its real part is a bounded solution.

1. Let

$$\sigma(r) = \frac{2q^2 - [n(n+1)/r]}{r} \qquad \text{so that} \qquad \sigma'(r) \le \frac{n(n+1)}{r^2}, \qquad r > 2.$$

Then

$$|\sigma(r)| \le E, \qquad \text{if } r \ge a \overset{\text{def}}{=} \max\left(2, \frac{2q^2}{E}, \frac{n(n+1)}{E}\right)$$

and (60) becomes $v'' + 2Ev = -\sigma v$. Multiply by $2v'$ and integrate over $[a, r]$ to get

$$((v')^2 + 2Ev^2)\,|_a^r = -2\int_a^r \sigma v v'\, ds = -\sigma v^2\,|_a^r + \int_a^r \sigma' v^2\, ds.$$

2. Conclude that for some constant A dependent on a and v:

$$2Ev^2(r) \le A + Ev^2(r) + n(n+1)\int_a^r \frac{v^2}{s^2}\, ds.$$

Thus $$Ev^2(r) \le A + n(n+1)\int_a^r \frac{v^2}{s^2}\, ds \overset{\text{def}}{=} \Psi(r) \Rightarrow \Psi(a) = A \ge 0.$$

3. Show that for $r \ge a$:

$$\Psi'(r) = \frac{n(n+1)}{r^2} v^2(r) \le \frac{n(n+1)}{Er^2}\Psi(r) = \frac{M}{r^2}\Psi(r), \text{ say,}$$

so that $(d/dr)(e^{-M/r}\Psi(r)) \le 0$ or $Ev^2(r) \le \Psi(r) \le e^{M/r}e^{-M/a}A \le A$, and this proves that for each $E > 0$, (60) has a bounded solution on $[0, \infty)$.

7.7*

Finite S–L Transforms

Consider the partial differential equation

$$\mathscr{L}(u) \overset{\text{def}}{=} (\tau u_x)_x - qu - \rho[Ku_{tt} + K_1 u_t] = 0 \tag{61}$$

for $a < x < b$ and $t > 0$, where τ, q, and ρ are given functions of x, while K and K_1

are constants. In this chapter we found formal solutions to special cases of this equation in series form

$$u(x, t) = \sum_{n=1}^{\infty} \hat{u}_n(t) y_n(x) \tag{62}$$

where the y_n are the eigenfunctions of some S–L problem associated with the separation equation

$$L(y) \stackrel{\text{def}}{=} (\tau y')' - qy = -\lambda \rho y, \qquad a < x < b. \tag{63}$$

We suppose that τ, q, and ρ meet the additional requirements set forth in Section 7.1, and that we have found eigenvalues λ_n with $\lambda_1 < \lambda_2 < \cdots$ and corresponding eigenfunctions y_n, $n = 1, 2, \ldots$, *ρ-orthonormalized* so that

$$\int_a^b \rho y_m y_n \, dx = \delta_{mn} \stackrel{\text{def}}{=} \begin{cases} 0, & m \neq n \\ 1, & m = n \end{cases}, \qquad m, n = 1, 2, \ldots . \tag{64}$$

Under these circumstances, assume that for some $t > 0$ the series in (62) converges uniformly on the *bounded* interval (a, b). Then by the usual arguments, it follows that for each $n = 1, 2, \ldots$,

$$\hat{u}_n(t) \stackrel{\text{def}}{=} \int_a^b \rho u(\cdot, t) y_n \, dx \tag{65}$$

is the nth S–L coefficient of the solution $u(\cdot, t)$. The function $\hat{u}$ defined by

$$\hat{u}(n, t) = \hat{u}_n(t), \qquad t > 0, \qquad n = 1, 2, \ldots$$

is called the **finite S–L transform** of u. The y_n are known, and we seek associated $\hat{u}_n$ that permit the series in (62) to satisfy given *nonhomogeneous* boundary conditions at a and/or at b.

Specifically, suppose that the S–L problem is regular (see Definition 7.0) and that the y_n are chosen to satisfy the Dirichlet endpoint conditions

$$y_n(a) = y_n(b) = 0. \tag{66}$$

Then we seek $\hat{u}_n$ so that the series (62) provides a formal solution to (61), satisfying the boundary conditions

$$u(a, t) = \alpha(t) \qquad \text{and} \qquad u(b, t) = \beta(t), \qquad t > 0, \tag{67}$$

with appropriate initial conditions

$$u(x, 0) = f(x) \qquad \text{and/or} \qquad u_t(x, 0) = h(x), \qquad a < x < b, \tag{67'}$$

for given functions α, β, f, and h. If series (62) converges uniformly on $[a, b]$ for some fixed $t > 0$, then

$$\alpha(t) = \beta(t) = u(a, t) = u(b, t) = 0.$$

However, if series (62) converges *nonuniformly*, then it is possible to have nonzero values for either

$$\lim_{x \searrow a} u(x, t) = \alpha(t) \qquad \text{or} \qquad \lim_{x \nearrow b} u(x, t) = \beta(t); \tag{68}$$

observe that such behavior requires an infinite series.

To obtain corresponding $\hat{u}_n$, we differentiate equation (65) formally with respect to t to get

$$\dot{\hat{u}}_n(t) = \int_a^{\ell} \rho u_t(\cdot, t) y_n \, dx$$
$$\text{and} \quad \ddot{\hat{u}}_n(t) = \int_a^{\ell} \rho u_{tt}(\cdot, t) y_n \, dx. \tag{69}$$

When equation (61) is used, we find that

$$\begin{aligned} K\ddot{\hat{u}}_n + K_1 \dot{\hat{u}}_n &= \int_a^{\ell} [(\tau u_x)_x - qu] y_n \, dx \\ &= \int_a^{\ell} (-\tau u_x y_n' - q u y_n) \, dx + (\tau u_x y_n)\Big|_a^{\ell} \\ &= \int_a^{\ell} [(\tau y_n')' - q y_n] u \, dx + [\tau u_x y_n - \tau u y_n']\Big|_a^{\ell}, \end{aligned}$$

so that, by (63),

$$K\ddot{\hat{u}}_n + K_1 \dot{\hat{u}}_n = -\lambda_n \int_a^{\ell} \rho u(\cdot, t) y_n \, dx - (\tau u y_n')\Big|_a^{\ell},$$

where we have successively integrated by parts the first term on the right and used (66) to eliminate some of the boundary conditions.

Finally, upon substituting (67), we see that $\hat{u}_n$ satisfies the ordinary nonhomogeneous linear differential equation

$$\ell_n(\hat{u}_n) \stackrel{\text{def}}{=} K\ddot{\hat{u}}_n + K_1 \dot{\hat{u}}_n + \lambda_n \hat{u}_n = \tau(a) y_n'(a)\alpha - \tau(\ell) y_n'(\ell)\beta, \tag{70}$$

together with the appropriate initial conditions obtained from (67′); viz.,

$$\hat{u}_n(0) = \int_a^{\ell} \rho f y_n \, dx, \qquad \dot{\hat{u}}_n(0) = \int_a^{\ell} \rho h y_n \, dx. \tag{70′}$$

Conversely, the ordinary linear equation (70) has a unique solution that meets conditions (70′) for all $t \geq 0$. Thus, in principle, the $\hat{u}_n$ can be determined from the given conditions, and then we can consider series (62) as a formal solution to our boundary value problem. Since the inhomogeneous conditions (67) can be met only with proper *nonuniform* convergence, we can anticipate that the differentiated series need not converge. Moreover, we have little assurance that either set of conditions (67) or (67′) will be satisfied. Despite these difficulties, solutions to some problems can be found in this manner.

The same transform methods may be used with the corresponding nonhomogeneous equation $\mathscr{L}(u) = F$, for a given function $F = F(x, t)$ (see Problem 7.7E), and to attack problems associated with singular S–L problems. (See [Mack].)

Example 9:

Suppose we want a solution to the problem shown in Figure 7.17 for the simple heat equation

$$\mathscr{L}(u) \stackrel{\text{def}}{=} u_{xx} - u_t = 0, \qquad 0 < x < \pi, \qquad t > 0.$$

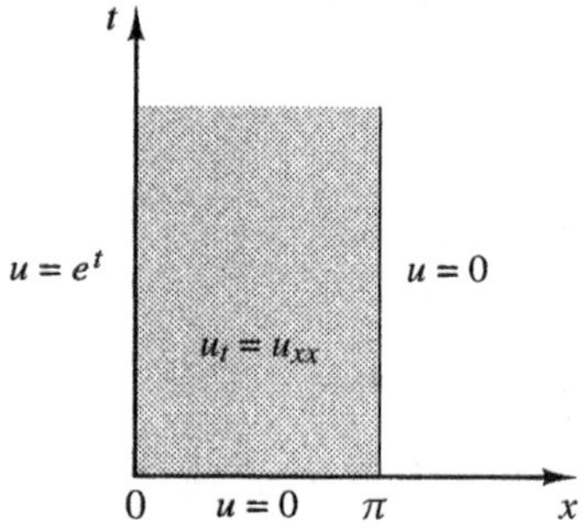

Figure 7.17

Here, $\rho = \tau = 1$, $q = 0$, $K = 0$, and $K_1 = 1$. If we substitute the normalized eigenfunctions

$$y_n(x) = \sqrt{\frac{2}{\pi}} \sin nx, \qquad \text{with} \qquad y_n'(x) = \sqrt{\frac{2}{\pi}} n \cos nx,$$

for $\lambda_n = n^2$, $n = 1, 2, \ldots$; in equation (70), we obtain the following:

$$\dot{\hat{u}}_n + n^2 \hat{u}_n = y_n'(0)\alpha - y_n'(\pi)\beta = \sqrt{\frac{2}{\pi}} n[\alpha - (-1)^n \beta]. \tag{71}$$

To have a solution $u = u(x, t)$ meeting the specific boundary conditions

$$u(0, t) = \alpha(t) = e^t \qquad \text{and} \qquad u(\pi, t) = \beta(t) = 0, \qquad t > 0,$$

and the homogeneous initial condition $u(x, 0) = f(x) = 0$, $0 < x < \pi$, we require a solution $\hat{u}_n$ of the first-order equation

$$\dot{\hat{u}}_n(t) + n^2 \hat{u}_n(t) = \sqrt{\frac{2}{\pi}} n e^t, \qquad t > 0 \tag{71'}$$

with the initial condition $\hat{u}_n(0) = 0$. It is easy to verify that $\hat{u}_n$ is given by

$$\hat{u}_n(t) = \sqrt{\frac{2}{\pi}} n e^{-n^2 t} \int_0^t e^{n^2 s} e^s \, ds = \sqrt{\frac{2}{\pi}} \frac{n}{n^2 + 1} (e^t - e^{-n^2 t}), \qquad n = 1, 2, \ldots .$$

Then, after substituting in (62) and rearranging, we obtain the formal solution

$$U(x, t) = \frac{2}{\pi} e^t \sum_{n=1}^{\infty} \frac{n \sin nx}{n^2 + 1} - \frac{2}{\pi} \sum_{n=1}^{\infty} \frac{n}{n^2 + 1} e^{-n^2 t} \sin nx.$$

Now, the second series converges uniformly and its termwise differentiated series converges uniformly when $t \geq t_0 > 0$. Hence, it is easy to see that the second series defines a solution of the heat equation that has zero boundary values, when $t > 0$. The first series exhibits the expected nonuniform convergence; however, using results from Example 2 of Section 1.3, we recognize it as the Fourier series of $(\pi\sigma/2) \sinh(\pi - x)$, $0 < x < \pi$, where $\sigma = 1/\sinh \pi$. Thus

$$U(x, t) = e^t \frac{\sinh(\pi - x)}{\sinh \pi} - \frac{2}{\pi} \sum_{n=1}^{\infty} \frac{n}{n^2 + 1} e^{-n^2 t} \sin nx, \qquad 0 < x < \pi$$

is an actual solution with the correct boundary values for $t > 0$, and through inspection $U(x, 0) = 0, 0 < x < \pi$. Once again, we see how one nonhomogeneous problem can be transformed into another. Here, $e^t\sigma \sinh(\pi - x)$ is a solution of the heat equation with the desired boundary values, while the remaining series is a solution with zero boundary values and the initial values $\sigma \sinh(\pi - x)$.

In this example, we were fortunate in being able to sum explicitly the part of the series that does not converge uniformly. When this is not possible, devices such as Abel's partial summation may be useful. Moreover, these linear problems may admit splitting into simpler ones.

Example 10:*

We can also utilize a finite transform approach to find a solution to Laplace's equation in the polar form

$$\mathscr{L}(u) \stackrel{\text{def}}{=} u_{\theta\theta} + r^2 u_{rr} + r u_r = 0$$

in the semidisk $\{0 < r < 1, 0 < \theta < \pi\}$ that meets nonhomogeneous boundary conditions such as those shown in Figure 7.18. Again we use the normalized eigenfunctions $y_n(\theta) = \sqrt{2/\pi}\sin n\theta$ to introduce the S-L transform

$$\hat{u}_n(r) = \int_0^{\pi} u(r, \cdot) y_n \, d\theta, \qquad n = 1, 2, \ldots .$$

Then, proceeding exactly as before, we find that $\hat{u}_n$ should satisfy the nonhomogeneous Euler equation

$$r^2\hat{u}_n'' + r\hat{u}_n' - n^2\hat{u}_n = \Big[u(\cdot, \theta)y_n' - u_\theta(\cdot, \theta)y_n\Big]_0^{\pi} = -\sqrt{\frac{2}{\pi}}\, rn, \qquad n = 1, 2, \ldots,$$

with $\hat{u}_n(1) = 0$ and $\hat{u}_n(0)$ finite. If we let $r = e^t$, we can convert this Euler equation to one in t with constant coefficients and obtain the bounded solutions

$$\hat{u}_1(r) = \frac{-1}{\sqrt{2\pi}} r \log r; \quad \hat{u}_n(r) = \sqrt{\frac{2}{\pi}}\frac{n}{n^2 - 1}(r - r^n), \quad n = 2, 3, \ldots .$$

The eigenfunction series

$$U(r, \theta) = \sum_{n=1}^{\infty} \hat{u}_n(r) y_n(\theta) = -\frac{r}{\pi}\log r \, \sin\theta + \frac{2}{\pi}\sum_{n=2}^{\infty}\frac{n(r - r^n)}{n^2 - 1}\sin n\theta$$

represents a formal solution to our problem. The series multiplying r is essentially that for $1 - (\theta/\pi)$ on $(0, \pi)$ so that it converges nonuniformly only near $\theta = 0$. (Recall the discussion of equation (41) in Section 1.8d.)

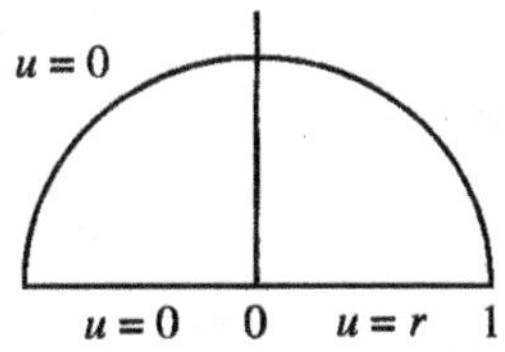

Figure 7.18

Multidimensional Problems

At the end of Section 7.1, we considered regular multidimensional S–L problems for a special case of the equation

$$L(\psi) \stackrel{\text{def}}{=} \nabla \cdot (\tau \nabla \psi) - q\psi = -\lambda \rho \psi, \qquad \text{in } V, \tag{72}$$

where V is a simple domain of $\mathbb{R}^d$ with boundary S. This equation arises in seeking product solutions $u(\mathbf{x}, t) = \psi(\mathbf{x})\hat{u}(t)$ of the equation

$$\mathscr{L}(u) \stackrel{\text{def}}{=} \nabla \cdot (\tau \nabla u) - qu - \rho[Ku_{tt} + K_1 u_t] = 0, \qquad \text{in } V \times (0, \infty), \tag{72$'$}$$

where τ, q, and ρ are functions of $\mathbf{x}$,and K and K_1 are constants. Observe that with proper choice of its parameters, (72$'$) provides versions of the heat equation, the (damped) wave equation, and the potential equation.

Suppose we have a sequence of eigenfunctions ψ_n to eigenvalues λ_n of (72$'$), where each $\psi_n \in C^1(\bar{V})$ with $\psi_n|_S = 0$, $n = 1, 2, \ldots$. If u is a solution of (72$'$), then operating formally as in the derivation of (70), we find that

$$\hat{u}_n(t) \stackrel{\text{def}}{=} \int_V \tau u(\cdot, t)\psi_n \, dV$$

satisfies the ordinary equation

$$\ell_n(\hat{u}_n) \stackrel{\text{def}}{=} K\ddot{\hat{u}}_n + K_1\dot{\hat{u}}_n + \lambda_n \hat{u}_n = -\int_S \left(\tau \frac{\partial \psi_n}{\partial n}\right)\bigg|_S \beta \, dS \tag{73}$$

where

$$\beta(\mathbf{s}, t) \stackrel{\text{def}}{=} \lim_{\mathbf{x} \to \mathbf{s}} u(\mathbf{x}, t), \qquad \mathbf{s} \in S. \tag{73$'$}$$

(See Problem 7.7G.) Thus to solve the nonhomogeneous boundary value problem

$$\mathscr{L}(u) = 0 \qquad \text{in} \qquad V \times (0, +\infty)$$

$$\text{with} \qquad u(\cdot, t)|_S = \beta(\cdot, t) \not\equiv 0, \quad t > 0 \tag{74}$$

and as required, the initial condition(s)

$$u(\mathbf{x}, 0) = f(\mathbf{x}), \qquad u_t(\mathbf{x}, 0) = h(\mathbf{x}), \qquad \mathbf{x} \in V, \tag{75}$$

we find the solution $\hat{u}_n$ to the equation

$$\ell_n(\hat{u}_n) = -\int_S \tau \frac{\partial \psi_n}{\partial n} \beta \, dS, \qquad t > 0,$$

with the corresponding initial value(s)

$$\hat{u}_n(0) = \int_V \rho f \psi_n \, dV, \qquad \dot{\hat{u}}_n(0) = \int_V \rho h \psi_n \, dV, \tag{75$'$}$$

for $n = 1, 2, \ldots$. Then, the series

$$U(\mathbf{x}, t) = \sum_{n=1}^{\infty} \hat{u}_n(t)\psi_n(\mathbf{x})$$

is a possible *formal* solution to the problem. It usually suffices to solve the problem with homogeneous initial conditions (75) and (75$'$), and then $\hat{u}_n$ is given by an explicit integral. (See Example 9 and Appendix B.1.) However, analysis of the resulting nonuniformly convergent series is still difficult, and in particular, it is not evident that U satisfies (72′). (See Section 9.6.)

Problem Set 7.7

7.7A Obtain a formal series solution to the problem in Example 9 when $\alpha(t) = 0$, $\beta(t) = \pi t$, $t > 0$, and $f(x) = x$, $0 < x < \pi$, and discuss its convergence.

7.7B Obtain a formal series solution to the wave equation

$$u_{xx} - u_{tt} = 0, \quad \text{in} \quad \{0 < x < \pi, \ t > 0\}$$

$$\text{with} \quad u(0, t) = e^{-t}, \quad u(\pi, t) = 0, \quad t > 0$$

$$\text{and} \quad u(x, 0) = x, \quad u_t(x, 0) = 0, \quad 0 < x < \pi,$$

and discuss its convergence.

7.7C Suppose that instead of (66), the eigenfunctions y_n satisfy $y_n'(a) = y_n'(b) = 0$.

1. Show that (70) should be replaced by $\ell_n(\hat{u}_n) = \tau(b)y_n(b)\beta - \tau(a)y_n(a)\alpha$ where now $\alpha(t) = u_x(a, t)$ and $\beta(t) = u_x(b, t)$.
2. Obtain a formal series solution to the problem $\mathscr{L}(u) \equiv u_{xx} - u_t = 0$, in $\{0 < x < \pi, \ t > 0\}$, with $u_x(0, t) = t$, $u_x(\pi, t) = e^{-t}$, $t > 0$; $u(x, 0) = \sin x$, $0 < x < \pi$.

7.7D 1. Verify the computations of all results obtained in Example 10.

2.* Find a formal solution to the problem of Example 10 under the mixed boundary conditions $u(r, 0) = r^2$, $u_\theta(r, \pi) = r$, $(0 < r < 1)$ and $u(1, \theta) = 1$, $(0 < \theta < \pi)$. (*Hint*: Take $y_n(\theta) = \sqrt{2/\pi}\sin(n\theta/2)$ and look for $\hat{u}_n$ in the form $\hat{u}_n(r) = A_n r^{n/2} + B_n r^2 + C_n r$, $n = 1, 3, 5, \ldots$.)

7.7E To find a formal series that satisfies $\mathscr{L}(u) = F$ in $\{a < x < b, \ t > 0\}$ where $\mathscr{L}(u)$ is defined in (61), show that under conditions (66), equation (70) becomes

$$\ell_n(\hat{u}_n) = \tau(a)y_n'(a)\alpha - \tau(b)y_n'(b)\beta - f_n$$

$$\text{where} \quad f_n(t) = \int_a^b F(\cdot, t)y_n \, dx, \qquad n = 1, 2, \ldots,$$

is the finite S–L transform of F/ρ.

7.7F Use the method in Problem 7.7E to obtain a formal solution of the non-homogeneous problem of Example 9, when $F(x, t) = xe^{-t}$.

7.7G Verify the results leading from equation (72) to (73′). Then formulate a multidimensional version of Proposition 7.1 for eigenfunctions governed by (72) and (11′), and explain how the divergence theorem could be used to establish it.

7.7H Can the tall-building problem (Example 5 of Section 5.6) be solved using the methods of this section? Explain.

8

Integral Transform Methods

Separation methods on linear boundary value problems frequently result in S–L eigenvalue problems on an interval. If the interval is finite, we can usually impose boundary conditions that eliminate all but a *sequence* of eigenvalues (and eigenfunctions) and solution by series superposition is feasible. However, when the interval is not finite, natural boundary conditions need not be so restrictive. For example, the simple eigenvalue problem

$$X'' + \omega^2 X = 0 \qquad \text{on } (0, \infty)$$

$$\text{with} \qquad X(0) = 0 \qquad \text{and} \qquad |X| \text{ bounded,}$$

admits the eigenfunction $X(x) = \sin \omega x$ for each $\omega > 0$.

In such cases, we might attempt solution by *integral* superposition and in this chapter we explore some of the consequences. We begin with Fourier sine and cosine transforms — the natural integral analogues of Fourier sine and cosine series — and use them to solve standard problems (Section 8.2) and to motivate the complex-valued Fourier transform (Section 8.3). In Section 8.4, we consider related integral transforms, in particular those of Laplace, Hankel, and Mellin, and examine associated representation formulas. Finally, in Section 8.5, we discuss uniqueness of solutions to simple problems in spatially unbounded domains.

The chapter concludes with an asymptotic analysis of Bessel's integral formula (and similar integral expressions) supplied by the method of stationary phase.

8.1

Sine and Cosine Transforms

In earlier chapters we encountered many boundary value problems that could be solved by representing a boundary function f (or an extension) as a Fourier series of the form

$$f(x) = \frac{c_0}{2} + \sum_{n=1}^{\infty} (c_n \cos \omega_n x + s_n \sin \omega_n x), \qquad -\ell < x < \ell, \qquad \textbf{(1)}$$

where

$$\left.\begin{matrix} c_n \\ s_n \end{matrix}\right\} = \frac{1}{\ell}\int_{-\ell}^{\ell} f(\xi) \left\{\begin{matrix} \cos \omega_n \xi \\ \sin \omega_n \xi \end{matrix}\right\} d\xi \tag{2}$$

with $\omega_n = n\pi/\ell$, $n = 0, 1, 2, \ldots$. This representation is valid when $f \in \hat{C}^1[-\ell, \ell]$, and it remains valid when f is sectionally smooth on $[-\ell, \ell]$, provided that in (1), $f(x)$ is replaced by $[f(x+) + f(x-)]/2$.

Fourier's Integral Formula

For corresponding problems of infinite extent, we would like to have an analogous representation formula for functions f defined on $\mathbb{R}$. To see how this formula might look, let's substitute (2) into (1) to get

$$f(x) = \frac{1}{2\ell}\int_{-\ell}^{\ell} f(\xi)\, d\xi + \frac{1}{\pi}\sum_{n=1}^{\infty} \Delta\omega\left[\int_{-\ell}^{\ell} f(\xi) \cos n\Delta\omega(\xi - x)\, d\xi\right], \tag{3}$$

where $\Delta\omega = \pi/\ell$, and ask what happens as $\ell \to +\infty$.

When f is in $\hat{C}^1[-\ell, \ell]$ for each $\ell > 0$ and $\int_{-\infty}^{\infty} |f(x)|\, dx < +\infty$, we see that

$$\left|\frac{1}{2\ell}\int_{-\ell}^{\ell} f(\xi)\, d\xi\right| \le \frac{1}{2\ell}\int_{-\infty}^{\infty} |f(\xi)|\, d\xi, \qquad \text{which} \to 0 \text{ as } \ell \to \infty;$$

thus, if we regard the infinite series in equation (3) as an approximation that improves as $\Delta\omega \to 0$, then for each fixed $x \in \mathbb{R}$, we should have

$$f(x) = \frac{1}{\pi}\int_0^{\infty} d\omega \int_{-\infty}^{\infty} f(\xi) \cos \omega(\xi - x)\, d\xi. \quad \textit{(Fourier's integral formula)} \tag{4}$$

This result was first obtained in 1811 by a similar limiting argument. It can also be expressed in the form

$$f(x) = \int_0^{\infty} [c(\omega) \cos \omega x + s(\omega) \sin \omega x]\, d\omega, \tag{5}$$

where

$$\left.\begin{matrix} c(\omega) \\ s(\omega) \end{matrix}\right\} = \frac{1}{\pi}\int_{-\infty}^{\infty} f(\xi) \left\{\begin{matrix} \cos \omega\xi \\ \sin \omega\xi \end{matrix}\right\} d\xi, \qquad \omega \ge 0. \tag{6}$$

Fourier's integral formula (4), together with (5) and (6), provides a complete formal analogy to (3), (1), and (2). Moreover, if f is only **absolutely integrable** over $\mathbb{R}$ in that

$$f \textit{ is integrable on each bounded interval, and } \int_{-\infty}^{\infty} |f(\xi)|\, d\xi < +\infty, \tag{6$'$}$$

then its **Fourier cosine transform** $c(\cdot)$ and its **Fourier sine transform** $s(\cdot)$ are defined for each $\omega \ge 0$, since the resulting improper integrals "converge." Then, the question is whether the integral

$$F(x) = \int_0^{\infty} [c(\omega) \cos \omega x + s(\omega) \sin \omega x]\, d\omega \tag{7}$$

exists, and if so, whether it equals $f(x)$, or possibly $[f(x+)+f(x-)]/2$, at a point of discontinuity. Here, we lack an analogue of orthogonality, but there is a Bessel inequality. (See Problem 8.1D.)

Let's defer consideration of these general questions until we have seen some examples. Note first that the requirement of absolute integrability eliminates many simple functions. For example, the functions $f(x) = 1, x, x^2, \sin x, \cos x, e^x, \log(1+|x|)$ and $1/(1+|x|)$ do not "die out" fast enough with large $|x|$ to make $\int_{-\infty}^{\infty} |f(x)|\,dx$ finite. $f(x) = 1/(1+x^2)$ is absolutely integrable and its Fourier sine transform $s(\omega) = 0$. (Why?) Its cosine transform cannot be calculated easily, but it can be obtained indirectly, as we shall see.

Example 1:

$f(x) = e^{-|x|}$ is an *even* function since $f(-x) = f(x)$. Hence, as in Proposition 1.4, we have that $s(\omega) = 0$, $\omega \geq 0$, while

$$c(\omega) = \frac{2}{\pi}\int_0^\infty e^{-\xi}\cos\omega\xi\,d\xi = \frac{2}{\pi}\lim_{\ell\to+\infty}\int_0^\ell e^{-\xi}\cos\omega\xi\,d\xi$$
$$= \frac{2}{\pi}\lim_{\ell\to+\infty}\frac{e^{-\ell}(\omega\sin\omega\ell - \cos\omega\ell)+1}{1+\omega^2} = \frac{2}{\pi}\frac{1}{1+\omega^2}, \tag{8}$$

since, e.g., $|e^{-\ell}\sin\omega\ell| \leq e^{-\ell}$, which $\to 0$, as $\ell \to +\infty$. Thus, we expect (4) to take the form

$$e^{-|x|} = \frac{2}{\pi}\int_0^\infty \frac{\cos\omega x}{1+\omega^2}\,d\omega, \qquad x \in \mathbb{R}, \tag{9}$$

and if this result is valid, it provides the *cosine* transform for $\tilde{f}(x) = 1/(1+x^2)$.

We can verify that (9) holds when $x = 0$, since

$$\int_0^\infty \frac{d\omega}{1+\omega^2} = \lim_{\ell\to\infty}\tan^{-1}\ell = \pi/2. \tag{10}$$

Example 2:

The function

$$f(x) = \begin{cases} e^{-x}, & x > 0 \\ 0, & x \leq 0 \end{cases} \tag{11}$$

has a jump discontinuity at 0. Clearly, as in the previous example

$$c(\omega) = \frac{1}{\pi}\int_0^\infty e^{-\xi}\cos\omega\xi\,d\xi = \frac{1}{\pi}\frac{1}{1+\omega^2}, \tag{12}$$

and similarly

$$s(\omega) = \frac{1}{\pi}\int_0^\infty e^{-\xi}\sin\omega\xi\,d\xi = \frac{1}{\pi}\frac{\omega}{1+\omega^2}. \tag{12'}$$

At $x = 0$, we expect from (5) that

$$\int_0^\infty c(\omega)\,d\omega = \frac{f(0+)+f(0-)}{2} = \frac{1}{2},$$

and this formula can be verified through (10). For other values of x, existence of the integral

$$\int_0^\infty s(\omega)\sin\omega x\,d\omega = \frac{1}{\pi}\int_0^\infty \frac{\omega}{1+\omega^2}\sin\omega x\,d\omega \tag{13}$$

is in question, but if our conjectured formula (5) holds, we can prove that

$$\frac{1}{\pi}\int_0^\infty \frac{\cos\omega x + \omega\sin\omega x}{1+\omega^2}\,d\omega = \begin{cases} e^{-x}, & x > 0 \\ 0, & x < 0. \end{cases} \tag{14}$$

It is seen that Fourier's integral formula, if admissible, permits the evaluation of certain infinite integrals in the same manner that the series representation formula of Theorem 1.6 provides evaluation of related infinite series.

Example 3:

The cosine transform of the function

$$f(x) \stackrel{\text{def}}{=} e^{-x^2}, \qquad x \in \mathbb{R},$$

whose bell-shaped graph is shown in Figure 8.1, is given by $c(\omega)$, where

$$\pi c(\omega) = \int_{-\infty}^\infty e^{-x^2}\cos\omega x\,dx = 2\int_0^\infty e^{-x^2}\cos\omega x\,dx, \tag{15}$$

since f is even. It is well known[†] that

$$c(0) = \frac{2}{\pi}\int_0^\infty e^{-x^2}\,dx = 1/\sqrt{\pi}.$$

For other values of ω, note that $\pi c(\omega)$ can be differentiated with respect to ω since the integral of the resulting integrand $-2xe^{-x^2}\sin\omega x$ converges *uniformly* with respect to ω. Then, partial integration gives

$$\pi c'(\omega) = \int_0^\infty -2xe^{-x^2}\sin\omega x\,dx = \underset{\to 0}{e^{-x^2}\sin\omega x\Big|_0^\infty} - \omega\int_0^\infty e^{-x^2}\cos\omega x\,dx$$

and therefore, in view of (15), $c'(\omega) = -(\omega/2)c(\omega)$.

This ordinary differential equation can be rewritten in the form

$$\frac{d}{d\omega}\left(e^{\omega^2/4}c(\omega)\right) = 0,$$

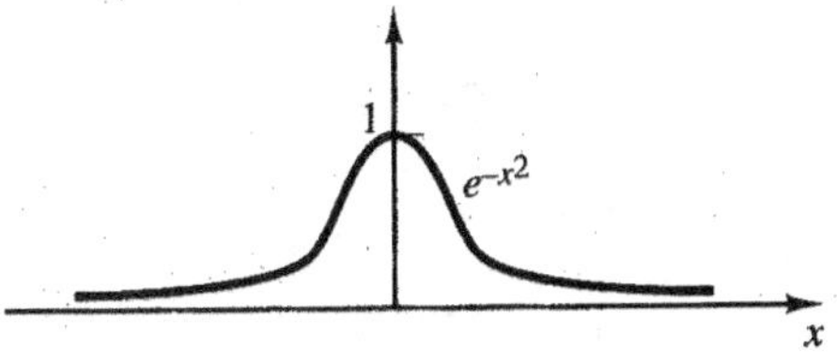

Figure 8.1

[†] An elementary proof is indicated in Problem 8.1F.

and integrated from 0 to ω to obtain

$$c(\omega) = c(0)e^{-\omega^2/4} = e^{-\omega^2/4}/\sqrt{\pi} = f(\omega/2)/\sqrt{\pi}.$$

Then, from (15) we get the useful formula

$$\int_0^\infty e^{-x^2} \cos \omega x \, dx = \frac{\sqrt{\pi}}{2} e^{-\omega^2/4} \tag{15'}$$

It follows easily that formula (5) holds for this f which is almost its own cosine transform.

Validity of Fourier's Integral Formula*

Now, let's show that formula (4) holds under reasonable conditions on f. Lemma 8.1 requires a rather long proof, but if it is accepted, the principal result (Theorem 8.2) follows easily. We establish (4) first when $x = 0$.

(8.1) Lemma *Assume that f is absolutely integrable on $\mathbb{R}$ and that at $x = 0$, f is continuous with derivatives from the right and from the left. Then,*

$$f(0) = \frac{1}{\pi}\int_0^\infty d\omega \int_{-\infty}^\infty f(\xi) \cos \omega\xi \, d\xi. \tag{16}$$

Proof:** If we express $f = f_E + f_O$ as in Section 1.4, we see that formally $\int_{-\infty}^\infty f_O(\xi) \cos \omega\xi \, d\xi = 0$, while from continuity, $f_O(0) = 0$. Thus it suffices to establish our lemma when f is an *even* function, in the form

$$\frac{\pi}{2} f(0) = \int_0^\infty d\omega \int_0^\infty f(\xi) \cos \omega\xi \, d\xi = I, \text{ say,} \tag{17}$$

and this is accomplished by a barrage of limit interchanges justified by appeals to uniform convergence. (Recall Section 0.3.)

(1) If I exists, then by definition,

$$\begin{aligned} I &= \lim_{\Omega\to\infty} \left(\int_0^\Omega d\omega \lim_{\ell\to\infty}\right) \int_0^\ell f(\xi) \cos \omega\xi \, d\xi \\ &= \lim_{\Omega\to\infty} \left(\lim_{\ell\to\infty} \int_0^\Omega d\omega\right) \int_0^\ell f(\xi) \cos \omega\xi \, d\xi, \end{aligned} \tag{18}$$

since the limit in ℓ is reached *uniformly* with respect to ω, permitting the limit interchange indicated. Our goal is to prove that

$$\lim_{\Omega\to\infty} \lim_{\ell\to\infty} I(\Omega, \ell) = \frac{\pi}{2} f(0), \tag{19}$$

where for large positive Ω and ℓ,

$$\begin{aligned} I(\Omega, \ell) &\stackrel{\text{def}}{=} \int_0^\Omega d\omega \int_0^\ell f(\xi) \cos \omega\xi \, d\xi \\ &= \int_0^\ell f(\xi) \, d\xi \int_0^\Omega \cos \omega\xi \, d\omega = \int_0^\ell \frac{f(\xi)}{\xi} \sin \Omega\xi \, d\xi. \end{aligned}$$

(2) Replacing $f(\xi)$ by $f(0)+[f(\xi)-f(0)]$, we see that

$$I(\Omega, \ell) = f(0)\int_0^\ell \frac{\sin \Omega\xi}{\xi}\, d\xi + \int_0^\ell \phi(\xi) \sin \Omega\xi \, d\xi, \tag{20}$$

$$\text{where} \qquad \phi(\xi) \overset{\text{def}}{=} \frac{f(\xi)-f(0)}{\xi}, \qquad \xi > 0,$$

and it remains to consider the limiting behavior of the integral terms in (20).

(3) First, our assumptions on f ensure that ϕ is absolutely integrable on $[0, \infty]$. Consequently, for each Ω,

$$\lim_{\ell\to\infty} \int_0^\ell \phi(\xi) \sin \Omega\xi \, d\xi$$

exists, and as above, this limit is reached *uniformly* in Ω. On the other hand, it will be shown in part 6 below that for each ℓ,

$$\lim_{\Omega\to\infty} \int_0^\ell \phi(\xi) \sin \Omega\xi \, d\xi = 0, \tag{21}$$

and we conclude that

$$\lim_{\Omega\to\infty} \lim_{\ell\to\infty} \int_0^\ell \phi(\xi) \sin \Omega\xi \, d\xi = 0, \tag{21$'$}$$

since the latter limits can be interchanged.

(4) Finally, for each $\Omega > 0$, if $t = \Omega\xi$, then

$$\lim_{\ell\to\infty} \int_0^\ell \frac{\sin \Omega\xi}{\xi}\, d\xi = \lim_{\ell\to\infty} \int_0^{\Omega\ell} \frac{\sin t}{t}\, dt = \int_0^\infty \frac{\sin t}{t}\, dt = \mathscr{S}, \tag{22}$$

where the number $\mathscr{S}$ exists and is independent of Ω. (See Problem 8.1C.)

(5) Therefore by (20), (21$'$), and (22), it follows that

$$\lim_{\Omega\to\infty} \lim_{\ell\to\infty} I(\Omega, \ell) = f(0)\,\mathscr{S},$$

so that from (18) and (19), I exists and

$$I = f(0)\,\mathscr{S}.$$

In particular, for the function $f(x) = e^{-|x|}$ of Example 1, we find that

$$\mathscr{S} = \mathscr{S} f(0) = I = \int_0^\infty \frac{d\omega}{1+\omega^2} = \frac{\pi}{2}. \tag{23}$$

Hence, (17) holds and our lemma is proved, provided that (21) is valid.

(6) To establish (21), note that for each $\epsilon > 0$, we can approximate ϕ on $[0, \ell]$ by a polynomial p so closely that

$$\int_0^\ell |\phi - p|\, d\xi < \epsilon.$$

(Recall Theorem 1.17.) Then using standard estimates and partial integration we see that for fixed ℓ:

$$\begin{aligned}\left|\int_0^\ell \phi(\xi)\sin\Omega\xi\,d\xi\right| &\le \int_0^\ell |\phi - p|\,d\xi + \left|\int_0^\ell p(\xi)\sin\Omega\xi\,d\xi\right| \\ &< \epsilon + \frac{1}{\Omega}\left|p(0) - p(\ell)\cos\Omega\ell + \int_0^\ell p'(\xi)\cos\Omega\xi\,d\xi\right| \\ &\le \epsilon + \frac{1}{\Omega}\left(|p(0)| + |p(\ell)| + \int_0^\ell |p'(\xi)|\,d\xi\right). \\ &< 2\epsilon, \qquad \text{if } \Omega \text{ is sufficiently large.}\end{aligned}$$

Since ϵ is arbitrary, (21) follows. ■

We can now establish our major result.

8.2 Theorem: *If f is absolutely integrable on $\mathbb{R}$, then at each point x where f has at most a jump discontinuity with derivatives from the right and from the left, Fourier's integral formula holds in the form*

$$\frac{1}{\pi}\int_0^\infty d\omega \int_{-\infty}^\infty f(\xi)\cos\omega(\xi - x)\,d\xi = \frac{f(x+) + f(x-)}{2}. \tag{24}$$

Proof: When $x = 0$, then as in the proof of Theorem 1.11, we observe that f_E is effectively continuous at 0 with limiting value $f_E(0) \stackrel{\text{def}}{=} [f(0+) + f(0-)]/2$ and derivatives from the right and from the left. In this case, the result follows from Lemma 8.1 (and the fact that $\int_{-\infty}^\infty f_O(\xi)\cos\omega\xi\,d\xi = 0$).

For the general case, let $\tilde{f}(t) = f(x+t)$, and note that when $\xi = x + t$:

$$\begin{aligned}\frac{1}{\pi}\int_0^\infty d\omega \int_{-\infty}^\infty f(\xi)\cos\omega(\xi - x)\,d\xi &= \frac{1}{\pi}\int_0^\infty d\omega \int_{-\infty}^\infty \tilde{f}(t)\cos\omega t\,dt \\ &= \frac{\tilde{f}(0+) + \tilde{f}(0-)}{2} = \frac{f(x+) + f(x-)}{2}.\end{aligned}$$ ■

Remarks: For validity of this result, it is not essential that f have derivatives from the right and from the left at x, only that the difference quotients $[f(x \pm t) - f(x\pm)]/t$ be absolutely integrable on some t-interval $(0, \delta]$. This is essentially what we used to prove Lemma 8.1. Here, we use "integrability" to mean possibly improper Riemann integrability, so that, for example, $f(x) = |x|^{-1/2}e^{-|x|}$ is admissible at $x = 0$. Similar questions arise when f is only Lebesque integrable on $\mathbb{R}$, but the answers are far more delicate. (See [Sog] for some recent results.)

Problem Set 8.1

8.1A Obtain the sine transform $s(\omega)$ for the function of Example 2.

8.1B Use (6) to obtain the sine and cosine transforms for each of the following functions f and appeal to Theorem 8.2 to evaluate the resulting F in (7).

1. $f(x) = \begin{cases} 1 - |x|, & |x| \leq 1 \\ 0, & |x| \geq 1 \end{cases}$

2. $f(x) = \begin{cases} x, & |x| < 2 \\ 0, & |x| > 2 \end{cases}$

3. $f(x) = \begin{cases} 1, & a < x < b \\ 0, & \text{otherwise,} \end{cases}$ for finite a, b

4. $f(x) = \begin{cases} \cos x, & 0 < x < \pi \\ 0, & \text{otherwise;} \end{cases}$ and conclude that $\displaystyle\int_0^\infty \frac{\omega \sin \omega\pi}{1 - \omega^2}\, d\omega = \frac{\pi}{2}$.

5. $f(x) = \begin{cases} e^{-x}, & x > 0 \\ -e^{x}, & x < 0, \end{cases}$ and conclude that

$$\int_0^\infty \frac{\omega \sin \omega x}{1 + \omega^2}\, d\omega = \int_0^\infty \frac{\cos \omega x}{1 + \omega^2}\, d\omega, \quad x > 0.$$

(See Example 2.)

6. $f(x) = e^{-x} \cos x, \quad x > 0,$ with even extension

7. $f(x) = e^{-x} \cos x, \quad x > 0,$ with odd extension

8.1C For $\Omega \geq \pi/2$, consider

$$S(\Omega) = \int_0^\Omega \frac{\sin t}{t}\, dt = S(\pi/2) + \int_{\pi/2}^\Omega \frac{\sin t}{t}\, dt.$$

1. Integrate the second term by parts to conclude that as $\Omega \to +\infty$,

$$S(\Omega) \to S(\pi/2) - \int_{\pi/2}^\infty \frac{\cos t}{t^2}\, dt = \mathcal{S}, \text{ (say)},$$

since the last integral is absolutely convergent. Why must $S(\ell\Omega) \to \mathcal{S}$ as $\Omega \to +\infty$, for each $\ell > 0$?

2. Argue that $|S(\Omega)| \leq S(\pi/2)$ for all $\Omega \geq 0$, so that $\mathcal{S} > 0$. (*Hint*: Sketch $(\sin t)/t$ and its integral, $S(\Omega)$, for $\Omega > 0$.)

8.1D 1. Using (3), with Bessel's inequality, show that for each $\ell > 0$ and $\Delta\omega = \pi/\ell$:

$$\sum_{n=1}^\infty \Delta\omega \left\{ \left[\frac{1}{\pi}\int_{-\ell}^{\ell} f(x) \cos(n\Delta\omega x)\, dx\right]^2 + \left[\frac{1}{\pi}\int_{-\ell}^{\ell} f(x) \sin(n\Delta\omega x)\, dx\right]^2 \right\}$$
$$\leq \frac{1}{\pi}\int_{-\ell}^{\ell} f^2(x)\, dx \leq \frac{1}{\pi}\int_{-\infty}^{\infty} f^2(x)\, dx.$$

2. Regard the series in part 1 as an approximation for an (improper) Riemann integral and conclude that as $\ell \to \infty$, *formally*, using (6), we obtain

$$\int_0^\infty [c^2(\omega) + s^2(\omega)]\, d\omega \leq \frac{1}{\pi}\int_{-\infty}^{\infty} f^2(x)\, dx$$

which is the integral analogue of Bessel's inequality.

3. Show that if Parseval's formula is used in part 1, instead of Bessel's *inequality*, then we obtain *equality* in part 2. Apply this result to $f(x) = e^{-|x|}$ in Example 1.

8.1E Suppose $f \in C^1(\mathbb{R})$ with f, f', and f'^2 absolutely integrable on $\mathbb{R}$.

1. If f is odd, bounded, and $f(0) = 0$, show that the sine transform s of f is also absolutely integrable on $\mathbb{R}$. (*Hint*: For $\omega \geq 1$, use partial integration, the Schwarz inequality, and Bessel's inequality for f'.)
2. When f is even, make a corresponding analysis for the cosine transform of f.

8.1F 1. For $R > 0$, set $I(R) = \int_0^R e^{-x^2}\,dx = \int_0^R e^{-y^2}\,dy$, and use polar coordinates with symmetry to show that

$$I^2(R) = 2\int_0^{\pi/4} d\theta \int_0^{R\sec\theta} re^{-r^2}\,dr = \frac{\pi}{4} - \int_0^{\pi/4} e^{-R^2\sec^2\theta}\,d\theta.$$

(See Figure 8.8.)

2. Since $\sec^2\theta \geq 1$, conclude that

$$\left|I^2(R) - \frac{\pi}{4}\right| \leq \frac{\pi}{4}e^{-R^2}, \text{ which} \to 0, \qquad \text{as } R \to \infty,$$

so that $$\int_0^{\infty} e^{-x^2}\,dx = \lim_{R\to\infty} I(R) = \frac{\sqrt{\pi}}{2}.$$

8.2

Applications

In Chapters 3 and 4 we used Fourier series methods to solve simple boundary value problems of the strip type. In these cases, boundary conditions on the sides of the strip produce S–L problems with a sequence of explicit eigenvalues and trigonometric eigenfunctions. When the strip is not of finite width, sine and cosine transforms can sometimes be used to solve corresponding problems. Let's illustrate this approach with a simple example for Laplace's equation and one for the heat equation.

Poisson's Formula for a Half-Plane

In planar electrostatics, charge distributed along the x-axis will produce a field of potential $u = u(x, y)$ in, say, the upper half-plane. u should be bounded and harmonic for $y > 0$, and it is reasonable to suppose that the behavior of $u(x, y)$ as $y \searrow 0$ is related to the charge distribution at x. (See Figure 8.2.)

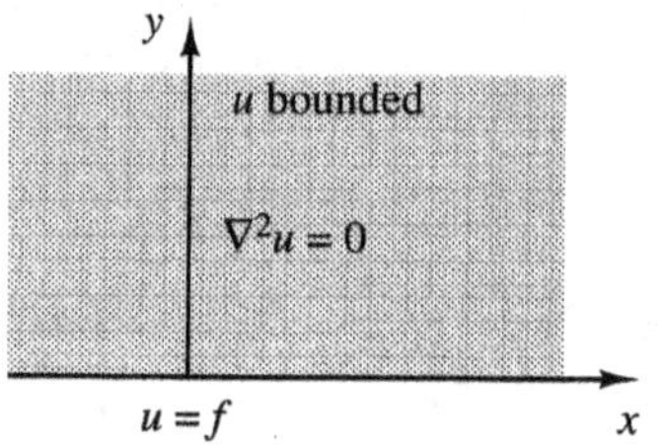

Figure 8.2

Example 4:

Let's find a function $u = u(x, y)$ that is bounded and harmonic in the half-plane $\{y > 0\}$ and that assumes given Dirichlet boundary values

$$u(x, 0) = f(x), \qquad -\infty < x < +\infty. \tag{25}$$

We need an appropriate solution u of Laplace's equation

$$\mathscr{L}(u) \stackrel{\text{def}}{=} u_{xx} + u_{yy} = 0. \tag{25'}$$

Solution in the separated form $u(x, y) = X(x)\,Y(y)$ requires a λ for which

$$X'' + \lambda X = 0; \qquad Y'' - \lambda Y = 0.$$

$\lambda = 0$ leads to the bounded solution $u_0 = \text{const.}$, while $\lambda = -\omega^2 < 0$ leads to $X(x) = ae^{\omega x} + be^{-\omega x}$, which is unbounded on $\mathbb{R}$ unless $a = b = 0$. Finally, $\lambda = \omega^2$ gives $X(x) = c \cos \omega x + s \sin \omega x$ and a corresponding $Y(y) = ae^{\omega y} + be^{-\omega y}$, which for $\omega > 0$ is unbounded in the upper half-plane unless $a = 0$. Thus, for arbitrary constants c and s, we have nontrivial solutions

$$u(x, y) = e^{-\omega y}(c \cos \omega x + s \sin \omega x)$$

for each $\omega \geq 0$, but we cannot further specify ω.

Instead, we postulate a solution in the integral form

$$U(x, y) = \int_0^\infty e^{-\omega y}\,[c(\omega) \cos \omega x + s(\omega) \sin \omega x]\, d\omega \tag{26}$$

for functions $c(\omega)$, $s(\omega)$ to be determined. U should be harmonic and bounded in the upper half-plane and, for $x \in \mathbb{R}$, U should satisfy the boundary condition

$$U(x, 0) = \int_0^\infty [c(\omega) \cos \omega x + s(\omega) \sin \omega x]\, d\omega = f(x). \tag{27}$$

If f is absolutely integrable on $\mathbb{R}$, we can take c and s to be its cosine and sine transforms as defined in (6). By careful study of the results in conjunction with Theorem 8.2, we can give conditions on f that make U of formula (26) a solution to our problem. (See Problem 8.1E.)

Now, let's try to obtain a nicer expression for our solution U. If we substitute formulas (6) into (26) and operate formally, we get

$$\begin{aligned} U(x, y) &= \frac{1}{\pi}\int_0^\infty e^{-\omega y}\, d\omega \int_{-\infty}^\infty f(\xi) \cos \omega(\xi - x)\, d\xi \\ &= \frac{1}{\pi}\int_{-\infty}^\infty f(\xi)\, d\xi \int_0^\infty e^{-\omega y} \cos \omega(\xi - x)\, d\omega, \end{aligned}$$

or

$$U(x, y) = \frac{y}{\pi}\int_{-\infty}^\infty \frac{f(\xi)}{(\xi - x)^2 + y^2}\, d\xi, \qquad y > 0. \tag{28}$$

In deriving (28) we assumed that the orders of integration are interchangeable and used standard methods to evaluate the resulting inner integral. (See Problem 8.2B.) But suppose we examine this formula independently of its derivation. Observe that for $y > 0$, a function $U(x, y)$ is defined by (28) when f is bounded and, say, continuous on $\mathbb{R}$; also, $|f| \leq M \Rightarrow |U| \leq M$. (See Problem 8.2B.) Under these conditions on f, we can use uniform convergence to justify differentiating across the integral sign as often as we please with respect to x or y. Consequently, for $y > 0$, U *is* harmonic, since for each fixed $\xi \in \mathbb{R}$:

$$\nabla^2 \left(\frac{2y}{(x-\xi)^2 + y^2} \right) = \nabla^2 \frac{\partial}{\partial y} \log[(x-\xi)^2 + y^2] = \frac{\partial}{\partial y} \nabla^2 \log[(x-\xi)^2 + y^2] = 0.$$

The boundary behavior of U as $y \searrow 0$ is set forth in the following.

(8.3) Theorem *(Poisson): Let f be bounded on $\mathbb{R}$ and suppose that f is integrable over each bounded interval. Then in the upper half-plane*

$$U(x, y) \stackrel{\text{def}}{=} \frac{y}{\pi} \int_{-\infty}^{\infty} \frac{f(\xi)\, d\xi}{(x-\xi)^2 + y^2}, \qquad y > 0 \tag{29}$$

is a bounded harmonic function that satisfies the boundary conditions

$$U(x, 0+) = \frac{f(x+) + f(x-)}{2} \tag{30}$$

at each x where $f(x+)$ and $f(x-)$ exist.

If f is continuous at x and $U(x, 0) \stackrel{\text{def}}{=} f(x)$, then U is continuous at $(x, 0)$.

Proof: It is only necessary to establish the limiting behavior, and we do so first at $x = 0$, assuming that f is an even function that is continuous at 0. Let M bound $|f|$, and for $\epsilon > 0$, choose $\delta > 0$ such that

$$|\xi| < \delta \Rightarrow |f(\xi) - f(0)| < \epsilon.$$

Observe that for $y > 0$:

$$U(0, y) - f(0) = \frac{2y}{\pi} \int_0^{\infty} \frac{[f(\xi) - f(0)]}{\xi^2 + y^2}\, d\xi.$$

Hence, standard estimates give

$$\begin{aligned} |U(0, y) - f(0)| &\leq \frac{2y}{\pi} \int_0^{\delta} \frac{|f(\xi) - f(0)|}{\xi^2 + y^2}\, d\xi + \frac{2y}{\pi} \int_{\delta}^{\infty} \frac{|f(\xi)| + |f(0)|}{\xi^2 + y^2}\, d\xi \\ &\leq \frac{2y\epsilon}{\pi} \int_0^{\infty} \frac{d\xi}{\xi^2 + y^2} + \frac{4M}{\pi} \tan^{-1} \frac{\xi}{y} \Big|_{\delta}^{\infty} \\ &= \epsilon + \frac{4M}{\pi} \left[\frac{\pi}{2} - \tan^{-1} \frac{\delta}{y} \right]. \end{aligned}$$

Now, as $y \searrow 0$, the bracketed expression approaches zero. Therefore, for sufficiently small $y > 0$,

$$|U(0, y) - f(0)| \leq 2\epsilon,$$

and since ϵ is arbitrary, it follows that

$$U(0, 0+) = f(0)$$

as desired. For general f and x, observe that by translation,

$$\int_{-\infty}^{\infty} \frac{f(\xi)}{(\xi - x)^2 + \xi^2}\, d\xi = \int_{-\infty}^{\infty} \frac{f(x+\xi)}{\xi^2 + y^2}\, d\xi$$

and apply the result just established to the *even part* of $\tilde{f}(\xi) \stackrel{\text{def}}{=} f(x+\xi)$, at $\xi = 0$. (Recall Section 1.4d.) Thus (30) holds, but it specifies the limiting behavior of U only along a line *normal* to the boundary. However, when f is continuous at x, we can modify our arguments to establish that U has the limit $f(x)$ along other paths in the half-plane that terminate at $(x, 0)$. The details are outlined in Problem 8.2M, and a similar argument is carried out in proving Theorem 8.5. ■

(8.4) Remarks:

(1) (29) is Poisson's formula for the half-plane. If f is bounded and continuous, then U is the *unique* bounded harmonic function in the half-plane with $U(\cdot, 0+) = f$. (See Theorem 8.6 in Section 8.5.)

(2) These results do not require that f be absolutely integrable and they hold independently of Theorem 8.2!

(3) The half-plane can be mapped conformally onto the disk and many other domains by *explicit* analytic functions whose boundary behavior is accessible (recall Section 4.4). Since the composite (harmonic) ∘ (analytic) is harmonic, we can obtain *explicit* Poisson-like formulas for such domains that provide solutions to Dirichlet problems. Their boundary behavior is analogous to that described in Theorem 8.3. In particular, we see that Poisson's formula for the unit disk must provide *radial limits* at every point, when the boundary function f is piecewise continuous, and limits along arbitrary paths that terminate at points of continuity of f.

The Age of the Earth

How old is our planet? In 1820, Fourier suggested how his methods might be used to answer this question, and his suggestion was pursued by Lord Kelvin in 1864. Suppose the Earth's interior was initially at a uniform temperature, U_0, and has only cooled subsequently. If we assume that the Earth's surface is flat, then it is reasonable to believe that the temperature $u(x, t)$ at depth x and time t obeys the one-dimensional heat equation of Section 3.3. For simplicity, assume that the Earth has constant thermal diffusivity k and that the temperature just beneath the surface has not varied appreciably from a reference value (chosen to be *zero*).

Example 5:

Let's find a solution to the heat equation (see also Figure 8.3, p. 334)

$$\mathscr{L}(u) = ku_{xx} - u_t = 0, \qquad \text{in } \{(x, t)\colon \; x > 0, t > 0\} \tag{31}$$

that satisfies *the boundary condition* $u(0, t) = 0, \qquad t > 0,$

and *the initial condition* $u(x, 0) = f(x), \qquad x > 0.$

Here k, the thermal diffusivity, is assumed constant, and we can take $k = 1$

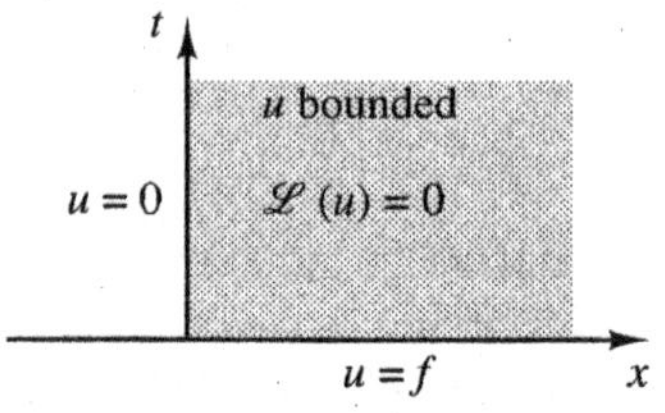

Figure 8.3

through the change of scale that replaces kt by t. (See Section 2.5.)

Separation in the form $u = XT$ yields *the eigenvalue problem*

$$X'' + \lambda X = 0 \qquad \text{and} \qquad T' = -\lambda T$$
$$X(0) = 0.$$

There is no mathematical basis for selecting a sign for λ; however,

$$T(t) = T(0)e^{-\lambda t}$$

and then the solution

$$u(x, t) = X(x)\,T(0)e^{-\lambda t}$$

is compatible with physical experience for a *temperature u* only when $\lambda \geq 0$ since in this situation we do not expect the temperature to increase with time. (See also the discussion of the maximum principle in Section 3.5.)

When $\lambda = 0$, then $X_0(x) = ax + b$ is unbounded unless $a = 0$; also, $X_0(0) = 0 \Leftrightarrow b = 0$. Therefore $X_0(x) = 0$. Finally, when $\lambda = \omega^2 (\omega > 0)$, we must take $X(x) = \sin \omega x$ to make $X(0) = 0$, but ω cannot be further specified. Instead, let's try a solution through *integral superposition* in the form

$$U(x, t) = \int_0^\infty e^{-\omega^2 t} s(\omega) \sin \omega x \, d\omega, \tag{32}$$

where $s(\omega)$ is to be determined if possible, so that $\mathscr{L}(U) = 0$, and $U(x, 0) = f(x)$. Observe that $U(0, t) = 0$, $t > 0$.

The boundary condition leads formally to the equation

$$f(x) = \int_0^\infty s(\omega) \sin \omega x \, d\omega, \qquad x > 0, \tag{33}$$

which suggests that we extend f as an *odd* function to all of $\mathbb{R}$. Then if f is absolutely integrable on $(0, \infty)$, comparison with (5) shows that $c(\omega) = 0$ and that the bounded function

$$s(\omega) = \frac{2}{\pi} \int_0^\infty f(\xi) \sin \omega\xi \, d\xi, \qquad \omega \geq 0, \tag{33$'$}$$

is the *sine transform* of f (as extended). Theorem 8.2 gives us conditions under which (33) is valid and U is a solution. (See Problem 8.1E.)

If we insert (33$'$) in (32), we obtain

$$U(x, t) = \frac{2}{\pi}\int_0^\infty e^{-\omega^2 t}\, d\omega\left(\int_0^\infty f(\xi) \sin \omega\xi \, d\xi\right) \sin \omega x$$
$$= \frac{1}{\pi}\int_0^\infty e^{-\omega^2 t}\, d\omega\left(\int_{-\infty}^\infty f(\xi) \cos \omega(x-\xi)\, d\xi\right),$$

since f is odd and $c(\omega) = (1/\pi)\int_{-\infty}^{\infty} f(\omega) \cos \omega\xi \, d\xi = 0$. Assuming that orders of integration can be interchanged we find that for $t > 0$:

$$U(x, t) = \frac{1}{\pi}\int_{-\infty}^\infty f(\xi)\, d\xi \int_0^\infty e^{-\omega^2 t} \cos \omega(x-\xi)\, d\omega$$

or
$$U(x, t) = \frac{1}{2\sqrt{\pi t}}\int_{-\infty}^\infty f(\xi) \exp\left\{-\frac{(x-\xi)^2}{4t}\right\} d\xi, \qquad t > 0 \qquad \textbf{(34)}$$

where we have used formula (15′) to evaluate the inner integral. (See Problem 8.2N.) With the substitution $\xi = x + 2\sqrt{t}\sigma$ (so $d\xi = 2\sqrt{t}\, d\sigma$), we can express our solution in the more attractive form

$$U(x, t) = \frac{1}{\sqrt{\pi}}\int_{-\infty}^\infty f(x + 2\sqrt{t}\sigma) e^{-\sigma^2}\, d\sigma. \qquad \textbf{(34')}$$

Before making a detailed study of these Poisson-like formulas, let's see what they provide for our original problem.

To estimate the age of the Earth, let $f(x) = U_0$, $x > 0$, have an *odd* extension to ensure that $U(0, t) = 0$, for $t > 0$. Then

$$f(x + 2\sqrt{t}\sigma) = \pm U_0 \qquad \text{if} \qquad \sigma \gtrless -\frac{x}{2\sqrt{t}},$$

and with this specification, and subsequent cancellation, (34′) gives

$$U(x, t) = \frac{2U_0}{\sqrt{\pi}}\int_0^{x/2\sqrt{t}} e^{-\sigma^2}\, d\sigma, \qquad t > 0. \qquad \textbf{(34'')}$$

It is straightforward to differentiate the last expression as required to show that $\mathscr{L}(U) = 0$ $(t > 0)$, and as $t \searrow 0$, we see that by (15′)

$$U(x, 0+) = \frac{2U_0}{\sqrt{\pi}}\int_0^\infty e^{-\sigma^2}\, d\sigma = U_0, \qquad x > 0.^\dagger$$

Hence U is the unique *bounded* solution to our problem. (See Theorem 8.7 in Section 8.5.)

Similarly,

$$U_x(0, t) = \frac{U_0}{\sqrt{\pi t}}\left(= \frac{U_0}{\sqrt{\pi k t}} \quad \text{when } k \neq 1\right).$$

The latter expression represents the spatial rate of increase in temperature at (or near) the Earth's surface; it can be determined empirically together with k, while U_0 can be found from the temperature of molten lava. According to this analy-

† However, as $(x, t) \to (0, 0)$ along the parabola defined by $x = \alpha\sqrt{t}$, $U(x, t) = U(\alpha\sqrt{t}, t)$ has a different limit for each $\alpha > 0$!

sis, the age of the Earth "now" assessed by the time of cooling should be approximately

$$t = \frac{1}{\pi k}\left(\frac{U_0}{U_x(0, t)}\right)^2,$$

which Kelvin estimated to be between 4×10^8 and 10^9 years. Although this estimate has been superseded by those obtained from radiocarbon dating techniques, it still represents a mathematical challenge to theological assertions about the date of creation. (See [Kor], [C-J], and [Wi].)

The Heat Kernel

Now, let's explore formulas (34) and (34′) more thoroughly. In general, $U(x, t)$ is defined by (34) for $t > 0$, provided that f is bounded on $\mathbb{R}$ and is, say, continuous. Moreover,

$$\text{if} \quad |f| \le M, \qquad \text{then} \qquad |U(x, t)| \le M, \qquad t > 0.$$

Using integral versions of the M-test arguments of Section 0.3, we can prove that U is C^∞ for $t > 0$, since differentiations across the integral sign are permitted. Consequently, from (31), we see that

$$\mathscr{L}(U) = \int_{-\infty}^{\infty} f(\xi)\,\mathscr{L}(\gamma)\,d\xi = 0, \qquad t > 0$$

since the **heat kernel**,

$$\gamma = \gamma(x, t; \xi, 0) \overset{\text{def}}{=} \frac{1}{2\sqrt{\pi t}} \exp\left\{-\frac{(x-\xi)^2}{4t}\right\}, \qquad t > 0$$

satisfies the heat equation, $\gamma_t = \gamma_{xx}$, for each fixed ξ. (Recall Problem 3.3A.)

The boundary behavior of U as $t \searrow 0$ is set forth in the following.

(8.5) Theorem: *Let f be bounded on $\mathbb{R}$ and suppose that f is integrable over each bounded interval. Then*

$$U(x, t) = \frac{1}{2\sqrt{\pi t}} \int_{-\infty}^{\infty} f(\xi) \exp\left\{-\frac{(x-\xi)^2}{4t}\right\} d\xi, \qquad t > 0,$$

is a bounded solution of the heat equation

$$U_t = U_{xx} \qquad \text{on} \qquad R \times (0, +\infty)$$

that satisfies the initial condition

$$U(x, 0+) = \frac{f(x+) + f(x-)}{2},$$

at each $x \in \mathbb{R}$ where $f(x+)$ and $f(x-)$ exist. If f is continuous at x and $U(x, 0) \overset{def}{=} f(x)$, then U is continuous at $(x, 0)$.

Proof:* Only the limiting behavior remains in question, and this can be attacked more easily using (34′). The proof of the first assertion is similar to that in Theorem 8.3, and it suffices to establish the second when $x = 0$. Given

$\epsilon > 0$, choose $\delta > 0$ such that

$$|\xi| < 2\delta \Rightarrow |f(\xi) - f(0)| < \epsilon, \qquad (\xi = x + 2\sqrt{t}\sigma).$$

Then if M bounds $|f|$, observe that by (15′)

$$U(x, t) - f(0) = \frac{1}{\sqrt{\pi}} \int_{-\infty}^{\infty} [f(\xi) - f(0)] e^{-\sigma^2} \, d\sigma,$$

so that by standard estimates,

$$|U(x, t) - f(0)| \le \frac{1}{\sqrt{\pi}} \int_{S_\delta} |f(\xi) - f(0)| e^{-\sigma^2} \, d\sigma + \frac{1}{\sqrt{\pi}} \int_{\tilde{S}_\delta} |f(\xi) - f(0)| e^{-\sigma^2} \, d\sigma,$$

where $S_\delta = \{\xi\colon |\xi| < 2\delta\}$ and $\tilde{S}_\delta = \{\xi\colon |\xi| \ge 2\delta\}$.

Now, if $\xi \in S_\delta$, then $|f(\xi) - f(0)| < \epsilon$

and if $\xi \in \tilde{S}_\delta$, then $|2\sqrt{t}\sigma| = |\xi - x| \ge 2\delta - |x| \ge \delta$, for $|x| \le \delta$.

Hence, if $|x| \le \delta$ and $t \le \delta^2/4$, then

$$|U(x, t) - f(0)| \le \frac{\epsilon}{\sqrt{\pi}} \int_{-\infty}^{\infty} e^{-\sigma^2} d\sigma + \frac{4M}{\sqrt{\pi}} \int_{\delta/2\sqrt{t}}^{\infty} e^{-\sigma\delta/2\sqrt{t}} \, d\sigma$$

$$\le \epsilon + \frac{4M}{\sqrt{\pi}} \frac{2\sqrt{t}}{\delta} \qquad \left(\text{since } \frac{\delta}{2\sqrt{t}} \ge 1\right).$$

Therefore, if $|x| \le \delta$ and $\sqrt{t} < \epsilon\delta\sqrt{\pi}/8M$, then $|U(x, t) - f(0)| < 2\epsilon$, for each $\epsilon > 0$; i.e., if $U(x, 0) \stackrel{\text{def}}{=} f(x)$, then U is continuous at $(x, 0)$.

For general x, apply the result just established to $\tilde{f}(\xi) \stackrel{\text{def}}{=} f(\xi + x)$ at $\xi = 0$. ■

Remarks:

1. When f is continuous, U is the unique bounded solution to the boundary value problem of Theorem 8.5 that is continuous for $y \ge 0$. (See Theorem 8.7 in Section 8.5.)
2. This result suggests that we regard $\gamma(x, t; \xi, 0)$ as the temperature at (x, t) generated by a one-dimensional heat source at $(\xi, 0)$ of unit strength. Then integral superposition of sources of distributed strength f along the ξ-axis produces at (x, t) the temperature $U(x, t)$.

Problem Set 8.2

8.2A 1. Use Poisson's formula to find a function $u = u(x, y)$ that is bounded and harmonic for $y > 0$, such that as $y \searrow 0$:

$$u(x, 0+) = \begin{cases} 1, & a < x < b \\ 0, & x < a, \text{ or } x > b, \end{cases}$$

where a and b are finite. What happens when $x = a$?, when $x = b$?

2. Is there an unbounded solution to this problem? more than one?
3. What happens to the solution if $b \to \infty$? if also $a \to -\infty$?

8.2B 1. If U is defined by (28) for $y > 0$, show that

$$|f| \le M \Rightarrow |U(x, y)| \le M.$$

2. Verify the integration used to obtain (28).

8.2C* Suppose f is bounded on $\mathbb{R}$ and *odd*. Obtain a formal Poisson-like solution to the following Neumann problem for the upper half-plane,

$$\nabla^2 u = 0, \qquad u \text{ bounded for } y > 0, \text{ and } u_y(x, 0+) = -f(x), x \in \mathbb{R}.$$

(*Hint*: $\int_{-\infty}^{\infty} f(\xi)\, d\xi \int_y^{\infty} dn/\eta = 0$ for $y > 0$.)

8.2D 1. Find a formal bounded solution in the semi-infinite strip,

$$V = \{(x, y) \colon x > 0,\ 0 < y < 1\}, \text{ to the problem}$$

$$\nabla^2 u = 0 \quad \text{in } V \quad \text{with} \quad \begin{cases} u_x(0, y) = 0, & 0 < y < 1 \\ u_y(x, 0) = 0, & x > 0 \end{cases}$$

$$\text{and} \quad u(x, 1) = f(x), \qquad x > 0,$$

in the integral form

$$U(x, y) = \int_0^{\infty} \gamma(\omega) \cosh \omega y \cos \omega x \, d\omega.$$

2. Obtain γ when $f(x) = e^{-x}$, $x > 0$, and interpret your answer physically.

8.2E In the strip V of Problem 8.2D, find a formal bounded solution and give a physical interpretation to the problem

$$\nabla^2 u = 0 \quad \text{in } V \quad \text{with} \quad \begin{cases} u(0, y) = 0, & 0 < y < 1 \\ u(x, 0) = 0, & x > 0 \end{cases}$$

$$\text{and} \quad u(x, 1) = f(x) = \begin{cases} 1, & 0 < x < 1 \\ 0, & x > 1 \end{cases}.$$

8.2F What changes are required in the analysis of Problem 8.2D if on $y = 1$ we prescribe

1. $u_y(x, 1) = f(x), \qquad x > 0$?
2. $u_y(x, 1) + u(x, 1) = f(x), \qquad x > 0$?

8.2G 1. Derive (34″) from (34′) when $f(x) = \pm U_0$, for $x \gtrless 0$, and show that $U_x(x, t) = (U_0/\sqrt{\pi t})e^{-x^2/4t}$, $t > 0$.

2. Conclude that $U_t = U_{xx}, \qquad t > 0$.

8.2H When $f(x) = \begin{cases} 1, & a < x < b \\ 0, & x < a \text{ or } x > b \end{cases}$ $\quad$ (a, b, finite),

express U as defined by (34′) in terms of the **error function** $(\text{erf})(z) \stackrel{\text{def}}{=} (2/\sqrt{\pi})\int_0^z e^{-\sigma^2} d\sigma$, $z \in \mathbb{R}$.

8.2I 1. Obtain a bounded formal solution U to the following problem.

$$\mathscr{L}(u) = u_{xx} - u_t = 0, \qquad \text{in} \qquad \{x > 0, t > 0\},$$

$$\text{with} \quad u_x(0, t) = 0, \qquad t > 0 \quad \text{and} \quad u(x, 0) = \frac{1}{1 + x^2}, \quad x > 0.$$

2. Is the associated $c(\omega)$ absolutely integrable? What can you conclude about U as an *actual* solution? Interpret your answer physically.

8.2J 1. Obtain a bounded formal solution U to the following problem.

$$\mathscr{L}(u) = u_{xx} - u_t = 0 \quad \text{in} \quad \{x > 0,\ t > 0\},$$

$$\text{with} \quad u(0, t) = 0, \quad t > 0, \quad \text{and} \quad u(x, 0) = e^{-x}, \quad x > 0.$$

2. Is the associated $s(\omega)$ absolutely integrable? What can you conclude about U?

8.2K 1. Obtain a bounded formal solution U to the problem

$$\mathscr{L}(u) = u_{xx} - u_t = 0, \text{ when } x \in \mathbb{R},\ t > 0,$$

$$\text{with } u(x, 0) = \begin{cases} e^{-x}, & x > 0 \\ 0, & x < 0 \end{cases}.$$

2. What can you conclude about U?

8.2L 1. Show that $u(x, y) = Y(y) \cos \omega x$ satisfies $u_x(0, y) = 0$

$$\text{and that } \mathscr{L}(u) \equiv \nabla^2 u - u = 0 \text{ if } Y'' - (\omega^2 + 1)Y = 0.$$

2. Conclude that $U(x, y) = \int_0^\infty \gamma(\omega) \sinh(\sqrt{\omega^2 + 1}\,y) \cos \omega x \, d\omega$ satisfies the equation of part 1, in the strip $0 < y < 1$, $x > 0$, with $u_x(0, y) = 0$, $0 < y < 1$, and $u(x, 0) = 0$, $x > 0$, if $\omega^2\gamma(\omega)$ is absolutely integrable.
3. How should γ be chosen to have (a) $U(x, 1) = f(x)$ or (b) $U_y(x, 1) = f(x)$, with f absolutely integrable on $(0, \infty)$?
4. Obtain U when $f(x) = e^{-x}$, $x > 0$.
5. Give some physical interpretations of this boundary value problem.

8.2M* To complete the proof of Theorem 8.3, assume that f is continuous at $\xi = 0$, and for given $\epsilon > 0$, replace δ by 2δ.

1. Explain why

$$|U(x, y) - f(0)| < \frac{\epsilon y}{\pi} \int_{-\infty}^{\infty} \frac{d\xi}{(x - \xi)^2 + y^2} + 2M \frac{y}{\pi} \int_{\tilde{S}_\delta} \frac{d\xi}{(x - \xi)^2 + y^2}$$

where $\tilde{S}_\delta = \{\xi\colon |\xi| \geq 2\delta\}$.

2. Evaluate the first integral and estimate the second when $|x| \leq \delta$ by noting that then $|\xi - x| \geq |\xi| - |x| \geq \delta$ if $\xi \in \tilde{S}_\delta$. Conclude that

$$|U(x, y) - f(0)| \leq \epsilon + \frac{4M}{\pi}\left[\frac{\pi}{2} - \tan^{-1}\frac{\delta}{y}\right],$$

and explain why $U(x, y) \to f(0)$ as $(x, y) \to (0, 0)$.

3. Explain why a similar result holds if f is continuous at x_0. (*Hint*: $\tilde{f}(\xi) \stackrel{\text{def}}{=} f(x_0 + \xi)$ is continuous at $\xi = 0$.)

8.2N Show how formula (34) follows when (15′) is used to evaluate $\int_0^\infty e^{-\omega^2 t} \cos \omega(x - \xi)\, d\omega$. (*Hint*: Let $s = \omega\sqrt{t}$.)

8.3

The Fourier Transform and Applications

When Euler's formula, $2\cos t = e^{it} + e^{-it}$, is used in (4), then Fourier's integral formula takes the form

$$f(x) = \frac{1}{2\pi}\int_0^\infty d\omega \int_{-\infty}^\infty f(\xi)\,[e^{i\omega(x-\xi)} + e^{-i\omega(x-\xi)}]\,d\xi$$

$$= \frac{1}{2\pi}\int_{-\infty}^\infty d\omega \int_{-\infty}^\infty f(\xi)e^{i\omega(x-\xi)}\,d\xi,$$

since $$\int_0^\infty d\omega \int_{-\infty}^\infty f(\xi)e^{-i\omega(x-\xi)}\,d\xi = \int_{-\infty}^0 d\omega \int_{-\infty}^\infty f(\xi)e^{i\omega(x-\xi)}\,d\xi,$$

after substitutions which effect the replacement of ω by $-\omega$.

Finally, $e^{-i\omega(x-\xi)} = e^{-i\omega x}e^{i\omega\xi}$, and we see that Fourier's integral formula can be stated

$$\left.\begin{aligned} f(x) &= \frac{1}{2\pi}\int_{-\infty}^\infty e^{-i\omega x}\hat{f}(\omega)\,d\omega, \qquad x \in \mathbb{R} \\ \text{where} \qquad \hat{f}(\omega) &\stackrel{\text{def}}{=} \int_{-\infty}^\infty e^{i\omega\xi}f(\xi)\,d\xi, \qquad \omega \in \mathbb{R} \end{aligned}\right\} \tag{35}$$

is called the **Fourier transform** of f.†

According to Theorem 8.2, this pair of formulas is valid for all $x \in \mathbb{R}$ if f is C^1 and absolutely integrable on $\mathbb{R}$. More generally, $\hat{f}$ is defined in (35) when f is absolutely integrable on $\mathbb{R}$, and then

$$\frac{f(x+) + f(x-)}{2} = \frac{1}{2\pi}\int_{-\infty}^{+\infty} e^{-i\omega x}\hat{f}(\omega)\,d\omega, \tag{35'}$$

at every x where f has derivatives from the right and from the left.

Example 6:

Using previous results, we can easily calculate the Fourier transform of the Gaussian distribution function

$$g(x) = \exp\left\{-\frac{(x-\mu)^2}{2\sigma^2}\right\}, \qquad x \in \mathbb{R}$$

† These formulas are sometimes replaced by equivalent ones that differ in their handling of the factor 2π and thereby provide competing definitions for the Fourier transform. Additional competition arises in versions in which i is replaced by $-i$.

for arbitrary constants $\sigma > 0$ and μ. Indeed if we let $x = \mu + \sqrt{2}\,\sigma\xi$, then

$$\begin{aligned}\hat{g}(\omega) &= \int_{-\infty}^{\infty} g(x)e^{i\omega x}\,dx = \sqrt{2}\,\sigma e^{i\omega\mu}\int_{-\infty}^{\infty} e^{-\xi^2} e^{i\omega\sqrt{2}\sigma\xi}\,d\xi \\ &= \sqrt{2}\sigma(e^{i\omega\mu})2\int_0^{\infty} e^{-\xi^2}\cos\omega\sqrt{2}\sigma\xi\,d\xi,\end{aligned}$$

since $e^{-\xi^2}$ is even. But this is essentially the *cosine transform* of $e^{-\xi^2}$, and from (15′), we see that

$$\hat{g}(\omega) = \sqrt{2\pi}\,\sigma e^{i\omega\mu}e^{-\sigma^2\omega^2/2}.$$

If we can differentiate f, then in general we find that

$$\begin{aligned}(f')\hat{}(\omega) &= \int_{-\infty}^{\infty} f'(x)e^{i\omega x}\,dx \\ &= f(x)e^{i\omega x}\Big|_{-\infty}^{\infty} - i\omega\int_{-\infty}^{\infty} f(x)e^{i\omega x}\,dx.\end{aligned}$$

Hence,

$$(f')\hat{}(\omega) = -i\omega\hat{f}(\omega),$$

provided that $f(x) \to 0$ as $|x| \to +\infty$. More generally, for $k = 1, 2, 3, \ldots,$

$$(f^{(k)})\hat{}(\omega) = (-i\omega)^{(k)}\,\hat{f}(\omega), \tag{36}$$

provided that for $j = 0, 1, 2, \ldots,\ k-1$: $f^{(j)}$ is absolutely integrable and $f^{(j)}(x) \to 0$ as $|x| \to +\infty$.

When we multiply $f(x)$ as given in (35) by $f(x)$ and integrate, we find that

$$\begin{aligned}\int_{-\infty}^{\infty} f^2(x)\,dx &= \frac{1}{2\pi}\int_{-\infty}^{\infty} f(x)\,dx\int_{-\infty}^{\infty} e^{-i\omega x}\hat{f}(\omega)\,d\omega \\ &= \frac{1}{2\pi}\int_{-\infty}^{\infty}\hat{f}(\omega)\,d\omega\int_{-\infty}^{\infty} e^{-i\omega x}f(x)\,dx\end{aligned}$$

if the orders of integration can be interchanged. The last inner integral is just $\overline{\hat{f}(\omega)}$ since f is real-valued, and in this manner we obtain **Parseval's formula**:

$$\int_{-\infty}^{\infty} f^2(x)\,dx = \frac{1}{2\pi}\int_{-\infty}^{\infty} |\hat{f}(\omega)|^2\,d\omega, \tag{37}$$

which is valid when f^2 is absolutely integrable on $\mathbb{R}$. Consequently, if f and g are absolutely integrable on $\mathbb{R}$, together with f^2 and g^2, we can derive **Plancherel's formula** (see Problem 8.3D):

$$\int_{-\infty}^{\infty} f(x)g(x)\,dx = \frac{1}{2\pi}\int_{-\infty}^{\infty}\hat{f}(\omega)\overline{\hat{g}(\omega)}\,d\omega. \tag{37′}$$

An attractive feature of the Fourier transform is the ease with which it extends to higher dimensions. Indeed, if f is C^1 and absolutely integrable over $\mathbb{R}^d$, then in Cartesian coordinates, we can show that

$$\left.\begin{aligned} f(\mathbf{x}) &= \frac{1}{(2\pi)^d}\int_{\mathbb{R}^d} e^{-i\boldsymbol{\omega}\cdot\mathbf{x}}\hat{f}(\boldsymbol{\omega})\,d\boldsymbol{\omega}, \qquad \mathbf{x} \in \mathbb{R}^d \\ \text{where}\qquad \hat{f}(\boldsymbol{\omega}) &= \int_{\mathbb{R}^d} e^{i\boldsymbol{\omega}\cdot\mathbf{x}}f(\mathbf{x})\,d\mathbf{x}, \qquad \boldsymbol{\omega} \in \mathbb{R}^d\end{aligned}\right\} \tag{38}$$

with $d\mathbf{x} = dx_1\,dx_2\cdots dx_d$ and $d\boldsymbol{\omega} = d\omega_1\,d\omega_2\cdots d\omega_d$. (See Problem 8.3G.)

If we apply (36) componentwise, we find that

$$(\nabla^2 f)\hat{}(\boldsymbol{\omega}) = -|\boldsymbol{\omega}|^2 \hat{f}(\boldsymbol{\omega}), \qquad \boldsymbol{\omega} \in \mathbb{R}^d \tag{38'}$$

under suitable conditions on f. (See Problem 8.3H.)

Spectral Analysis of a Function

Equations (35) recall the complex form of a Fourier series used in Section 1.7. In the simplest case where an integrable function f of period 2π is specified on $\mathbb{R}$, the Fourier series generated by f can be expressed in the form

$$F(x) = \sum_{n=-\infty}^{\infty} \hat{f}(n) e^{-inx}. \tag{39}$$

Here, for each integer n,

$$\hat{f}(n) \overset{\text{def}}{=} \frac{1}{2\pi} \int_{-\pi}^{\pi} f(t) e^{int}\, dt \tag{39'}$$

provides a complex measure of how much of the base component $\cos(nt - \alpha)$ is present in the "signal" f. Indeed, if for some $m = 0, 1, 2, \ldots$, $f(t) = A\cos(mt - \alpha)$ for a phase α, then of course $F(x) = f(x)$ so that

$$\hat{f}(m) = \frac{A}{2} e^{i\alpha}, \hat{f}(-m) = \frac{A}{2} e^{-i\alpha}, \qquad \text{and otherwise } \hat{f}(n) = 0.$$

Hence, both A and α may be recovered for the given m from knowledge of $\hat{f}$ for that m.

More generally, a given f is said to have the discrete Fourier **spectrum** $\hat{f}$ defined above, and then the representation

$$f(x) = \sum_{n=-\infty}^{\infty} \hat{f}(n) e^{-inx}$$

is said to be achieved by **spectral synthesis.**†

Similarly, when f is defined and absolutely integrable on $\mathbb{R}$, its Fourier transform $\hat{f}$ in (35) is said to provide a **spectral resolution** of f; the natural question is whether f can be recovered from a corresponding spectral synthesis. However, since pure signals $f(t) = \cos(\omega t - \alpha)$ are *not* absolutely integrable, this terminology is less readily explained. In practice, the Fourier transform of an electronic signal f is obtained by integrating f against the pure signals above over a relatively large but finite time interval as the frequency ω and phase α are varied. Thus to some extent, the earlier interpretation applies.

(a) Sampling and band-limited signals: When transmitted electrically, human speech may be regarded as a smooth signal f whose spectrum is negligible above frequencies exceeding 20,000 cps. If we suppose that $\hat{f}(\omega) = 0$,

† The origin of this terminology is rather curious. When light from distant stars is directed through a prism or diffraction grating, the resulting *color spectrum* provides a visual image of the Fourier transform of the incoming signal. The term *spectrum* as employed in mathematics derives from this association. (See [Wal].)

when $|\omega| \geq \ell$ for some positive ℓ, then the signal f is said to be *band limited,* and synthesis produces the following general interpolation or **sampling formula**:

$$f(x) = \sum_{n=-\infty}^{\infty} f\left(\frac{n\pi}{\ell}\right) \frac{\sin(n\pi - \ell x)}{n\pi - \ell x}. \tag{40}$$

This result makes it possible to recover a telephone conversation from transmitted "bits" and so to use the same line to send hundreds of messages simultaneously.

(To derive this sampling formula, observe that at $x = n\pi/\ell$, (35′) gives

$$f\left(\frac{n\pi}{\ell}\right) = \frac{1}{2\pi}\int_{-\ell}^{\ell} e^{-i\omega n\pi/\ell} \hat{f}(\omega)\, d\omega = \frac{\ell}{\pi}\,\hat{e}_n,$$

where $\hat{e}_n$ is the nth complex Fourier coefficient of $\hat{f}$ on $[-\ell, \ell]$ in the representation $\hat{f}(\omega) \sim \sum_{n=-\infty}^{\infty} \hat{e}_n e^{i\omega n\pi/\ell}$, $(|\omega| < \ell)$. It follows that for other values of x, (35′) gives

$$\begin{aligned} f(x) &= \frac{1}{2\pi}\int_{-\ell}^{\ell} e^{i\omega x}\left(\sum_{n=-\infty}^{\infty} \hat{e}_n e^{i\omega n\pi/\ell}\right) d\omega \\ &= \sum_{n=-\infty}^{\infty} \hat{e}_n \frac{1}{2\pi}\int_{-\ell}^{\ell} e^{i\omega(n\pi - \ell x)/\ell}\, d\omega \\ &= \sum_{n=-\infty}^{\infty} \left(\frac{\ell}{\pi}\hat{e}_n\right) \frac{\sin(n\pi - \ell x)}{n\pi - \ell x}, \end{aligned}$$

as required, where we justify the above interchange of summation and integration by appealing to Parseval's formula.)

A signal with jump discontinuities cannot be band limited, since contributions of $\hat{f}(\omega)$ for arbitrarily large ω are required to represent f near points of discontinuity. (Recall Gibbs' behavior in the discrete case discussed in Section 1.8d.) In fact, a smooth band-limited signal f is analytic (hence C^{∞}) since this is true of its representing integral $(1/2\pi)\int_{-\ell}^{\ell} e^{-i\omega x} \hat{f}(\omega)\, d\omega$. Consequently unless $f = 0$, f cannot *also* be "time-limited" in that for some $T > 0$: $|x| \geq T \Rightarrow f(x) = 0$; this observation is of practical concern, since it implies that integration over a large but finite time interval cannot determine the spectrum of a band-limited signal. The search for acceptable compromise has motivated important research in Fourier analysis. (See [D–M].)

(b) The uncertainty principle: In quantum mechanics, a particle of unit mass in rectilinear motion under a potential field V is said to be in the stationary (real) state ψ when ψ is an eigenfunction to the eigenvalue, $-2E/\hbar^2$, for the reduced Schrödinger equation

$$L(\psi) \stackrel{\text{def}}{=} \psi'' - \frac{2}{\hbar^2} V(x)\psi = -\frac{2}{\hbar^2} E\psi, \qquad -\infty < x < +\infty.$$

We also suppose that ψ is normalized in that $\int_{-\infty}^{\infty} \psi^2\, dx = 1$. E is the energy of the state and $2\pi\hbar$ is Planck's constant. (Recall the discussion of the hydrogen atom in Section 7.6.)

A significant consequence of a quantum-mechanical model for this particle is that its position x and momentum p cannot be specified simultaneously. Indeed, in the formalism of quantum physics:

$$x \text{ corresponds to the state operator } Q \quad \text{where} \quad Q\psi = x\psi$$

and

$$p \text{ corresponds to the state operator } P \quad \text{where} \quad P\psi = -\hbar i\psi'.$$

Any such (self-adjoint) operator S on states ψ is termed an *observable* and its mean in the real state ψ is defined by the integral

$$\int_{-\infty}^{\infty} \psi(S\psi)\, dx.$$

In particular, in state ψ, the particle has *mean position* $\mu = \int_{-\infty}^{\infty} x\psi^2\, dx$, assumed finite. μ is seen to be the probabilistic mean of a random variable X (of position) with density function ψ^2, and X has standard deviation σ from this mean where

$$\sigma^2 = \int_{-\infty}^{\infty} (x-\mu)^2 \psi^2\, dx. \tag{41}$$

Similarly, in state ψ, the particle has *momentum* with *mean*

$$-\hbar i \int_{-\infty}^{\infty} \psi\psi'\, dx = \frac{\hbar}{2\pi}\int_{-\infty}^{\infty} \omega\,|\hat{\psi}|^2\, d\omega = 0,$$

assuming that Plancherel's formula (37′) applies. (See Problem 8.3E.) From (37),

$$\frac{1}{2\pi}\int_{-\infty}^{\infty} |\hat{\psi}|^2\, d\omega = \int_{-\infty}^{\infty} \psi^2\, dx = 1;$$

thus $|\hat{\psi}|^2/2\pi$ is the probability density for a random variable $\hat{X}$ with mean *zero* and standard deviation $\hat{\sigma}$, where

$$\hat{\sigma}^2 = \frac{1}{2\pi}\int_{-\infty}^{\infty} \omega^2\,|\hat{\psi}|^2\, d\omega = \int_{-\infty}^{\infty} (\psi')^2\, dx. \tag{41′}$$

In fact, $\hbar\hat{X}$ may be regarded as the momentum in state ψ, since if $\psi \in C^n$ and $\int_{-\infty}^{\infty} (\psi^{(n)})^2\, dx < +\infty$, then by (37′) and (36)

$$\int_{-\infty}^{\infty} \psi(P^n\psi)\, dx = (-i\hbar)^n \int_{-\infty}^{\infty} \psi\psi^{(n)}\, dx = \frac{\hbar^n}{2\pi}\int_{-\infty}^{\infty} \omega^n\,|\hat{\psi}|^2\, d\omega, \qquad n = 0, 1, 2, \ldots;$$

i.e., the ψ-averages of the powers of P are obtained as corresponding moments of $\hbar\hat{X}$, so that $\hbar\hat{X}$ and P have the same probabilistic behavior.

σ and $\hat{\sigma}$ are related through the inequality

$$\sigma\hat{\sigma} \geq 1/2, \tag{42}$$

which implies that one is small iff the other is correspondingly large. This mathematical fact, called the **uncertainty principle**, yields the Heisenberg inequality of quantum physics in the form

$$\Delta x \Delta p \geq \hbar/2$$

concerning the average uncertainties Δx and Δp in specification of position and momentum.

To establish (42), let's suppose $x\psi^2(x) \to 0$ as $|x| \to +\infty$. Then upon partial integration, we find that

$$\frac{1}{2} = \frac{1}{2}\left(\int_{-\infty}^{\infty} \psi^2(x)\, dx\right) = \tfrac{1}{2}(x-\mu)\underbrace{\psi^2(x)\Big|_{-\infty}^{\infty}}_{0} - \int_{-\infty}^{\infty} [(x-\mu)\psi]\psi'\, dx$$

$$\leq \left(\int_{-\infty}^{\infty} (x-\mu)^2\psi^2(x)\, dx\right)^{1/2} \left(\int_{-\infty}^{\infty} \psi'(\xi)^2\, d\xi\right)^{1/2} = \sigma\hat{\sigma},$$

where we used the Schwarz inequality and our previous expressions for σ and $\hat{\sigma}$ given in (41) and (41′) respectively.

This result, which is valid for more general ψ, indicates globally why a function and its Fourier transform cannot both be confined to small intervals. Much work has been done recently to derive corresponding local results for more general transformations. (See [D–M] and [Fef].)

Poisson's Formula for the Wave Equation

In Section 5.6 we derived d'Alembert's solution of the initial value problem for the one-dimensional wave equation. Higher-dimensional analogues can be obtained from the Fourier transform of (38), and the computations can be carried through in the physically important case of dimension three.

*Example 7:**

Let's find a bounded solution $u(\mathbf{x}, t)$ to the wave equation

$$u_{tt} = c^2\nabla^2 u, \qquad \mathbf{x} \in X = \mathbb{R}^d, \qquad t \in \mathbb{R} \tag{43}$$

that assumes continuously the initial values

$$u(\mathbf{x}, 0) = 0, \text{ and } u_t(\mathbf{x}, 0) = h(\mathbf{x}), \qquad \mathbf{x} \in X. \tag{44}$$

To do so, we suppose h and $u(\cdot, t)$ to be absolutely integrable over $X = \mathbb{R}^d$, for each fixed t. Then, for each $\boldsymbol{\omega} \in X$, we can introduce the Fourier transform

$$\hat{u}(\boldsymbol{\omega}, t) = \int_X e^{i\boldsymbol{\omega}\cdot\mathbf{x}} u(\mathbf{x}, t)\, d\mathbf{x}. \tag{45}$$

Operating formally, we find from (44) that when $t = 0$

$$\left.\begin{aligned} &\hat{u}(\boldsymbol{\omega}, 0) = 0, \\ \text{and} \quad &\hat{u}_t(\boldsymbol{\omega}, 0) = \hat{h}(\boldsymbol{\omega}) = \int_X e^{i\boldsymbol{\omega}\cdot\mathbf{x}} h(\mathbf{x})\, d\mathbf{x}. \end{aligned}\right\} \tag{45'}$$

Similarly, using (43), we see that $\hat{u}$ should satisfy the equation

$$\begin{aligned} \hat{u}_{tt} &= \int_X e^{i\boldsymbol{\omega}\cdot\mathbf{x}} u_{tt}(\mathbf{x}, t)\, d\mathbf{x} \\ &= c^2 \int_X e^{i\boldsymbol{\omega}\cdot\mathbf{x}} \nabla^2 u(\mathbf{x}, t)\, d\mathbf{x}. \end{aligned}$$

Then, from (38′), we get

$$\hat{u}_{tt} = -c^2\rho^2\hat{u}, \qquad \text{where} \qquad \rho = |\boldsymbol{\omega}|, \tag{46}$$

if we assume that both u and $|\nabla u|$ die out as $|\mathbf{x}| \to \infty$, so that the boundary terms arising in partial integrations vanish in the limit.

For fixed ω, this last equation may be regarded as an ordinary differential equation in t, and with initial conditions (45′) it has the well-known solution

$$\hat{u}(\omega, t) = \hat{h}(\omega)\,\frac{\sin \rho c t}{\rho c}, \qquad \text{if } \rho = |\omega| \neq 0. \tag{46′}$$

Thus, we now know $\hat{u}$ explicitly, and we can use the inversion formula of (38) to recover u from its Fourier transform as follows.

$$u(\mathbf{x}, t) = \frac{1}{c(2\pi)^d}\int_X e^{-i\omega\cdot\mathbf{x}}\hat{h}(\omega)\,\frac{\sin \rho c t}{\rho}\,d\omega,$$

or by (45′),

$$u(\mathbf{x}, t) = \frac{1}{c(2\pi)^d}\int_X (e^{-i\omega\cdot\mathbf{x}})\,\frac{\sin \rho c t}{\rho}\,d\omega \int_X e^{i\omega\cdot\xi}h(\xi)\,d\xi.$$

Interchanging orders of integration, and replacing ξ by $\xi + \mathbf{x}$, we get

$$u(\mathbf{x}, t) = \frac{1}{c(2\pi)^d}\int_X h(\mathbf{x}+\xi)\,d\xi \int_X (e^{i\omega\cdot\xi})\,\frac{\sin \rho c t}{\rho}\,d\omega. \tag{47}$$

At this point, calculations are facilitated by introducing $r = |\xi|$ with associated "polar" coordinates, and the results are dimensionally dependent. For $d = 1$, we recover d'Alembert's formula (see Problem 8.3I), but for $d = 2$, the analysis is more complicated. Surprisingly, when $d = 3$, it can be carried out (see Problem 8.3J) and this gives us **Poisson's mean-value formula** of 1818

$$u(\mathbf{x}, t) = \frac{t}{4\pi}\int_{S_1} h(\mathbf{x} + ct\sigma)\,dS, \qquad t > 0, \tag{48}$$

where σ denotes a point on S_1, the surface of the unit sphere in $\mathbb{R}^3$.

This derivation of Poisson's formula relies on several formal mathematical operations of questionable validity. Instead of attempting to justify each of these, we can show directly that the function $u(\mathbf{x}, t)$ so defined provides a solution to the problem. We see readily that it satisfies the given initial conditions (44), since $\int_{S_1} 1\,dS = 4\pi$, and with some work we can prove that it is a solution of the wave equation when h is C^2. Then, since u_t also satisfies this equation, it follows that for $t > 0$

$$u(\mathbf{x}, t) \overset{\text{def}}{=} \frac{\partial}{\partial t}\left[\frac{t}{4\pi}\int_{S_1} f(\mathbf{x} + ct\sigma)\,dS\right] + \frac{t}{4\pi}\int_{S_1} h(\mathbf{x} + ct\sigma)\,dS \tag{49}$$

will satisfy the three-dimensional wave equation and the initial conditions

$$\begin{aligned} u(\mathbf{x}, 0) &= f(\mathbf{x}) \\ u_t(\mathbf{x}, 0) &= h(\mathbf{x}) \end{aligned}, \qquad \text{when } f \text{ and } h \text{ are in } C^2(\mathbb{R}^3);$$

u depends continuously on these conditions when $0 \le t \le T$ ($T < +\infty$). (See Problem 8.3K.)

In Section 6.4 we saw that the three-dimensional wave equation controls the propagation of sound in space. Therefore, formula (49) supports **Huyghen's principle** that at time $t > 0$, we hear only what was generated initially a distance

ct from us. In spatial dimension $d \geq 2$ this *sharp* principle for the wave equation holds iff d is odd, and then there is a formula analogous to (49) that can be used to obtain a solution in dimension $d-1$ by the "method of descent." (See [Ga], [B-C].) The principle is invalid when $d = 2$; this is evidenced by the series of ripples caused by dropping a stone into a small pond and is confirmed by the solution obtained in Problem 8.3M. The anomalous situation in dimension one is indicated by d'Alembert's formula, which shows that the sharp principle holds iff the initial velocity $u_t(x, 0) \equiv 0$.

Problem Set 8.3

8.3A Find the Fourier transform of the function

$$f(x) = \begin{cases} 1, & a \leq x \leq b \\ 0, & \text{otherwise} \end{cases}$$

where a and b are finite. Can this f be recovered from its transform? Explain and make similar analyses of the other functions of Problem 8.2B.

8.3B Let f be absolutely integrable on $\mathbb{R}$ with sine and cosine transforms, s and c, respectively.

1. Show that $\hat{f}(\omega) = \pi c(\omega)$, if f is even and that $\hat{f}(\omega) = \pi i s(\omega)$, if f is odd.
2. Conclude that in general $\hat{f}(\omega) = \pi[c(\omega) + is(\omega)]$, and use this fact to find the Fourier transform of the function of Example 2 in Section 8.1.

8.3C Show that the Fourier series generated by a function f can be written in the complex form of (39).

8.3D If f, f^2, g, and g^2 are absolutely integrable on $\mathbb{R}$, use (37) and the Schwarz inequality to prove (37′). (*Hint*: See Problem 1.5F.)

8.3E If f, f^2, f', and f'^2 are absolutely integrable and $f(\pm\infty) = 0$, show that $\int_{-\infty}^{\infty} \omega |\hat{f}|^2(\omega)\, d\omega = 0$.

8.3F If f, f^2, g, and g^2 are absolutely integrable on $\mathbb{R}$, we can define the *convolution* of f and g by

$$(f * g)(x) = \int_{-\infty}^{\infty} f(\xi)\, g(x - \xi)\, d\xi, \qquad x \in \mathbb{R}.$$

1. Show that $g * f = f * g$.
2. Use formal manipulations to show that $(f * g)\hat{} = \hat{f}\hat{g}$.

8.3G

1. Let $\mathbf{x} = (x_1, x_2)$ and $\boldsymbol{\omega} = (\omega_1, \omega_2)$ be points in $\mathbb{R}^2$. Show that if f is continuous and absolutely integrable over $\mathbb{R}^2$, then

$$\hat{f}(\boldsymbol{\omega}) = \int_{-\infty}^{\infty} \int_{-\infty}^{\infty} e^{i\boldsymbol{\omega}\cdot\mathbf{x}} f(\mathbf{x})\, dx_1\, dx_2$$

has the formal inverse $f(\mathbf{x}) = (2\pi)^{-2} \int_{-\infty}^{\infty} \int_{-\infty}^{\infty} \hat{f}(\boldsymbol{\omega}) e^{-i\boldsymbol{\omega}\cdot\mathbf{x}}\, d\omega_1\, d\omega_2$. Give conditions on f that will validate this result. (*Hint*: Consider componentwise.)

2. Explain how to generalize this analysis to $\mathbb{R}^3$ and to $\mathbb{R}^d$, for $d > 3$.

8.3H 1. If $f \in C^1(\mathbb{R}^2)$, show that $(f_{x_1})\hat{}(\boldsymbol{\omega}) = -i\omega_1 \hat{f}(\boldsymbol{\omega})$, $\boldsymbol{\omega} \in \mathbb{R}^2$, provided that f is absolutely integrable on $\mathbb{R}^2$ and dies out appropriately as $|\mathbf{x}| \to \infty$.

2. If $f \in C^2(\mathbb{R}^2)$, give nontrivial conditions on f under which formula (38′) holds.

8.3I 1. When $d = 1$ in Example 7, show that for $c = 1$, the formal solution (47) is

$$u(x, t) = \frac{1}{2\pi}\int_{-\infty}^{\infty} h(x+\xi)\, d\xi \int_{-\infty}^{\infty} e^{i\omega\xi}\,\frac{\sin \omega t}{\omega}\, d\omega.$$

2. Use trigonometric identities and substitutions to show that the *inner* integral vanishes unless $\xi^2 \le t^2$, in which case it equals $2\int_0^\infty [(\sin s)/s]\, ds = \pi$. (See (22) and (23) in Section 8.1.)

8.3J 1. When $d = 3$ in Example 7, show that the formal solution is given by

$$u(\mathbf{x}, t) = \lim_{R\to\infty} \frac{1}{c(2\pi)^2}\int_X h(\mathbf{x}+\boldsymbol{\xi})\, d\boldsymbol{\xi} \int_0^R \rho \sin \rho c t\, d\rho \int_0^\pi e^{ir\rho\cos\psi} \sin\psi\, d\psi,$$

where ψ is the angle between $\boldsymbol{\omega}$ and $\boldsymbol{\xi}$, and $r = |\boldsymbol{\xi}|$. (*Hint*: Express $\boldsymbol{\omega}$ in spherical coordinates (ρ, ψ, ν).)

2. Evaluate the inner integral to obtain

$$u(\mathbf{x}, t) = \frac{1}{2c\pi^2} \lim_{R\to\infty}\int_0^R \sin \rho c t\, d\rho \int_X h(\mathbf{x}+\boldsymbol{\xi})\,\frac{\sin r\rho}{r}\, d\boldsymbol{\xi}$$

$$= \frac{1}{2c\pi^2}\int_0^{2\pi} d\theta \int_0^\pi \sin\phi\, d\phi \int_0^\infty \sin\rho c t\, d\rho \int_0^\infty h(\mathbf{x}+r\boldsymbol{\sigma}) r \sin r\rho\, dr,$$

where $\boldsymbol{\sigma}$ has spherical coordinates $(1, \phi, \theta)$.

3. Conclude that as in the text

$$u(\mathbf{x}, t) = \frac{t}{4\pi}\int_{S_1} h(\mathbf{x}+ct\boldsymbol{\sigma})\, dS.$$

(*Hint*: Use the inversion formula for the *sine* transform of $h(\mathbf{x}+r\boldsymbol{\sigma})r$.)

8.3K When $c = 1$, denote the previous integral by v; i.e.,

$$v(\mathbf{x}, t) = \int_{S_1} h(\mathbf{x}+t\boldsymbol{\sigma})\, dS, \qquad t > 0,\ \mathbf{x} \in \mathbb{R}^3.$$

1. Show that if h is C^2, then

$$v_t(\mathbf{x}, t) = \int_{S_1} \nabla h(\mathbf{x}+t\boldsymbol{\sigma})\cdot\mathbf{n}\, dS = t^{-2}\int_{S_t(\mathbf{x})} \nabla h\cdot\mathbf{n}\, dS = t^{-2}\int_{V_t(\mathbf{x})} \nabla^2 h\, dV,$$

where $V_t(\mathbf{x})$ is the sphere of radius t centered at $\mathbf{x}$ with boundary $S_t(\mathbf{x})$. (*Hint*: Use the divergence theorem.)

2. Express the last integral in spherical coordinates (centered at $\mathbf{x}$), and differentiate once more to obtain

$$v_{tt} = -\frac{2}{t}v_t + t^{-2}\int_{S_t(\mathbf{x})} \nabla^2 h\, dS = -\frac{2}{t}v_t + \nabla^2 v.$$

3. Conclude that for $t > 0$, $u(\mathbf{x}, t) = tv(\mathbf{x}, t)$ satisfies the standard wave equation (for $c = 1$) with $u(\mathbf{x}, 0) = 0$ and $u_t(\mathbf{x}, 0) = 4\pi h(\mathbf{x})$.
4. Verify that when h is C^3, then $w = u_t = (tv)_t$ also satisfies this wave equation with $w(\mathbf{x}, 0) = 4\pi h(\mathbf{x})$ and $w_t(\mathbf{x}, 0) = 0$. Show how this result leads to (49).

8.3L 1. When $\mathbf{x} \in \mathbb{R}^3$, show that the **retarded potential**

$$u(\mathbf{x}, t) = \frac{1}{4\pi}\int_0^t \tau\, d\tau \int_{S_1} F(\mathbf{x} + \tau\boldsymbol{\sigma}, t - \tau)\, dS$$

solves the inhomogeneous wave equation $u_{tt} - \nabla^2 u = F$, $t > 0$ when $F = F(\mathbf{x}, t)$ is prescribed and C^2, with $u(\mathbf{x}, 0) = u_t(\mathbf{x}, 0) = 0$. (*Hint*: See Problem 5.6I and equation (48).)

2. Verify that

$$u(\mathbf{x}, t) = \frac{1}{4\pi}\int_{V_t} \frac{F(\boldsymbol{\xi}, t - \tau)}{|\mathbf{x} - \boldsymbol{\xi}|}\, d\boldsymbol{\xi}$$

where V_t is the sphere of radius t centered at the origin.

8.3M 1. If $h(\mathbf{x}) = h(x, y)$ only, in Problem 8.3K, and $(\alpha, \beta, \gamma) \in S_1$ (the surface of the unit sphere in $\mathbb{R}^3$), then show that

$$u(x, y, t) = \frac{t}{4\pi}\int_{S_1} h(x + t\alpha, y + t\beta)\, dS$$

is a solution of the *two*-dimensional wave equation $u_{tt} = \nabla^2 u$ with $u(x, y, 0) = 0$ and $u_t(x, y, 0) = h(x, y)$.

2. Verify that with coordinates $\xi = t\alpha$, $\eta = t\beta$,

$$u(x, y, t) = \frac{1}{2\pi}\iint_{D_t} \frac{h(x + \xi, y + \eta)}{\sqrt{t^2 - \xi^2 - \eta^2}}\, d\xi\, d\eta,$$

where D_t is the *disk* of radius t centered at the origin.

3. Write an integral that solves the corresponding nonhomogeneous problem $v_{tt} = \nabla^2 v + F$ in $\mathbb{R}^3$, with $v(x, y, 0) = v_t(x, y, 0) = 0$. (*Hint*: See Problems 5.6I and 8.3L.)

8.3N Show that when h and u are as in Problem 8.3M, then for constant a, $v(x, y, z, t) = e^{-az}u(x, y, t)$ satisfies the equation $v_{tt} = \nabla^2 v + a^2 v$, with $v(x, y, z, 0) = 0$ and $v_t(x, y, z, 0) = e^{-az}h(x, y)$.

8.4*

Other Integral Transforms

Local behavior of linear partial differential equations and their fundamental solutions can be obtained by examining the Fourier transform of the equations. (See [Ho], Tre].) However, solutions of specific boundary value problems for these equations require transforms that are tailored to the problems. In this section we look at a few of the more important integral transformations: those of Fourier, Laplace, Hankel, and Mellin. For each we need a supply of explicit transforms of known functions and theoretical assurance that an inversion formula exists. As we shall see, the latter formula can sometimes be found by manipulating Fourier's integral formula, although other approaches are conceivable.

In Section 7.7 we saw that finite integral transforms can provide solutions $u = u(x, t)$ to boundary value problems involving the equation

$$\mathscr{L}(u) \overset{\text{def}}{=} (\tau u_x)_x - qu - \rho[Ku_{tt} + K_1 u_t] = 0, \tag{50}$$

on finite intervals (a, b). The results are expressed in terms of appropriate solutions $y(x, \lambda)$ to an S–L problem for the equation

$$L(y) \overset{\text{def}}{=} (\tau y')' - qy = -\lambda \rho y. \tag{51}$$

In some instances, solutions to equation (51) are valid on all of $\mathbb{R}$, and it is natural to ask whether corresponding integral transforms might be useful in solving problems on unbounded intervals. For simplicity, we just consider bounded solutions for $x \geq 0$, $t \geq 0$, but the approach is valid more generally.

Again, we suppose that "enough" solutions $y(x, \lambda)$ to (51) are at hand and consider for positive ρ the associated **S–L transform**

$$\hat{u}(t, \lambda) \overset{\text{def}}{=} \int_0^\infty \rho u(\cdot, t) y(\cdot, \lambda)\, dx, \tag{52}$$

assuming that this integral converges absolutely.

Proceeding formally, with differentiations, substitutions, and partial integrations exactly as before we obtain, for each t, that

$$K\ddot{\hat{u}} + K_1\dot{\hat{u}} + \lambda \hat{u} = \tau(u_x y - uy')\big|_{x=0}^{x=\infty} \tag{53}$$

$$= 0, \tag{53'}$$

when, e.g., $\tau y(x) \to 0$ as $x \searrow 0$, $u(0, t) = 0$, and the terms on the right in (53) vanish as $x \to +\infty$.

If we want a solution to (50) with $u(x, 0) = f(x)$ prescribed, then from (52), we see that

$$\hat{u}(0, \lambda) = \int_0^\infty \rho f y(\cdot, \lambda)\, dx = \hat{f}(\lambda), \text{ say,} \tag{54}$$

and, in principle, we can find solutions to (53′) that satisfy (54) and approach zero as $t \to +\infty$. Thus, we can suppose that $\hat{u}(t; \lambda)$ is known for each admissible λ.

The obvious questions concern how (or whether) u may be recovered from its transform (52). In particular, is it possible that a solution can be obtained by superposition in the integral form

$$U(x, t) = \int y(x, \lambda)\, \hat{u}(t, \lambda)\, d\mu(\lambda), \tag{55}$$

for some integration element $d\mu$ over some λ-interval?

If we simply differentiate (55) as required, we see that, formally, for $t > 0$, $x > 0$

$$\mathscr{L}(U) = \int \{[(\tau y')' - qy]\hat{u}(t, \cdot) - \rho(x)(K\ddot{\hat{u}} + K_1\dot{\hat{u}})y\} \, d\mu(\lambda)$$

$$= -\rho(x) \int (K\ddot{\hat{u}} + K_1\dot{\hat{u}} + \lambda\hat{u}) y(x, \lambda) \, d\mu(\lambda)$$

$$= 0, \text{ under the conditions used to derive } (53').$$

Hence U as defined by (55) is a formal solution to equation (50). Moreover, $U(0, t) = 0$, $t > 0$, if we require $y(0, \lambda) = 0$. Finally, to have $U(x, 0) = f(x)$, then in view of (54), we want

$$f(x) = \int y(x, \lambda)\, \hat{u}(0, \lambda) \, d\mu(\lambda) = \int y(x, \lambda) \hat{f}(\lambda) \, d\mu(\lambda), \tag{56}$$

$$\text{where} \quad \hat{f}(\lambda) = \int_0^\infty \rho y(\cdot, \lambda) f \, d\xi \tag{56'}$$

defines the eigenfunction transform of f. If we can determine $d\mu$ (and the associated λ-interval) to satisfy (56), then, in principle, formula (55) provides a formal solution to our problem.

(a) Fourier sine transform: Let's use this approach to find a bounded solution $u = u(x, t)$ to the wave equation

$$\mathscr{L}(u) = u_{xx} - u_{tt} = 0, \ x > 0, \ t > 0,$$

with $\quad u(0, t) = 0, \ t > 0; \quad$ and $\quad u(x, 0) = f(x), \ x > 0.$

In this case, (51) becomes

$$L(y) = y'' = -\lambda y$$

and with $y(0, \lambda) = 0$, we can take $y(x; \lambda) = \sin \sqrt{\lambda} x$, for $\lambda > 0$. Then we wish to find $d\mu(\lambda)$ so that (56) becomes

$$f(x) = \int \sin \sqrt{\lambda} x \, d\mu(\lambda) \int_0^\infty f(\xi) \sin \sqrt{\lambda} \xi \, d\xi. \tag{57}$$

By comparison with (33) and (33′), we see that we want

$$d\mu(\lambda) = \frac{2}{\pi} d(\sqrt{\lambda}), \qquad 0 < \lambda < +\infty.$$

Then, with $\omega = \sqrt{\lambda}$, formula (57) takes the more familiar form

$$f(x) = \frac{2}{\pi} \int_0^\infty \sin \omega x \, d\omega \int_0^\infty f(\xi) \sin \omega \xi \, d\xi,$$

which is valid when f is continuous and absolutely integrable on $\mathbb{R}$, with derivatives from the right and left at each $x \in \mathbb{R}$. From (53′) it follows that for $\lambda > 0$, the transform $\hat{u}(t, \lambda)$ satisfies the equation

$$\ddot{\hat{u}} + \omega^2 \hat{u} = 0, \qquad \text{with} \qquad \hat{u}(0, \lambda) = \int_0^\infty f(\xi) \sin \omega \xi \, d\xi = \hat{f}(\lambda).$$

This problem from ordinary differential equations has the solution

$$\hat{u}(t, \lambda) = \hat{f}(\lambda) \cos \omega t + s(\lambda) \sin \omega t, \tag{58}$$

where $s(\lambda)$ is arbitrary. If we take $s(\lambda) = 0$, we see that

$$U(x, t) = \frac{2}{\pi}\int_0^\infty \sin \omega x \hat{f}(\omega^2) \cos \omega t \, d\omega = \frac{2}{\pi}\int_0^\infty \hat{f}(\omega^2) \sin \omega x \cos \omega t \, d\omega$$

should supply a complete solution to the original problem. Observe that it can be written in the d'Alembert form (see Section 5.6)

$$U(x, t) = \frac{1}{\pi}\int_0^\infty \hat{f}(\omega^2) \sin \omega(x - t) \, d\omega + \frac{1}{\pi}\int_0^\infty \hat{f}(\omega^2) \sin \omega(x + t) \, d\omega.$$

$$\text{Here, } U_t(x, 0) = -\frac{2\omega}{\pi}\int_0^\infty \hat{f}(\omega^2) \sin \omega x \sin \omega t \, d\omega \bigg|_{t=0} = 0,$$

but, in principle, $s(\lambda)$ in (58) could be chosen to make U_t satisfy a different condition at $t = 0$. (See Problem 8.4A.)

Similar analysis generates the Fourier cosine transform. (See Problem 8.4B.)

(b) Hankel transforms: Water rushing through a long straight circular pipe produces sounds that might be described by axially symmetric solutions of the standardized wave equation in $\mathbb{R}^3$ (Section 6.4). Thus, as in Example 5 of Section 7.4, we need solutions $u = u(r, t)$ of the equation

$$\mathscr{L}(u) \stackrel{\text{def}}{=} (ru_r)_r - ru_{tt} = 0 \qquad \text{for} \qquad r > 0, t > 0, \tag{59}$$

and to obtain them, we look for eigenfunctions $y = y(r, \lambda)$ of the equation

$$L(y) \stackrel{\text{def}}{=} (ry')' = -\lambda ry, \qquad r > 0. \tag{60}$$

For $\lambda > 0$, the substitution, $x = \sqrt{\lambda}r$, transforms equation (60) into Bessel's equation of order zero, and we obtain the solutions

$$y(r, \lambda) = J_0(\omega r) \qquad \text{for} \qquad r \geq 0, \lambda = \omega^2 \geq 0.$$

Thus from (56) and (56′), to have $u(r, 0) = f(r)$, we must find $d\mu(\lambda)$ to make

$$f(r) = \int J_0(\omega r) \, d\mu(\lambda) \int_0^\infty \xi J_0(\omega\xi) f(\xi) \, d\xi. \tag{61}$$

By use of the two-dimensional Fourier transform in (38) and Bessel's integral representation formula

$$J_m(x) = \frac{1}{\pi}\int_0^\pi \cos(m\theta - x \sin \theta) \, d\theta, \qquad m = 0, 1, 2, \ldots, \tag{62}$$

it can be shown that (61) holds when $d\mu(\lambda) = \omega \, d\omega$ for $\omega > 0$. (See Problems 8.4C and D and Section 8.6.) Therefore,

$$f(r) = \int_0^\infty \omega J_0(\omega r) \, d\omega \int_0^\infty \xi J_0(\omega\xi) f(\xi) \, d\xi,$$

if f is continuous with $\sqrt{r} f(r)$ absolutely integrable on $(0, \infty)$ and with right and left derivatives at each $r > 0$.

Thus, we define the **Hankel transform of order 0** of f by

$$\check{f}(\omega) = \int_0^\infty \xi J_0(\omega\xi) f(\xi) \, d\xi$$

and, as in the previous example, take

$$\check{u}(t, \omega) = \check{f}(\omega) \cos \omega t.$$

Then from (55) we see that

$$U(r, t) = \int_0^\infty \omega J_0(\omega r)\, \check{u}(t, \omega)\, d\omega$$

$$= \int_0^\infty \omega \check{f}(\omega) J_0(\omega r) \cos \omega t \, d\omega$$

provides a formal solution to $\mathscr{L}(u) = 0$ for $r > 0$, $t > 0$, with $U(r, 0) = f(r)$ and $U_t(r, 0) = 0$.

Although it is nontrivial to calculate $\check{f}$ for any function $f \neq 0$, tables of Hankel transforms are available, and this method can be carried out in a few cases of interest. (See [Sn].)

Through the same techniques, we can also establish that if f has a **Hankel transform of order m** $= 0, 1, 2, \ldots$ given by

$$\check{f}_m(\omega) = \int_0^\infty \xi f(\xi) J_m(\xi\omega)\, d\xi, \qquad \omega \in \mathbb{R},$$

then under the above conditions,

$$f(r) = \int_0^\infty \omega \check{f}_m(\omega) J_m(r\omega)\, d\omega, \qquad r > 0. \tag{63}$$

Therefore, for each $m = 0, 1, \ldots$, $U_m(r, t) = \int_0^\infty \check{f}_m(\omega) J_m(r\omega) \cos \omega t \, d\omega$ provides a formal solution to the problem

$$\mathscr{L}_m(u) \stackrel{\text{def}}{=} (ru_r)_r - \frac{m^2}{r} u - ru_{tt} = 0 \tag{63$'$}$$

$$\text{with} \qquad u(r, 0) = f(r) \qquad \text{and} \qquad u_t(r, 0) = 0.$$

These results supply *non*-axisymmetric solutions of the two-dimensional wave equation $\nabla^2 u - u_{tt} = 0$, since from periodicity (in θ), we get elementary product solutions $u(r, \theta, t) = u_m(r, t)(c_m \cos m\theta + s_m \sin m\theta)$ where u_m satisfies (63$'$). Hence we look for solutions in the form

$$U(r, \theta, t) = \sum_{m=0}^{\infty} U_m(r, t)(c_m \cos m\theta + s_m \sin m\theta),$$

$$\text{where} \qquad U(r, \theta, 0) = F(r, \theta) = \sum_{m=0}^{\infty} \check{f}_m(r)(c_m \cos m\theta + s_m \sin m\theta),$$

and now we would use a combination of the above methods and Fourier series considerations to find the $\check{f}_m$ and thereby U_m.

In the preceding transforms, λ is positive. Positivity is not essential, as the next transforms show, but there are complications.

(c) Laplace transform: To solve Laplace's equation

$$\mathscr{L}(u) \stackrel{\text{def}}{=} u_{xx} + u_{tt} = 0$$

in the quarter-plane $\{x > 0, t > 0\}$; with

$$u(x, 0) = f(x) \text{ and } u(x, t) \to 0 \text{ as } x \to +\infty,$$

we seek appropriate solutions to the equation

$$L(y) \stackrel{\text{def}}{=} y'' = -\lambda y.$$

If we take $\lambda > 0$, this leads to $y(x, \lambda) = \sin \sqrt{\lambda}(x - \alpha)$, which cannot approach zero as $x \to +\infty$. However if we take $\lambda = -s^2$ for real $s > 0$, then $y(x; s) = e^{-sx}$ is an admissible eigenfunction.

Therefore, we define the **Laplace transform** of f by

$$\tilde{f}(s) = \int_0^\infty e^{-s\xi} f(\xi)\, d\xi, \tag{64}$$

and, as in (56), we look for a $d\mu(s)$ that gives us

$$f(x) = \int e^{-sx} \tilde{f}(s)\, d\mu(s). \tag{65}$$

From (35) and (64) we see that $\tilde{f}(s)$ may be regarded as the Fourier transform $\hat{f}$ of f ($\equiv 0$ on $(-\infty, 0)$) evaluated at $\omega = is$. Hence, we might expect a related inversion formula.

In fact, if f is absolutely integrable on $(0, +\infty)$ and continuous, with right and left derivatives, then

$$f(x) = \frac{1}{2\pi} \int_{-\infty}^{\infty} e^{-i\omega x} \hat{f}(\omega)\, d\omega = \frac{1}{2\pi} \int_{-\infty}^{\infty} e^{-i\omega x} \tilde{f}(-i\omega)\, d\omega.$$

Upon substituting $s = -i\omega$ or $\omega = is$, we find that

$$f(x) = \frac{1}{2\pi i} \int_{-i\infty}^{i\infty} e^{sx} \tilde{f}(s)\, ds = \frac{1}{2\pi i} \int_{-i\infty}^{+i\infty} e^{-sx} \tilde{f}(-s)\, ds, \tag{66}$$

where these last integrals are evaluated along the *imaginary axis* in the complex plane.

Remarks: This last feature complicates the use of the Laplace transform. However, because of the exponential factor in (64), the Laplace transform $\tilde{f}$ can be defined for a function f that is *not* absolutely integrable on $(0, +\infty)$. It suffices if for large $\xi > 0$, $|f(\xi)| \leq Me^{a\xi}$ for some constants M, a. Then when $A > a$, $g(\xi) = e^{-A\xi} f(\xi)$ *is* absolutely integrable on $\mathbb{R}$ if $f(\xi) = 0$, $\xi < 0$; also, g has the Fourier transform $\hat{g}(\omega) = \tilde{f}(A - i\omega)$. (Why?) Hence if f is continuous with right and left derivatives at x, then by (35′),

$$e^{-Ax} f(x) = \frac{1}{2\pi} \int_{-\infty}^{\infty} e^{-i\omega x} \tilde{f}(A - i\omega)\, d\omega,$$

or with $s = A - i\omega$:

$$f(x) = \frac{1}{2\pi i} \int_{A-i\infty}^{A+i\infty} e^{sx} \tilde{f}(s)\, ds, \tag{67}$$

where again, the limits on the last integral indicate how it is to be evaluated.

Extensive tables of Laplace transforms and their inverses have been obtained (see, for example, [Do]), which facilitate applications.

Suppose that f itself is absolutely integrable and that (66) applies. Comparison with (56) shows that we still do not have what we wished. However, in this case equation (53′) reduces to

$$-\ddot{\tilde{u}} - s^2\tilde{u} = 0,$$

and its solution with $\tilde{u}(0, s) = \tilde{f}(s)$ is given by

$$\tilde{u}(t, s) = \tilde{f}(s) \cos st.$$

Then

$$U(x, t) \stackrel{\text{def}}{=} \frac{1}{2\pi i}\int_{-i\infty}^{+i\infty} e^{-sx}\tilde{f}(-s) \cos st \, ds$$

will supply a formal solution to our problem, with $U_t(x, 0) = 0$.

If we use Euler's formula to express

$$\cos st = \frac{e^{ist} + e^{-ist}}{2},$$

then from (66) we see that

$$U(x, t) = \tfrac{1}{2}\left[f(x + it) + f(x - it)\right], \tag{68}$$

assuming that the right side is defined. This result, which resembles d'Alembert's formula of Section 5.6, suggests an important analogy. The formal substitution $\tau = it$ transforms Laplace's equation

$$u_{xx} + u_{tt} = 0$$

into the wave equation

$$\mathscr{u}_{xx} - \mathscr{u}_{\tau\tau} = 0,$$

whose solution has the d'Alembert form

$$\mathscr{u}(x, \tau) = \phi(x + \tau) + \psi(x - \tau).$$

When $\mathscr{u}(s, 0) = u(s, 0) = f(x)$ and $\mathscr{u}_\tau(x, 0) = -iu_t(x, 0) = 0$, then,

$$\mathscr{u}(x, \tau) = \frac{f(x + \tau) + f(x - \tau)}{2}.$$

Thus we could have anticipated (68), or more generally, that in the plane, harmonic functions $u = u(x, y)$ must take the form

$$u(x, y) = \phi(x + iy) + \psi(x - iy)$$

for appropriate functions ϕ and ψ. It can be shown that ϕ and ψ are complex analytic functions of $z = x + iy$ and $\bar{z} = x - iy$, respectively, giving further evidence of the link between planar harmonic functions and complex analytic functions. (Recall Section 4.4.)

(d) Mellin transform: To find harmonic functions $u = u(r, \theta)$ in polar coordinates in the quarter-plane $\{0 < \theta < \pi/2\}$ with $u(r, 0) = f(r)$, we seek solutions of Laplace's equation in the form

$$\mathcal{L}(u) \stackrel{\text{def}}{=} (ru_r)_r + \frac{1}{r}\, u_{\theta\theta} = 0.$$

Hence, we need eigenfunctions $y(r, \lambda)$ of the equation

$$L(y) \stackrel{\text{def}}{=} (ry')' = -\frac{\lambda}{r}\, y, \qquad \text{or} \qquad r^2 y'' + ry' + \lambda y = 0.$$

This equidimensional equation has solutions

$$y(r; s) = r^s \qquad \text{or} \qquad r^{-s}, \qquad \text{for } \lambda = -s^2 \neq 0,$$

which will be of the form $r^{i\omega}$ or $r^{-i\omega}$ if $\lambda = \omega^2 > 0$.

If $s > 0$, only $y(r; s) = r^s$ is finite at $r = 0$, and since $\rho(r) = 1/r$, we introduce the **Mellin transform** of f given by

$$\breve{f}(s) = \int_0^\infty r^{s-1} f(r)\, dr = \int_0^\infty e^{(s-1)\log r} f(r)\, dr. \tag{69}$$

If we substitute $r = e^\xi$, and $g(\xi) = f(e^\xi)$, then we see that

$$\breve{f}(s) = \int_{-\infty}^\infty e^{\xi s} g(\xi)\, d\xi, \qquad s > 0,$$

provided that this integral exists. Now, if g is absolutely integrable on $(-\infty, \infty)$ and is continuous with right and left derivatives, then for $s = i\omega$, it has the Fourier transform (35)

$$\hat{g}(\omega) = \int_{-\infty}^\infty e^{i\omega\xi} g(\xi)\, d\xi = \breve{f}(i\omega), \qquad \omega \in \mathbb{R},$$

and inverse

$$f(e^x) = g(x) = \frac{1}{2\pi}\int_{-\infty}^\infty e^{-i\omega x}\, \breve{f}(i\omega)\, d\omega = \frac{1}{2\pi i}\int_{-i\infty}^{+i\infty} e^{-sx}\, \breve{f}(s)\, ds.$$

With $r = e^x$, the inverse of (69) is given by

$$f(r) = \frac{1}{2\pi i}\int_{-i\infty}^{+i\infty} r^{-s}\breve{f}(s)\, ds = \frac{1}{2\pi i}\int_{-i\infty}^{+i\infty} r^s\, \breve{f}(-s)\, ds. \tag{70}$$

Hence (as in the previous example), we solve $-\ddot{\breve{u}} - s^2\breve{u} = 0$, with $\breve{u}(0, s) = \breve{f}(s)$, to obtain $\breve{u}(\theta, s) = \breve{f}(s)\cos s\theta$ and see that

$$U(r, \theta) = \frac{1}{2\pi i}\int_{-i\infty}^{i\infty} r^s\, \breve{f}(-s)\cos s\theta\, ds$$

will provide a formal solution to our problem, with $U_\theta(r, 0) = 0$.

The strong requirement that $f(e^x)$ be absolutely integrable on $(-\infty, \infty)$ may be relaxed by moving the line of integration as in the remarks following (66).

Problem Set 8.4

8.4A To have a formal solution $U = U(x, t)$ to the equation

$$\mathcal{L}(u) = u_{xx} - u_{tt} = 0 \qquad \text{in} \qquad \{x > 0,\ t > 0\}$$

satisfying $u(0, t) = 0$, $t > 0$ and $u(x, 0) = f(x)$, $u_t(x, 0) = h(x)$, $x > 0$, how should $s(\lambda)$ in (58) be chosen? What is the resulting U?

8.4B How is the argument in Problem 8.4A changed if we want a solution u with $u_x(0, t) = 0$?

8.4C (*Bessel's integral formula*) Let $y(x) = \int_0^\pi \cos(x \sin\theta)\, d\theta$, $x \in \mathbb{R}$.

1. Show that $y(0) = \pi$ and $y'(0) = 0$.
2. Verify that

$$-[xy'(x)]' = \int_0^\pi \cos(x \sin\theta) x \sin^2\theta\, d\theta + \int_0^\pi \sin(x \sin\theta)\sin\theta\, d\theta = xy(x).$$

(*Hint*: Integrate the *second* term by parts.)

3. Conclude that $y(x) = \pi J_0(x)$, which gives (62) for $m = 0$.
4. Verify that if $y_m(x) = \int_0^\pi \cos(m\theta - x\sin\theta)\, d\theta$, then

$$xy_{m+1}(x) - my_m(x) + xy_m'(x) = \int_0^\pi \cos(m\theta - x\sin\theta)(x\cos\theta - m)\, d\theta$$
$$= 0, \text{ if } m \text{ is an integer.}$$

5. Conclude that (62) holds for $m = 0, 1, 2, \ldots$. (*Hint*: Use the first recursion formula of Problem 7.4F.)

8.4D* (*The Hankel inversion formula for fixed* $m = 0, 1, 2, \ldots$)

1. Use (62) to show that

$$\check{f}_m(\rho) \stackrel{\text{def}}{=} \int_0^\infty rf(r)J_m(r\rho)\, dr = \int_0^\infty r\, dr \int_{-\pi}^\pi d\theta\, g(x_1, x_2)e^{-i\rho x_2} = \hat{g}(0, -\rho),$$

where $\hat{g}$ is the Fourier transform of $g(x_1, x_2) = f(r)e^{im\theta}/2\pi$, relative to *polar coordinates* (r, θ) such that $x_1 = r\cos\theta$ and $x_2 = r\sin\theta$.

2. By similar means, verify that, in general,

$$\hat{g}(\omega_1, \omega_2) = \check{f}_m(\rho)e^{im[\phi + (\pi/2)]}$$

when expressed in polar coordinates ρ, ϕ where $\omega_1 = \rho\cos\phi$, $\omega_2 = \rho\sin\phi$.

3. Conclude that under reasonable conditions on f, we should have

$$f(r)e^{im\theta} = (2\pi)^{-1}\int_{-\infty}^\infty \int_{-\infty}^\infty \hat{g}(\omega_1, \omega_2)e^{-i\boldsymbol{\omega}\cdot\mathbf{x}}\, d\omega_1\, d\omega_2$$
$$= (2\pi)^{-1}\int_0^\infty \rho\, d\rho \int_{-\pi}^\pi d\phi\, \check{f}_m(\rho)e^{im[\phi + (\pi/2)]}e^{-i\rho r\cos(\phi - \theta)},$$

so that when $\theta = \pi/2$, we obtain

$$f(r) = (2\pi)^{-1}\int_0^\infty \rho\check{f}_m(\rho)\, d\rho \int_{-\pi}^\pi \cos(m\phi - \rho r\sin\phi)\, d\phi,$$

and now (63) follows from (62).

8.4E Let $f(r) = \begin{cases} r^m, & 0 < r < a \\ 0, & r > a \end{cases}$.

1. Use the second recurrence relation in Problem 7.4F to show that f has the Hankel transform

$$\check{f}_m(\rho) = \rho^{-1}a^{m+1}J_{m+1}(\rho a), \qquad \rho > 0, \qquad m = 0, 1, 2, \ldots.$$

2. What integral follows from the inversion formula (63)?
3. When $m = 0$, obtain the corresponding formal solution $U(r, t)$ to the problem of (59) with $u(r, 0) = f(r)$ and $u_t(r, 0) = 0$.
4. What would you change in order to have a solution with $u_t(r, 0) = h(r)$?

8.4F 1. Verify that when absolutely integrable, $g(x) = e^{-Ax} f(x)$ has the Fourier transform $\hat{g}(\omega) = \tilde{f}(A - i\omega)$ where $\tilde{f}$ is the Laplace transform of f defined as in (64).
2. Show that $-\tilde{f}'(s)$ is the Laplace transform of $xf(x)$. How would you obtain the Laplace transform of $x^k f(x)$? when $k = 2, 3, \ldots$? when $k = -1$?
3. When f is C^1, show that for $s > A$ the Laplace transform of f' is $s\tilde{f}(s) - f(0)$. What is the transform of f''? of $f^{(k)}$ (assuming sufficient differentiability of f)?

8.4G In Problem 8.3F, let $h = f * g$. Write h when $f(x) = g(x) = 0$, for $x < 0$, and relate its Laplace transform to those of f and g.

8.5

Uniqueness of Solutions in Unbounded Regions

In the preceding sections we used integral transforms to find formal solutions to standard boundary value problems in unbounded regions. Even when these are actual solutions to the problems it does not follow that they are unique. For example, $u_0(x, y) = 0$, $u_1(x, y) = y$, and $u_2(x, y) = xy$ satisfy Laplace's equation in the upper half-plane $(y > 0)$ and vanish on the boundary $(y = 0)$; the associated boundary value problem does not have a unique solution, in a sense, because we have not imposed boundary requirements "at infinity." Observe that among these solutions only u_0 is bounded, and in fact it is the only *bounded* solution to this problem, as we shall prove below. Similarly, we shall establish that there is at most one *bounded* solution to the heat equation $u_t = u_{xx}$ in the half-plane $t > 0$, with prescribed initial values on $(-\infty, \infty)$. Finally, we prove uniqueness for solution of the corresponding problem for the half-plane $x > 0$.

The Laplace Equation

(8.6) Theorem: *There is at most one bounded solution $u = u(x, y)$ to the boundary value problem*

$$\nabla^2 u = u_{xx} + u_{yy} = F \quad \textit{in} \quad \{-\infty < x < +\infty,\ y > 0\}$$
$$\textit{with } u(x, 0) = f(x) \textit{ prescribed}, \ -\infty < x < +\infty. \tag{71}$$

Proof: The difference u of two solutions is *harmonic* and bounded for $y > 0$ and has as $y \searrow 0$, the continuous boundary values $u(x, 0) = 0$, $x \in \mathbb{R}$. Let's extend u

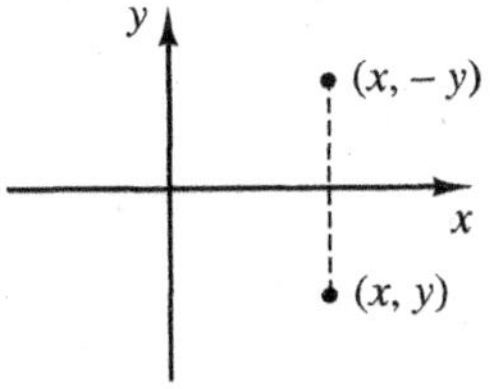

Figure 8.4

to $\{y < 0\}$ by *reflection* (see Figure 8.4) so that

$$u(x, y) \overset{\text{def}}{=} -u(x, -y), \qquad y < 0.$$

The resulting function is bounded and continuous in $\mathbb{R}^2$, and it is harmonic! For $y < 0$, harmonicity follows from the fact that

$$u_{yy}(x, y) = -u_{yy}(x, -y) = u_{xx}(x, -y) = -u_{xx}(x, y),$$

but when $y = 0$, we must work harder.

For $R > 0$, set $f(t) = u(R \cos t, R \sin t)$. Then by Example 12 of Section 3.5, the function

$$U(r, \theta) = \frac{1}{2\pi} \int_{-\pi}^{\pi} \frac{(R^2 - r^2) f(t)}{|(R, t) - (r, \theta)|^2} \, dt \tag{72}$$

is harmonic in the disk $\{r < R\}$ of Figure 8.5 and continuous on the closure $\{r \le R\}$, and it has the *same* boundary values as u. However for $-\pi < t < \pi$: $\cos(\theta - t) = \cos(\theta + t)$ when $\theta = 0$ and $\theta = \pi$, while from reflection $f(-t) = -f(t)$. It follows that $U(r, 0) = U(r, \pi) = 0$, for $r \le R$, and thus $U - u$ is harmonic in either the upper or the lower open semidisk and continuous on its closure with zero boundary values. Consequently, $U = u$ in either semidisk by Theorem 3.5, and from continuity we conclude that $u = U$ is harmonic within a disk of arbitrarily large radius R. We also know from the hypothesis and construction that u is bounded (say, $|u(x, y)| \le M$).

Now we can show that u is constant, and hence $u \equiv u(x, 0) = 0$. Observe that in polar coordinates

$$v(r, \theta) \overset{\text{def}}{=} u(r \cos \theta, r \sin \theta) + M \ge 0,$$

and v is also harmonic. Hence within each disk $\{r < R\}$, v admits Poisson integral

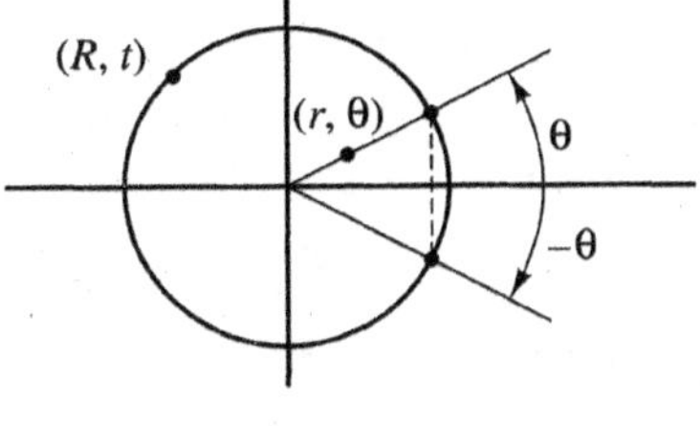

Figure 8.5

representation in terms of its boundary values on that disk in the form

$$v(r,\theta) = \frac{1}{2\pi}\int_{-\pi}^{\pi} \frac{(R^2 - r^2)v(R, t)\,dt}{|(R, t) - (r, \theta)|^2}, \qquad 0 \le r < R. \tag{73}$$

In particular, $v(0, \theta) = (1/2\pi)\int_{-\pi}^{\pi} v(R, t)\,dt$, and it is geometrically evident that for all t, θ:

$$R - r \le |(R, t) - (r, \theta)| \le R + r, \qquad 0 \le r < R.$$

Since $v \ge 0$, we can estimate the integral in (73) to obtain **Harnack's inequalities**

$$v(r,\theta) \le \frac{R+r}{R-r}\frac{1}{2\pi}\int_{-\pi}^{\pi} v(R, t)\,dt = \frac{R+r}{R-r}v(0,\theta) \qquad (0 \le r < R)$$

and **(73')**

$$v(r,\theta) \ge \frac{R-r}{R+r}v(0,\theta),$$

which provide growth estimates for harmonic functions $v \ge 0$. Now $v(0, \theta) = u(0, \theta) + M = M$, and if we let $R \to +\infty$, in the previous inequalities we see that for fixed r:

$$M = \lim_{R\to\infty} \frac{R-r}{R+r} M \le v(r,\theta) \le \lim_{R\to\infty}\frac{R+r}{R-r} M = M$$

so that $v \equiv M$ or $u \equiv 0$ as desired. ∎

Remarks: In the plane, it is somewhat simpler to obtain this result by exploiting the intimate relations between harmonic and complex analytic functions. (See [Ne].) However, the arguments indicated here have natural extensions to harmonic functions in higher dimensions, and so to solutions of other elliptic equations, in "half" space. There are also some corresponding results for the Neumann and Robin problems in these and a few related regions. The analysis for the general region is complicated as always by the potential irregularity of the boundary, although uniqueness arguments are less affected than those for existence.

In proving Theorem 8.6, it was supposed that u is continuous for $y \ge 0$. However, through Lebesgue integral methods, it can be shown that uniqueness also holds for solutions to the problem in which (71) is replaced by the weaker condition $u(x, 0+) = f(x)$, $x \in \mathbb{R}$.

The Heat Equation

(8.7) Theorem: *Let V be an unbounded domain in $\mathbb{R}^d$ with boundary ∂V. Then there is at most one bounded solution $u = u(\mathbf{x}, t)$ to the boundary value problem*

$$\mathcal{L}(u) \stackrel{\text{def}}{=} \nabla^2 u - u_t = F, \qquad \text{in} \qquad \mathcal{V} = V \times (0, \infty)$$

$$\textit{with} \qquad u(\mathbf{x}, 0) = f(\mathbf{x}), \qquad \mathbf{x} \in V \tag{74}$$

$$\textit{and} \qquad u(\mathbf{x}, t) = h(\mathbf{x}), \qquad \mathbf{x} \in \partial V, \qquad t > 0 \textit{ prescribed.}$$

Proof: u, the difference of two solutions, is bounded ($|u(\mathbf{x}, t)| \leq M$, say), satisfies the homogeneous equation in $\mathscr{V}$, and assumes *continuously* the initial/boundary values of zero. For each $R > 0$, let $V_R = V \cap \{|\mathbf{x}| < R\}$ and observe that the everywhere-continuous function

$$p_R(\mathbf{x}, t) = \frac{M}{R^2}(|\mathbf{x}|^2 + 2td) \tag{74'}$$

is a *positive* solution of $\mathscr{L}(u) \equiv 0$ in $\mathscr{V}$, such that when $|\mathbf{x}| = R$: $M \leq p_R(\mathbf{x}, t)$. From the maximum principle (Theorem 3.3) it follows that $u - p_R \leq 0$ in $V_R \times (0, \infty)$; thus, for *fixed* $\mathbf{x}$, t:

$$u(\mathbf{x}, t) \leq p_R(\mathbf{x}, t) \qquad \text{if} \qquad |\mathbf{x}| < R, \qquad t > 0.$$

Now, if we let $R \to \infty$, we see from (74′) that $u(\mathbf{x}, t) \leq 0$, and upon replacing u by $-u$ in this argument, that $u \equiv 0$ in $\mathscr{V}$, as desired. ■

Remarks: It is simple to modify the proof just given to cover the heat operator $\mathscr{L}(u) = k\nabla^2 u - u_t$ where k is a positive constant, and similar results are available for more general parabolic operators (see [Fr]). Moreover, uniqueness for the heat equation may still be established in certain cases for solutions that grow rapidly with $|\mathbf{x}|$ or t. (See Problem 8.5B and [P-W].) Finally, when V is $\mathbb{R}^d$, uniqueness of solution also holds for the problem in which condition (74) is replaced by $u(\mathbf{x}, 0+) = f(\mathbf{x})$, $\mathbf{x} \in \mathbb{R}^d$. (See [Wi].)

We conclude this section with a proof that the bounded solution of the one-dimensional heat equation problem in Example 1 of Section 3.1 is unique. This does not seem to follow immediately from earlier results, and we need some facts about the Fourier integral solution $p = p(x, t)$ of the quarter-plane problem indicated in Figure 8.6.

From (34″) in Example 5, we see that p is given by

$$p(x, t) = \frac{2M}{\sqrt{\pi}} \int_0^{x/2\sqrt{t-T}} e^{-\sigma^2}\, d\sigma, \qquad x > 0, \qquad t > T. \tag{75}$$

p is positive in the quarter-plane, and since $e^{-\sigma^2} \leq 1$, we have the easy estimate

$$p(x, t) \leq \frac{M}{\sqrt{\pi}} \frac{x}{\sqrt{t-T}}, \qquad t > T. \tag{76}$$

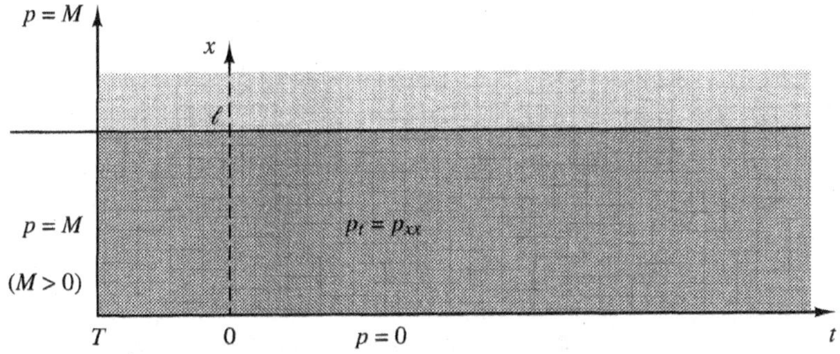

Figure 8.6

(8.8) Theorem: *There is at most one bounded solution $u = u(x, t)$ to the boundary value problem for the heat equation $u_t = u_{xx}$ in the half-plane $x > 0$, with prescribed values*

$$u(0, t) = f(t), \qquad t \in \mathbb{R}.$$

Proof: Suppose that $f = 0$ and that $|u(x, t)| \le M < +\infty$. We want to prove that $u(x, t) \equiv 0$, $x > 0$.

First observe that for $0 < x < \ell < +\infty$:

$$v(x, t) \stackrel{\text{def}}{=} u(x, t) - \frac{M}{\ell} x$$

satisfies the same heat equation with

$$v(0, t) = 0, \qquad v(\ell, t) = u(\ell, t) - M \le 0,$$

$$\text{and} \qquad v(x, T) \le M \qquad \text{for any finite } T.$$

Now p from (75) is a *nonnegative* solution of the heat equation in the strip $\{0 < x < \ell, t > T\}$, and so, by Theorem 3.3 (the maximum principle), $v - p$ cannot have an interior maximum value in this strip. This means that the value of $v(x, t) - p(x, t)$ cannot exceed the largest bound that it can approach as $(x, t) \to (x_0, t_0) \in \mathscr{A}_T$, where $\mathscr{A}_T$ is the boundary of this strip. If we examine v and p, we see that $v - p$ has *nonpositive* boundary limits at each point $(x_0, t_0) \in \mathscr{A}_T$ except when $(x_0, t_0) = (0, T)$. At this exceptional point, v has the limit zero, and therefore $v - p$ has zero as an upper bound.

Thus, in the strip, by (76),

$$v(x, t) \le p(x, t) \le \frac{M}{\sqrt{\pi}} \frac{x}{\sqrt{t - T}},$$

and this holds for any $T < t$. If we keep t fixed and let $T \to -\infty$, it follows that

$$u(x, t) - \frac{M}{\ell} x = v(x, t) \le 0, \qquad \text{or} \qquad u(x, t) \le \frac{M}{\ell} x, \qquad 0 < x < \ell.$$

However, if we now hold x fixed and let $\ell \to +\infty$, we see that $u(x, t) \le 0$, $x > 0$. Similarly $-u(x, t) \le 0$, $x > 0$, so that $u(x, t) \equiv 0$, $x > 0$. ∎

Remark: For this problem, uniqueness of solution also holds for a solution u that grows no faster than x^c for a fixed $c < 1$. However, the solutions, $u(x, t) = ax$, show that uniqueness fails when $c = 1$.

Problem Set 8.5

8.5A *1.* Suppose u is harmonic and bounded in the quarter-plane $\{x > 0, y > 0\}$ with the continuously assumed boundary values

$$u(x, 0) = 0, \quad x \ge 0$$

$$\text{and} \qquad u(0, y) = 0, \quad y \ge 0.$$

Conclude that u can be extended to a bounded harmonic function in $\{y > 0\}$, with $u(x, 0) = 0$, and hence $u \equiv 0$. (*Hint*: Use reflection.)

2. Formulate a theorem similar to Theorem 8.6 for this quarter-plane.
3. How would you establish a similar result in another quarter-plane? another half-plane? (*Hint*: Harmonicity is preserved under rotations and translations.)

8.5B* We want to prove that $u = 0$ is the only solution $u = u(x, t)$ of the equation $\mathscr{L}(u) \stackrel{\text{def}}{=} u_{xx} - u_t = 0$ in the half-plane $\{t > 0\}$ when for some $M > 0$, $|u(x, t)| \le Me^{|x|+t}$, and u assumes (continuously) the initial values

$$u(x, 0) = 0,\ x \in \mathbb{R}.$$

Choose a fixed $T > 0$ and for $0 < t < T$, let

$$v(x, t) = \sqrt{T - t}\, \exp[-x^2/4(T - t)]u(x, t).$$

1. Show that v satisfies the equation

$$v_t - v_{xx} = \frac{x}{T - t} v_x, \qquad 0 < t < T,$$

with $v(x, 0) = v(x, T) = 0$, $x \in \mathbb{R}$.

2. Verify that $|v(x, t)| \le M\sqrt{T}\, e^T e^{|x|} e^{-x^2/4T}$, $0 \le t \le T$ and that the equation in part 1 is a special case of (30) in Section 3.5.
3. Conclude that v satisfies the *maximum and minimum* principles in each box

$$\mathscr{V}_{R,T} = \{|x| \le R,\ 0 \le t \le T\}$$

so that for *fixed* (x, t) in this box:

$$|v(x, t)| \le M\sqrt{T} e^R e^{-R^2/4T}, \qquad \text{which} \to 0 \text{ as } R \to \infty,$$

and hence $u(x, t) \equiv 0$ as desired.

4. How much greater growth of u with x or t is permitted if this proof is to remain valid?

8.6*

Bessel's Integral Formula and Asymptotic Developments

In Sections 7.4 and 8.4 we saw the role that Bessel functions play in solving boundary value problems involving the Laplacian in cylindrical coordinates. F.W. Bessel (1784–1846) was the director of the astronomical observatory in Königsberg, and his own interest in these functions (from 1816 onward) was motivated partly by their application to his field.

Example 8:

Let's consider how to predict the location of a planet moving around the sun. According to Kepler's laws (which Newton derived from the inverse square

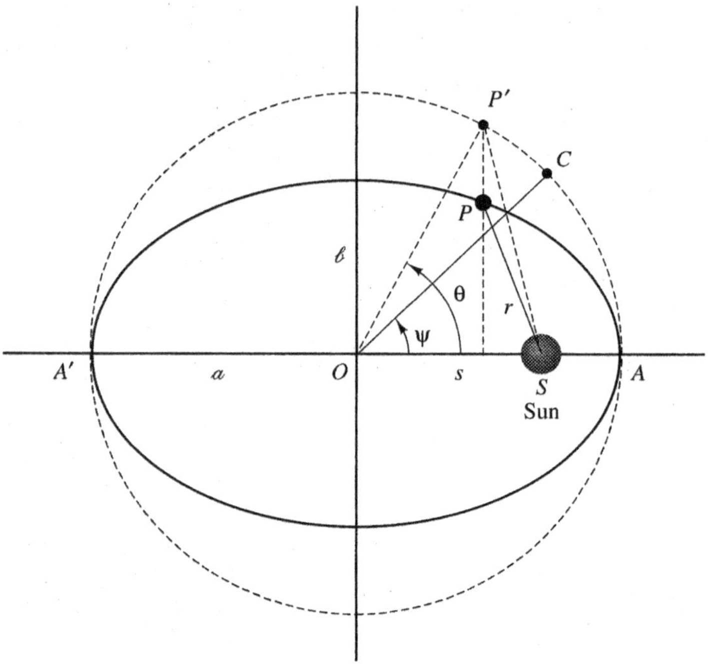

Figure 8.7

law of gravitation), *a planet P moves in a planar elliptical orbit with period* $2T$ *around the sun S located at a focus of the ellipse,* as shown in Figure 8.7. Also, *the area within the ellipse swept out by the radius vector SP during a time interval* Δt *is proportional to* Δt.

Suppose that the ellipse with center O has a horizontal major axis of length $2a$ with endpoints A, A', and on a concentric circle of radius a, let P' be the point vertically nearest to P. Finally, imagine a point C that traces the circle with *constant speed* and at times $t = 0$ and T coincides with P at the perihelion A and aphelion A', respectively, of the planet. Then at an intervening time t, the central angle AOC is given by $\psi = \pi t/T$ so that the position of C is known. But Kepler needed to know where P (or P') is located in relation to C.

From the law of areas stated above, we can deduce **Kepler's equation**

$$\psi = \theta - \epsilon \sin \theta, \tag{77}$$

where $\epsilon = s/a$ is the eccentricity of the ellipse and θ is the angle AOP'. (See Problem 8.6A.) We would like to invert Kepler's equation and obtain θ as a function of ψ. Since, by symmetry, $\theta - \psi$ is an *odd* function of ψ with period 2π, in 1770, Lagrange proposed representing it by the sine series

$$\theta - \psi = \sum_{n=1}^{\infty} s_n \sin n\psi, \qquad 0 \le \psi \le \pi. \tag{78}$$

Then as usual, for $n = 1, 2, \ldots,$

$$s_n = \frac{2}{\pi}\int_0^\pi (\theta - \psi)\sin n\psi\, d\psi,$$

and after partial integration we obtain

$$s_n = \frac{2}{\pi n}\int_0^\pi \cos n\psi\, d(\theta - \psi),$$

since $\theta - \psi$ vanishes when $\psi = 0$ and when $\psi = \pi$. But $\int_0^\pi \cos n\psi\, d\psi = 0$, $n = 1, 2, \ldots$, and from (77) it follows that

$$s_n = \frac{2}{\pi n}\int_0^\pi \cos n\psi\, d\theta = \frac{2}{\pi n}\int_0^\pi \cos(n\theta - n\epsilon \sin\theta)\, d\theta. \qquad \textbf{(78')}$$

This led Bessel to introduce the functions

$$J_m(x) = \frac{1}{\pi}\int_0^\pi \cos(m\theta - x\sin\theta)\, d\theta, \qquad m = 0, 1, 2, \ldots, \qquad x \in \mathbb{R}, \qquad \textbf{(79)}$$

and derive for them the now-familiar properties characterizing Bessel functions.[†] Formula (79) is now known as **Bessel's integral formula**, and when it is used in (78′), we see that $s_n = (2/n)\, J_n(n\epsilon)$, $n = 1, 2, \ldots$.

Thus θ can be represented by the *uniformly convergent* series

$$\theta = \psi + 2\sum_{n=1}^{\infty} \frac{J_n(n\epsilon)}{n}\sin n\psi, \qquad \textbf{(80)}$$

which involves Bessel functions of *all* integer orders $n = 1, 2, \ldots$.[‡] Lagrange obtained power series expressions for its coefficients, and he also derived a similar series for r, the distance from S to P, based on the geometric fact that

$$r = a(1 - \epsilon\cos\theta). \qquad \textbf{(81)}$$

(See [Wat] and Problem 8.6B.)

Asymptotic Considerations

The integral formula (79) that Bessel introduced in 1818 has several advantages over the usual power series representation of J_m when $m = 0, 1, 2, \ldots$. First, it reduces the calculation of values of $J_m(x)$ to integration of elementary functions, which can now be carried out numerically with great precision. Second, it provides an immediate proof that $|J_m(x)| \le 1$. But perhaps its most significant contribution is the following expression for the asymptotic behavior of $J_m(x)$ for large x:

$$J_m(x) = \sqrt{\frac{2}{\pi x}}\cos\left(x - \frac{\pi}{4} - \frac{m\pi}{2}\right) + O\left(\frac{1}{x}\right). \qquad \textbf{(82)}$$

Here, the symbol $O\left(\frac{1}{x}\right)$ (read "big oh of $1/x$") indicates a function that is bounded

[†] Formula (79) is invalid for nonintegral values of m, even for $m = 1/2$. For integral values, see Problem 8.4C.

[‡] From (77), $d\psi/d\theta = 1 - \epsilon\cos\theta > 0$ (since $\epsilon < 1$) so that θ is a strictly increasing C^1 function of ψ and $\theta - \psi$ is C^1. Now use Theorem 1.7.

in absolute value by some constant times $(1/x)$ when x is sufficiently large. A statement equivalent to (82) is that there exist positive constants a and A such that

$$\left| J_m(x) - \sqrt{\frac{2}{\pi x}} \cos\left(x - \frac{\pi}{4} - \frac{m\pi}{2}\right) \right| \leq \frac{A}{x}, \qquad x \geq a. \tag{82'}$$

We can obtain formula (82) from a more general asymptotic result.

Method of Stationary Phase

Using Euler's formula $\cos t = \mathcal{R}_e\{e^{it}\}$, we can express (79) in the form

$$\pi J_m(x) = \mathcal{R}_e \int_0^\pi e^{imt} e^{-ix \sin t}\, dt, \qquad x > 0. \tag{83}$$

Therefore, let's try to obtain asymptotic behavior for the more general integral

$$F(x) = \int_a^b f(t)\, e^{ix\phi(t)}\, dt, \tag{84}$$

where the interval $[a, b]$ is bounded, the **amplitude** $f \in C^2[a, b]$, and the **phase** $\phi \in C^3[a, b]$. (ϕ is real valued, but f may be complex valued.)

Now, if $h \overset{\text{def}}{=} f/\phi' \in C^1[a, b]$, then

$$f(t)\, e^{ix\phi(t)} = -\frac{i}{x} \frac{f(t)}{\phi'(t)} \frac{d}{dt}(e^{ix\phi(t)}) = -\frac{i}{x} h(t) \frac{d}{dt} e^{ix\phi(t)},$$

and partial integration of (84) gives

$$F(x) = -\frac{i}{x}\left[\int_a^b h'(t) e^{ix\phi(t)}\, dt - h(t)\, e^{ix\phi(t)}\Big|_a^b\right].$$

The bracketed term is bounded since $|e^{ix\phi(t)}| \leq 1$, and we would conclude that $F(x) = O(\frac{1}{x})$ for large positive x. We can do this unless there are points in $[a, b]$ at which ϕ' vanishes; i.e., points of **stationary phase**. If there are only a finite number of such points in $[a, b]$, then by further partitioning, if necessary, we can express (84) as a sum

$$F(x) = \sum_{k=1}^{K} \int_{a_k}^{b_k} f(t)\, e^{ix\phi(t)}\, dt, \tag{84'}$$

where in each interval $[a_k, b_k]$, ϕ' vanishes only at *one* endpoint. Therefore, *it suffices to discover how the integral in* (84) *behaves when* ϕ' *vanishes only at* a *or only at* b.

Suppose $\phi'(a) = 0$ and $\phi''(a) > 0$. Then by adding and subtracting to f and ϕ their respective values at a it is not difficult to show as above that

$$F(x) = f(a)\, e^{ix\phi(a)} \int_a^b e^{ix[\phi(t)-\phi(a)]}\, dt + O\left(\frac{1}{x}\right). \tag{85}$$

Next, since $\phi'(t) > 0$ on $(a, b]$, ϕ is strictly increasing on $[a, b]$. If we make the substitutions

$$s^2 = \phi(t) - \phi(a), \qquad c^2 = \phi(b) - \phi(a), \qquad \text{and} \qquad 2s\, ds = \phi'(t)\, dt,$$

the last integral becomes

$$\int_0^c e^{ixs^2}\left(\frac{dt}{ds}\right) ds = \left(\frac{dt}{ds}\right)_0 \int_0^c e^{ixs^2}\, ds + \int_0^c e^{ixs^2}\left[\frac{dt}{ds} - \left(\frac{dt}{ds}\right)_0\right] ds.$$

Here, by l'Hôpital's rule,

$$\left(\frac{dt}{ds}\right)_0 = \lim_{s\searrow 0}\left(\frac{dt}{ds}\right) = \left[\lim_{t\searrow a}\frac{4(\phi(t)-\phi(a))}{\phi'(t)^2}\right]^{1/2} = \frac{\sqrt{2}}{\sqrt{\phi''(a)}}$$

and by partial integration, it can be shown that the *second* integral on the right is $O(\frac{1}{x})$. With the substitution $u = s\sqrt{x}$ (so $du = \sqrt{x}\, ds$), the first integral on the right becomes $x^{-1/2}E(c\sqrt{x})$, where, as we shall show presently,

$$E(x) \overset{\text{def}}{=} \int_0^x e^{iu^2}\, du = \frac{\sqrt{\pi}}{2}\, e^{i\pi/4} + O\left(\frac{1}{x}\right). \tag{86}$$

Therefore, if we set $A(x) = f(a)\,|\,\phi''(a)\,|^{-1/2} e^{ix\phi(a)}$, we see that

$$F(x) = A(x)\sqrt{\frac{\pi}{2x}}\, e^{\pm i\pi/4} + O\left(\frac{1}{x}\right), \qquad \text{if } \phi''(a) \gtrless 0. \tag{87}$$

(When $\phi'(a) = 0$ but $\phi''(a) < 0$, we reverse the sign of s^2 and so of i in (86).) If b is the only point where ϕ' vanishes and $\phi''(b) \neq 0$, then (87) holds with a replaced by b. Each interior stationary-phase point of $[a, b]$ will appear as both an upper and a lower endpoint of subintervals in (84′) and hence will contribute twice to the asymptotic development of $F(x)$. We have sketched a proof for the following.

(8.9) Theorem: *Suppose that $[a, b]$ is bounded, that $f \in C^2[a, b]$, and $\phi \in C^3[a, b]$, and that $\phi'(t) \neq 0$ except at a finite set of points $t_k \in [a, b]$ where $\phi''(t_k) \neq 0$, $k = 1, 2, \ldots, K$. For each such k, set*

$$T_k(x) = f(t_k)\,|\,\phi''(t_k)\,|^{-1/2} e^{ix\phi(t_k)}\begin{cases}(+1), & \text{if } \phi''(t_k) > 0\\ (-i), & \text{if } \phi''(t_k) < 0\end{cases}.$$

Then for large $x > 0$:

$$\int_a^b f(t)\, e^{ix\phi(t)}\, dt = \sqrt{\frac{\pi}{2x}}\, e^{i\pi/4}\sum_{k=1}^{K}\delta_k T_k(x) + O\left(\frac{1}{x}\right) \tag{88}$$

where $\delta_k = 2$ or 1 according to whether $t_k \in (a, b)$ or not. ■

Now we can get the asymptotic behavior of J_m.

(8.10) Theorem: *For each $m = 0, 1, \ldots$ and large $x > 0$,*

$$J_m(x) = \sqrt{\frac{2}{\pi x}}\left[\cos\left(x - \frac{\pi}{4} - \frac{m\pi}{2}\right) - \frac{M}{2x}\sin\left(x - \frac{\pi}{4} - \frac{m\pi}{2}\right)\right] + O\left(\frac{1}{x^{5/2}}\right), \tag{89}$$

where $M = m^2 - \frac{1}{4}$.

Proof: In (88) we take $f(t) = e^{imt}$, $\phi(t) = -\sin t$, and note that on $[0, \pi]$, $\phi'(t) = -\cos t$ vanishes only at the *interior* point $t_1 = \pi/2$ where $\phi''(t_1) = 1$.

Hence, we set $T_1(x) = e^{im\pi/2}\, e^{-ix}$ and $\delta_1 = 2$, so that (83) yields the estimate

$$\pi J_m(x) = \sqrt{\frac{\pi}{2x}}\,\mathcal{R}_e\{2\,e^{i\pi/4}\,e^{im\pi/2}\,e^{-ix}\} + O\left(\frac{1}{x}\right)$$

and this is equivalent to (82). Then, we can use (82) to identify the constants in the formula

$$J_m(x) = \frac{A}{\sqrt{x}}\left[\cos(x-\alpha) - \frac{M}{2x}\sin(x-\alpha)\right] + O\left(\frac{1}{x^{5/2}}\right)$$

derived by elementary means at the end of Section 7.4. Through comparison, we see that $A = \sqrt{2/\pi}$ and $\alpha = (m\pi/2) + (\pi/4)$, and this gives (89). By the earlier method, asymptotic results may be successively improved as desired. ■

Remark: The asymptotic behavior of the zeros of J_m is taken up in Problem 9.3K.

Asymptotic behavior of $\boldsymbol{E(x) = \int_0^x e^{iu^2}\,du}$**:** We still need to establish (86). Observe that we can express $E^2(x)$ as follows:

$$E^2(x) = \int_0^x e^{iu^2}\,du \int_0^x e^{iv^2}\,dv = \int_0^x \int_0^x e^{i(u^2+v^2)}\,du\,dv$$

$$= 2\int_0^{\pi/4} d\theta \int_0^{x\sec\theta} re^{ir^2}\,dr = i\frac{\pi}{4} - i\int_0^{\pi/4} e^{ix^2\sec^2\theta}\,d\theta,$$

if we perform the integration over the square $\{0 \le u \le x,\ 0 \le v \le x\}$ in polar coordinates (r, θ) and incorporate the symmetry of the integrand with respect to the line $u = v$ as shown in Figure 8.8. In the last integral let $w = \tan\theta$ so that $\sec^2\theta = 1 + w^2$ and $d\theta = dw/(1 + w^2)$. Then we get

$$E^2(x) = \frac{i\pi}{4} - ie^{ix^2}\int_0^1 \frac{e^{ix^2w^2}}{1+w^2}\,dw$$

$$= \frac{i\pi}{4} - ie^{ix^2}\int_0^1 e^{ix^2w^2}\,dw + O\left(\frac{1}{x^2}\right) \tag{90}$$

by a repetition of the argument used to obtain (85). (See Problem 8.6D.) Finally, if we let $u = xw$ and use the definition of E, we see that E satisfies asymptotically

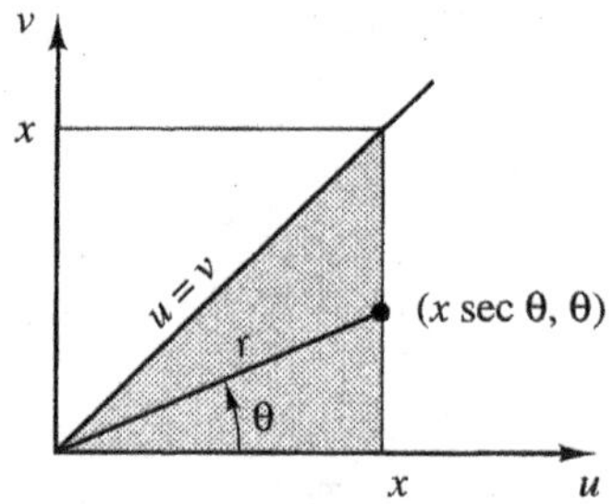

Figure 8.8

the *quadratic equation*

$$E^2(x) = \frac{i\pi}{4} - \frac{ie^{ix^2}}{x} E(x) + O\left(\frac{1}{x^2}\right).$$

Upon completing the square we get

$$\left[E(x) + \frac{ie^{ix^2}}{2x}\right]^2 = \frac{i\pi}{4} + O\left(\frac{1}{x^2}\right) = \frac{\pi}{4} e^{i\pi/2}\left[1 + O\left(\frac{1}{x^2}\right)\right]$$

and after taking square roots of both sides, we find that

$$E(x) = \int_0^x e^{iu^2}\,du = \pm\frac{\sqrt{\pi}}{2} e^{i\pi/4} - \frac{ie^{ix^2}}{2x} + O\left(\frac{1}{x^2}\right). \tag{91}$$

As $x \to \infty$, we deduce that

$$\int_0^\infty (\cos u^2 + i \sin u^2)\,du = \int_0^\infty e^{iu^2}\,du = \pm\frac{\sqrt{\pi}}{2} e^{i\pi/4} = \pm\sqrt{\frac{\pi}{8}}(1+i)$$

and the sign is that of the real integral

$$\int_0^\infty \sin u^2\,du = \frac{1}{2}\int_0^\infty \frac{\sin v}{\sqrt{v}}\,dv,$$

(which exists). But the last integrand is positive on $(0, \pi)$ and alternates in sign with decreasing mean amplitude on successive intervals of length π. Consequently, the last integral is *positive* and (91) holds with the + sign; this gives (86). We also see that

$$\int_0^\infty \cos u^2\,du = \int_0^\infty \sin u^2\,du = \sqrt{\frac{\pi}{8}} \tag{92}$$

and thereby evaluate these **Fresnel integrals.**

Problem Set 8.6

8.6A *1.* Use Kepler's law of areas and Figure 8.7 to infer that

$$\frac{t}{2T} = \frac{\psi}{2\pi} = \frac{\text{Area } \widehat{SAP}}{\pi a b} = \frac{\text{Area } \widehat{SAP'}}{\pi a^2}$$

where b is the length of the semiminor axis, so that the ellipse is obtained from the circle through the mapping

$$x_1 = x, \qquad y_1 = \frac{b}{a} y.$$

2. Show that

$$\text{Area } \widehat{SAP'} = (a^2\theta/2) - \text{Area } SOP' = (a^2\theta/2) - (s/2)(a \sin\theta)$$

and conclude that (77) follows.

8.6B *1.* In Figure 8.7, let O' be the point where the line from P' to P meets the axis AA', and show that the segments OO' and $O'P$ have lengths $a\cos\theta$ and $b\sin\theta$, respectively.

2. Use the Pythagorean theorem to obtain (81).
3. Why is it appropriate to express r as the series

$$r = \frac{c_0}{2} + \sum_{n=1}^{\infty} c_n \cos n\psi?$$

4. Conclude that

$$\frac{r}{a} = 1 + \frac{\epsilon^2}{2} - 2\epsilon \sum_{n=1}^{\infty} \frac{J_n'(n\epsilon)\cos n\psi}{n}.$$

(*Hint*: Differentiate (79).)

8.6C* With θ and ψ as in Figure 8.7, and $0 \leq \epsilon < 1$:

1. show that $[1/(1 - \epsilon \cos\theta)] = 1 + 2\sum_{n=1}^{\infty} J_n(n\epsilon)\cos n\psi$.
2. conclude that $1/\sqrt{1-\epsilon^2} = 1 + 2\sum_{n=1}^{\infty} J_n^2(n\epsilon)$.

8.6D In the analysis of $E(x)$:

1. verify the explicit expressions obtained for $E^2(x)$ through use of polar coordinates.
2. show how (90) follows from examination of the integral in the preceding line. (*Hint*: $[(1+w^2)^{-1} - 1]/w$ is well behaved at $w = 0$.)
3.* if $f \in C(1, \infty)$ is complex valued and $\lim_{x\to\infty} f^2(x) = A^2 \neq 0$ for some complex number A, why must $\lim_{x\to\infty} f(x) = A$ or $\lim_{x\to\infty} f(x) = -A$? (Why is this relevant to the derivation of (91)?)

8.6E Differentiate (79) and use Theorem 8.10 to obtain an asymptotic result for $J_m'(x)$.

8.6F Apply Theorem 8.10 to find asymptotic developments for each of the following integrals for $x > 0$:

1. $\displaystyle\int_{-1}^{1} \frac{1}{1+t^2} e^{ixt^2}\, dt,$ 2. $\displaystyle\int_{0}^{4} \frac{1}{1+t} e^{ix\cos t}\, dt,$

3. $\displaystyle\int_{0}^{2\pi} (\sin^2 t)\, e^{ix\sin t}\, dt,$ 4. $\displaystyle\int_{-\pi/2}^{\pi} \frac{e^{ix\sin t}}{1+t^2}\, dt,$

8.6G* 1. If $f \in C^2[0, \ell]$, explain why $f(t) = f(0) + tg(t)$ where $g \in C^1[0, \ell]$. Then if $\phi \in C^2[0, \ell]$ with $\phi'(0) = 0$, $\phi''(0) > 0$, and $\phi'(t) > 0$ for $t > 0$, explain why $h(t) = [f(t) - f(0)]/\phi'(t)$ gives a function in $C^1[0, \ell]$. (*Hint*: Use l'Hôpital's rule.)

2. With ϕ as above and $a = 0$, verify that the indicated substitutions lead from (84) to (85).

9

Sturm–Liouville Problems

In this chapter, we reexamine the Sturm–Liouville (S–L) eigenvalue problems introduced in Chapter 7. In particular, we want to find conditions that guarantee S–L series representations for useful classes of functions. Our results apply to the regular problems of Section 7.1, as well as to the singular problems from Chapter 7 associated with Bessel's equation of order m

$$(xy')' - \frac{m^2}{x} y = -\lambda xy, \qquad \text{on } (0, 1) \tag{1}$$

and Legendre's equation of order m

$$[(1 - x^2)y']' - \frac{m^2}{1 - x^2} y = -\lambda y, \qquad \text{on } (-1, 1) \tag{2}$$

for $m = 0, 1, 2, \ldots$.

We use the terminology from Section 7.1, and throughout the chapter we simply assume the existence of enough eigenfunctions and eigenvalues to achieve our goals. For regular problems, existence can be established in general (see [In]), and for singular problems involving (1) or (2), existence is assured by series solution as in Sections 7.4–7.6. The question of whether there are "enough" eigenfunctions and eigenvalues is taken up in our discussion of completeness in Section 9.2.

The bulk of this chapter is devoted to one-dimensional problems. We begin by identifying problems that have nonnegative eigenvalues and then (in Section 9.2) show how this leads to series representation with uniform convergence on compact (closed and bounded) subintervals. In Section 9.3 we will use Rayleigh's principle to characterize eigenvalues and gain further information about the nature of the eigenfunctions. Then, in Section 9.4, we will obtain asymptotic estimates for eigenfunctions to large eigenvalues that are strong enough to furnish useful pointwise convergence results in Section 9.5. The chapter concludes with a brief look at the situation in higher dimensions and an application to wave-guide design.

The study of this class of problems was initiated in 1835 by J. C. F. Sturm (1803–1855), who was interested in the applications to heat conduction, and by

J. Liouville (1809–1882), who also made significant contributions to complex analysis, Hamiltonian mechanics, and number theory. These mathematicians were located in Paris at the Sorbonne and College de France, respectively, and their long collaboration resulted in methods to establish existence and distribution of eigenvalues for such problems and the oscillatory character of the eigenfunctions. However, satisfactory demonstration that large classes of functions could be represented by convergent series of the eigenfunctions was not available until Kneser's work of 1904. Then, in 1905 David Hilbert (1862–1943), the most accomplished mathematician of his era, showed that an equivalent integral-operator formulation of an S–L problem in terms of its Green function (see Appendix B.2) is more tractable analytically. This supplied a means of establishing both existence and completeness of eigenfunctions, and it ushered in our century of functional analysis. (See [Rei] for a detailed historical account.)

9.1

Positive S–L Problems

For our purposes, a Sturm–Liouville (S–L) eigenvalue problem consists of a second-order linear differential equation for $y = y(x)$ in the "self-adjoint" form.

$$L(y) \equiv (\tau y')' - qy = -\lambda \rho y, \qquad \text{on } (a, b), \tag{3}$$

together with a subspace $\mathscr{S}$ of continuous functions meeting regular or singular boundary requirements which (as discussed in Remark 7.1′) guarantee that

$$y, v \in \mathscr{S} \Rightarrow (\tau y'v)(e) = (\tau v'y)(e), \text{ at each endpoint } e. \tag{3′}$$

Here τ, q, and ρ are given continuous functions on a *fixed* open interval (a, b) in which τ is C^1 and $\rho\tau$ is C^2. We assume that τ and ρ are positive with finite endpoint limits from within (a, b), and we will use $\langle a, b\rangle$ to denote the interval obtained by adjoining to (a, b) the *regular* endpoints of the problem. In particular, for a *regular problem*, we see that $\langle a, b\rangle = [a, b]$ and is *compact* (closed and bounded).

(9.1) Definition: *$\mathscr{S}$ is the set of functions $y \in \hat{C}^1\langle a, b\rangle$, for which both y and y' are bounded,*† $\|y\| \overset{\text{def}}{=} (\int_a^b \rho y^2\,dx)^{1/2}$ *is finite, and $B_e(y) = 0$ at each regular endpoint e.* (See Definition 7.0.)

Because of the Schwarz inequality in the form

$$\int_a^b \rho\,|y|\,|v|\,dx \le \|y\|\,\|v\|, \tag{4}$$

it follows that the inner product

$$(y, v) \overset{\text{def}}{=} \int_a^b \rho y v\,dx, \tag{4′}$$

is finite when both y and v are in $\mathscr{S}$ and that $\mathscr{S}$ *is* a subspace. (Problem 9.1E.)

† Specifically, a continuous y is in $\hat{C}^1\langle a, b\rangle$ when $y'(x)$ is defined at all except a finite number of points $x \in \langle a, b\rangle$, and $y \in \hat{C}^1[a', b']$ for each compact interval $[a', b'] \subseteq \langle a, b\rangle$. (Recall Definition 1.5.)

(9.2) Definition: *Each nonzero* $y \in \mathcal{S} \cap C^2$ *which satisfies equation* (3) *for some* $\lambda \in \mathbb{R}$ *is called an* **eigenfunction** *to the* **eigenvalue** λ *of the corresponding* S–L *problem. It is* **normalized** *when also* $\| y \| = 1$.

For a given S–L problem, when $y \in \mathcal{S}$, we can introduce†

$$I(y, y) \stackrel{\text{def}}{=} \int_a^b (\tau y'^2 + q y^2)\, dx - (\tau y' y)\Big|_a^b \tag{5}$$

and use our boundary requirements to show that

$$I(y, y) = \int_a^b (\tau y'^2 + q y^2)\, dx + A(\tau y^2)(a) + B(\tau y^2)(b) \tag{5'}$$

(see Problem 9.1D), for suitable regular endpoint constants A and/or B, where we delete each term evaluated at a singular endpoint or at a regular endpoint at which y or y' vanishes. Moreover, if $y \in \mathcal{S} \cap C^2$, then by partially integrating the $(\tau y')y'$ term in (5) and using (3), we find that

$$I(y, y) = -\int_a^b L(y) y\, dx. \tag{6}$$

In particular, when y is a *normalized* eigenfunction to the eigenvalue λ, it follows that

$$\lambda = \lambda \int_a^b \rho y^2\, dx = I(y, y), \tag{6'}$$

and this result in conjunction with (5′) affords valuable estimates of the eigenvalues.

Positive Problems

An S–L problem that has only nonnegative eigenvalues is simpler to investigate, and from (6′) we see that this occurs if the S–L problem is **positive** in that

$$y \in \mathcal{S} \Rightarrow I(y, y) \geq 0. \tag{7}$$

Then, by examining (5′), we see that an S–L problem is positive if it is **fully positive** in that

$$q \geq 0, \qquad \text{and in (5')}, \qquad A \geq 0 \qquad \text{and } B \geq 0. \tag{7'}$$

Now, without changing the eigenfunctions, a regular problem can always be replaced by an equivalent regular problem that is positive and has only positive eigenvalues. (The arguments are outlined in Problems 9.1H and 9.1I*.) However, in Section 7.2, we encountered a regular problem that is positive but not fully positive.‡ To simplify our presentation, we *assume* hereafter that *singular problems are fully positive*.

† According to our assumptions, $I(y, y)$ is always defined but in a singular case (where $q \geq 0$ is presupposed) it might have the value $+\infty$.

‡ The constant B used in Section 7.2 is the *negative* of that in (5′).

(9.3) Proposition: *For a positive S–L problem, each eigenvalue λ is ≥ 0; moreover, $\lambda = 0$ is an eigenvalue of a fully positive problem iff $q \equiv 0$, and $y = 1$ is an associated eigenfunction.*

Proof: If y is a normalized eigenfunction to λ, then, from (6′), $\lambda = I(y, y) \geq 0$. Moreover, $\lambda = 0 \Rightarrow I(y, y) = 0$, and under the conditions (7′) of full positivity, this implies that $\int_a^b \tau y'^2\,dx = 0$, so that $y' \equiv 0$. Hence $y(x)$ is a constant, which we may take to be 1. But then also $0 \leq \int_a^b qy^2\,dx = 0 \Rightarrow q \equiv 0$. The converse assertion follows from (3). ∎

For any S–L problem, suppose that $y, v \in \mathcal{S}$ while both $I(y, y)$ and $I(v, v)$ are finite. Then since $|yv| \leq y^2 + v^2$ and $|y'v'| \leq y'^2 + v'^2$, we see that $I(y, v)$ is finite, where, as in (5),

$$I(y, v) \stackrel{\text{def}}{=} \int_a^b (\tau y'v' + qyv)\,dx - (\tau y'v)\Big|_a^b = I(v, y); \tag{8}$$

$$\text{if also } y \in C^2, \text{ then } I(y, v) = -\int_a^b L(y)v\,dx. \tag{8′}$$

The symmetric relation $I(y, v) = I(v, y)$ in (8) follows immediately from (3′), and the result in (8′) is obtained by partially integrating the $(\tau y')v'$ term.

Example 1:

The temperature distribution $u = u(x, t)$ in a long thin circular rod of length ℓ and nonuniform properties is governed by the equation

$$\nu u_t = (Ku_x)_x - qu, \qquad 0 < x < \ell, \qquad t > 0, \tag{9}$$

where ν and K are positive functions on $[0, \ell]$. (See Figure 9.1.) $q = q(x)$ is a function that allows heat transfer through the lateral surface, and the rod is assumed to be surrounded by a medium at reference temperature zero. (See Problem 3.1C.)

Suppose that the ends are insulated so that

$$u_x(0, t) = u_x(\ell, t) = 0, \qquad t > 0,$$

and an initial temperature distribution $u(x, 0) = f(x)$ is prescribed. Then since ν, K, and q are functions only of x, separation of variables leads to elementary product solutions $u(x, t) = y(x)e^{-\lambda t}$ provided that y is an eigenfunction to the

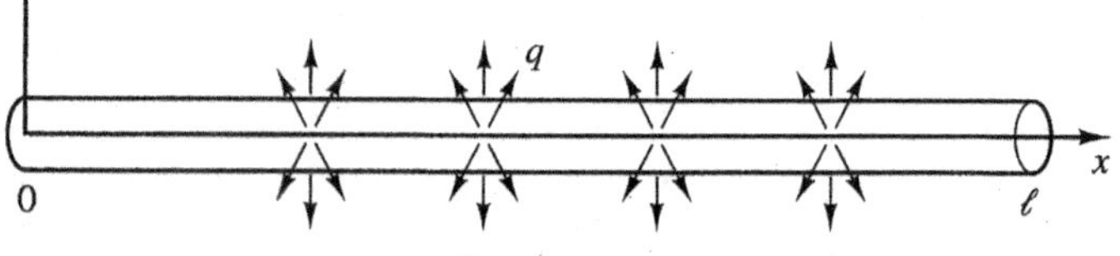

Figure 9.1

eigenvalue λ of the *regular* S–L problem

$$\begin{aligned} (Ky')' - qy &= -\lambda \nu y, \qquad 0 < x < \ell \\ \text{with} \qquad y'(0) &= y'(\ell) = 0. \end{aligned} \tag{10}$$

Note that when $q \geq 0$, the problem is fully positive. In general, the sign of q determines the heat-transfer conditions at the lateral surface and influences the location of the eigenvalues. If $q > 0$, heat is lost by radiation to the surroundings and by Proposition 9.3, all eigenvalues are positive. If $q = 0$, the surface is insulated, and 0 is an eigenvalue by inspection. Finally, if $q < 0$, heat is gained rather than lost at the surface, and negative eigenvalues can occur. (See Problem 7.1B.)

In any case, suppose we obtain a sequence of eigenvalues λ_n and associated eigenfunctions y_n, $n = 1, 2, \ldots$. Then we would try to solve our problem with the formal series

$$U(x, t) = \sum_{n=1}^{\infty} c_n y_n(x) e^{-\lambda_n t} \tag{11}$$

and look for coefficients c_n that satisfy the initial condition

$$U(x, 0) = f(x) = \sum_{n=1}^{\infty} c_n y_n(x), \qquad 0 < x < \ell. \tag{12}$$

For $K = \nu = 1$ and $q = 0$, we found eigenfunctions $y_n(x) = \cos(n\pi x/\ell)$ to eigenvalues $\lambda_n = (n\pi/\ell)^2$, $n = 0, 1, 2, \ldots$, and the resulting questions of representing f were examined in Chapter 1 through consideration of Fourier series. But in the general situation above, we do not know any y_n or λ_n explicitly, nor whether there are "enough" such y_n to provide representation of given functions f. Apparently, we must build an infinite-storeyed structure without having a single one of its floors to stand upon, along the vague outlines suggested by a Fourier series blueprint. That such construction cannot be simple is evident; that it can be accomplished is one of the triumphs of modern analysis.

Problem Set 9.1

9.1A Classify each of the following problems as regular or not, and indicate those that are fully positive.

1. $y'' = -\lambda(1 + x^2)y$ on $(0, 1)$ with $y(0) = y(1) = 0$.
2. $(xy')' - y = -e^x \lambda y$ on $(1, 2)$ with $y(1) = y'(2) = 0$.
3. $(xy')' - y = -e^x \lambda y$ on $(0, 1)$ with $y(1) = 0$, and y, y' bounded.
4. $[(\sin x)y']' - x^2 y = -\lambda y$ on $(0, \pi)$ with y, y' bounded.
5. $[(\sin x)y']' + x^2 y = -\lambda y$ on $(\pi/4, 3\pi/4)$ with $y(\pi/4) = y(3\pi/4) = 0$.
6. $[(1 + x^2)y']' + 4y = -\lambda y$ on $(0, 1)$ with $y(0) = 0, y'(1) = y(1)$.
7. $y'' = -(\lambda/x^2)y$ on $(1, 2)$ with $y(1) = y'(2) = 0$.
8. $x^2 y'' + xy' = -\lambda y$ on $(1, e)$ with $y(1) = y(e) = 0$.
9. $((1 - x^2)y')' = -\lambda y$ on $(0, 1)$ with $y(0) = 0$, and y, y' bounded.
10. $[(1/x^2)y']' - y = -e^{-x}\lambda y$ on $(1, \infty)$ with $y(1) = 0$, and y, y' bounded.

11. $-[(1/(x^2+1))y']'+y=e^{-x^2}\lambda y$ on $(-\infty, \infty)$ with y, y' bounded.

12. $y''-x^2y=-\lambda y$ on $(0, 1)$ with $y'(0)-y(0)=0=y'(1)+y(1)$.

13. $y''-x^2y=-\lambda y$ on $(0, 1)$ with $y'(0)=y(0)$ and $y'(1)=y(1)$.

9.1B Determine the eigenvalues and corresponding eigenfunctions for

$$y''=-\lambda y, \qquad 0<x<1, \qquad \text{with } y(0)+y'(0)=0 \text{ and } y(1)=0$$

and show that there is no negative eigenvalue. Is zero an eigenvalue?

9.1C In one dimension, the diffusion-convection equation for the concentration $u=u(x, t)$ of a dye for water in a trench is $u_t=(\kappa u_x)_x-(vu)_x$ (see Problem 3.1G), for a positive diffusion coefficient function $\kappa=\kappa(x)$ and a drift velocity function $v=v(x)$, $0<x<\ell$. Let $u(x, 0)=f(x)$ be given for $0<x<\ell$.

1. Show that $u(x, t)=y(x)e^{-\lambda t}$ is a solution provided that
$$(\kappa y')'-(vy)'=-\lambda y, \; 0<x<\ell.$$
2. When $v(x)=3$, put this equation in the form of (3). For absorbing ends where the end concentrations $u(0, t)=u(\ell, t)=0$, show that the resulting S–L problem is positive. If $\kappa=1$, give a formal series solution of the initial value problem.
3. Give some other end conditions that retain the positivity of the problem in part 2. What happens if $v(x)=-3$?
4. Analyze the corresponding problem when $v\in C^1[0, \ell]$ is not constant, and show that the equation in part 1 may be written as
$$(P\kappa y')'-Pv'y=-\lambda Py,$$
where $P(x)=\exp\{-\int_0^x (v/\kappa)\,d\xi\}$.

9.1D
1. If $y, v\in C^1[a, b]$ and $B_a(y)=0=B_a(v)$ at the regular endpoint a of an S–L problem, then show that
$$\lim_{x\searrow a}(\tau y'v)(x)=\lim_{x\searrow a}(\tau v'y)(x).$$
2. Conclude that $\lim_{x\searrow a}(\tau y'y)(x)=A(\tau y^2)(a)$ for some constant A.
3. Verify the partial integration used to establish (6).

9.1E Show that (4) follows from the fact that if any compact interval $[a', b']\subseteq\langle a, b\rangle$, then

$$0\le\int_{a'}^{b'}\rho(x)\,dx\int_{a'}^{b'}\rho(t)\,dt\,[y(x)v(t)-y(t)v(x)]^2.$$

Indicate why $\mathscr{S}$ is a subspace.

9.1F The damped unforced vibrations of a nonuniform stretched string give rise to a deflection function $u(x, t)$ that satisfies the equation (from Section 5.4)

$$\rho u_{tt}=-2\delta u_t+(\tau u_x)_x \quad \text{in} \quad 0<x<\ell,\; t>0,$$
$$\text{with} \quad u(0, t)=u(\ell, t)=0.$$

1. When δ is a constant but $0 < \tau \in C^1[0, \ell]$, show that $u(x, t) = y(x)\,T(t)$ is a solution, provided that y is an eigenfunction of a regular S–L problem. Why must its eigenvalues be positive?
2. Assume that these eigenvalues λ_n and associated normalized eigenfunctions y_n are known for $n = 1, 2, \ldots$. Explain how you would find corresponding functions $T_n(t)$ to solve (formally) the initial value problem when

$$u(x, 0) = 0 \quad \text{and} \quad u_t(x, 0) = f(x), \qquad 0 < x < \ell.$$

3. How would the formal analysis change if the end condition at ℓ is replaced by $u_x(\ell, t) = Bu(\ell, t)$ for some constant $B \le 0$? what if $B > 0$?
4.* Explain how in part 3 you would handle a forcing term $F(x, t)$ on the right side of the original equation. (*Hint*: See Problem 5.6I.)

9.1G* According to the analysis in Problem 5.3F, the free transverse vibrations of the tapered clamped beam shown in the figure occur at frequency ω in a mode shape $y = y(x)$ that satisfies the following equation (for $\lambda = \omega^2$):

$$L(y) \stackrel{\text{def}}{=} (\mu y'')'' = \lambda \rho y, \qquad 0 < x < \ell,$$

$$\text{with} \qquad y(0) = y'(0) = y(\ell) = y'(\ell) = 0,$$

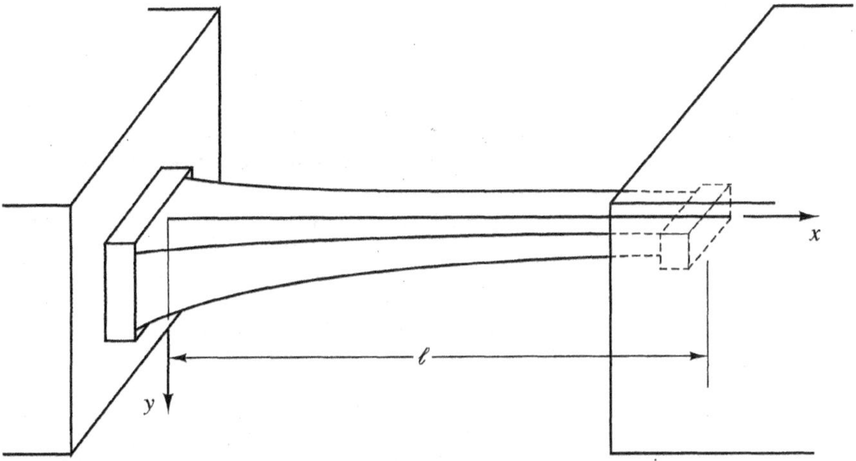

where μ and ρ are given positive functions on $[0, \ell]$, having μ'' and ρ continuous. We wish to develop an S–L framework for this problem.

1. Use the inner product $[y, v] = \int_0^\ell yv\,dx$ to conclude that formally,

$$[Ly, v] - [y, Lv] = 0,$$

when both y and $v \in \mathscr{C}^4[0, \ell]$ and each satisfies the above boundary conditions. (*Hint*: Integrate by parts as required.)
2. Use the inner product $(y, v) = \int_0^\ell \rho yv\,dx$ and show that eigenfunctions corresponding to distinct eigenvalues are ρ-orthogonal.
3. Introduce $I(y, y) = \int_0^\ell \mu(y'')^2\,dx$ on

$$\mathscr{S} = \{y \in C^4[0, \ell] : y(0) = y'(0) = y(\ell) = y'(\ell) = 0\},$$

and conclude that each eigenvalue λ is nonnegative.

4. Could $\lambda = 0$ be an eigenvalue? Explain.
5. What can you say about the eigenvalues λ for the same beam under the cantilever conditions of part 2c in Problem 5.3F?

9.1H Suppose that $(yy')(a) = (yy')(b) = 0$ for $\mathcal{S}$ of a regular problem in which q may have negative values. Let $m = \min_{a \leq x \leq b} [q(x)/\rho(x)]$.

1. Show that m is a lower bound for the eigenvalues of the problem. (*Hint*: Use (6) and (6′), and remember that ρ is positive on $[a, b]$.)
2. Conclude that the eigenvalues $\tilde{\lambda}$ of $\tilde{L}(y) \overset{\text{def}}{=} L(y) + m\rho y = -\tilde{\lambda}\rho y$ are given by $\tilde{\lambda} = \lambda - m \geq 0$.
3. Show that the problem for $\tilde{L}$ on $\mathcal{S}$ is positive.

9.1I* For a regular problem on (a, b), the Robin condition $y'(a) = Ay(a)$ contributes the term $A(\tau y^2)(a)$ to (5′).

1. If $y \in \mathcal{S}$ and $\| y \| = 1$, use the mean value theorem to show that

$$1 = \int_a^b \rho y^2 \, dx = \ell \rho(\xi) y^2(\xi), \qquad \text{for some } \xi \in (a, b),$$

where $\ell = b - a$.

2. Use part 1 to show that

$$y^2(a) \geq \frac{1}{\ell \rho(\xi)} - 2\int_a^\xi |y||y'| \, dx \geq -\frac{1}{\ell \rho_0} - \frac{2}{\sqrt{\tau_0 \rho_0}} Y,$$

$$\text{where } Y = \left(\int_a^b \tau y'^2 \, dx \right)^{1/2},$$

if ρ_0 and τ_0 denote the minimum values of ρ and τ respectively on $[a, b]$. (*Hint*: Use (4).)

3. Obtain the estimate $A(\tau y^2)(a) \geq -2CY - C$ for a suitably large problem-constant C.
4. With m as in Problem 9.1H, use (6) to show that

$$\begin{aligned} I(y, y) &\geq Y^2 + m - 2CY - C + B(\tau y^2)(b) \\ &\geq Y^2 - 2CY - C, \end{aligned}$$

for perhaps a larger constant C, so that for this new C

$$I(y, y) \geq (Y - C)^2 - C^2 - C \geq -\tilde{m} + 1, \qquad \text{if } \tilde{m} = C^2 + C + 1.$$

5. Conclude that if $\tilde{L}(y) = L(y) - \tilde{m}\rho y$, then $\tilde{I}(y, y) = I(y, y) + \tilde{m} \geq 1$, so that the modified problem for $\tilde{L}(y) = -\tilde{\lambda}\rho y$ on $\mathcal{S}$ is regular and positive, and each of its eigenvalues $\tilde{\lambda}$ is ≥ 1. In fact, $\tilde{\lambda} = \lambda + \tilde{m}$, where λ is an eigenvalue of the original problem. The eigenfunctions are the same for both problems.

9.2

Completeness and S–L Series

According to Theorem 7.2, a *regular* S–L problem has a sequence of normalized eigenfunctions $y_1, y_2, \ldots$ such that for each $f \in \mathcal{S}$:

$$\sum_{n=1}^{\infty} c_n^2 = \int_{a}^{b} \rho f^2 \, dx, \qquad \textbf{(Parseval's formula)} \quad \textbf{(13)}$$

where

$$c_n = (f, y_n) = \int_{a}^{b} \rho f y_n \, dx, \qquad n = 1, 2, \ldots . \qquad \textbf{(13')}$$

More generally, we say that any S–L problem has a **complete** sequence of *normalized* eigenfunctions $y_1, y_2, \ldots$ when Parseval's formula holds for each $f \in \mathscr{S}$. Then, by standard approximation arguments, it is straightforward to extend (13) to each f that has finite norm $\| f \|$. Moreover, then the problem cannot have an eigenfunction y ρ-orthogonal to each y_n, $n = 1, 2, \ldots$ since from (13), we would conclude that $\int_a^b \rho y^2 \, dx = 0$, which means that y vanishes identically on (a, b). Consequently, by Proposition 7.1, the problem has *only* the distinct eigenvalues $\lambda_1, \lambda_2, \ldots$, where for each $n = 1, 2, \ldots, \lambda_n$ is the eigenvalue corresponding to y_n. We will say that the eigenvalue sequence $\{ \lambda_n \}_1^{\infty}$ is **regular** when it is *unbounded* and $\lambda_1 < \lambda_2 < \cdots$. It follows that there is at most a *finite* number of *negative* eigenvalues in a regular sequence, and there is a least positive integer P such that $n \geq P \Rightarrow \lambda_n > 0$.

Approximation by S–L Sums

For further insight, suppose that $\{y_n\}_1^{\infty}$ is any sequence of normalized eigenfunctions of an S–L problem to eigenvalues forming a *regular* sequence $\{ \lambda_n \}_1^{\infty}$. Then we can consider how well an f of finite norm is *approximated* by the Nth S–L sum

$$F_N = \sum_{n=1}^{N} c_n y_n, \qquad N = 1, 2, \ldots . \qquad \textbf{(14)}$$

To answer this question we will try to mimic the Fourier-series arguments from Chapter 1, especially those used to provide mean-square approximation (Section 1.8a) and uniform approximations (Theorem 1.14). In what follows we indicate the principal signposts along a path leading to our goal.

First, using ρ-orthogonality (Proposition 7.1) it is straightforward to establish that[†]

$$0 \leq \| f - F_N \|^2 = \int_{a}^{b} \rho f^2 \, dx - \sum_{n=1}^{N} c_n^2, \qquad \textbf{(15)}$$

and we see that the eigenfunction sequence $\{ y_n \}_1^{\infty}$ is complete iff

$$\| f - F_N \| \to 0 \qquad \text{as } N \to \infty \qquad \text{for each } f \in \mathscr{S}.^{\ddagger} \qquad \textbf{(16)}$$

(See Problem 9.2A.)

Moreover, if $\tilde{F}_N = \sum_{n=1}^{N} \tilde{c}_n y_n$, then by similar computations, we find that

$$\| f - F_N \|^2 = \| f - \tilde{F}_N \|^2 - \sum_{n=1}^{N} (c_n - \tilde{c}_n)^2 \leq \| f - \tilde{F}_N \|^2, \qquad \textbf{(17)}$$

[†] You should try to relate these expressions to ones introduced geometrically in Section 0.1e.

[‡] When we use (16) with (14) to define completeness for an arbitrary sequence of functions from $\mathscr{S}$, we find that the completeness of an eigenfunction sequence is not affected by normalization of its functions. See Problem 9.2F.

which guarantees that the Nth S–L sum provides the *best approximation-in-norm* to f among all sums of the form $\tilde{F}_N$. (See Problem 9.2B.)

Now, as $N \to \infty$ in (15), we get

$$\sum_{n=1}^{\infty} c_n^2 \le \int_a^b \rho f^2\,dx, \qquad \textit{(Bessel's inequality)} \quad \textbf{(18)}$$

and in proving Theorem 1.14, Bessel's inequality for f' was used to examine uniform approximation of f by its Fourier sums. For our S–L problem, we have an analogous inequality in the following lemma.

(9.4) Lemma: *If $f \in \mathscr{S}$ and* (13) *holds, then*

$$\sum_{n=1}^{\infty} \lambda_n c_n^2 \le I(f, f). \qquad \textbf{(19)}$$

Proof: Assume first that the problem is positive, and note that for each $N = 1, 2, \ldots,$

$$f_N \stackrel{\text{def}}{=} f - F_N = f - \sum_{n=1}^{N} c_n y_n \in \mathscr{S}$$

(since $\mathscr{S}$ is a subspace), and f_N is ρ-orthogonal to each y_n for $n \le N$. (Why?) Thus by (5) and (8)

$$\begin{aligned} 0 \le I(f_N, f_N) &= I(f, f) - I(F_N, f) - I(F_N, f_N) \\ &= I(f, f) - \sum_{n=1}^{N} \lambda_n c_n^2 \le I(f, f), \end{aligned} \qquad \textbf{(20)}$$

since as in (8′),

$$I(F_N, f) = -\int_a^b L(F_N) f\,dx = \sum_{n=1}^{N} \lambda_n c_n \int_a^b \rho y_n f\,dx = \sum_{n=1}^{N} \lambda_n c_n^2$$

is finite, while $I(F_N, f_N) = 0$, by a similar computation. As $N \to \infty$, we obtain the desired inequality. (See Problem 9.2C.)

Therefore the inequality holds when the problem is positive, and, in particular, it holds for our singular problems. But each regular problem on a subspace $\mathscr{S}$ can be replaced by the equivalent regular problem on $\mathscr{S}$ for the equation

$$\tilde{L}(y) \stackrel{\text{def}}{=} L(y) - \tilde{m}\rho y = -(\lambda + \tilde{m})\,\rho y, \qquad \textbf{(21)}$$

which is *positive* with positive eigenvalues $\tilde{\lambda}_n = \lambda_n + \tilde{m}$, $n = 1, 2, \ldots$ if the positive *constant* $\tilde{m}$ is sufficiently large. (See Problem 9.1I*.) The associated eigenfunctions are the same for both problems; hence, if $f \in \mathscr{S}$, we can use (19) for the modified problem to get

$$\sum_{n=1}^{\infty} (\lambda_n + \tilde{m}) c_n^2 \le \tilde{I}(f, f) = I(f, f) + \tilde{m} \int_a^b \rho f^2\,dx.$$

If in addition (13) holds, we recover (19). ∎

Now, $I(f, f) < +\infty$ for each $f \in \mathscr{S}_0$, where

$$\mathscr{S}_0 \overset{\text{def}}{=} \left\{ f \in \hat{C}^1\langle a, b\rangle: \quad I(f) \overset{\text{def}}{=} \int_a^b [\tau f'^2 + (|q| + \rho) f^2]\, dx < +\infty \right.$$
$$\left. \text{and } f \text{ has the zero boundary limits (if any) required in } \mathscr{S} \right\}, \tag{22}$$

and by standard approximations we can extend (19) to $\mathscr{S}_0$. (See Problem 9.2G*.)

S–L Series

The functions in $\mathscr{S}_0$ provide nice candidates for S–L series representation, even in singular cases. In fact, we have the following theorem.

(9.5) Theorem: *Suppose that an S–L problem on (a, b) has a complete sequence of normalized eigenfunctions $\{y_n\}_1^\infty$ associated with a regular eigenvalue sequence $\{\lambda_n\}_1^\infty$. If*

$$\Lambda^+ \overset{\text{def}}{=} \sum_{n=P}^{\infty} \frac{1}{\lambda_n} < +\infty, \tag{23}$$

then each $f \in \mathscr{S}_0$ has the representation

$$f(x) = \sum_{n=1}^{\infty} c_n y_n(x), \qquad x \in \langle a, b\rangle, \tag{24}$$

$$\text{where} \qquad c_n = (f, y_n) = \int_a^b \rho f y_n \, dx, \qquad n = 1, 2, \ldots . \tag{24'}$$

The series converges absolutely and uniformly on each compact interval $[a', b'] \subseteq \langle a, b\rangle$.

Proof: To establish this theorem we will use the nontrivial fact that for $x \in [a', b'] \subseteq \langle a, b\rangle$, $[y_n(x)]^2 \le M'$, where M' is a positive constant *independent* of n. (See Theorem 9.21.) Then, for $N \ge P$, Cauchy's inequality gives the estimate

$$\begin{aligned}
\left[\sum_{n=N}^{\infty} c_n y_n(x)\right]^2 &\le \left[\sum_{n=N}^{\infty} (\sqrt{\lambda_n}\,|c_n|)\left(\frac{|y_n(x)|}{\sqrt{\lambda_n}}\right)\right]^2 \\
&\le \left(\sum_{n=N}^{\infty} \lambda_n c_n^2\right) \sum_{m=N}^{\infty} \frac{y_m^2(x)}{\lambda_m} \\
&\le \left(\sum_{n=N}^{\infty} \lambda_n c_n^2\right) M' \sum_{m=N}^{\infty} \frac{1}{\lambda_m}, \qquad \text{if } x \in [a', b'] \\
&\le \left(\sum_{n=N}^{\infty} \lambda_n c_n^2\right) M' \Lambda^+.
\end{aligned}$$

The last series converges when $f \in \mathscr{S}_0$. (Why?) It follows that on $\langle a, b\rangle$ the series $\sum_{n=1}^{\infty} c_n y_n$ converges absolutely with sum F; moreover, the convergence is uniform on each compact subinterval $[a', b']$ so that F is continuous. It remains to prove that $F = f$ on $\langle a, b\rangle$, and it is here that completeness is used.

Observe that as in the above inequality, for $N \geq P$,

$$[F(x) - F_N(x)]^2 = \left| \sum_{n=N+1}^{\infty} c_n y_n(x) \right|^2 \leq M'\Lambda^+ \sum_{n=N}^{\infty} \lambda_n c_n^2, \qquad \text{if } x \in [a', b'].$$

Thus for fixed a' and b':

$$\int_{a'}^{b'} \rho(F - F_N)^2\, dx \leq M'\Lambda^+ \left(\sum_{n=N}^{\infty} \lambda_n c_n^2 \right) \int_{a'}^{b'} \rho\, dx, \text{ which } \to 0, \qquad \text{as } N \to \infty. \tag{25}$$

If we use the inequality $(u + v)^2 \leq 2u^2 + 2v^2$, we find that

$$0 \leq \int_{a'}^{b'} \rho(f - F)^2\, dx = \int_{a'}^{b'} \rho(f - F_N + F_N - F)^2\, dx$$

$$\leq 2 \int_{a}^{b} \rho(f - F_N)^2\, dx + 2 \int_{a'}^{b'} \rho(F - F_N)^2\, dx, \text{ which } \to 0, \qquad \text{as } N \to \infty,$$

since in the last line the first integral approaches zero by (16) and our assumption that the y_n are complete, while the second does so by (25). It follows that

$$\int_{a'}^{b'} \rho(f - F)^2\, dx = 0 \qquad \text{so that} \qquad \rho(f - F)^2(x) \equiv 0 \qquad \text{on } [a', b'],$$

$$\text{or since } \rho > 0\text{:} \quad f(x) = F(x), \qquad x \in [a', b'].$$

But each $x \in \langle a, b\rangle$ is in some compact subinterval $[a', b']$, and this establishes the representation. ∎

Observe that without completeness, the result cannot hold since the remaining hypotheses are satisfied when we delete λ_1 and y_1. However, $y_1 \in \mathscr{S}_0$, and it surely *cannot* be represented as $y_1 = \sum_{n=2}^{\infty} c_n y_n$, since, by Proposition 7.1, $c_n = (y_1, y_n) = 0$, $n = 2, 3, \ldots$.

Any regular problem has a complete sequence of eigenfunctions associated with a regular sequence of eigenvalues (see [Wein]), and in Section 9.3 we will see that the eigenvalues satisfy (23). Thus this theorem guarantees the uniform representation result stated in Theorem 7.2. However, each singular problem requires individual consideration since it might not admit *any* eigenfunctions. Let us look at two important cases.

Bessel Series

We know that the functions $J_0(\omega_n x)$, $u = 1, 2, \ldots$ used to form Bessel series in Section 7.4 are ρ-orthogonal on (0, 1) for $\tau(x) = \rho(x) = x$, and it can be shown that their normalized versions are complete. (See [Wein] and [Hi].) We also see that the eigenvalues $\lambda_n = \omega_n^2$ form a regular sequence, and since $\omega_n \approx n\pi - (\pi/4)$ for $n \geq 5$, we conclude that $\Lambda^+ = \sum_{n=1}^{\infty} \omega_n^{-2} < +\infty$. For this singular problem, $\langle 0, 1\rangle = (0, 1]$, and we can take

$$\mathscr{S}_0 = \{ f \in \hat{C}^1(0, 1]\text{:} \quad f \text{ and } f' \text{ are bounded and } f(1) = 0 \}. \tag{26}$$

Moreover, each normalized eigenfunction takes the form $y_n(x) = j_n J_0(\omega_n x)$, and if we use the fact that

$$\int_0^1 x J_0^2(\omega_n x)^2 \, dx = \tfrac{1}{2}[J_0'(\omega_n)]^2, \qquad n = 1, 2, \ldots$$

(see Problem 7.4E), the normalizing constant j_n is given by $j_n = \sqrt{2}/|J_0'(\omega_n)|$. Then, from Theorem 9.5, we obtain our principal result.

(9.6) Proposition: *Each $f \in \mathscr{S}_0$ of (26) has the Bessel series representation*

$$f(x) = \sum_{n=1}^{\infty} C_n J_0(\omega_n x), \qquad 0 < x \le 1 \tag{27}$$

where $J_0(\omega_n) = 0$ and

$$C_n = \frac{2}{[J_0'(\omega_n)]^2} \int_0^1 x f(x) J_0(\omega_n x) \, dx, \qquad n = 1, 2, \ldots . \tag{27'}$$

The series converges absolutely and uniformly on each compact interval $[a, 1] \subseteq (0, 1]$. ■

Corresponding results can be obtained for the series involving Bessel functions of order $m \neq 0$ and for problems involving other endpoint conditions. (See [Pi] and [Wein].)

Legendre Series

An important case where completeness can be established directly is that of the Legendre polynomials on $(-1, 1)$, which were introduced in Section 7.5. Let us first use (39) from that section to define the *normalized polynomials*

$$p_n(x) = \sqrt{\frac{2n+1}{2}} \, P_n(x), \qquad n = 0, 1, 2, \ldots , \tag{28}$$

Then a given $f \in C[-1, 1]$ has coefficients

$$c_n = \int_{-1}^{1} f p_n \, dx, \qquad n = 0, 1, 2, \ldots \tag{29}$$

and generates the polynomial *approximants*

$$F_N(x) = \sum_{n=0}^{N} c_n p_n(x), \qquad N = 1, 2, \ldots . \tag{30}$$

(9.7) Lemma: *If $f \in C[-1, 1]$, then $\int_{-1}^{1} (f - F_N)^2 \, dx \to 0$ as $N \to \infty$.*

Proof: For each $\epsilon > 0$, by the Weierstrass approximation theorem (1.17), f can be uniformly approximated on $[-1, 1]$ by a polynomial P_ϵ for which

$$\int_{-1}^{1} (f - P_\epsilon)^2 \, dx < \epsilon.$$

However, P_ϵ is of *some* degree N and since each p_n is exactly of degree n, it follows that

$$P_\epsilon = \sum_{n=0}^{N} \tilde{c}_n p_n,$$

for *some* choice of coefficients $\tilde{c}_1, \tilde{c}_2, \ldots, \tilde{c}_N$. Therefore, exactly as in the proof of (17) we see from (30) that

$$\int_{-1}^{1} (f - F_N)^2\, dx \le \int_{-1}^{1} (f - P_\epsilon)^2\, dx < \epsilon,$$

and since ϵ is arbitrary, we have the desired conclusion. ∎

Using standard approximation arguments, we can extend the conclusion of the lemma to any f Riemann integrable on $[-1, 1]$ and establish completeness of the $\{p_n\}$ (or equivalently, of the $\{P_n\}$). It follows that Legendre's equation of order zero

$$[(1 - x^2)y']' = -\lambda y \qquad \text{on } (-1, 1)$$

cannot have any integrable eigenfunctions except scalar multiples of the P_n. Moreover, $\lambda_n = n(n+1)$, $n = 0, 1, \ldots$ provides a regular eigenvalue sequence satisfying (23). Finally, if $f \in C^1[-1, 1]$, then (19) gives the inequality

$$\sum_{n=1}^{\infty} \lambda_n c_n^2 \le I(f, f) = \int_{-1}^{1} \tau f'^2\, dx < +\infty, \tag{31}$$

which extends to $\mathscr{S}_0 = \{f \in \hat{C}^1(-1, 1) : I(f) < +\infty\}$. Thus Theorem 9.5 is applicable as stated, and if we return to the original P_n, we obtain the following result.

(9.8) Theorem: *If $f \in \hat{C}^1(-1, 1)$ and $\int_{-1}^{1} (\tau f'^2 + f^2)\, dx < +\infty$, where $\tau(x) = 1 - x^2$, then f has the Legendre series representation*

$$f(x) = \sum_{n=0}^{\infty} C_n P_n(x),$$

where

$$C_n = (n + \tfrac{1}{2}) \int_{-1}^{1} f P_n\, dx, \qquad n = 0, 1, 2, \ldots .$$

The series converges absolutely and uniformly on each interval $[a, b] \subseteq (-1, 1)$.† ∎

Using a method patterned after the proof of Theorem 1.11, it is possible to obtain equally strong pointwise convergence results at an interior point x of jump discontinuity, where the series then has the value $[f(x+) + f(x-)]/2$. (See Problems 9.5E and F.) Corresponding results for the associated Legendre functions of Section 7.6 can be derived. (See Problem 9.2E.)

Orthogonal Polynomials

The Legendre polynomials give us one system of polynomials that is complete and ρ-orthogonal on an interval and that provides useful series representation for functions on the interval.

---

† If $f' \in \hat{C}^1[-1, 1]$, the series converges absolutely and uniformly on $[-1, 1]$. See Problem 7.5F.

As examples of other systems, we have the following.

(a) ***Tchebycheff polynomials:*** $T_n(x) = 2^{1-n} \cos(n \cos^{-1} x)$, which are ρ-orthogonal on $(-1, 1)$ for $\rho(x) = (1 - x^2)^{-1/2}$, and satisfy the equation

$$(\rho^{-1} y')' = -n^2 \rho y, \qquad n = 0, 1, 2, \ldots . \tag{32}$$

These polynomials have best uniform approximation properties. (See Problem 1.9I.)

(b) ***Hermite polynomials:*** $H_n(x) = (-1)^n e^{x^2} (d^n/dx^n)(e^{-x^2})$, which are ρ-orthogonal on $(-\infty, \infty)$ for $\rho(x) = e^{-x^2}$, and satisfy the equation

$$(\rho y')' = -2n \rho y, \qquad n = 0, 1, 2, \ldots . \tag{33}$$

These polynomials arise in analyzing the quantum-mechanical model of a linear oscillator. (See [Ja].)

(c) ***Laguerre polynomials:*** $L_n(x) = e^x (d^n/dx^n)(x^n e^{-x})$, which are ρ-orthogonal on $(0, \infty)$ for $\rho(x) = e^{-x}$, and satisfy the equation

$$(x \rho y')' = -n \rho y, \qquad n = 0, 1, 2, \ldots . \tag{34}$$

These polynomials help describe the eigenstates of the quantum-mechanical model for the hydrogen atom. (See Problem 7.6L.).

Observe that the eigenvalues of (b) and (c) do *not* satisfy (23). For more on these and similar systems, see [C-H] and [Ja].

Problem Set 9.2

9.2A To establish (15):

1. show that $\| f - F_N \|^2 = (f - F_N, f - F_N) = (f, f) - (f, F_N) - (f - F_N, F_N)$.
2. then show that $(f - F_N, F_N) = 0$, while $(f, F_N) = \sum_{n=1}^{N} c_n^2$. (*Hint*: $(f - F_N, y_n) = 0$, for $n \le N$.)

9.2B To establish (17):

1. use $f - \tilde{F}_N = (f - F_N) + (F_N - \tilde{F}_N)$ to show that $\| f - \tilde{F}_N \|^2 = \| f - F_N \|^2 + \| F_N - \tilde{F}_N \|^2$, since $(f - F_N, F_N - \tilde{F}_N) = 0$, by the hint in Problem 9.2A.
2. show that $\| F_N - \tilde{F}_N \|^2 = \sum_{n=1}^{N} (c_n - \tilde{c}_n)^2$, using (15) with $f = \sum_{n=1}^{N} (c_n - \tilde{c}_n) y_n$.

9.2C Verify the computations indicated in (20).

9.2D Consider an S–L problem defined by (3) and Definition 9.1.

1. If $\| f \| < \infty$ and $\| \tilde{f} \| < \infty$, conclude that $\| f \pm \tilde{f} \| < +\infty$. (*Hint*: $(u \pm v)^2 \le 2u^2 + 2v^2$. Then show that $\| f \pm \tilde{f} \| \le \| f \| + \| \tilde{f} \|$ [the triangle inequality] follows from (4).)
2. If the problem has a complete set of normalized eigenfunctions, show that $\sum_{n=1}^{\infty} c_n \tilde{c}_n = (f, \tilde{f})$. (*Hint*: $2f\tilde{f} = (f + \tilde{f})^2 - f^2 - \tilde{f}^2$.)

9.2E When $m = 1$, the associated Legendre functions $P_n^1(x) = \sqrt{1-x^2}\,P_n'(x)$ of Section 7.6 are eigenfunctions of Legendre's equation of order 1 corresponding to the eigenvalues $\lambda_{1n} = n(n+1)$, $n = 1, 2, \ldots$.

1. Using Theorem 1.17 and the fact that P_n' is a polynomial of degree $n-1$, explain how to show that $\{P_n^1\}$ is complete.
2. Define $\mathscr{S}_0$ and use Theorem 9.5 to formulate a representation result for these P_n^1.
3. How would these results change for the associated Legendre functions $\{P_n^2\}$? for the functions $\{P_n^m\}$?

9.2F Let Y_n be eigenfunctions to distinct eigenvalues λ_n, $n = 1, 2, \ldots$ for an S–L problem, and suppose that to each f with $\| f \| < +\infty$ and each $\epsilon > 0$, there is some

$$\tilde{F}_N = \sum_{n=1}^{N} \tilde{c}_n Y_n \qquad \text{with} \qquad \| f - \tilde{F}_N \|^2 < \epsilon.$$

Show that if y_n is a normalized eigenfunction to λ_n, then the y_n are complete. (*Hint*: Use equation (17).)

9.2G* To extend Lemma 9.4, we need to show that if $f \in \mathscr{S}_0$ of (22) for a *positive* problem, then for each $N = 1, 2, \ldots$,

$$\sum_{n=1}^{N} \lambda_n c_n^2 \le I(f, f) = \int_a^b (\tau f'^2 + q f^2)\, dx + A f^2(a) + B f^2(b),$$

where we have used (5′) with $A\tau(a)$ replaced by A and $B\tau(b)$ replaced by B. Proof is required only when $\mathscr{S} \subsetneq \mathscr{S}_0$.

1. If $I(f-y) < \epsilon$ for some $y \in \mathscr{S}$ with S–L coefficients $\tilde{c}_n$, $n = 1, 2, \ldots$, show that $\sum_{n=1}^{N}(c_n - \tilde{c}_n)^2 \le \epsilon$. (*Hint*: Use (17).)
2. By a variant of (4), show also that
$$\int_a^b \tau y'^2\, dx \le \int_a^b \tau f'^2\, dx + \epsilon + 2\sqrt{\epsilon I(f)}$$
and similarly
$$\int_a^b |q| y^2\, dx \le \int_a^b |q| f^2\, dx + \epsilon + 2\sqrt{\epsilon I(f)}.$$
3. If $A = B = 0$ in a positive problem, argue that for some constant M:
$$\sum_{n=1}^{N} \lambda_n \tilde{c}_n^2 \le I(y, y) \le I(f, f) + \sqrt{\epsilon} M$$
so that as $\epsilon \to 0$, by part 1, each $\tilde{c}_n \to c_n$, and hence $\sum_{n=1}^{N} \lambda_n c_n^2 \le I(f, f)$ if we can approximate f by y appropriately. In this case, modify f near *regular* endpoints as necessary to obtain $y \in \mathscr{S}$, with $I(f-y) < \epsilon$. (*Hint*: If $y'(a) = 0$ is required, then set $y(x) = f(a')$, $a < x \le a'$, where a' is so near a that the Schwarz inequality estimate
$$(f-y)^2(x) \le |x - a'| \left| \int_{a'}^{x} (f' - y')^2\, d\xi \right|, \qquad a < x \le a'$$
controls the inequality.)

4. If only $A \neq 0$, then the approximation in part 1 must be made with $Ay^2(a) \leq Af^2(a) + M\sqrt{\epsilon}$. Explain why this is possible, and how the argument in part 3 is affected. What should we do if also $B \neq 0$?

9.2H 1. If $\int_0^1 xf^2\,dx < +\infty$, then $\int_0^1 \tilde{f}^2\,dx < +\infty$, where $\tilde{f}(x) = \sqrt{x}f(x)$. Use Theorem 1.14 (Parseval's formula for Fourier series) with standard approximation arguments to establish that

$$\int_0^1 \tilde{f}^2(x)\,dx = \sum_{n=1}^{\infty} s_n^2$$

where $s_n = 2\int_0^1 \tilde{f}(x)\sin n\pi x\,dx$, $n = 1, 2, \ldots$.

2. How does the result in part 1 show completeness of the eigenfunctions

$$y_n(x) = \sqrt{\frac{2}{x}}\sin n\pi(1-x) = (-1)^{n+1}\sqrt{\frac{2}{x}}\sin n\pi x, \qquad n = 1, 2, \ldots,$$

for the Bessel problem of order $m = 1/2$. (See Example 7 in Section 9.4.)

3. State a representation theorem for this problem using Theorem 9.5. Can you obtain stronger results from Section 1.8?

9.3

The Rayleigh Principle

While investigating natural vibrations of a nonuniform stretched string in 1870, Lord Rayleigh discovered an important energy identity. To obtain it, suppose the string is stretched over the x-interval $[a, b]$ with local density $\rho = \rho(x)$ and tension $\tau = \tau(x)$. Then free vibrations at frequency ω (of period $c = 2\pi/\omega$) have mean potential energy per cycle given as in Section 5.1 by

$$\bar{P} = \frac{1}{2c}\int_0^c dt \int_a^b \tau u_x^2\,dx = -\frac{1}{2c}\int_0^c dt \int_a^b u(\tau u_x)_x\,dx \tag{35}$$

and mean kinetic energy per cycle

$$\bar{T} = \frac{1}{2c}\int_0^c dt \int_a^b \rho u_t^2\,dx = -\frac{1}{2c}\int_0^c dt \int_a^b u(\rho u_t)_t\,dx. \tag{36}$$

The boundary terms produced by the partial integrations vanish in (35) because $u(a, t) = u(b, t) = 0$ and in (36) because $u(x, 0) = u(x, c)$ and $u_t(x, 0) = u_t(x, c)$, from periodicity.

Thus $\bar{P} = \bar{T}$, since $(\tau u_x)_x = (\rho u_t)_t$. (Why?) Moreover, when $u(x, t) = y(x)\cos\omega t$, then

$$\bar{P} = \frac{1}{2c}\int_0^c \cos^2\omega t\,dt \int_a^b \tau(y')^2\,dx = \frac{1}{4}\int_a^b \tau(y')^2\,dx$$

and

$$\bar{T} = \frac{\omega^2}{2c}\int_0^c \sin^2\omega t\,dt \int_a^b \rho y^2\,dx = \frac{\omega^2}{4}\int_a^b \rho y^2\,dx,$$

since $\displaystyle\int_0^c \sin^2 \omega t\, dt = \int_0^c \cos^2 \omega t\, dt = \frac{c}{2}$, when $\omega = 2\pi/c$.

Hence

$$\omega^2 = \left(\frac{\displaystyle\int_a^b \tau y'^2\, dx}{\displaystyle\int_a^b \rho y^2\, dx} \right) = R(y), \text{ say.}$$

In particular, this relation holds for the fundamental, or lowest, natural frequency ω_1 and its associated mode shape y_1. Rayleigh argued that vibrations in other modes y could be induced by appropriate constraints whose effect was to "stiffen" the system and raise the fundamental frequency, ω. Thus we should expect that if $y \in C^1[a, b]$, with $y(a) = y(b) = 0$, then

$$\omega_1^2 = R(y_1) \le \omega^2 = R(y). \tag{37}$$

This inequality, known as **Rayleigh's principle**, is the basis of most subsequent work in the subject. Observe that for the above problem the **Rayleigh ratio** (or quotient) can be expressed as

$$R(y) = \frac{I(y, y)}{\displaystyle\int_a^b \rho y^2\, dx} \tag{38}$$

if we use (5).

Assume that an S–L problem specified by (3) and Definition 9.1 has a *complete* sequence of normalized eigenfunctions $y_1, y_2, \ldots$ associated with a *regular* sequence of eigenvalues $\lambda_1, \lambda_2, \ldots$. Then we can establish Rayleigh's principle for λ_1 and supply an analogous characterization of each λ_n.

(9.9) Definition: *Let $\mathscr{S}_1$ be the class of functions in $\mathscr{S}_0$ which do not vanish identically on (a, b), and for each $N = 2, 3, \ldots$, let $\mathscr{S}_N$ be the class of functions in $\mathscr{S}_1$ that are ρ-orthogonal to y_n when $n < N$.*

Observe that $\mathscr{S}_N \subsetneq \mathscr{S}_{N-1} \subsetneq \cdots \subsetneq \mathscr{S}_1 \subsetneq \mathscr{S}_0$ (which was defined in (22)) and that the functions in these sets are the survivors as successive constraints are placed on the functions of $\mathscr{S}_0$. Moreover, $y_n \in \mathscr{S}_n \sim \mathscr{S}_{n+1}$, $n = 1, 2, \ldots, N-1$.

(9.10) Proposition (*Rayleigh's principle*):†

$$\lambda_N = R(y_N) = \min_{y \in \mathscr{S}_N} R(y), \; N = 1, 2, \ldots .$$

Proof: That $\lambda_N = R(y_N)$ follows from (6′). If $y \in \mathscr{S}_N$ has S–L coefficients c_n, $n = 1, 2, \ldots$, then $c_n = 0$ if $n < N$ (Why?), and from Parseval's formula (13), we see that

$$\sum_{n=N}^{\infty} c_n^2 = \int_a^b \rho y^2\, dx. \tag{39}$$

† This characterization of eigenvalues was first proposed by H. Weber in 1869.

Similarly, $\lambda_n \geq \lambda_N$ when $n \geq N$, and from (19) it follows that

$$\lambda_N \sum_{n=N}^{\infty} c_n^2 \leq \sum_{n=N}^{\infty} \lambda_n c_n^2 \leq I(y, y).$$

Thus, by (38),

$$\lambda_N \leq R(y). \qquad \blacksquare$$

This result provides a means of defining the eigenvalues successively and approximating them from above. Moreover, using variational methods, it can be shown that for regular problems, functions that actually minimize at each stage must be eigenfunctions to the eigenvalue so defined. (See, e.g., [Tr], App. 6.)

Nodes of Eigenfunctions

We can use Rayleigh's principle to obtain precise information about the **nodes** of the eigenfunctions y_n, those points $x \in (a, b)$ where $y_n(x) = 0$. At each node, y_n' cannot vanish (Why?) so that y_n changes sign.

(9.11) Proposition: *The nth eigenfunction y_n of the regular S–L problem*

$$L(y) = -\lambda \rho y \qquad \text{on } [a, b], \qquad \text{with } y(a) = y(b) = 0,$$

has exactly $n - 1$ nodes in (a, b) and it changes sign at each node.

Proof: Suppose that y_n has at least $J - 1$ nodes, x_j, partitioning the interval $[a, b]$ into subintervals $I_j = [x_{j-1}, x_j]$, $j = 1, 2, \ldots, J$, where $J \geq 1$, and $x_0 = a$, $x_J = b$. Then for each $j = 1, 2, \ldots, J$, the function

$$\phi_j \overset{\text{def}}{=} \begin{cases} y_n, & \text{in } I_j \\ 0, & \text{otherwise} \end{cases} \quad \text{is in } \mathcal{S}_0. \tag{40}$$

Observe that for any constants, $c_j, j = 1, 2, \ldots, J$,

$$\phi \overset{\text{def}}{=} \sum_{j=1}^{J} c_j \phi_j \in \mathcal{S}_0,$$

and we can select these c_j to make ϕ nonzero and ρ-orthogonal to $y_1, \ldots, y_{J-1}$. (See the proof of Proposition 9.14.) Thus $\phi \in \mathcal{S}_J$, so that by Proposition 9.10 $\lambda_J \leq R(\phi)$. By actual computation, $R(\phi) = \lambda_n$. (See Problem 9.3J.) Hence $J \leq n$ or $J - 1 \leq n - 1$, so that y_n has no more than $n - 1$ nodes.

In particular, y_1 cannot have *any* nodes in (a, b), so that it is nonvanishing there, and we will suppose it to be *negative*.

We know also that y_2 cannot have more than one node. Observe that the Wronskian,

$$W_{12} = \tau(y_1 y_2' - y_2 y_1'), \tag{41}$$

has, by (3) and cancellation, the derivative

$$W_{12}' = y_1[(\tau y_2')' - q y_2] - y_2[(\tau y_1')' - q y_1] = -(\lambda_2 - \lambda_1)\rho y_1 y_2, \tag{41'}$$

which carries the sign of y_2. (Why?)

If y_2 is assumed nonvanishing and positive on (a, b), then W_{12} is *strictly* increasing on $[a, b]$ contradicting the fact that by (3') $W_{12}(a) = W_{12}(b) = 0$.

Hence, y_2 has exactly one node in (a, b); i.e., *y_2 has one node strictly between the successive zeros a and b of y_1*. Observe that this argument is still valid when $y_2(a)$ or $y_2(b)$ is positive since $y_1'(a) < 0$ and $y_1'(b) > 0$. (See the proof of Theorem B.3 in Appendix B.)

Next, let c denote the node of y_2 so that on each of the subintervals (a, c) and (c, b), y_2 is a nonvanishing eigenfunction that vanishes at the endpoints. If also y_3 does not vanish on one of these subintervals, we would have a contradiction upon considering W_{23} defined in analogy to W_{12}. Hence, y_3 has exactly two nodes — one between each pair of successive zeros of y_2, and this argument evidently extends inductively to each $n = 1, 2, \ldots$. ■

Remarks: On the interval (a, c) as above, y_2 is a *nonvanishing* eigenfunction for the regular problem $L(y) = -\lambda\rho y$ with $y(a) = y(c) = 0$. Hence, it must be an eigenfunction for the *least* eigenvalue on this interval; i.e., λ_2 is the least eigenvalue for this same problem on the interval (a, c) (and also on the interval (c, b)). Similarly, λ_n is the least eigenvalue for the problem on each of the n subintervals of (a, b) between successive zeros of y_n.

If the simple boundary conditions of Proposition 9.11 are replaced by another set satisfying conditions of Definition 9.1, we can still apply the above Wronskian comparisons on subintervals to get the following.

(9.12) Corollary: *In general, y_n must have at least one node between successive zeros of y_j, for each $j < n$.* ■

As a result, we might expect the Legendre polynomials, or any system of orthogonal polynomials associated with a positive S–L problem, to have increasing numbers of zeros in the intervals defining the polynomials. (A more precise result is established in Problem 9.3C.) We can also obtain insight about the distribution of zeros of J_0. (See Problem 9.3D.) However, we should also recall the situation in Example 3 of Section 7.2 with $B > 1$, where $y_0(x) = \beta_0 \sinh \omega_0 x$ does not have successive zeros and $y_1(x) = \beta_1 \sin \omega_1 x$ has no node in (0, 1).

Maximin Principle

In practice, the eigenfunctions y_n cannot be determined with sufficient numerical accuracy to make the orthogonality conditions defining $\mathcal{S}_N$ meaningful. Thus in estimating λ_N much effort has been directed toward circumventing the use of $y_1, y_2, \ldots, y_{N-1}$.

(9.13) Lemma: *If $y = \sum_{m=1}^{N} c_m y_m$ and some constant $c_m \neq 0$, $m = 1, 2, \ldots, N$, then $R(y) \leq \lambda_N$.*

Proof: Note that $R(y)$ *is* finite. Indeed, by substitution

$$R(y) = \frac{\sum_{m,n=1}^{N} \gamma_{mn} c_m c_n}{\sum_{m,n=1}^{N} \delta_{mn} c_m c_n},$$

where
$$\delta_{mn} = 0 \text{ if } m \neq n, \text{ and } \delta_{nn} = 1;\ m, n, = 1, 2, \ldots \tag{42}$$

and by (8′) and (3):

$$\gamma_{mn} \overset{\text{def}}{=} I(y_n, y_m) = -\int_a^b y_n L(y_m)\,dx = \lambda_m \int_a^b \rho y_n y_m\,dx = \lambda_m \delta_{mn},$$
$$m, n, = 1, 2, \ldots, N.$$

Thus

$$R(y) = \left(\frac{\sum_{n=1}^{N} c_n^2 \lambda_n}{\sum_{n=1}^{N} c_n^2} \right) \leq \lambda_N \left(\frac{\sum_{n=1}^{N} c_n^2}{\sum_{n=1}^{N} c_n^2} \right) = \lambda_N$$

since $\lambda_n \leq \lambda_N$ when $n \leq N$. ■

Now, for each $n = 1, 2, \ldots, N-1$, let ϕ_n be *any* integrable function with $\int_a^b \rho \phi_n^2\,dx < +\infty$, and let Φ_N denote the class of functions in $\mathscr{S}_1$ that are ρ-orthogonal to these ϕ_n.

(9.14) Proposition: $\min_{y \in \Phi_N} R(y) \leq \lambda_N, \quad N = 1, 2, \ldots.$

Proof: We can find constants c_m not all zero such that y of Lemma 9.13 is in Φ_N since the necessary $N-1$ equations in the N unknown c_m; viz.,

$$0 = \int_a^b \rho \phi_n y\,dx = \sum_{m=1}^{N} c_m \int_a^b \rho \phi_n y_m\,dx,\ n = 1, 2, \ldots, N-1,$$

always have such a solution. Thus $y \in \Phi_N$ and $R(y) \leq \lambda_N$. ■

It can be shown that the minimum in Proposition 9.14 is achieved. (See [W-S].) If $\phi_n = y_n$, $n \leq N$, we have equality in Proposition 9.14 and it follows that

$$\lambda_N = \max_{\{\phi_1, \ldots, \phi_{N-1}\}} \min_{y \in \Phi_N} R(y). \tag{43}$$

This result, known as the **Courant–Weyl maximin principle**,† gives a characterization of λ_N that is independent of eigenfunctions. It can be used, for example, to verify that the eigenvalues decrease when we relax conditions at b from $y(b) = 0$ to $y'(b) = -By(b)$ $(B \geq 0)$. Such behavior is to be expected from Rayleigh's principle. (See Problem 9.3F.)

Eigenvalue Comparisons

On physical grounds, we expect the natural frequencies of a vibrating string to decrease as the tension τ is decreased and/or the density ρ is increased, assuming that the length and the boundary conditions are unchanged. Using the above techniques, we can establish more comparisons between eigenvalues of S–L problems that have *complete sets of eigenfunctions and regular sequences of eigenvalues.*

† In fact, it was first formulated by H. Weyl in 1911. (See [W-S].)

Suppose that we have such S–L problems for equations

$$L(y) \stackrel{\text{def}}{=} (\tau y')' - qy = -\lambda \rho y, \qquad \text{on } (a, b), \tag{44}$$

and

$$\tilde{L}(y) \stackrel{\text{def}}{=} (\tilde{\tau} y')' - \tilde{q}y = -\tilde{\lambda} \tilde{\rho} y, \qquad \text{on } (a, b), \tag{44'}$$

with respective boundary-requirement sets $\mathscr{S}$ and $\tilde{\mathscr{S}}$ of Definition 9.1. Under these circumstances, we have the following result.

(9.15) Proposition: *If* (1) $\tilde{\tau} \le \tau$, $\tilde{q} \le q$, $\rho \le \tilde{\rho}$, and $\tilde{\mathscr{S}} = \mathscr{S}$ *and* (2) *in* (5′), *both A and B are nonnegative, then the eigenvalues of* (44) *and* (44′) *satisfy the inequalities* $\tilde{\lambda}_N \le \lambda_N$, $N = 1, 2, \ldots$.

Remarks: When conditions (1) hold, conditions (2) can be relaxed somewhat; for example, A can be negative provided that $\tau(a) = \tilde{\tau}(a)$. Moreover, eigenvalue comparison is possible in some cases where $\mathscr{S} \neq \tilde{\mathscr{S}}$. (See Problem 9.3F.)

Proof: Let $\tilde{y}_n$ denote a normalized eigenfunction to $\tilde{\lambda}_n$ of (44′), $n = 1, 2, \ldots$. Then for a given $N > 1$ we can construct a nonzero y of the form

$$y = \sum_{m=1}^{N} c_m y_m$$

exactly as in the proof of Proposition 9.14 which is $\tilde{\rho}$-orthogonal to each $\tilde{y}_n$ when $n < N$. Moreover,

$$\int_a^b \tilde{\rho} y^2 \, dx < +\infty$$

(since $\mathscr{S} = \tilde{\mathscr{S}}$) so that with obvious notation, by Proposition 9.10,

$$\tilde{\lambda}_N \le \tilde{R}(y) = \frac{\tilde{I}(y, y)}{\int_a^b \tilde{\rho} y^2 \, dx}.$$

But also by the hypothesized monotonicity we see from (5′) that

$$\tilde{I}(y, y) \le I(y, y), \qquad \text{while} \qquad \int_a^b \rho y^2 \, dx \le \int_a^b \tilde{\rho} y^2 \, dx.$$

Hence, using Lemma 9.13, we conclude that

$$\tilde{\lambda}_N \le \tilde{R}(y) \le R(y) \le \lambda_N, \qquad \text{if } N > 1;$$

when $N = 1$, we have $\tilde{\lambda}_1 \le \tilde{R}(y_1) \le R(y_1) = \lambda_1$. ∎

Example 2 (*The zeros of* J_m):

The singular problem for Bessel's equation of order $m \ge 0$

$$L_m(y) \stackrel{\text{def}}{=} (xy')' - \frac{m^2 y}{x} = -\lambda xy \qquad \text{on } (0, 1], \tag{45}$$

with $y(1) = 0$ and y and y' bounded, has a complete sequence of normalized eigenfunctions, y_{mn}, to eigenvalues λ_{mn}, $n = 1, 2, \ldots$ forming a regular sequence. (See [Hi], [Wein], and [Wat].)

Within normalization on (0, 1], the only bounded solutions and hence the only eigenfunctions of the above equation are the Bessel functions $J_m(\sqrt{\lambda}x)$, so that when $x = 1$, we have $J_m(\sqrt{\lambda_{mn}}) = 0$, $n = 1, 2, \ldots$. For each such λ_{mn}, $J_m(\sqrt{\lambda_{mn}}x)$ gives a corresponding eigenfunction for our problem. We wish to investigate the distribution of these eigenvalues λ_{mn}, or equivalently that of the positive zeros of J_m.

Clearly $q(x) = m^2/x$ increases with m, so that by Proposition 9.15, for each $n = 1, 2, \ldots$ we have

$$\lambda_{\tilde{m}n} \leq \lambda_{mn}, \qquad \text{when} \qquad 0 \leq \tilde{m} \leq m.$$

Moreover when $m = 1/2$, it will be shown in Example 7 that

$$y_n(x) = \sqrt{\frac{2}{x}} \sin nx, \qquad x \in (0, 1] \tag{46}$$

is an eigenfunction to the eigenvalue $\lambda_n = n^2\pi^2$, $n = 1, 2, \ldots$ and that these eigenfunctions are complete. (See Problem 9.2H.) Hence for $m \geq 1$:

$$\lambda_{0n} \leq n^2\pi^2 \leq \lambda_{mn}, \qquad n = 1, 2, \ldots .$$

Now let $\omega_{mn} = \sqrt{\lambda_{mn}}$ denote the nth zero of J_m. Since $J_0' = -J_1$ we see that J_0 and J_1 cannot vanish simultaneously, and moreover, by Rolle's theorem, that between successive zeros of J_0, there must be a zero of J_1. However, from the preceding estimates, this behavior is only possible if

$$\omega_{01} \leq \pi \leq \omega_{11} < \omega_{02} \leq 2\pi \leq \omega_{12} \ldots .$$

Consequently, both J_0 and J_1 have precisely one zero in each indicated interval of length π, and therefore,

$$\lim_{n\to\infty} \frac{\lambda_{0n}}{n^2} = \lim_{n\to\infty} \frac{\lambda_{1n}}{n^2} = \pi^2.$$

This argument can be extended successively to each J_m because (from Problem 7.4F)

$$[x^{-m} J_m(x)]' = -x^{-m} J_{m+1}(x), \qquad m = 0, 1, 2, \ldots .$$

For instance, when $m = 1$, Rolle's theorem gives

$$\omega_{11} < \omega_{21} < \omega_{2,n} < \omega_{1,n+1} \leq (n+2)\pi.$$

Therefore

$$\lim_{n\to\infty} \frac{\lambda_{2n}}{n^2} = \lim_{n\to\infty} \frac{\omega_{2n}^2}{n^2} = \pi^2. \tag{47}$$

We can summarize our conclusions as follows.

(9.16) Proposition: *For $m = 0, 1, 2, \ldots$, let $\omega_{mn} = \sqrt{\lambda_{mn}}$, $n = 1, 2, \ldots$ denote the successive positive zeros of J_m. These numbers satisfy the inequalities*

$$\left.\begin{aligned} &(n-1)\pi < \omega_{0n} \leq n\pi \leq \omega_{mn} < (n+m)\pi, \qquad (m \geq 1) \\ &\omega_{m,n} < \omega_{m+1,n} < \omega_{m,n+1} \end{aligned}\right\} \tag{48}$$

so that

$$\lim_{n\to\infty} \frac{\lambda_{mn}}{n^2} = \pi^2. \tag{49}$$

■

Remark: More precise estimates show that for $m = 0, 1, 2, \ldots$:

$$\left|\omega_{mn} - \left(n - \frac{1}{4} + \frac{m}{2}\right)\pi\right| \le \frac{|M|}{n\pi}, \quad n = 1, 2, \ldots \tag{50}$$

where $M = m^2 - \frac{1}{4}$. (See [Wat] and Problem 9.3K.)

Estimates of Eigenvalues of Regular Problems

One possible source of comparison-equations (44) or (44′) is that where the coefficient functions are appropriately chosen constants. For a *regular problem*, constant bounds are available for τ, q, and ρ in the form

$$\left.\begin{array}{l} 0 < \underline{\tau} \le \tau(x) \le \bar{\tau} \\ \underline{q} \le q(x) \le \bar{q} \\ \underline{\rho} \ge \rho(x) \ge \bar{\rho} > 0 \end{array}\right\} x \in [a, b]. \tag{51}$$

From these constants, we can calculate the ratios

$$\underline{T} = \frac{\underline{\tau}}{\underline{\rho}}, \qquad \bar{T} = \frac{\bar{\tau}}{\bar{\rho}} \qquad \text{and} \qquad \underline{Q} = \frac{\underline{q}}{\underline{\rho}}, \qquad \bar{Q} = \frac{\bar{q}}{\bar{\rho}} \tag{52}$$

and use them in the following.

(9.17) Theorem: *Suppose that the* S–L *problem of* (3) *and Definition* 9.1 *is regular and that $\mathscr{S}$ requires the vanishing of yy' at both ends. Then the eigenvalues λ_n satisfy the inequalities*

$$\left[\frac{s_n\pi}{b - a}\right]^2 \underline{T} + \underline{Q} \le \lambda_n \le \left[\frac{s_n\pi}{b - a}\right]^2 \bar{T} + \bar{Q}, \qquad n = 1, 2, \ldots \tag{53}$$

where $s_n = n - 1 + (s/2)$, and s denotes the number of endpoints at which the functions in $\mathscr{S}$ must vanish.

Proof: The boundary conditions imply that we can take $A = B = 0$ in (5′), and Proposition 9.15 shows that

$$\underline{\lambda}_N \le \lambda_N \le \bar{\lambda}_N, \qquad N = 1, 2, \ldots, \tag{54}$$

where the $\underline{\lambda}_N$ are the eigenvalues of the S–L problem for the *same* boundary set $\mathscr{S}$ and the differential equation

$$\underline{L}(y) \overset{\text{def}}{=} (\underline{\tau} y')' - \underline{q} y = -\underline{\lambda}\underline{\rho} y, \qquad \text{or} \qquad \underline{T} y'' + (\underline{\lambda} - \underline{Q}) y = 0,$$

while the $\bar{\lambda}_N$ are those for the equation

$$\bar{L}(y) \overset{\text{def}}{=} (\bar{\tau} y')' - \bar{q} y = -\bar{\lambda}\bar{\rho} y, \qquad \text{or} \qquad \bar{T} y'' + (\bar{\lambda} - \bar{Q}) y = 0.$$

It suffices to assume that $a = 0$, so that $b - a = b$. These S–L problems on $[0, b]$ are necessarily regular, and at each end, y or y' must vanish. We see that in case $y(0) = y(b) = 0$, the eigenvalues are such that

$$\frac{\lambda_n - \underline{Q}}{\underline{T}} = \frac{\bar{\lambda}_n - \bar{Q}}{\bar{T}} = \frac{n^2\pi^2}{b^2}.$$

Thus $\underline{\lambda_n} = \underline{T}\frac{n^2\pi^2}{b^2} + \underline{Q}$ and $\bar{\lambda}_n = \bar{T}\frac{n^2\pi^2}{b^2} + \bar{Q}, \qquad n = 1, 2, \ldots;$

when $y'(0) = y'(b) = 0$, n is replaced by $n - 1$, and in the remaining mixed cases, n is replaced by $n - \frac{1}{2}$. The estimates in (53) follow. ∎

Remarks: Corresponding estimates can be derived for each regular problem, but comparison eigenvalues for the more general endpoint conditions permitted in Definition 9.1 are not available explicitly. However, further comparison can be achieved by invoking Rayleigh's principle concerning the increase of eigenvalues with "stiffness." (See Problem 9.3F.) Indeed, the observation that $n - 1 \le n - \frac{1}{2} \le n$ confirms this Rayleigh behavior for the eigenvalues of the three constant-coefficient problems used in proving the theorem, since successive pinning of the ends stiffens the system.

Example 3:

The regular problem for the equation

$$(e^x y')' - (1 + x)^{-2} y = -\lambda y \qquad \text{on } (0, 1), \tag{55}$$

with $y(0) = y(1) = 0$, is fully positive, but $y_0(x) = 1$ is not an eigenfunction. Consequently, all eigenvalues are positive. To use Theorem 9.17, note that on $[0, 1]$, $\tau(x) = e^x$ and $\rho(x) = 1$ have bounds

$$\underline{\tau} = 1,\ \bar{\tau} = e, \qquad \text{and} \qquad \underline{\rho} = \rho(x) = \bar{\rho} = 1,$$

so that $\underline{T} = 1$, $\bar{T} = e$, in (52). However, for q, we have

$$\underline{q} = \tfrac{1}{4} \le q(x) = (1 + x)^{-2} \le 1 = \bar{q}, \qquad 0 \le x \le 1,$$

so that $\underline{Q} = \frac{1}{4}$ and $\bar{Q} = 1$, in (52).

Hence from Theorem 9.17 the eigenvalues λ_n satisfy the estimates

$$n^2\pi^2 + \tfrac{1}{4} \le \lambda_n \le n^2\pi^2 e + 1, \qquad n = 1, 2, \ldots .$$

Observe that when we divide through by n^2, we get

$$\pi^2 + \frac{1}{4n^2} \le \frac{\lambda_n}{n^2} \le \pi^2 e + \frac{1}{n^2},$$

and for large enough n, we can make the lower and upper bounds for λ_n/n^2 as near π^2 and $\pi^2 e$, respectively, as we wish, but we cannot "squeeze" λ_n/n^2 any tighter using these estimates. However, in the next section we learn that in fact

$$\lim_{n\to\infty} \frac{\lambda_n}{n^2} = \frac{\pi^2}{\ell^2}, \qquad \text{where } \ell = 2(1 - e^{-1/2}). \tag{56}$$

(See Example 5.)

*Example 4:**

If $\ell < +\infty$, then an S–L problem for the equation

$$y'' - qy = -\lambda y \qquad \text{on } (0, \ell) \tag{57}$$

is *regular* (with $\tau = \rho = 1$) when $q \in C[0, \ell]$.

If $(yy')(0) = (yy')(\ell) = 0$, we have the following eigenvalue estimates from Theorem 9.17:

$$\left[\frac{\mathscr{s}_n \pi}{\ell}\right]^2 + \underline{q} \le \lambda_n \le \left[\frac{\mathscr{s}_n \pi}{\ell}\right]^2 + \bar{q}, \qquad n = 1, 2, \ldots .$$

Let's see if we can improve these results for large n (or equivalently, large $\mathscr{s}_n$). First, multiply through by $[\ell/\mathscr{s}_n \pi]^2$ and set $M = (|\underline{q}| + |\bar{q}|)\ell^2/\pi^2$. Then, we get the weaker (but simpler) inequalities

$$1 - \frac{M}{\mathscr{s}_n^2} \le \frac{\lambda_n}{\mathscr{s}_n^2}\frac{\ell^2}{\pi^2} \le 1 + \frac{M}{\mathscr{s}_n^2}.$$

Finally, if $\mathscr{s}_n^2 > M$ and $\lambda_n = \omega_n^2$, take square roots to see that

$$1 - \frac{M}{\mathscr{s}_n^2} \le \sqrt{1 - \frac{M}{\mathscr{s}_n^2}} \le \frac{\omega_n \ell}{\mathscr{s}_n \pi} \le \sqrt{1 + \frac{M}{\mathscr{s}_n^2}} \le 1 + \frac{M}{\mathscr{s}_n^2},$$

and if we now multiply through by $\mathscr{s}_n \pi$, we conclude that for $n > 1$

$$|\omega_n \ell - \mathscr{s}_n \pi| \le \frac{M\pi}{\mathscr{s}_n} \le \frac{2\pi M}{n} \tag{58}$$

since $\mathscr{s}_n \ge n - 1 \ge n/2$. Hence, for large n, $\omega_n \ell \approx \mathscr{s}_n \pi$ and in the limit as $n \to \infty$, $\omega_n/n \to \pi/\ell$ since $\mathscr{s}_n/n = [1 - (1/n) + (\mathscr{s}/n)] \to 1$; i.e.,

$$\lim_{n\to\infty} \frac{\lambda_n}{n^2} = \frac{\pi^2}{\ell^2}. \tag{59}$$

The limiting result (59) is like (49) and is valid even when Robin conditions govern this regular S–L problem. However, our asymptotic estimate (58) holds in such cases only within possible replacement of every $\mathscr{s}_n$ by $\mathscr{s}_{n+1}$. (See Problem 9.3L.)

Problem Set 9.3

9.3A Use Rayleigh's principle (Proposition 9.10) on the given $\tilde{y}$ to obtain an upper bound to the least eigenvalue λ_1 for each of the following problems.

1. $y'' = -\lambda y$ on $[0, 1]$, with $y(0) = y(1) = 0$; $\tilde{y}(x) = x(1 - x)$.
2. $y'' = -\lambda xy$ on $[0, 1]$, with $y(0) = y'(1) = 0$; $\tilde{y}(x) = \sin(\pi x/2)$.
3. $(xy')' - xy = -2\lambda y$ on $(0, 1]$, with $y'(1) = 0$; $\tilde{y}(x) = 1$.
4. $y'' + xy = -\lambda y$ on $[0, 1]$, with $y(0) = 0$ and $y'(1) = y(1)$; $\tilde{y}(x) = x$.

9.3B Find upper and lower bounds to λ_n as in Example 3 for the problems

1. $(1 + x)y'' + y' = -\lambda e^x y$ on $(0, 1)$, with $y(0) = y(1) = 0$.
2. $2y'' - xy = -\lambda e^x y$ on $(0, 1)$, with $y'(0) = y'(1) = 0$.

9.3C The nth Legendre polynomial can be represented in the form

$$P_n(x) = a_n(x - x_1)(x - x_2)\cdots(x - x_n),$$

for some constant a_n and *distinct* zeros $x_1, x_2, \ldots, x_n \in \mathbb{C}$, so indexed that $x_1, x_2, \ldots, x_J$ are its nodes in $(-1, 1)$. Obviously, $J \leq n$, but we wish to prove that $J = n$. Suppose $J < n$ and consider $P(x) = (x - x_1)(x - x_2)\cdots(x - x_J)$.

1. Argue that with correct choice of a_n,

$$P(x)P_n(x) \geq 0 \qquad \text{on } (-1, 1) \qquad \text{so that } \int_{-1}^{1} PP_n\,dx > 0.$$

2. But P_n is orthogonal to any polynomial of lesser degree (Why?), a contradiction.
3. How would this argument change for the Tchebycheff polynomials? the Hermite polynomials? the Laguerre polynomials? (See the end of Section 9.2.)

9.3D Let ω_n denote the successive positive zeros of J_0, $n = 1, 2, \ldots$ so that $y_n(x) = J_0(\omega_n x)$ is an nth eigenfunction to the S–L problem in Example 2, on $(0, 1)$.

1. Why is $\omega_1/\omega_2 < \omega_2/\omega_3 < \omega_2$? (*Hint*: Consider y_2, y_3.)
2. What can you conclude about ω_4 by means of similar estimates?

9.3E (*Monotonicity of* λ_n *with interval*) Suppose that $(\tilde{a}, \tilde{b}) \subseteq (a, b)$, and we compare the regular eigenvalue problem of Proposition 9.11 to the *same* problem on the smaller interval, with eigenfunctions $\tilde{y}_n$. Show that as with a shortened guitar string, the associated eigenvalue $\tilde{\lambda}_n \geq \lambda_n$, $n = 1, 2, \ldots$. (*Hint*: Each $\tilde{y}_n$ with $\tilde{y}_n(\tilde{a}) = \tilde{y}_n(\tilde{b}) = 0$ can be extended to a function in $\mathcal{S}_0$ by setting it equal to zero outside $(\tilde{a}, \tilde{b})$.)

9.3F Compare the eigenvalues λ_n for the regular problem of Proposition 9.11 with those $\tilde{\lambda}_n$ for the same equation, but the regular boundary conditions

$$\tilde{y}(a) = 0 \text{ and } \tilde{y}'(b) = -B\tilde{y}(b), \qquad \text{for } B \geq 0,$$

defining a set $\tilde{\mathcal{S}}$, with $\tilde{\mathcal{S}}_1 \supseteq \mathcal{S}_1$.

1. Show that $\tilde{\lambda}_1 < \lambda_1$. (*Hint*: $\tilde{R}(y_1) = R(y_1)$.)
2. Show that $\tilde{\lambda}_N < \lambda_N$ for $N = 1, 2, \ldots$. (*Hint*: For given functions $\phi_1, \ldots, \phi_{N-1}$: $y \in \Phi_N \Rightarrow \tilde{R}(y) = R(y)$.)
3. Verify that $\tilde{\lambda}_n$ increases to λ_n as $B \to \infty$. Why is this consistent with Rayleigh's principle?

9.3G In Problem 9.3F, show that:

1. $\tilde{y}_n$ cannot have more than $n - 1$ nodes, by comparing with y_n (which cannot).
2. $\tilde{y}_n$ must have $n - 1$ nodes, by comparing with $\tilde{y}_k$ for lesser k.
3. $\lambda_1 < \tilde{\lambda}_2$ (since $\tilde{\lambda}_2 < \lambda_1$ leads to contradiction).
4. $\tilde{\lambda}_1 < \lambda_1 < \tilde{\lambda}_2 < \lambda_2 < \cdots$ (a separation theorem).

9.3H (*Sturm separation theorem*) The Wronskian analysis in Proposition 9.11 has a more general form:

Let y_k be a solution of $L_k(y) \overset{\text{def}}{=} (\tau y')' + p_k y = 0$ on (a, b) for $k = 1, 2$, where $p_1(x) \le p_2(x)$.

1. Define W_{12} as in (41) and conclude that $W_{12}' = (p_1 - p_2) y_1 y_2$.
2. Prove that unless $p_1 = p_2$, y_2 must have a node between successive zeros of y_1.

9.3I For the problem of Proposition 9.11 we know that the "first" normalized eigenfunction y_1 is, in general, nonvanishing on (a, b) and so can be presumed *positive*. Let u be a linearly independent function in $\mathscr{S} \cap C^2$ that is positive on (a, b), and set

$$u_1 \overset{\text{def}}{=} [\tau(u' y_1 - y_1' u)]'.$$

1. Show that $\int_a^b u_1 \, dx = 0$, so that u_1 must have both positive and negative values.
2. For $x \in (a, b)$, use (3) to obtain
$$\lambda_1 = U(x) + \left(\frac{u_1}{\rho u y_1}\right)(x), \qquad \text{where } U \overset{\text{def}}{=} -\frac{L(u)}{\rho u}.$$
3. Use the information in part 1 to conclude that
$$\min_{x \in (a,b)} U(x) \le \lambda_1 \le \max_{x \in (a,b)} U(x).$$
4. For the simple problem $L(y) \equiv y'' = -\lambda y$ on (0, 1) with $y(0) = y(1) = 0$, take $u(x) = x(1 - x)$, and use part 3 to estimate $\lambda_1 (= \pi^2)$ from *below*.

9.3J When ϕ_j is defined by (40) so that

$$\phi = \sum_{j=1}^{J} c_j \phi_j \in \mathscr{S}_J,$$

verify that $R(\phi) = \lambda_n$. (*Hint*: Observe that, in general, $\phi_i \phi_j = \phi_i' \phi_j = \phi_i' \phi_j' = 0$, $i \ne j$.)

9.3K* (*Asymptotic estimates for the zeros of J_m*)

1. Combine the results of Proposition 9.16 and Theorem 8.10 to conclude that for $m = 0$, $\omega_{0n} = \omega_n = n\pi - (\pi/4) + \delta_n$,
$$\text{where } 0 \le \delta_n < \pi \text{ and } \tan \delta_n = -\cot\left(\omega_n - \frac{\pi}{4}\right) = \frac{1}{8\omega_n} + O\left(\frac{1}{n^2}\right),$$
so that for large n:
$$\omega_n = n\pi - \frac{\pi}{4} + \frac{1}{8n\pi} + O\left(\frac{1}{n^2}\right).$$
(See Problem 7.2D.)
2. Make a similar analysis for the zeros of J_1 and for those of J_m for $m = 2, 3, \ldots$ to show that for all m,
$$\omega_{mn} = n\pi + \frac{m\pi}{2} - \frac{\pi}{4} - \frac{M}{2n\pi} + O\left(\frac{1}{n^2}\right), \qquad \text{where } M = m^2 - \frac{1}{4}.$$

9.3L* Let λ_1 be the least eigenvalue for a regular problem involving the equation $y'' = -\lambda y$ on $(0, \ell)$, and for $n > 1$, let $\lambda_n = \omega_n^2$.

1. Under the conditions of Example 3 of Section 7.2 (where $\ell = 1$) show that

$$\omega_n = \tilde{s}_n \pi \approx s_n \pi, \quad \text{where for some } M \quad |\tilde{s}_n - s_n| \leq M/n, \quad n > M.$$

2. Obtain the same conclusion (when $\ell = 1$) under the conditions of Problem 7.2D, when $A \neq 0$, $B = 0$.

3.* For the general case in Problem 7.2D where $AB \neq 0$, show that $\omega_n = \tilde{s}_n \pi/\ell$, where either the previous inequality holds or

$$|\tilde{s}_n - s_{n+1}| \leq M/n, \qquad n > M.$$

4. Explain why the arguments in Example 4 carry through in the above cases when $\tilde{s}_n$ replaces s_n throughout. Then explain why (59) holds as stated, while (58) holds as stated except when $s = 0$, in which case it is sometimes necessary to replace s_n by s_{n+1}.

9.4

Asymptotic Considerations

The conviction that a nonuniform string should not vibrate too differently from an appropriate uniform string leads to the following.

(9.18) Conjecture: *The eigenvalues and eigenfunctions of the general S–L problem of* (3) *and Definition* 9.1 *should not behave too differently from those of a corresponding "uniform" problem, where the differential equation is* $\ddot{v} + \lambda v = 0$, *with solutions* $v(t) = A \sin(\sqrt{\lambda}\, t + \alpha)$, *and for large n, there are eigenvalues* λ_n *approximately proportional to* n^2. ■

This conjecture is substantially true for large λ and the asymptotic approximations used to verify it provide the basis for the representation results of this chapter. Let us first show how to replace a given S–L problem by an equivalent problem for which $\tau = \rho = 1$.

The Liouville Transformation

If y and $\lambda = \omega^2$ satisfy equation (3) and, in addition,

$$\ell \stackrel{\text{def}}{=} \int_a^b \sqrt{\frac{\rho}{\tau}}\, d\xi < +\infty, \quad \text{set } p = (\tau\rho)^{1/4}, \tag{60}$$

and, following Liouville, introduce new independent and dependent variables

$$t = \int_a^x \sqrt{\frac{\rho}{\tau}}\, d\xi \quad \text{and} \quad v(t) = p(x)y(x). \tag{60'}$$

Then, using the dot to denote t-differentiation, we find that v satisfies the equivalent equation

$$\ddot{v}+\omega^2 v = rv, \qquad \text{on } (0, \ell), \tag{61}$$

where

$$r = r(t) = \frac{\ddot{P}}{P}(t) + \frac{q}{\rho}(x), \qquad \text{if } P(t) \stackrel{\text{def}}{=} p(x). \tag{61'}$$

Observe that the transformed equation (61) is indeed that for an S–L problem with $\tau = \rho = 1$. Moreover,

$$\int_0^\ell v^2\, dt = \int_a^b \rho y^2\, dx = 1,$$

when y is a normalized eigenfunction to λ since under (60′), $dt/dx = \sqrt{\rho/\tau} = \rho/p^2$. (See Problem 9.4C.)

Eigenvalues of Regular Problems

Now, if the original problem is *regular*, then ℓ is finite, and since

$$v(t) = (py)(x) \qquad \text{and} \qquad \dot{v}(t) = (p'y + y'p)\sqrt{\frac{\tau}{\rho}}(x), \tag{62}$$

it is seen that a regular problem on (a, b) is transformed into a new regular problem on $(0, \ell)$. Also, the new eigenfunctions v will vanish at the same number s of endpoints as the old. Finally, (61) is essentially the equation we studied in Example 4, and these observations set the stage for our first reward from the Liouville transformation.

(9.19) Proposition: *If the original S–L problem is regular, then its eigenvalues $\lambda_n = \omega_n^2$ satisfy*

$$\text{either} \qquad |\omega_n \ell - s_n \pi| \le M/n, \qquad \text{for } n > M \tag{63}$$

$$\text{or} \qquad |\omega_n \ell - s_{n+1} \pi| \le M/n, \qquad \text{for } n > M \tag{63'}$$

where M is some problem-constant, $s_n = n - 1 + (s/2)$, $n = 1, 2, \ldots$, and (63′) is discarded unless $s = 0$.

In any case, $\lim_{n\to\infty} \lambda_n/n^2 = \pi^2/\ell^2$, where $\ell = \int_a^b \sqrt{\rho/\tau}\, dx$. (64)

Proof: It suffices to consider the transformed problem and call upon the results from Example 4, with $q = r$ and $b = \ell$. (See also Problem 9.3L.). ■

Remark: For each specific problem with $s = 0$, there is only one valid choice (63) or (63′), but it is determined by boundary conditions of the transformed problem, and for example, the condition $y'(a) = 0$ is transformed under (62) into the Robin condition $v'(0) = Ev(0)$, where usually $E \ne 0$.

Example 5:

In the regular problem of Example 3, we had $\tau(x) = e^x$ and $\rho(x) = 1$, on $(0, 1)$ so that

$$\ell = \int_0^1 \sqrt{\frac{\rho}{T}}\,dx = \int_0^1 e^{-x/2}\,dx = 2(1 - e^{-1/2})$$

Consequently, (56) holds. Moreover, because the boundary conditions $y(0) = y(1) = 0$ give $v(0) = v(\ell) = 0$, then $\sigma = 2$, and from (63), we conclude that $\omega_n \approx n\pi/\ell$ for large n.

Eigenfunction Behavior

Observe that for large $\omega > 0$, at each $t \in (0, \ell)$ the term $(rv)(t)$ in (61) is negligible in comparison with $\omega^2 v(t)$, and so we might expect the associated solutions to be well approximated by $A \sin(\omega t + \alpha)$. However, for useful results, we need estimates of the approximation expressed in terms of positive finite **problem-constants**, denoted M, that do not depend on a particular choice of v or ω (when $\omega \geq 1$). (Recall the constant M in Example 4.)

Using formula (1) of Proposition B.1, we can "solve" (61) to obtain the following Volterra integral equation for v:

$$\left.\begin{aligned} v(t) &= A \sin(\omega t + \alpha) + \frac{V(t)}{\omega}, \\ \text{where}\quad V(t) &\stackrel{\text{def}}{=} \int_0^t \sin\omega(t-s)\,(rv)(s)\,ds, \end{aligned}\right\} \quad 0 < t < \ell \tag{65}$$

for constants $A > 0$ and α that depend on ω and the boundary requirements.

Our task is to derive a priori estimates of A, α, and ω, and we consider first how normalization affects choice of the positive amplitude constant A.

(9.20) Proposition: *For $\omega > 1 + (2\pi/\ell)$, let v be a solution of (65) with $\int_0^\ell v^2\,dt = 1$.*

(1) If ℓ and $R \stackrel{\text{def}}{=} \int_0^\ell r^2\,dt$ are finite, then

$$\left| v(t) - \sqrt{\frac{2}{\ell}} \sin(\omega t + \alpha) \right| \leq \frac{M}{\omega}, \qquad 0 \leq t \leq \ell, \tag{66}$$

for some problem-constant M.

(2) If $[0', \ell'] \subseteq (0, \ell)$, then

$$|v(t)| \leq M', \qquad 0' \leq t \leq \ell', \tag{66'}$$

for some problem-constant M' dependent on $0'$ and ℓ'.

Proof:* Upon rearranging (65) and squaring, we find that

$$\frac{A^2}{2}(1 - \cos 2(\omega t + \alpha)) = v^2(t) - \frac{2}{\omega} v(t)\,V(t) + \frac{1}{\omega^2} V^2(t).$$

Suppose ℓ is finite and choose $\ell' \in (\ell - \pi/\omega, \ell)$ such that

$$\int_0^{\ell'} \cos 2(\omega t + \alpha)\,dt = 0.$$

Then integration of the first equation over $[0, \ell']$ gives the key result

$$\ell' \frac{A^2}{2} = \int_0^{\ell'} v^2\,dt - \frac{2}{\omega}\int_0^{\ell'} vV\,dt + \frac{1}{\omega^2}\int_0^{\ell'} V^2\,dt. \tag{67}$$

(1) Next, using the Schwarz inequality and the assumed normalization of v to estimate V in (65), we have

$$V^2(t) \le \left(\int_0^t |rv|\,ds\right)^2 \le \int_0^{\ell} r^2\,ds \int_0^{\ell} v^2\,ds = R, \qquad t \in [0, \ell]$$

and similarly, we find that

$$\left|\int_0^{\ell'} vV\,dt\right| \le \left(\int_0^{\ell} v^2\,dt \int_0^{\ell} V^2\,dt\right)^{1/2} \le \sqrt{R\ell}.$$

Consequently, for $\omega \ge 1$, (67) admits the estimate

$$\left|\ell' \frac{A^2}{2} - \int_0^{\ell'} v^2\,dt\right| \le \frac{2}{\omega}\sqrt{R\ell} + \frac{1}{\omega^2} R\ell \le \frac{2}{\omega}(\sqrt{R\ell} + R\ell).$$

Now for large ω, $\ell' \approx \ell$ so that $\int_0^{\ell'} v^2\,dt \approx 1$, and we see that if R is finite, then

$$\ell \frac{A^2}{2} \approx 1 \qquad \text{or} \qquad A \approx \sqrt{\frac{2}{\ell}}.$$

In fact, by techniques similar to those above, it can be shown that

$$\left|A - \sqrt{\frac{2}{\ell}}\right| \le \frac{M}{\omega} \tag{67$'$}$$

for a problem-constant M. (See Problem 9.4G.) Then from the triangle inequality and (65) we obtain (66) as follows:

$$\left|v(t) - \sqrt{\frac{2}{\ell}}\sin(\omega t + \alpha)\right| \le |v(t) - A\sin(\omega t + \alpha)| + \left|A - \sqrt{\frac{2}{\ell}}\right|$$

$$\times |\sin(\omega t + \alpha)| \le \frac{|V(t)|}{\omega} + \frac{M}{\omega} \le \frac{\sqrt{R} + M}{\omega}, \qquad t \in [0, \ell].$$

(2) Note that for $\omega \ge 1$, this inequality yields the bound $|v(t)| \le M$, $t \in [0, \ell]$, assuming that ℓ and R are finite. To obtain a corresponding result on $[0', \ell'] \subseteq (0, \ell)$, when ℓ or R is infinite, set $v = c'v'$, where c' is a positive constant, and $\int_{0'}^{\ell'} v'^2\,dt = 1$. Then $(c')^2 = \int_{0'}^{\ell'} v^2\,dt \le 1$ so that $|v| = c'|v'| \le |v'|$. We can replace 0 by $0'$ in (65), and since $R' \overset{\text{def}}{=} \int_{0'}^{\ell'} r^2\,dt$ *is* finite, we see that

$$|v(t)| \le M', \qquad t \in [0', \ell']$$

as desired. ■

Remarks: *Suppose that (66) holds*: Observe that in (65), $V(0) = \dot{V}(0) = 0$. Therefore $v(0) = 0 \Rightarrow A \sin \alpha = 0$ and in (66) we would take $\alpha = 0$, (or $\alpha = \pi$ which corresponds to replacing v by $-v$). Similarly, if $\dot{v}(0) = Ev(0)$ for some problem-constant E, we see that

$$|\cos \alpha| = \left|\frac{E \sin \alpha}{\omega}\right| \leq \frac{|E|}{\omega},$$

so that for large ω, $\alpha \approx \pi/2$ (or $\alpha \approx -\pi/2$). In fact, in (66) we can replace α by $\pi/2$ (or by $-\pi/2$). (See Problem 9.4H.)

Upon restoring the original variables we obtain the major results of this section.

(9.21) Theorem: *The normalized eigenfunctions y to large eigenvalues $\lambda = \omega^2$ of an S–L problem of* (3) *and* Definition 9.1 *have bounds*

$$|y(x)| \leq M', \qquad a' \leq x \leq b' \tag{68}$$

on each compact interval $[a', b'] \subseteq \langle a, b \rangle$, where M' is a problem-constant dependent on a' and b'.

Proof: If we define $t(x) = \int_{a'}^{x} \sqrt{\rho/\tau}\, d\xi$ and $\ell' = t(b')$, then (68) follows from the second result in Proposition 9.20. ■

(9.22) Theorem: *If the problem in Theorem 9.21 is regular, set $\ell = t(b)$ where $t(x) = \int_a^x \sqrt{\rho/\tau}\, d\xi$, $x \in [a, b]$. Then for every sufficiently large integer n, there is a normalized eigenfunction y_n to the eigenvalue ω_n^2 such that*

$$\left| y_n(x) - \sqrt{\frac{2}{\ell}}[(\tau\rho)(x)]^{-1/4} \sin(\omega_n t(x) + \alpha) \right| \leq \frac{M}{\omega_n}, \qquad x \in [a, b] \tag{69}$$

for problem-constants M and α. ($\alpha = 0$ or $\pi/2$ according to whether or not $\mathcal{S}$ requires its functions to vanish at a, and ω_n can be replaced by $s_n \pi/\ell$ or $s_{n+1}\pi/\ell$ as in Proposition 9.19.)

Proof: Regularity of the problem ensures that ℓ is finite and that in (61′) r is bounded on $[0, \ell]$. Hence $R = \int_0^{\ell} r^2\, dt < +\infty$, so that (66) is applicable and (69) is just a restatement. Moreover the S–L problem for v of (61) will also be regular, and clearly $y(a) = 0$ iff $v(0) = 0$. Consequently, the selection of α is just that discussed in the remarks above, and if (63) or (63′) holds, the replacement of ω_n can be justified by an easy application of the triangle inequality. ■

Remarks: If the S–L problem is *singular* with a *regular* eigenvalue sequence $\lambda_1, \lambda_2, \ldots$ and y_n is an eigenfunction to $\lambda_n = \omega_n^2$, with $\int_{a'}^{b'} \rho y_n^2\, dx = 1$, then (69) holds on the compact interval $[a', b'] \subseteq \langle a, b \rangle$ with $t(x) \stackrel{\text{def}}{=} \int_{a'}^{x} \sqrt{\rho/\tau}\, d\xi$ and $\ell = t(b')$. When a is a regular endpoint of the problem, we take $a' = a$, and choose $\alpha = 0$ or $\pi/2$ in (69) according to whether or not $\mathcal{S}$ requires that $y(a) = 0$. If only b is regular, then in (69), set $t(x) = \int_x^{b} \sqrt{\rho/\tau}\, d\xi$, $\ell = t(a')$, and select $\alpha = 0$ or $\pi/2$ according to whether or not $\mathcal{S}$ requires that $y(b) = 0$.

Example 6:

For the regular problem

$$y'' - (1+x)^{-2} y = -\lambda y \qquad \text{on } (0, 1)$$
$$\text{with} \qquad y(0) = y(1) = 0 \qquad (\text{so } \mathscr{s} = 2),$$

we see that since $\rho = \tau$, then $t(x) = \int_0^x 1\, d\xi = x$, and $\ell = 1$. In Theorem 9.22, we take $\alpha = 0$ and predict that for every large integer n there will be a normalized eigenfunction y_n with the uniform approximation

$$y_n(x) \sim \sqrt{2} \sin n\pi x \qquad \text{on } [0, 1].$$

We hope that the eigenfunctions for this problem will be as effective in providing series representations for functions f on $(0, 1)$ as are their trigonometric approximants.

Observe that these asymptotic approximations would be unchanged if in the differential equation, $q(x) = (1+x)^{-2}$ is replaced by *any* function $q \in C[0, 1]$. However, the choice of q will be reflected in how large we must take n to achieve approximation within prescribed accuracy. (See [In].)

Example 7:

For fixed $m = 0, 1, 2, \ldots$ consider *Bessel's equation* (1):

$$(xy')' - \frac{m^2}{x} y = -\lambda x y, \qquad \text{on } (0, 1). \tag{70}$$

Associated S–L problems are singular since $\tau(x) = \rho(x) = x \to 0$ as $x \to 0$. Suppose we require that each eigenfunction y vanishes at the regular endpoint 1. Here, $\langle 0, 1 \rangle = (0, 1]$, and since $p(x) = (x^2)^{1/4} = \sqrt{x}$, then for $x \in (0, 1)$ we define $v = v(t) = \sqrt{x}\, y(x)$, where

$$t = t(x) = \int_x^1 \sqrt{\frac{\rho}{\tau}}\, d\xi = \int_x^1 d\xi = 1 - x, \qquad \text{and} \qquad \ell = 1.$$

If we complete the Liouville substitutions (60′), equation (70) takes the form (see also Problem 7.4L)

$$\ddot{v} + \omega^2 v = \frac{4m^2 - 1}{4(1-t)^2} v, \qquad \text{on } [0, 1). \tag{70′}$$

In particular, when $m = 1/2$, the equation simplifies to

$$\ddot{v} + \omega^2 v = 0,$$

and with $v(0) = 0$, its only eigenfunctions have the form

$$v(t) = A \sin \omega t.$$

Therefore, in this case, a normalized eigenfunction *is*

$$y(x) = \sqrt{\frac{2}{x}} \sin \omega(1 - x) \tag{71}$$

which is finite at $x = 0$ iff $\sin \omega = 0$. Consequently, when $m = 1/2$, the eigenvalues for the S–L problem are

$$\lambda_n = n^2\pi^2, \qquad n = 1, 2, \ldots, \tag{71'}$$

with associated eigenfunctions

$$y_n(x) = \sqrt{\frac{2}{x}} \sin n\pi(1-x) = \pm\sqrt{\frac{2}{x}} \sin n\pi x. \tag{71''}$$

In this special case, the asymptotic results are exact.

If $m \neq 1/2$ and $a \in (0, 1)$, we can use (69) on $[a, 1]$ to see that for each large eigenvalue λ_n, *some* associated eigenfunction y_n *normalized over* $[a, 1]$ admits the asymptotic approximation

$$y_n(x) \approx \sqrt{\frac{2}{x\ell}} \sin\left(\sqrt{\lambda_n}(1-x)\right), \qquad \text{for } 0 < a \leq x \leq 1.$$

Here, $\ell = 1 - a$, and we have set $\alpha = \alpha_n = 0$, since $y_n(1) = 0$.

We cannot obtain information from this approximation concerning behavior of $y_n(x)$ as $x \searrow 0$, since we must permit n to increase to attain the same order of accuracy as $a \searrow 0$. In the process, we lose control of any specific y_n. However, for this problem, we can take $y_n(x) = j_n J_m(\omega_{mn}x)$ where $J_m(\omega_{mn}) = 0$, $n = 1, 2, \ldots,$ and choose j_n to satisfy the normalization condition

$$j_n^2 \int_0^1 x J_m^2(\omega_{mn}x)\, dx = 1.$$

From results of Problems 7.4E and F, we know that

$$j_n = \frac{\sqrt{2}}{|J_m'(\omega_{mn})|} = \frac{\sqrt{2}}{|J_{m+1}(\omega_{mn})|}, \qquad n = 1, 2, 3, \ldots;$$

then using asymptotic results for J_m and J_{m+1} obtained in Theorem 8.10 we conclude that for $M = m^2 - \frac{1}{4}$ and $x \geq a > 0$:

$$y_n(x) = \sqrt{\frac{2}{x}}\left[\cos\left(\omega_{mn}x - \frac{\pi}{4} - \frac{m\pi}{2}\right) - \frac{M}{2\omega_{mn}x}\sin\left(\omega_{mn}x - \frac{\pi}{4} - \frac{m\pi}{2}\right)\right] + O\left(\frac{1}{n^{5/2}}\right) \tag{72}$$

where the last term depends on a. (See Problem 9.4I.)

Example 8:

For *Legendre's equation*

$$[(1-x^2)y']' - \frac{m^2}{1-x^2}y = -\lambda y \qquad \text{on } (-1, 1), \tag{73}$$

$$\text{where } m = 0, 1, 2, \ldots,$$

S–L problems are necessarily singular since $\tau(x) = 1 - x^2 \to 0$ as $|x| \to 1$. However, in Sections 7.5 and 7.6, we found eigenfunctions $P_n{}^m(x)$ to eigenvalues $\lambda_{mn} = n(n+1)$, $n \geq m$.

Here,

$$\lim_{n\to\infty} \frac{\lambda_{mn}}{n^2} = \lim_{n\to\infty} \frac{n(n+1)}{n^2} = 1,$$

which formally agrees with (64) since $\rho(x) = 1$ and

$$\ell = \int_{-1}^{1} \sqrt{\frac{\rho}{\tau}}\, dx = \int_{-1}^{1} \frac{dx}{\sqrt{1-x^2}} = \sin^{-1} x \Big|_{-1}^{1} = \pi.$$

Note also that $\langle -1, 1\rangle = (-1, 1)$.

Let's consider first the simpler case $m = 0$, where an eigenfunction to $\lambda_n = n(n+1)$, is given by the *normalized* Legendre polynomial

$$p_n(x) = \sqrt{\frac{2n+1}{2}}\, P_n(x), \qquad n = 0, 1, 2, \ldots .$$

From (68), we see that if $0 < \mathscr{b} < 1$, then $[-\mathscr{b}, \mathscr{b}] \subseteq (-1, 1)$ so that

$$|p_n(x)| \le M(\mathscr{b}), \qquad |x| \le \mathscr{b},$$

if $n \ge N(\mathscr{b})$. But then, for perhaps a larger constant $M(\mathscr{b})$, this estimate must hold for *all* $n = 0, 1, 2, \ldots$. Consequently,

$$|P_n(x)| \le \sqrt{\frac{2}{2n+1}}\, M(\mathscr{b}), \qquad |x| \le \mathscr{b} < 1, \tag{73$'$}$$

and for convergence analysis of Legendre series, this result is a substantial improvement on the bound $|P_n(x)| \le 1$, given in Section 7.5.

It remains to apply Theorem 9.22. Since neither endpoint is regular, it is better to define

$$t = \int_0^x \sqrt{\frac{1}{1-\xi^2}}\, d\xi = \sin^{-1} x,$$

and set $\ell = \sin^{-1} \mathscr{b}$ for $\mathscr{b}$ near 1. Then we can repeat the asymptotic analysis on the interval $[-\mathscr{b}, \mathscr{b}] \subseteq (-1, 1)$. For the Legendre polynomials $\tilde{p}_n$ normalized over $[-\mathscr{b}, \mathscr{b}]$, we find the following asymptotic approximations:

$$\tilde{p}_n(x) \approx \sqrt{\frac{1}{\ell}}\, \frac{1}{(1-x^2)^{1/4}} \sin\left[(n+\tfrac{1}{2}) \sin^{-1} x + \alpha_n\right], \qquad |x| \le \mathscr{b}$$

where we have also approximated $\sqrt{\lambda_n} = \sqrt{n(n+1)}$ by $n + \frac{1}{2}$. Since $\tilde{p}_n$ has the parity of n, it follows that $\tilde{p}_n(0) = 0$ if n is odd, while $\tilde{p}_n'(0) = 0$ when n is even. Consequently, we can take $\alpha_n = 0$ or $\pi/2$, according to whether n is odd or even, and obtain the following results:

$$\begin{aligned} n \text{ odd:}\quad & \tilde{p}_n(x) \approx \pm \frac{1}{\sqrt{\ell}}\, \frac{1}{(1-x^2)^{1/4}} \sin\left[(n+\tfrac{1}{2}) \sin^{-1} x\right] \\ & \hspace{18em} |x| \le \mathscr{b} = \sin \ell \\ n \text{ even:}\quad & \tilde{p}_n(x) \approx \pm \frac{1}{\sqrt{\ell}}\, \frac{1}{(1-x^2)^{1/4}} \cos\left[(n+\tfrac{1}{2}) \sin^{-1} x\right] \end{aligned} \tag{74}$$

where for each $\mathscr{b}$ near 1, the approximation is uniformly valid within $M(\mathscr{b})/n$.

Here, the $\tilde{p}_n$ are specific normalized eigenfunctions, whereas our asymptotic approximants are only selected within a choice of α. Thus matching can only be guaranteed within a sign.

For even m, these asymptotic approximations also apply to appropriately normalized versions $\tilde{p}_n^m$ of the associated Legendre functions P_n^m introduced in Section 7.6. For odd m, they apply with a reversal in parity on n. This is evident when we recall that $P_n^m(x) = (1 - x^2)^{m/2} P_n^{(m)}(x)$ has the parity of n when m is even, and the parity of $n + 1$ when m is odd.

Problem Set 9.4

9.4A Use Proposition 9.19 to obtain information about the eigenvalues of the regular problems in Problem 9.1A.

9.4B Use Theorem 9.22 to obtain asymptotic information for the regular problems of Problem 9.1A.

9.4C 1. Using (60′), show that when $P(t) = p(x(t))$: $\dot{v} = P\sqrt{(\tau/\rho)}\,y' + y\dot{P} = (\tau y'/P) + y\dot{P}$.

2. Then, establish that $\ddot{v} = [P(\tau y')'/\rho] + y\ddot{P}$.
3. Conclude that (3) becomes (61).

9.4D 1. In the proof of Proposition 9.20, we used the estimate $V^2(t) \le \int_0^\ell r^2\,ds = R$. If $t \le \ell$, show how this follows from (4) and the assumed normalization.

2. Use the same methods to establish that $\int_0^{\ell'} |vV|\,dt \le \sqrt{R\ell'}$, when $0 \le \ell' \le \ell$.

9.4E Consider the regular S–L problem

$$y'' = -\lambda(1+x)^{-2}y, \qquad 0 < x < 1,$$
$$\text{with } y(0) = y(1) = 0.$$

1. Take $t(x) = \int_0^x d\xi/(1+\xi)$ in Theorem 9.22, and show that for each large integer n there is a normalized eigenfunction

$$y_n(x) \sim \sqrt{\frac{2}{\log 2}(1+x)}\,\sin\left[\frac{n\pi}{\log 2}\log(1+x)\right] = \tilde{y}_n(x), \text{ say.}$$

2. Verify by substitution that $\tilde{y}_n$ is an actual eigenfunction corresponding to the eigenvalue

$$\lambda_n = \left(\frac{n\pi}{\log 2}\right)^2 + \frac{1}{4}, \qquad n = 1, 2, \ldots .$$

3. Explain what will change in the asymptotic formula for $\tilde{y}_n$ if we change the boundary conditions to $y(0) = y'(1) = 0$.

9.4F 1. Differentiate (65) to obtain

$$\frac{\dot{v}(t)}{\omega} = A\cos(\omega t + \alpha) + \frac{1}{\omega}\dot{V}(t),$$

where $\dot{V}(t) = \int_0^t \cos\omega(t-s)(rv)(s)\,ds$.

2.* How should the proof of Proposition 9.20 be modified to derive the estimate

$$\left|\frac{\dot{v}(t)}{\omega} - \sqrt{\frac{2}{\ell}}\cos(\omega t + \alpha)\right| \leq \frac{M}{\omega}$$

when ℓ and R are both finite?

3. What are the resulting asymptotic approximations for y' corresponding to those in Theorem 9.22?

9.4G (*Improved asymptotic estimates*)

1. To obtain (67′), show first that from (67) we have

$$\frac{A^2}{2}\ell' \leq 1 + \frac{2}{\omega}\sqrt{R\ell'} + \frac{R\ell'}{\omega^2}$$

so that $A^2/2 \leq M'$ independently of v or ω (if $\omega \geq 1$).

2. Then integrate over $[0, \ell]$ and use the fact that $\int_0^\ell v^2\,dt = 1$ to obtain

$$\left|\frac{A^2}{2}\ell - 1\right| \leq \frac{A^2}{2}(\ell - \ell') + \frac{2}{\omega}\sqrt{R\ell} + \frac{R\ell}{\omega^2} \leq \frac{M}{\omega}.$$

3. Conclude that

$$\sqrt{\frac{2}{\ell}}\left|A - \sqrt{\frac{2}{\ell}}\right| \leq \left|A^2 - \frac{2}{\ell}\right| \leq \frac{2M}{\ell\omega}$$

and that this gives (67′) for an appropriate constant M.

9.4H 1. Use the mean value theorem to show that

$$|\sin x - \sin \xi| \leq |x - \xi|, \qquad \text{for } x, \xi \in \mathbb{R}.$$

2. Verify graphically that if $\delta \in \mathbb{R}$, then

$$|\delta| \leq \frac{\pi}{2} \Rightarrow \frac{2}{\pi}|\delta| \leq |\sin \delta| \leq |\delta|.$$

3. For the S–L problem of (61) on $(0, \ell)$, assume that regularity at 0 requires $\dot{v}(0) = Ev(0)$. Differentiate (65) and evaluate as necessary to get

$$\cos \alpha = E\frac{\sin \alpha}{\omega} \Rightarrow \left|\sin\left(\alpha - \frac{\pi}{2}\right)\right| \leq \frac{|E|}{\omega}.$$

Use the triangle inequality and the preceding estimates to verify that for v (or for $-v$), (66) holds with $\alpha = \pi/2$.

9.4I Show that if $J_m(\omega_{mn}) = 0$, $n = 1, 2, \ldots$, then for large n:

$$J_{m+1}(\omega_{mn}) = \sqrt{\frac{2}{\pi\omega_{mn}}} + O\left(\frac{1}{n^{3/2}}\right),$$

and explain how (72) in Example 7 follows.

9.5*

Representation of Discontinuous Functions

One of the most remarkable features of a Fourier series is its ability to represent functions that exhibit jump discontinuities (Theorem 1.11). Thus far, our analysis of an S–L series has been confined to conditions ensuring local uniform convergence (Theorems 7.2 and 9.5) and this precludes discontinuities in the sum. However, if we combine our results for Fourier series with the asymptotic estimates from Section 9.4, we can show that, in general, an S–L series of a function f behaves like a Fourier series for the same function. In particular, at an interior point x where f has a jump discontinuity with derivatives from the left and from the right, the S–L series should have the sum $[f(x+)+f(x-)]/2$.

Let's look at the regular case first, following [In], and then indicate possible extensions to singular problems. Fortunately, it suffices to characterize the behavior of the corresponding series for the standardized *positive* problem of (61),

$$\ddot{v}+\omega^2 v = rv \qquad \text{on } [0, \ell],$$

with regular boundary conditions as in Definition 9.1 where, by a scale change, we may suppose that $\ell = \pi$. For if we assume that this problem has normalized eigenfunctions $v_n(t)$ to eigenvalues $0 \le \lambda_n = \omega_n^2$, $n = 1, 2, \ldots$, then by (60) and (60′), $v_n(t) = p(x)y_n(x)$, where each y_n is an eigenfunction of the original problem. It is an easy exercise to show that if f is integrable on $[a, b]$, and $\tilde{f}(t) = p(x)f(x)$, then $\tilde{f}$ and f have the *same* S–L coefficients c_n, for the respective eigenfunctions v_n and y_n, $n = 1, 2, \ldots$. (See Problem 9.5A.) Moreover, $p = (\rho\tau)^{1/4}$ is continuous and positive so that the convergence behavior of these series will be the same at corresponding points

$$x \text{ and } t = \int_a^x \sqrt{\frac{\rho}{\tau}}\, d\xi, \qquad \text{for } x \in [a, b].$$

Next, we recall from Proposition 9.20 that for all large n:

$$v_n(t) = \sqrt{\frac{2}{\pi}} \sin(\omega_n t + \alpha_n) + \frac{M_n(t)}{\omega_n}, \tag{75}$$

and from Theorem 9.22 that for regular problems we can replace α_n by α, where either $\alpha = 0$ or $\alpha = \pi/2$. Finally, except for the cases where $a = 1$, we can reindex the ω_n if necessary and replace ω_n by n. Thus, except when $a = 1$, there is some v_n with

$$v_n(t) = \sqrt{\frac{2}{\pi}} \sin(nt + \alpha) + \frac{M_n(t)}{n}, \tag{75′}$$

where independently of n,

$$|M_n(t)| \le M < +\infty \qquad \text{for all } t \in [0, \pi].^\dagger \tag{75″}$$

† These estimates can be used to show that the sequence $\{v_n\}$ (and hence the sequence $\{y_n\}$) is complete. See [B-R].

If we iterate (65), then after considerable work (taken up in Problems 9.5G–I), we obtain the second-order asymptotic estimate†

$$v_n(t) = \sqrt{\frac{2}{\pi}}\sin(nt+\alpha) + \frac{R(t)}{\sqrt{2\pi}\,n}\cos(nt+\alpha) + \frac{M_n(t)}{n^2}, \tag{76}$$

where (75″) holds and R is a continuous function dependent only on r and the constants used in defining $\mathscr{S}$. Equation (76) may be regarded as defining $M_n(t)$ for *all* $n = 1, 2, \ldots$. What is significant is that this can be accomplished using functions M_n satisfying (75″). If $\alpha = 0$, define $v_0(t) = 0$ (when $\alpha = \pi/2$, v_0 is the "first" eigenfunction of the problem). Now, for $\alpha = 0$ or $\pi/2$, the trigonometric functions

$$\begin{aligned}\phi_n(t) &\overset{\text{def}}{=} \sqrt{\frac{2}{\pi}}\sin(nt+\alpha), \qquad n = 1, 2, \ldots \\ \phi_0(t) &= \frac{1}{\sqrt{\pi}}\sin\alpha\end{aligned} \tag{77}$$

are orthogonal on $[0, \pi]$, and in each case provide terms for Fourier series representation. Therefore, if $\tilde{f}$ is Riemann integrable on $[0, \pi]$, it is natural to compare the partial sum

$$V_N(t) = \sum_{n=0}^{N} c_n v_n(t) = \int_0^{\pi} \tilde{f}(s)\left[\sum_{n=0}^{N} v_n(s)\,v_n(t)\right] ds \tag{78}$$

of the S–L series with the corresponding partial sum

$$\Phi_N(t) = \sum_{n=0}^{N} \tilde{c}_n \phi_n(t) = \int_0^{\pi} \tilde{f}(s)\left[\sum_{n=0}^{N} \phi_n(s)\,\phi_n(t)\right] ds \tag{79}$$

of the associated Fourier series, where

$$c_n = \int_0^{\pi} \tilde{f} v_n\, ds \qquad \text{and} \qquad \tilde{c}_n = \int_0^{\pi} \tilde{f}\phi_n\, ds, \quad n = 0, 1, 2, \ldots .$$

From equations (76) and (77), together with simple trigonometric identities, we find that

$$\begin{aligned}\Psi_N(s,t) &\overset{\text{def}}{=} \sum_{n=0}^{N}[v_n(s)v_n(t) - \phi_n(s)\phi_n(t)] = [v_0(s)v_0(t) - \phi_0(s)\phi_0(t)] \\ &\quad + \frac{R(s)-R(t)}{2\pi}\sum_{n=1}^{N}\frac{\sin n(s-t)}{n} - (\cos 2\alpha)\,\frac{R(s)+R(t)}{2\pi}\sum_{n=1}^{N}\frac{\sin n(s+t)}{n} \\ &\quad + M_N(s,t)\end{aligned} \tag{80}$$

where $|M_N(s,t)| \le M < +\infty$, for $s, t \in [0, \pi]$, and $N = 1, 2, \ldots$.

Now $|R(t)| \le R < +\infty$, and if we use the nontrivial fact that for all $x \in \mathbb{R}$, and $N = 1, 2, \ldots$

$$\left|\sum_{n=1}^{N}\frac{\sin nx}{n}\right| \le S < +\infty \tag{81}$$

† Assuming that $r \in C^1[0, \pi]$. Otherwise, see Problem 9.5J.

(Problem 9.5D), we see that each term on the right in (80) is bounded independently of s or t or N; i.e.,

$$|\Psi_N(s, t)| \leq \Psi < +\infty, \tag{82}$$

for some problem-constant Ψ. (To obtain the corresponding results in the remaining cases where $\mathscr{s} = 1$, we omit the terms where $n = 0$ and replace by $n - \frac{1}{2}$ each $n \geq 1$ that is *not* a subscript or a summation index. The associated version of (80) holds and (82) follows.)

(9.23) Proposition: *For the above regular problem of v on* $[0, \pi]$, *if* $\tilde{f}$ *is Riemann integrable on* $[0, \pi]$, *then its S–L series* $\sum_{n=0}^{\infty} c_n v_n$ *converges precisely as does its associated Fourier series* $\sum_{n=0}^{\infty} \tilde{c}_n \phi_n$, *and the sums, when defined, are the same.*

Proof: We know by Theorems 1.14 and 7.2 that within $\epsilon > 0$, $\tilde{f}$ may be approximated by a function $\tilde{v} \in \mathscr{S}_0$ (in that $\int_0^\pi |\tilde{f} - \tilde{v}|\, dt < \epsilon$), for which *both* the S–L series and the associated Fourier series *converge uniformly* to $\tilde{v}$. Consequently, if $\tilde{v}$ replaces $\tilde{f}$ in (78) and (79) and the results are denoted $\tilde{V}_N$ and $\tilde{\phi}_N$, respectively, we see from (80) that as $N \to \infty$:

$$\int_0^\pi \tilde{v}(s)\, \Psi_N(s, t)\, ds = \tilde{V}_N(t) - \tilde{\phi}_N(t) \to 0 \textit{ uniformly} \text{ on } [0, \pi].$$

Therefore, by obvious estimates

$$|V_N(t) - \Phi_N(t)| \leq \int_0^\pi |\Psi_N(s, t)|\, |(\tilde{f} - \tilde{v})(s)|\, ds + \left| \int_0^\pi \Psi_N(s, t)\tilde{v}(s)\, ds \right|$$

$$\leq \Psi\epsilon + \epsilon, \qquad \text{if } N \geq N(\epsilon) \qquad (\text{for } \tilde{v}). \tag{83}$$

Hence, as $N \to \infty$, it follows that V_N and Φ_N have the *same* limits (if any) and the convergence is uniform or not on the same subintervals. ■

If f is sectionally smooth on $[0, \pi]$, the convergence behavior of the associated Fourier series can be inferred from Theorem 1.6. In particular, when $\mathscr{s} = 0$, then in (77), $\alpha = \pi/2$, and the associated (cosine) series converges uniformly on those closed subintervals of $[0, \pi]$ on which $\tilde{f}$ is continuous. To obtain such behavior in the other cases, we must restrict the endpoint behavior of $\tilde{f}$. Accordingly we introduce

$$\mathscr{S}_0{}^* = \{f\colon f \textit{ is sectionally smooth on } \langle a, b\rangle \textit{ and } f \textit{ has the zero endpoint limits required by } \mathscr{S}.\} \tag{84}$$

(9.24) Theorem: *For a regular S–L problem of* (3) *and* Definition 9.1, *let* $f \in \mathscr{S}_0{}^*$. *Then the S–L series for f converges pointwise with sum* $[f(x+) + f(x-)]/2$ *at each* $x \in (a, b)$, *and uniformly to f on each closed subinterval of* $[a, b]$ *in which f is continuous. Its partial sums exhibits Gibbs' behavior near each point of discontinuity of f.*

Proof: When $\mathscr{s} = 2$, then $\tilde{f}(0+) = \tilde{f}(\pi-) = 0$ so that $\tilde{f}$ has an odd extension of period 2π without jump discontinuity at $-\pi$, 0, or π. Hence, the associated (sine) series for $\tilde{f}$ converges as required. When $\mathscr{s} = 1$ and, say, $\alpha = 0$, then $f(0+) = 0$, and the associated series $\sum_{n=1}^{\infty} c_n \sqrt{(2/\pi)} \sin(n-\frac{1}{2})t$ converges as required; this is most easily seen through replacing $t/2$ by ξ and defining $\tilde{f}(2\xi) = \tilde{f}(2\pi - 2\xi)$ for $\xi \in ((\pi/2), \pi)$. The remaining case where $\mathscr{a} = 1$ and $\alpha = \pi/2$ can be reduced to the last one if t is replaced by $\pi - t$. The partial sums mimic the Gibbs' behavior of the associated Fourier series. ■

(9.25) Remarks:

(1) The endpoint restrictions on f are clearly necessary for the second conclusion of Theorem 9.24. However, if f is only sectionally smooth on $[a, b]$, then both conclusions apply to (a, b) and *its* closed subintervals.

(2) For a singular problem, we need information concerning the asymptotic behavior of α_n, ω_n and normalizing constants in order to attempt a corresponding analysis. Even when such knowledge is available it can be extremely difficult to proceed, as we illustrate in Appendix C with the Bessel problem of Example 7. For series of Legendre polynomials, a direct proof of an analogous theorem modeled on that for Fourier series in Section 1.8 is outlined in Problems 9.5E and F. (See also the articles of Hobson in Volume 7 of *Proceedings of the London Mathematical Society*.)

(3) Contour integration methods provide an elegant approach to validating S–L series expansions for even more general f in both regular and singular cases. (See [Ti].)

Problem Set 9.5

9.5A Use substitutions (60′) to verify that, in general, f and $\tilde{f}(t) = p(x)f(x)$ have the same S–L coefficients with respect to the eigenfunctions y_n and $v_n(t) = p(x)y_n(x)$.

9.5B When $\alpha = 0$ and $\phi_n(t) = \sqrt{(2/\pi)} \sin nt$, $n = 1, 2, \ldots$, use (76) for $v_n(s)$ and $v_n(t)$ to obtain (80).

9.5C Explain what would change when $\alpha = \pi/2$ in Problem 9.5B.

9.5D 1. For $0 \le x \le \pi$, use (32′) from Section 1.8 to show that

$$S_N(x) = \sum_{n=1}^{N} \frac{\sin nx}{n} = \int_0^x \left[\sum_{n=1}^{N} \cos nt\right] dt = \frac{1}{2}\int_0^x \left[\frac{\sin(N+\frac{1}{2})t}{\sin t/2} - 1\right] dt$$

$$= \int_0^x \frac{\sin Nt}{t}\, dt + \frac{1}{2}\int_0^x (\sin Nt)\left[\cot\frac{t}{2} - \frac{2}{t}\right] dt - \frac{x}{2} + \frac{\sin Nx}{N}.$$

2. Argue that the last integral in part 1, which we denote by $s_N(x)$, is essentially the Nth sine coefficient of the uniformly bounded Riemann integrable function

$$h(x, t) = \begin{cases} \cot\dfrac{t}{2} - \dfrac{2}{t}, & 0 \le t \le x \\ 0, & x < t \le \pi \end{cases}$$

so that by Bessel's inequality, $s_N^2(x) \le (2/\pi) \int_0^\pi h^2(x, t)\, dt \le H$, for some suitably large constant H.

3. Finally,

$$\int_0^x \frac{\sin Nt}{t}\,dt = \int_0^{Nx} \frac{\sin u}{u}\,du$$

(Why?) and this last integral is bounded independently of Nx as we saw in Problem 8.1C. Why can we conclude that for some constant S, $|S_N(x)| \leq S$ for all $x \in [-\pi, \pi]$, and hence for all x?

9.5E* (*Pointwise convergence of Legendre series*)
Let $C_n = (n + \frac{1}{2}) \int_{-1}^{1} fP_n\,dt$, $n = 1, 2, \ldots$ be the Legendre coefficients for a (Riemann) integrable f on $[-1, 1]$.

1. Use Bessel's inequality (18) to show that $c_n = (n + \frac{1}{2})^{-1/2} C_n \to 0$ as $n \to \infty$. Then, recall from Example 8 that at each $x \in (-1, 1)$, the normalized polynomials $p_n(x) = \sqrt{n + \frac{1}{2}} P_n(x)$ are bounded in absolute value independently of n, so that $c_n p_n(x) = C_n P_n(x) \to 0$ as $n \to \infty$.
2. Set $K_N(t, x) = (t - x) \sum_{n=0}^{N} (2n + 1) P_n(t) P_n(x)$, and show that

$$F_N(x) \stackrel{\text{def}}{=} \sum_{n=0}^{N} C_n P_n(x) = \frac{1}{2} \int_{-1}^{1} f(t) \frac{K_N(t, x)}{t - x}\,dt.$$

and that when $f \equiv 1$, then $1 = \frac{1}{2} \int_{-1}^{1} [K_N(t, x)/(t - x)]\,dt$.

3. At a fixed $x \in (-1, 1)$ where f is continuous with right and left derivatives, the function $\tilde{f}(t) = (f(t) - f(x))/(t - x)$ is integrable, with Legendre coefficients $\tilde{C}_n$, $n = 0, 1, 2, \ldots$. Conclude that

$$\begin{aligned} F_N(x) - f(x) &= \frac{1}{2} \int_{-1}^{1} \tilde{f}(t) K_N(t, x)\,dt \\ &= \frac{2N + 2}{2N + 3} \tilde{C}_{N+1} P_N(x) - \frac{2N + 2}{2N + 1} \tilde{C}_N P_{N+1}(x) \to 0 \end{aligned}$$

as $N \to \infty$ (by part 1); i.e., $f(x) = \sum_{n=0}^{\infty} C_n P_n(x)$. (*Hint*: Use (39′) of Section 7.5 twice (once for t and once for x) to show that $K_N(t, x) = (N + 1)[P_{N+1}(t) P_N(x) - P_N(t) P_{N+1}(x)]$.)

4. Prove that if $|x| \leq b < 1$, then the series

$$F(x) = \sum_{n=1}^{\infty} \frac{C_n P_n(x)}{n} = \sum_{n=1}^{\infty} \frac{c_n p_n(x)}{n}$$

converges absolutely and uniformly. (*Hint*: Use Cauchy's inequality.)

9.5F (*Legendre series at a point of discontinuity*)

1. Use the results of Problem 7.5D to show that if $N \geq 1$ and $x \in (-1, 1)$, then

$$\begin{aligned} \int_x^1 \frac{K_N(t, x)}{t - x}\,dt &= 1 - x + \sum_{n=1}^{N} P_n(x)[P_{n-1}(x) - P_{n+1}(x)] \\ &= 1 - P_N(x) P_{N+1}(x) \to 1 \qquad \text{as } N \to +\infty. \end{aligned}$$

(Why?)

2. Let f be as in Problem 9.5E and suppose that $f(x+)$ exists for some fixed $x \in (-1, 1)$ where f has a derivative from the right. Show that

$$\frac{1}{2}\int_x^1 \frac{f(t)-f(x+)}{t-x} K_N(t, x)\, dt \to 0 \qquad \text{as } N \to +\infty$$

by setting $f(t) \equiv 0$ for $t \le x$.

3. Conclude that if in part 2, $f \equiv 0$ for $t \le x$, then

$$F_N(x) - \frac{f(x+)}{2}[1 - P_N(x)P_{N+1}(x)] \to 0 \qquad \text{as } N \to +\infty,$$

$$\text{or} \qquad F_N(x) \to \frac{f(x+)}{2}$$

(from part 1).

4. Recall that $\int_{-1}^{1} [K_N(t, x)/(t-x)]\, dt = 2$, and explain how to prove that if in part 2, f also has a derivative from the left at x, then its Legendre series converges at x with sum $[f(x+) + f(x-)]/2$. (*Hint*: Split the integral at x.)

9.5G** (*Second-order asymptotic estimates when* $r \in C^1[0, \pi]$)
To obtain (76), in case $\sigma = 2$, where $v(0) = v(\pi) = 0$:

1. Take $A = 1$ and $\alpha = 0$ in (65) and iterate to get

$$v(t) = \sin \omega t + \frac{1}{\omega}\int_0^t r(s) \sin \omega(t-s)\left[\sin \omega s + \frac{V(s)}{\omega}\right] ds$$
$$= \sin \omega t - \frac{\cos \omega t}{2\omega} R(t) + \frac{V_1(t)}{\omega^2},$$

where $R(t) = \int_0^t r(s)\, ds$, while V_1 involves integrals of V and $\dot{r}$ over $[0, t]$ so that $V_1(0) = 0$, and V_1 is bounded on $[0, \pi]$.

2. Set $v(\pi) = 0$ in part 1 to get that

$$\tan \omega\pi = \frac{R(\pi)}{2\omega} - \frac{V_1(\pi)}{\cos \omega\pi}\frac{1}{\omega^2},$$

and conclude that $\omega\pi = n\pi + \delta_n$ where

$$\left|\delta_n - \frac{R(\pi)}{2\omega}\right| \le \frac{M}{n^2},$$

for some constant M.

3. Thus

$$v(t) = v_n(t) = \sin nt - \frac{\tilde{R}(t)}{2n}\cos nt + \frac{M_n(t)}{n^2}, \quad \text{for } \tilde{R}(t) = R(t) - \frac{R(\pi)}{\pi},$$

$$\text{and} \qquad a_n^2 \overset{\text{def}}{=} \int_0^\pi v_n^2(t)\, dt = \frac{\pi}{2} + \frac{M_n}{n^2};$$

conclude that $v_n(t)/a_n$ is a normalized eigenfunction satisfying (76) for an appropriate function R.

9.5H* To obtain the analogue of (76), when $\sigma = 1$ where $\dot{v}(0) = E_0 v(0)$ and $v(\pi) = 0$:

1. Take A and α in (65) so that

$$v(t) = \cos \omega t + \frac{E_0}{\omega} \sin \omega t + \frac{V(t)}{\omega}$$

satisfies the first condition and iterate to get

$$v(t) = \cos \omega t + \frac{E_0}{\omega} \sin \omega t + \frac{\cos \omega t}{2\omega} R(t) + \frac{V_1(t)}{\omega^2},$$

for appropriate functions R and V_1.

2. Set $v(\pi) = 0$ and conclude that for some constant M

$$\cos \omega\pi = -\frac{E_0}{2\omega} \sin \omega\pi + \frac{M}{\omega^2},$$

which leads to $\omega = \omega_n = (n - \frac{1}{2}) + (\delta_n/\pi)$ and

$$v_n(t) = \cos (n - \tfrac{1}{2})t - \frac{R_0(t)}{2n} \sin (n - \tfrac{1}{2})t + \frac{M_n(t)}{n^2}.$$

3. Normalize to get (76) with n replaced by $n - \frac{1}{2}$.

9.5I** To establish (76) when $\alpha = 0$ where $\dot{v}(0) = E_0 v(0)$ and $\dot{v}(\pi) = E_1 v(\pi)$:

1. Differentiate v in Problem 9.5H1 and evaluate at π as required to get $\tan \omega\pi = (R_1/\omega) + (M/\omega^2)$ for constants R_1, M.
2. Substitute and normalize to obtain (76).

9.5J*** If r is only continuous on $[0, \pi]$:

1. Explain how to modify the results in Problem 9.5G systematically to obtain (76) with the term $(1/n\sqrt{2\pi}) \int_0^t r(\xi) \cos n(t - 2\xi)\, d\xi$ added.
2. Then show that (80) still holds for some function M_N involving a finite sum of similar integral terms with a *combined integrand* to which (81) applies. Conclude that (82) remains valid.
3. Explain how to modify the results in Problems 9.5H and I to reach the same conclusions.

9.5K The following results are needed in Appendix C.

1. Show that

$$\sin\left(\frac{x}{4}\right) \sum_{n=1}^{N} \frac{\cos nx}{n} = \frac{\sin\left(N + \dfrac{1}{2}\right)x - \sin \dfrac{x}{2}}{4 \cos\left(\dfrac{x}{4}\right)}$$

is uniformly bounded independently of N when $|x| \leq 2\pi(1 - \delta)$ for some fixed $\delta \in (0, 1)$.

2. If $\sigma_n = (n - \frac{1}{4})\pi s - (\pi/4)$ and $\tau_n = (n - \frac{1}{4})\pi t - (\pi/4)$, conclude that

$$\sum_{n=1}^{N} \frac{\sin (\sigma_n - \tau_n)}{n} = \sum_{n=1}^{N} \frac{\sin (n - \frac{1}{4})\, \pi(s - t)}{n}$$

is bounded independently of N, when $s, t \in [\delta, 1]$.

3. Draw a similar conclusion for the sum $\sum_{n=1}^{N} [\sin(\sigma_n + \tau_n)]/n$. (*Hint*: Let $t = 2 - u$.)
4. Use the last results to bound the sums $\sum_{n=1}^{N} (\sin \sigma_n \cos \tau_n)/n$.

9.6

Multidimensional Problems

Our model for generating one-dimensional S–L problems is a nonhomogeneous string vibrating under various boundary conditions. Suppose that instead of a nonhomogeneous string, we consider the free vibrations of a nonhomogeneous *membrane*. Then, as in Section 7.1, we are led to seek solutions $\psi = \psi(\mathbf{x})$ to the Helmholtz equation (for $\lambda = \omega^2$),

$$\nabla \cdot (\tau \nabla \psi) = -\lambda \rho \psi, \tag{85}$$

in a bounded domain V, under various homogeneous conditions related to those of Theorem 5.4 of Section 5.2.

Moreover, in quantum mechanics, the spatial component ψ for the wave function of a single particle satisfies the equation

$$\nabla^2 \psi - q\psi = -\lambda \rho \psi, \tag{86}$$

for appropriate functions q and ρ.

Each of these is seen to be a special case of the equation

$$L(\psi) \equiv \nabla \cdot (\tau \nabla \psi) - q\psi = -\lambda \rho \psi, \tag{87}$$

which is the natural analogue of (3).

Assuming that V is a simple domain in $\mathbb{R}^d$ with boundary S it is surprisingly straightforward to obtain corresponding formal extensions to previous results from this chapter. The divergence theorem of Section 0.2 provides partial integration for **Lagrange's identity**

$$\begin{aligned}\int_V [uL(v) - vL(u)]\, dV &= \int_V [u\nabla \cdot (\tau \nabla v) - v\nabla \cdot (\tau \nabla u)]\, dV \\ &= \int_S \tau \left(u\frac{\partial v}{\partial n} - v\frac{\partial u}{\partial n}\right) dS,\end{aligned} \tag{88}$$

which is valid for all sufficiently smooth u and v. In particular, we can introduce

$$\left.\begin{aligned}I(u, u) &\stackrel{\text{def}}{=} \int_V (\tau |\nabla u|^2 + qu^2)\, dV - \int_S \tau \left(u\frac{\partial u}{\partial n}\right) dS \\ &= \lambda \int_V \rho u^2\, dV, \qquad \text{if } u \text{ satisfies (87),}\end{aligned}\right\} \tag{89}$$

and conclude that if τ, q, and ρ are nonnegative in V, while $\int_S \tau u(\partial u/\partial n)\, dS \leq 0$, then each *eigenvalue* λ is nonnegative. Let's put boundary requirements[†] on the problem that make the right side of (88) vanish whenever u, v are in a special

[†] Such as those of (11′) or (12) in Section 7.1.

subspace $\mathscr{S}$. Then, if $u, v \in \mathscr{S}$, we see that

$$\left.\begin{aligned} I(u, v) &\overset{\text{def}}{=} \int_V (\tau \nabla u \cdot \nabla v + quv)\, dV - \int_S \tau u \frac{\partial v}{\partial n}\, dS = I(v, u) \\ &= \lambda \int_V \rho uv\, dV, \qquad \text{if } u \text{ satisfies (87).} \end{aligned}\right\} \tag{90}$$

Thus, the eigenfunctions ψ, $\tilde{\psi}$ corresponding to distinct eigenvalues λ, $\tilde{\lambda}$, respectively, are ρ-orthogonal in that (see Problem 9.6A),

$$\int_V \rho\psi\tilde{\psi}\, dV = 0. \tag{91}$$

We cannot, however, prove that the eigenfunction associated with a given eigenvalue λ is unique within ρ-normalization and sign. In fact, as is evident from physical considerations of symmetry, if a square membrane with sides in the coordinate directions (x, y) can vibrate freely at frequency $\sqrt{\lambda}$ in a mode that has a cylindrical shape with respect to the x-axis, then it should vibrate freely at the *same* frequency in a mode that has the same cylindrical shape —but with respect to the y-axis. (See Figure 9.2.) These modes cannot be described by functions that are scalar multiples of each other.

Fortunately, in most cases there can be only a *finite* number of linearly independent eigenfunctions associated with a given λ, and they can be supposed mutually ρ-orthogonal by using the Gram–Schmidt process (see Problem 7.5H). Consequently, a multidimensional S–L problem might have eigenvalues $\lambda_1, \lambda_2, \ldots$ where $\lambda_1 < \lambda_2 = \lambda_3 = \lambda_4 < \lambda_5 = \lambda_6 < \lambda_7 = \cdots$.

To each λ_n, there is an eigenfunction ψ_n. If we assume these ψ_n to be ρ-orthonormal and suitably complete, then the remainder of the representation arguments can be duplicated without change. In particular, we have the following inequality

$$\sum_{n=1}^{\infty} \lambda_n \left(\int_V \rho u \psi_n\, dV \right)^2 \leq I(u, u) \tag{92}$$

(valid for those $\hat{C}^1$ functions u meeting the boundary requirements).† This yields

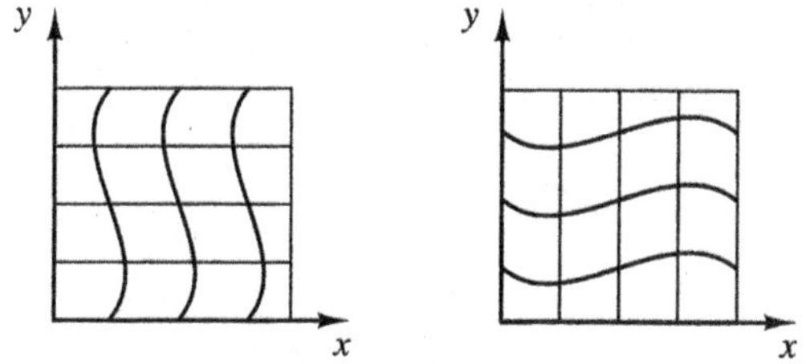

Figure 9.2

† A function $u \in C(\bar{V})$ is in $\hat{C}^1(\bar{V})$ when for some finite K, $u \in C^1(\bar{V}_k)$, $k = 1, 2, \ldots, K$, and $\bar{V} = \cup_{k=1}^{K} \bar{V}_k$, where the V_k are *disjoint* domains.

Rayleigh's principle:

$$\text{For } N = 1, 2, \ldots: \qquad \lambda_N \le R(u) \overset{\text{def}}{=} \frac{I(u,u)}{\int_V \rho u^2\, dV}, \tag{92$'$}$$

when $u \in \mathscr{S}$ *is sufficiently smooth and* ρ*-orthogonal to* $\psi_1, \psi_2, \ldots, \psi_{N-1}$.

(See Problem 9.6C.) Maximin characterizations of eigenvalues follow as before, and there are associated Ritz–Galerkin approximations that have been developed extensively in recent years through multidimensional finite-element methods. (See [S–F].)

Domain Comparison of Eigenvalues

As we saw in Section 9.3, the Rayleigh principle affords comparison between eigenvalues of related problems, and there are multidimensional versions of those results. In particular, for the "same" problem we expect smaller domains to have larger eigenvalues reflecting Rayleigh's initial observation that increasing stiffness is accompanied by higher natural frequencies of vibration. We will illustrate this for the Dirichlet problem in a simple domain $\tilde{V}$ that contains a simple subdomain V and its boundary, as shown in Figure 9.3.

Let $\tilde{\lambda}_n$, $\tilde{\psi}_n$, and $\tilde{R}$ denote the eigenvalues, normalized eigenfunctions, and Rayleigh quotient for the problem on $\tilde{V}$ with homogeneous Dirichlet conditions and let λ_n, ψ_n, R be their correspondents for the *same* equation on V with homogeneous Dirichlet conditions. By extending each ψ_n to be zero outside V, it may be regarded as a $\hat{C}^1$ function on $\tilde{V}$, and by the usual argument, for each N, we can find $u = \sum_{n=1}^{N} c_n \psi_n \ne 0$, which is ρ-orthogonal to each $\tilde{\psi}_n$, $n = 1, 2, \ldots, N-1$. Moreover, $\tilde{R}(u) = R(u)$. (Why?) Hence by Rayleigh's principle we have as conjectured (see Problem 9.6C):

$$\tilde{\lambda}_N \le \tilde{R}(u) = R(u) = \frac{\sum_{n=1}^{N} \lambda_n c_n^2}{\sum_{n=1}^{N} c_n^2} \le \lambda_N. \tag{93}$$

Since in simple cases, we can solve the Dirichlet problem in a few "nice" domains, this comparison can be quite valuable. By similar techniques, we can

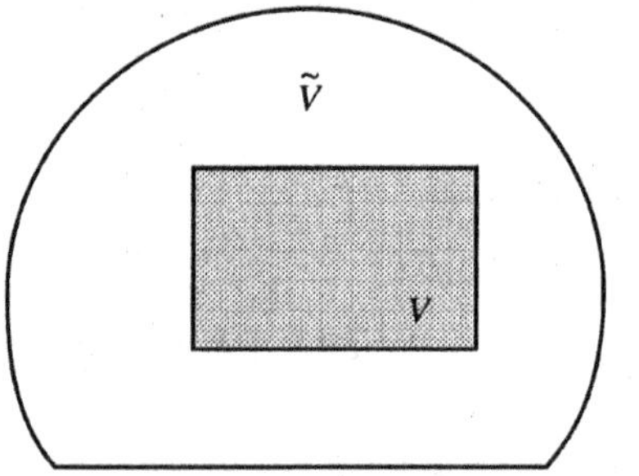

Figure 9.3

extend the proof of the first part of Proposition 9.11 to show that the nodes of ψ_n cannot subdivide V into more than n subdomains; in particular, an eigenfunction ψ_1 to λ_1 must be of one sign in V. If ψ_1 is *nonvanishing* in V, then each other eigenfunction ψ_n *must* change sign in V since $\int_V \rho\psi_1\psi_n\,dV = 0$, $n > 1$, and thus λ_1 *is nondegenerate*. However, the number of nodally determined subregions of V need *not* increase with n. (For these and related results, see Volume I of [C–H], the classic reference work on this subject.)

Membrane Eigenvalues

For a planar membrane of uniform properties, we can assume that both the density ρ and the tensile resistance τ are constant. Then natural vibrations will occur at frequency ω in modes whose shape $\psi = \psi(x, y)$ satisfies the Helmholtz equation,

$$\nabla^2\psi = -\frac{\omega^2}{c^2}\psi, \qquad \text{where } c^2 = \tau/\rho. \tag{94}$$

By the scale change $\bar{x} = x/c$, $\bar{y} = y/c$, we can obtain the same equation for $\bar{\psi}(\bar{x}, \bar{y}) = \psi(x, y)$ with $c^2 = 1$, and we shall simply assume that $c^2 = 1$ in (94). Then with $\lambda = \omega^2$, we seek nontrivial solutions to

$$\nabla^2\psi = -\lambda\psi, \tag{95}$$

in a simple domain V with boundary S, under the Dirichlet condition $\psi|_S = 0$.

The admissible λ and ψ for this problem are called the **membrane eigenvalues and eigenfunctions** of the domain V. As we indicated above, λ_1 is nondegenerate, since ψ_1 is nonvanishing in V. (See Problem 9.6G.) It is also known that $\lambda_2 < 3\lambda_1$ and $\lambda_3 < 4\lambda_1$; however, unless V is convex, little is known about the eigenfunctions associated with λ_n for $n \geq 2$. (see [Sch], pp. 82–93.)

Fortunately, independently of series considerations, we can estimate the natural frequencies of vibration for uniform membranes stretched over a simple domain $\tilde{V}$ (and certain other bounded domains). For $\tilde{V}$ contains a rectangle of sides a, b with a^2/b^2 irrational. From (93) it follows that, properly indexed, the eigenvalues $\tilde{\lambda}_{mn}$ for $\tilde{V}$ satisfy

$$\tilde{\lambda}_{mn} \leq \lambda_{mn} = \pi^2\left(\frac{m^2}{a^2} + \frac{n^2}{b^2}\right), \tag{96}$$

where we have incorporated the results from Example 4 of Section 7.3. Moreover, $\tilde{V}$ is contained in a larger rectangle of sides A, B, with A^2/B^2 irrational. Hence, for $m, n = 1, 2, \ldots$

$$\frac{m^2}{A^2} + \frac{n^2}{B^2} \leq \frac{\tilde{\lambda}_{mn}}{\pi^2} \leq \frac{m^2}{a^2} + \frac{n^2}{b^2}. \tag{97}$$

By choosing rectangles that best approximate $\tilde{V}$, we can improve such estimates. (See Problems 9.6D and E.)

Can two different (i.e., noncongruent) drumheads have the same set of natural frequencies? Or, "Can we hear the shape of a drum?" The last question, posed by M. Kac in 1966, remained open until quite recently. The answer seems to be no! See pages 134–137 of the *Bulletin of the American Mathematical Society*, Vol. 27, No. 1, July, 1992, where it is argued that the two domains shown in Figure 9.4 have the same set of membrane eigenvalues.

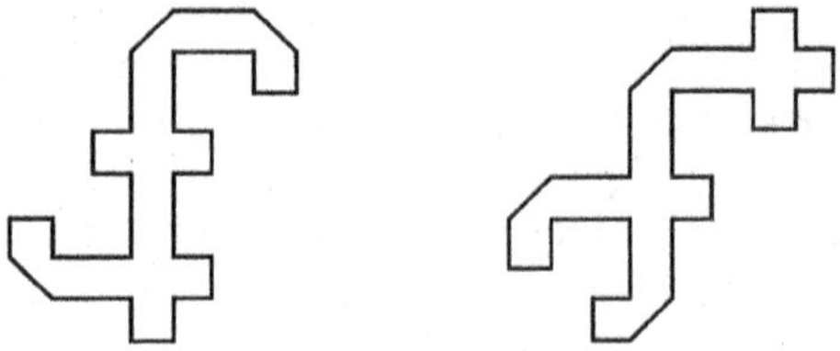

Figure 9.4†

Acoustic Waveguides

The transmission of sound along a straight tunnel of constant planar cross section V produces a pressure p that satisfies the wave equation

$$p_{tt} = c^2 \nabla^2 p \tag{98}$$

where c, the speed of sound within the tunnel, is considered constant. At a fixed point on the tunnel wall there is no normal flow and from Euler's momentum equation (22) of Section 6.4b, it follows that

$$-\nabla p \cdot \mathbf{n} = 0$$

where $\mathbf{n}$ is the outward directed unit normal vector at the point. (Assume that V is a simple domain with boundary S.)

We choose Cartesian coordinates in which the tunnel is located along the positive z-axis with one end at $z = 0$ as shown in Figure 9.5. Then it is natural to attempt separation in the form

$$p(x, y, z, t) = \psi(x, y)\, Z(z) \cos c\omega t$$

corresponding to prescribed oscillatory pressure excitation at the end $z = 0$.

We find easily that we require

$$\frac{\nabla^2 \psi}{\psi} = -\left(\omega^2 + \frac{Z''}{Z}\right) = -\nu, \text{ say,} \tag{99}$$

so that ψ should satisfy the Helmholtz equation $\nabla^2 \psi = -\nu\psi$ in V and the Neumann condition $(\partial\psi/\partial n)|_S = 0$.

The admissible ν for this problem are called the **Neumann eigenvalues** of the

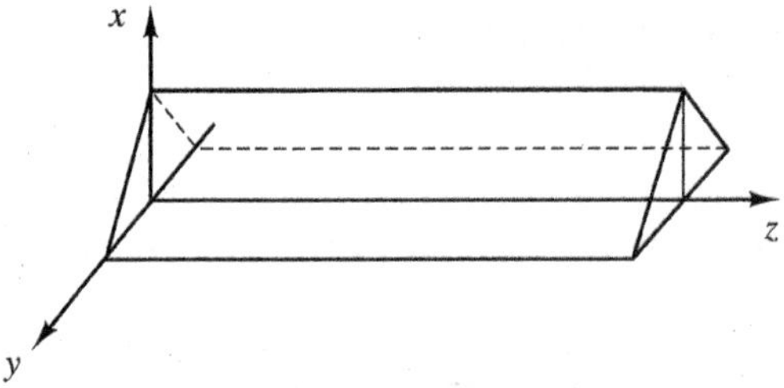

Figure 9.5

† Reprinted from "One Cannot Hear the Shape of a Drum," by Carolyn Gordon, David L. Webb, and Scott Wolpert, *Bulletin (New Series) of the American Mathematical Society*, Volume 27, Number 4, July, 1992, p. 135, by permission of the American Mathematical Society.

domain V. They are nonnegative (Why?) and it is seen that $\nu_1 = 0$ is an eigenvalue to the nonvanishing eigenfunction $\psi_1 \equiv 1$, which is therefore unique within a constant factor. (See Problem 9.6G.) Moreover, when ordered $0 = \nu_1 < \nu_2 \leq \nu_3 \ldots$, then for $k = 1, 2, \ldots$ it follows from the maximin principle that $\nu_k < \lambda_k$, the corresponding membrane eigenvalue (see Problems 9.6C and 9.3F), and, for example, it is known that $\nu_2 < \lambda_1$. (See [Sch, 121–133].)

For each ν_k a solution to our problem is possible for a corresponding Z_k that satisfies the equation

$$Z_k'' + (\omega^2 - \nu_k) Z_k = 0. \tag{100}$$

However, for the associated pressures to be propagated along the tunnel, we require

$$\nu_k \leq \omega^2.$$

(Otherwise, the pressures change exponentially with z.)

In particular, if we have an infinitely long tunnel (or **waveguide**) with *rectangular* cross section $V = \{0 < x < a, 0 < y < a/2\}$, then the ν_k take the form

$$\nu_{mn} = \frac{\pi^2}{a^2}(m^2 + 4n^2), \tag{101}$$

with associated $\psi_{mn}(x, y) = \cos(m\pi x/a) \cos(2n\pi y/a)$, m, $n = 0, 1, 2, \ldots$ (see Section 7.3). Propagation of an oscillatory excitation at $z = 0$, at "frequency" ω can occur only in those (complex) modes

$$\phi_{mn}(x, y, z) = \psi_{mn}(x, y) \exp\left(iz\sqrt{\omega^2 - \nu_{mn}}\right) \tag{102}$$

where $\omega \geq (\pi/a)\sqrt{m^2 + 4n^2}$, m, $n = 0, 1, 2, \ldots$, and these depend on the actual size of a. There will always be traveling-wave solutions of the complex form

$$p_0(x, y, z, t) = \exp i\omega(z - ct)$$

and these will be the only ones to propagate unless $\omega \geq \pi/a$. If $\pi/a \leq \omega < 2\pi/a$, then we add those of the form

$$p_1(x, y, z, t) = \cos\frac{\pi x}{a} \exp i\omega\left(z\sqrt{1 - \frac{\pi}{a\omega}} - ct\right), \qquad \text{etc.}$$

Each other term will die out exponentially as z increases, so that at large distances along this *acoustic wave guide*, we will "hear" only a finite part of the initial signal. For example, suppose we stretch a membrane over the end $z = 0$ and force it to oscillate at frequency $\omega \in [\pi/a, \sqrt{5}\pi/a)$ in the mode shape

$$f(x, y) = xy\left(1 - \frac{x}{a}\right)\left(1 - \frac{y}{2a}\right).$$

This gives rise to a proportional pressure signal of which only the terms ψ_{00}, ψ_{01}, ψ_{10}, and ψ_{11} in the representation

$$\Psi(x, y) = \sum_{m,n=0}^{\infty} c_{mn}\psi_{mn}(x, y) \tag{103}$$

will persist along the guide. But here $c_{10} = c_{11} = 0$ and similarly $c_{01} = 0$ (since $\int_0^a x[1 - (x/a)] \cos(\pi x/a)\, dx = 0$) so that only the traveling-wave term p_0 survives at large distances.

These results indicate what must be considered in designing wave guides for the propagation of acoustic or electromagnetic signals. In particular, effective transmission at a given frequency depends on both the size of the guide and its shape in relation to that of the signal at excitation. (See [E–H].)

Problem Set 9.6

9.6A 1. Verify equations (90) for $u, v \in \mathscr{S}$, and when u satisfies (87).

2. Conclude that if $\tilde{u}$ also satisfies (87) for the eigenvalue $\tilde{\lambda} \neq \lambda$, then (91) holds.

9.6B Suppose that for $n = 1, 2, \ldots, N$, u_n is a normalized eigenfunction to λ_n for (87) with the Dirichlet condition $u|_{\partial V} = 0$ defining $\mathscr{S}$. (Assume V is simple and that $\partial u/\partial n$ is well defined and bounded.)

1. For $f \in \mathscr{S}$ set

$$c_n = \int_V \rho f u_n \, dV, \qquad n = 1, 2, \ldots, N$$

(assuming that $\int_V \rho u_n u_m \, dV = \delta_{mn}$). Conclude that $I(f_N, f_N) \leq I(f, f)$, where $f_N = f - \sum_{n=1}^{N} c_n u_n$.

2. State formal analogues of Lemma 9.4 and Theorem 9.5. How would you define $I(f)$? $\mathscr{S}_0$?

9.6C 1. Discuss Rayleigh's principle for the previous problem. How is $\mathscr{S}_N$ defined?

2. Show how (93) follows from (90).

3. Formulate a maximin characterization of λ_n analogous to that in (43).

9.6D 1. Use (97) to obtain upper and lower estimates of the eigenvalues λ_{mn} for the vibrations of a uniform elliptical membrane with axes of length 1 and π. (Treat the elliptical domain as though it were simple and squeeze your estimates as tightly as you can.)

2. Use (97) to obtain corresponding estimates for a circular membrane of radius 1. Compare with the actual values obtained from the zeros $\sqrt{\lambda_{mn}}$ of J_m, $n = 1, 2, \ldots$ for $m = 0, 1, 2$. (See [J–E].)

9.6E 1. Verify that natural vibrations of a triangular drumhead with edges $y = \pm ax$, $y = \pi$, can occur in modes

$$u_n(x, y) = \sin 2ny - 2 \sin ny \cos nax, \qquad n = 1, 2, \ldots$$

that satisfy the equation $\nabla^2 u_n = -\mu_n u_n$ for some $\mu_n > 0$.

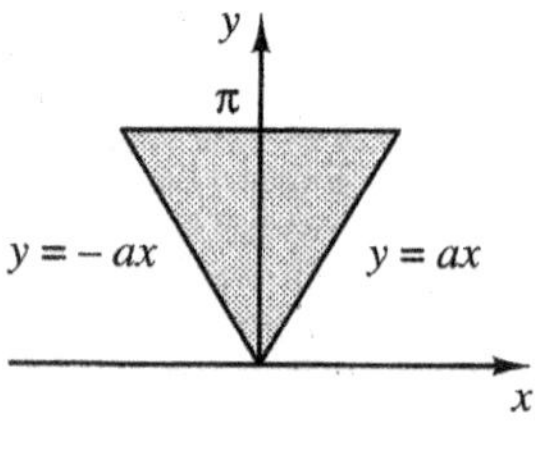

2. Show that $a = \pm\sqrt{3}$ (defining an equilateral triangle) is the only possibility for these u_n. What are the associated eigenvalues μ_n? (See M. Pinsky, *The Eigenvalues of an Equilateral Triangle*, SIAM Journal of Mathematical Analysis, (1980), 819–827.)

3. Show that $u_1(x, y) > 0$ in the triangle so that $\mu_1 = \lambda_1(= ?)$ is the least eigenvalue for this problem.

4.* Compare these μ_n with certain corresponding eigenvalues of squares of sides π and $\pi/2$ as in (97), for values $n = 1, 2, 3, 4, 5$. Can you draw any conclusions as to whether there are "more" triangular mode shapes and/or eigenvalues than those obtained above? Can you supply one? many? (*Hint*: Use symmetry.)

9.6F

1. Show that under the scale change $\tilde{\mathbf{x}} = \ell\mathbf{x}$, which stretches a plane domain V onto a domain $\tilde{V}$, the corresponding membrane eigenvalues are related by $\tilde{\lambda}_n = \lambda_n/\ell^2$, $n = 1, 2, \ldots$.

2. Verify that

$$u_1(x, y) = \sin 2x \sin y - \sin x \sin 2y$$

is a positive eigenfunction for a uniform triangular membrane with boundaries $y = 0$, $y = x$, $x = \pi$. Find λ_1, the least eigenvalue.

3. Compare δ, the least membrane eigenvalue of the sector $\Delta = \{0 < r < 1, 0 < \theta < \pi/4\}$, with those for triangles. What can you conclude about J_4?

9.6G Let ν_1 be an eigenvalue for the problem of (87) in a simple domain V with $(\partial u/\partial n)|_{\partial V} = 0$. If an associated eigenfunction $u = u_1$ is positive in V, we want to show that it is unique within a constant factor; i.e., ν_1 is nondegenerate.

1. Any other eigenfunction may be expressed as $u_2 = uv$, where $v = u_2/u$ is C^2 in V, since $u = u_1 > 0$. Verify that in V:

$$\begin{aligned}\tau|\nabla(uv)|^2 &= \tau(u^2|\nabla v|^2 + v^2|\nabla u|^2) + \tau u\nabla u \cdot \nabla(v^2)\\ &= \tau u^2|\nabla v|^2 + \nabla\cdot(\tau v^2 u\nabla u) - v^2 u\nabla\cdot(\tau\nabla u).\end{aligned}$$

2. Then use (87) for $\lambda = \nu_1$ to get

$$\tau|\nabla u_2|^2 + qu_2^2 = \tau u^2|\nabla v|^2 + \nu_1\rho u_2^2 - \nabla\cdot(\tau u_2 v\nabla u).$$

3. Finally, apply the divergence theorem to show that (89) gives

$$I(u_2, u_2) = \int_V (\tau u^2|\nabla v|^2 + \nu_1\rho u_2^2)\, dV \geq \nu_1 \int_V \rho u_2^2\, dV$$

with equality iff $\nabla v = \mathbf{0}$ in V.

4. If u_2 is an eigenfunction to $\nu_2 = \nu_1$, conclude that $u_2 = cu_1$ for some constant c.

5. How does this argument change if we require $[(\partial u/\partial n)+u]|_{\partial V}=0$ instead of a Neumann condition at the boundary?

9.6H (*Critical design for a nuclear reactor*) In the theory of reactor design, the thermal flux u for a critical system must be a solution to the Helmholtz equation

$$\nabla^2 u + \lambda u = 0$$

subject to the condition that the flux go to zero at the effective boundaries of the reactor. Here, the *geometric buckling* is the lowest eigenvalue for the equation and its boundary conditions.

1. Explain how to use the result of Problem 7.3I to estimate the least eigenvalue λ_1 of the Dirichlet problem

$$\nabla^2 u + \lambda u = 0 \qquad \text{in } V$$

with $u|_{\partial V}=0$, when $V=\{\mathbf{x}\in\mathbb{R}^3\colon |\mathbf{x}|<1\}$.

2.* Find the actual eigenvalues by separation.

3.* If $V=\{(x,y,z)\in\mathbb{R}^3\colon 0<x<a,\ 0<y<b,\ 0<z<c\}$, show how λ_1 is related to a, b, c.

4. For $\lambda_1 = 37.2\times 10^{-4}\,\text{cm}^{-2}$ (a uranium-beryllium system), find the size of a critical cubical reactor.

9.6I 1. Explain how to estimate the eigenvalues λ of the positive problem $L(u)\equiv \nabla\cdot(\tau\nabla u)-qu=-\lambda\rho u$ in V with $u|_{\partial V}=0$ where V is the planar domain shown in the figure while $\tau(x,y)=2+(x^2+y^2)$, $q(x,y)=\sin x\sin y$, and $\rho(x,y)=2-(x^2+y^2)$. (*Hint*: Compare with corresponding eigenvalues for simpler coefficients in simpler domains using Rayleigh's principle.)

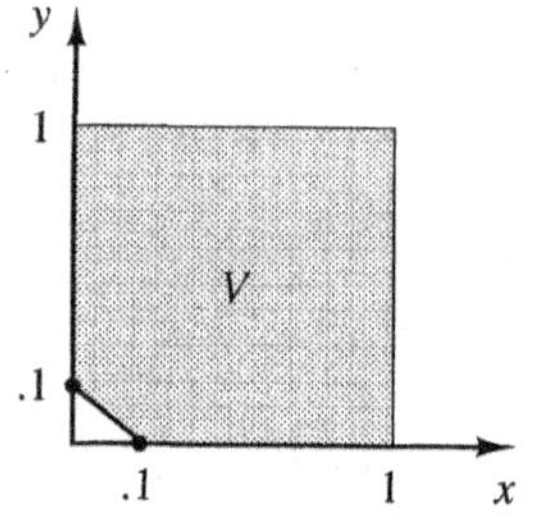

2.* Carry out your program.

9.6J The propagation of waves in the region between two parallel planes $\{-\infty<x,y<+\infty,\ 0\le z\le \ell\}$ is governed by the boundary value problem

$$\Phi_{tt}=c^2\nabla^2\Phi$$

$$\text{with } \Phi=0 \qquad \text{at} \qquad z=0,\ z=\ell.$$

Find the lowest frequency ω of a monochromatic wave that can propagate horizontally between the two planes.

9.6K Consider the problem of standing acoustic waves in a sphere satisfying

$$\nabla^2\Phi + k^2\Phi = 0 \quad \text{when} \quad x^2 + y^2 + z^2 < 1$$

$$\text{with } \Phi = 0 \quad \text{when} \quad x^2 + y^2 + z^2 = 1.$$

Find the lowest nonzero k for a spherically symmetric mode of the form $\Phi = \Psi(r)/r$ where $r^2 = x^2 + y^2 + z^2$.

Appendix A
Existence and Compactness

Sometimes we can establish existence of a special point in a set, or of a solution to a differential equation, by exhibiting a candidate with the required properties. Otherwise we may have to construct the desired object as the limit of an approximating sequence. If a sequence does not converge, it may still provide usable limits of convergent *subsequences*. For example, the sequence of numbers $(-1)^n$, $n = 0, 1, 2, \ldots$ does not converge, but its subsequences $(-1)^n$, $n = 0, 2, 4, \ldots$ and $(-1)^n$, $n = 1, 3, 5, \ldots$ do converge — albeit with different limits. A sequence of elements a_n, $n = 1, 2, \ldots$ provides many distinct subsequences, each of the form a_{n_j}, $j = 1, 2, \ldots$ for *some* choice of integers $n_j < n_{j+1}$, $j = 1, 2, \ldots$. Observe that if the sequence itself has limit a_0, then each of these subsequences converges to a_0 since $n_j \geq j$. On the other hand, the sequence $a_n = n$, $n = 1, 2, \ldots$ has no finite limit nor does any of its subsequences.

Definition: *A subset of* $\mathbb{R}^d$ *is said to be* **compact** *when each sequence of its points provides a convergent subsequence whose limit point is in the set.* ∎

In particular, in $\mathbb{R}^d$, a compact set is *closed* in that it contains the limit point of each convergent sequence of its points, and we show below that it is *bounded.* In Section A.2, we will see that conversely, a *closed and bounded subset of* $\mathbb{R}^d$ *is compact.* But first, let's see what makes compact sets so useful.

A.1 Continuity and Compactness

To appreciate the value and character of arguments involving compactness, look at the following results.

(A.1) Theorem: *If a real-valued function f is defined and continuous on a compact set $K \subseteq \mathbb{R}^d$, then*

(1) f assumes (finite) maximum and minimum values at some points in K, and K is bounded.

(2) f is uniformly continuous on K.

Proof: For part 1 observe that we can at least find a sequence $\mathbf{x}_n \in K$, $n = 1, 2, \dots$ for which $f(\mathbf{x_n}) \nearrow \mu$, the least upper bound of values of f on K. ($\mu \in \mathbb{R} \cup \{+\infty\}$.) Since K is compact, we can find a *subsequence* $\mathbf{x}_{n_1}, \mathbf{x}_{n_2}, \mathbf{x}_{n_3} \dots \mathbf{x}_{n_j}$ of these $\mathbf{x}_n$'s converging to a limit $\mathbf{x}_0$, as $j \to \infty$. Moreover, $\mathbf{x}_0 \in K$ since K is compact. But f is continuous at $\mathbf{x}_0$, so that

$$f(\mathbf{x}_0) = \lim_{j \to \infty} f(\mathbf{x}_{n_j}) = \mu;$$

thus μ *is* finite and f assumes its maximum value (μ) at $\mathbf{x}_0$. The proof that f assumes its minimum value is similar. In particular, the continuous function $f(\mathbf{x}) = |\mathbf{x}|$ is bounded on K which means that K is bounded.

For part 2 we wish to show that for each $\epsilon > 0$, there is a $\delta = \delta(\epsilon)$ such that for *any* points $\mathbf{x}, \tilde{\mathbf{x}} \in K$

$$|\mathbf{x} - \tilde{\mathbf{x}}| < \delta \Rightarrow |f(\mathbf{x}) - f(\tilde{\mathbf{x}})| < \epsilon.$$

(See Section 0.1c.) Suppose this fails for some $\epsilon_0 > 0$. Then we can find pairs of points $\mathbf{x}_n$, $\tilde{\mathbf{x}}_n$ in K, with

$$|\mathbf{x}_n - \tilde{\mathbf{x}}_n| < \frac{1}{n} \text{ while } |f(\mathbf{x}_n) - f(\tilde{\mathbf{x}}_n)| \geq \epsilon_0,\ n = 1, 2, \dots .$$

By compactness of K as above, a subsequence $\mathbf{x}_{n_j} \to x_0 \in K$, as $j \to \infty$. Then

$$|\tilde{\mathbf{x}}_{n_j} - \mathbf{x}_0| \leq |\mathbf{x}_{n_j} - \mathbf{x}_0| + |\mathbf{x}_{n_j} - \tilde{\mathbf{x}}_{n_j}| \leq |\mathbf{x}_{n_j} - \mathbf{x}_0| + \frac{1}{n_j}, \text{ which} \to 0, \qquad \text{as } j \to \infty.$$

Hence, $\tilde{\mathbf{x}}_{n_j} \to \mathbf{x}_0$ as $j \to \infty$. But then for all j:

$$0 < \epsilon_0 \leq |f(\mathbf{x}_{n_j}) - f(\tilde{\mathbf{x}}_{n_j})| \leq |f(\mathbf{x}_{n_j}) - f(\mathbf{x}_0)| + |f(\tilde{\mathbf{x}}_{n_j}) - f(\mathbf{x}_0)|,$$

so that the right side cannot approach zero as $j \to \infty$, contradicting the fact that f is continuous at $\mathbf{x}_0$. ■

Now, let's find some compact sets.

A.2

Compact Sets in $\mathbb{R}^d$†

To establish the compactness of the bounded closed interval $[a, b] \subseteq \mathbb{R}$, it will be enough to prove that the unit interval $I = [0, 1]$ is compact. Indeed, a scale change and translation permits us to regard $[a, b]$ as the unit interval in another

† This section is reprinted with permission from John L. Troutman, *Variational Calculus with Elementary Convexity* (*UTM*), New York: Springer-Verlag (1983), pp. 327–329.

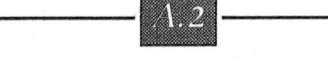

coordinate system, and convergence is clearly preserved under these elementary operations.

(A.2) Lemma: *$I = [0, 1]$ is compact.*

Proof: Each $x \in I$ has the *binary* expansion $x = \sum_{k=1}^{\infty} b_k 2^{-k}$, where $b_k = 0$ or 1, $k = 1, 2, \ldots$, and in case of ambiguity, we choose the expansion that terminates; e.g., we take $2^{-(m-1)}$ instead of $\sum_{k=m}^{\infty} 2^{-k}$, $m = 2, \ldots$.

Now, for each $n = 1, 2, \ldots$, suppose that $x_n \in I$ has the binary expansion $x_n = \sum_{k=1}^{\infty} b_{nk} 2^{-k}$. Then we can select a subsequence $\tilde{x}_1, \tilde{x}_2, \ldots$ of these x_n that converges to a limit point $\tilde{x} \in I$, in the following manner:

If there is an infinite set of the x_n with *first* binary coefficient $b_{1n} = 0$, choose $\tilde{x}_1$ to be that x_n from this set with *least* index n_1. Otherwise, there is an infinite set of the x_n with $b_{1n} = 1$, and we take $\tilde{x}_1$ to be that among them with least index. In either case, let $\tilde{b}_1$ be the first binary coefficient of $\tilde{x}_1$.

Next, let $\tilde{x}_2$ be that x_n of least index $n_2 > n_1$ for which $b_{1n} = \tilde{b}_1$ and $b_{2n} = 0$, if there is an infinite set of such x_n; otherwise, for which $b_{1n} = \tilde{b}_1$ and $b_{2n} = 1$. Let $\tilde{b}_2$ be the *second* binary coefficient of $\tilde{x}_2$.

This selection process can be continued indefinitely, since at each stage we consider only those remaining x_n that form an infinite set. There results a *subsequence* $\tilde{x}_m = x_{n_m}$ and an associated sequence of binary coefficients $\tilde{b}_m$, $m = 1, 2, \ldots$, determining the *number* $\tilde{x} \overset{\text{def}}{=} \sum_{k=1}^{\infty} \tilde{b}_k 2^{-k} \in I$.

But by construction, $\tilde{x}$ and $\tilde{x}_m$ have the *same* initial m binary coefficients $\tilde{b}_1, \tilde{b}_2, \ldots, \tilde{b}_m$, and the remaining binary coefficients cannot exceed 1. Hence, by an easy estimate, $|\tilde{x} - \tilde{x}_m| \le \sum_{k=m}^{\infty} 2 \cdot 2^{-k} = 4 \cdot 2^{-m}$, which $\to 0$ as $m \to \infty$. ■

> ***Remarks:*** The selection process used in proving Lemma A.2 may be visualized by successively bisecting (sub)intervals of I and retaining at each stage the *left*-most interval that contains an infinite set of the x_n. The $\tilde{b}_k$ index these selections, and, more important, define the limit point $\tilde{x}$ as the sum of a series of positive terms. We are assuming here (and elsewhere) that the set of real numbers is complete in that it contains a sum for each convergent series of its elements. For further discussion of this somewhat subtle point, see [Ru].

If we consider *successively* each component of a sequence $\mathbf{x}_n \in B = \{\mathbf{x} \in \mathbb{R}^d : a_j \le x_j \le b_j, j = 1, 2, \ldots, d\}$, $n = 1, 2, \ldots$, then it is straightforward to establish compactness of the *closed* box B. [The proven compactness of $[a_1, b_1]$ guarantees a convergent sequence of the *first* components of the $\mathbf{x}_n$ with a limit point $\tilde{x}_1 \in [a_1, b_1]$. Then from *this subsequence* of the second components, we may extract a convergent subsequence with limit point $\tilde{x}_2 \in [a_2, b_2]$. Continuing *successively*, we obtain a dth sub-sub-sub-sequence $\{\tilde{\mathbf{x}}_n\}_{n=1}^{\infty}$, say, for which the dth components converge to a limit point $\tilde{x}_d \in [a_d, b_d]$, while each of the other components converges to its previous limit. (Why?) It follows by an easy estimate that $\lim_n \tilde{\mathbf{x}}_n = \tilde{\mathbf{x}} = (\tilde{x}_1, \tilde{x}_2, \ldots, \tilde{x}_d) \in B$.]

(A.3) Theorem: *Each closed and bounded set $K \subseteq \mathbb{R}^d$ is compact.*

Proof: Since K is bounded, then $K \subseteq B$ for some (large) closed box B. If $\mathbf{x}_n \in K \subseteq B$, $n = 1, 2, \ldots$, the compactness of B guarantees a convergent subsequence $\{\mathbf{x}_{n_k}\}$, with limit point $\mathbf{x}_0 \in B$. But K is closed and therefore it contains the limit of each of its convergent sequences. Hence $\mathbf{x}_0 \in K$, so that K *is* compact. ■

Appendix B Ordinary Linear Differential Equations

In this appendix we present first a few facts about solutions $y = y(x)$ of second-order ordinary differential equations of the form

$$y'' + py' + qy = f, \tag{1}$$

where p, q, and f are given continuous functions on some fixed *compact* interval $[a, b]$. In Section B.2 we consider the Green function for associated two-point boundary value problems but postpone until Section B.3 a discussion of existence of solutions for these equations and for general first-order linear systems.

B.1 Second-Order Equations

The general solution to (1) may be expressed as

$$y = c_1 y_1 + c_2 y_2 + y_f$$

for constants c_1 and c_2, where y_f is *any* solution of (1) on $[a, b]$, and y_1 and y_2 are solutions on $[a, b]$ of the associated homogeneous equation

$$y'' + py' + qy = 0 \tag{1'}$$

that are linearly independent. Linear independence is assured if the Wronskian

$$W = y_1 y_2' - y_1' y_2 \tag{2}$$

is nonvanishing on $[a, b]$ because then neither y_k can be a constant multiple of the other; in fact,

$$\left(\frac{y_1}{y_2}\right)' = -\frac{W}{y_2^2} \tag{2'}$$

will be zero if $y_1 = cy_2$ or if $y_2 = cy_1$, for some constant c. Moreover, the Wronskian of linearly independent solutions *is* nonvanishing since $W' = -pW$, so that $W(x) = C \exp(-\int^x p(t)\,dt)$ and $C \neq 0$, for otherwise y_1/y_2 is constant by (2′).

When the coefficient functions p and q are constant, the components of the general solution can be determined. The most direct approach is to observe that with the operator notation $D^n = d^n/dx^n$, $n = 1, 2, \ldots$, equation (1) can be written

$$(D^2 + pD + q)y = f$$

and factored in the form

$$(D - \beta)(D - \alpha)y = f$$

where $\alpha = (-p + \sqrt{p^2 - 4q})/2$ and $\beta = (-p - \sqrt{p^2 - 4q})/2$.

If $\alpha = r + is$, set $e^{\alpha x} = e^{rx}e^{isx}$ so that

$$De^{\alpha x} = \alpha e^{\alpha x} \qquad \text{and} \qquad (D - \alpha)y = e^{\alpha x}D(e^{-\alpha x}y).$$

Then the factored equation can be replaced by

$$e^{\beta x}D[e^{-\beta x}e^{\alpha x}D(e^{-\alpha x}y)] = f,$$

which can be solved by inverting the indicated operations successively.

When $\alpha \neq \beta$, we get

$$y(x) = Ae^{\alpha x} + Be^{\beta x} + e^{\alpha x}\int^x e^{-\alpha\xi}e^{\beta\xi}\,d\xi \int^\xi e^{-\beta t}f(t)\,dt.$$

Finally, partial integration with $u(\xi) = \int^\xi e^{-\beta t}f(t)\,dt$ eliminates the double integral and after rearrangement, we find that

$$y(x) = Ae^{\alpha x} + Be^{\beta x} + \int^x \frac{e^{\alpha(x-\xi)} - e^{\beta(x-\xi)}}{\alpha - \beta} f(\xi)\,d\xi, \qquad \alpha \neq \beta.$$

Similarly,

$$y(x) = Ae^{\alpha x} + Bxe^{\alpha x} + \int^x e^{\alpha(x-\xi)}(x - \xi)f(\xi)\,d\xi, \qquad \alpha = \beta.$$

Here A and B are arbitrary constants (possibly complex) and the last integrals may be made definite if desired by integrating from, say, a to x.

For the most common equations with constant coefficients, the results are as follows.

(B.1) Proposition: *Let ω and δ be nonzero real constants. If $f \in C[a, b]$ and $y \in C^2[a, b]$ then*

(1) $y'' + \omega^2 y = f \Rightarrow y(x) = c\cos\omega x + s\sin\omega x + (1/\omega)\int_a^x \sin\omega(x - \xi)f(\xi)\,d\xi.$

(2) $y'' - \omega^2 y = f \Rightarrow y(x) = c\cosh\omega x + s\sinh\omega x + (1/\omega)\int_a^x \sinh\omega(x - \xi)f(\xi)\,d\xi.$

(3) $y'' + 2\delta y' + \delta^2 y = f \Rightarrow y(x) = e^{-\delta x}\{(c + sx) + \int_a^x e^{\delta\xi}(x - \xi)f(\xi)\,d\xi\}.$

(4) $y'' + 2\delta y' + \omega^2 y = f$ where $\beta = |\delta^2 - \omega^2|^{1/2} \neq 0: \Rightarrow$

$$y(x) = e^{-\delta x}\left\{(c\cos\beta x + s\sin\beta x) + (1/\beta)\int_a^x e^{\delta\xi}\sin\beta(x-\xi)\,f(\xi)\,d\xi\right\}, \text{ if } |\omega| > |\delta|,$$

$$y(x) = e^{-\delta x}\left\{(c\cosh\beta x + s\sinh\beta x) + (1/\beta)\int_a^x e^{\delta\xi}\sinh\beta(x-\xi)\,f(\xi)\,d\xi\right\}, \text{ if } |\omega| < |\delta|$$

for some real constants c and s. ■

By differentiation, it is easy to verify that each function y defined in these expressions does satisfy the relevant differential equation, and by proper selection of c and s, y and y' can be specified at a point in $[a, b]$.

When the coefficients p and q in (1) are not constant, then Theorem B.6 of Section B.3 guarantees existence of a unique solution y on $[a, b]$ with specified values of $y(a)$ and $y'(a)$, but y cannot usually be expressed in simple form. Let's first consider the homogeneous equation (1′) which has the trivial solution $y(x) = 0$.

Suppose that u is a solution of equation (1′) on $[a, b]$ that is nonvanishing. Then (by **reduction of order**) we can find a *linearly independent* solution in the form $y = vu$, where v is a nonconstant solution of the linear equation

$$v'' + \left(p + \frac{2u'}{u}\right)v' = 0.$$

Now, this equation is first-order in v', and it can be integrated. For appropriate integration constants, we get $u^2v' = P$, where $P(x) = \exp\left(-\int_a^x p(t)\,dt\right)$, so that one solution is given by

$$y_1(x) = u(x)\int_a^x \frac{P(t)}{u^2(t)}\,dt, \qquad x \in [a, b]. \tag{3}$$

Observe that $y_1(a) = 0$, except possibly in a limiting situation in which $u(a) = 0$.

Finally, when a pair of linearly independent solutions y_1 and y_2 of (1′) has been found, then (by **variation of parameters**) we can obtain a particular solution to equation (1) in the form

$$y_f(x) = \int_a^x \left[\frac{y_2(x)y_1(\xi) - y_1(x)y_2(\xi)}{W(\xi)}\right] f(\xi)\,d\xi. \tag{3'}$$

A function y_f *is* defined by this formula since W in (2) is nonvanishing (Why?), and it is straightforward to verify that y_f is a solution of (1). (A similar computation is carried out at the end of Section B.2.) Clearly, $y_f(a) = 0$, and similarly, $y_f'(a) = 0$; however, by adding to y_f an appropriate linear combination of y_1 and y_2, we can obtain the unique solution y having other specified values of y and y' at a (or at some other point x_0 in $[a, b]$).

Analytic Solutions

The homogeneous equation (1′)

$$y'' + p(x)y' + q(x)y = 0$$

arises frequently in the form

$$R(x)y'' + P(x)y' + Q(x)y = 0, \tag{4}$$

where P, Q, and R are polynomials or other functions that are real-analytic in a common neighborhood of a point $x = a$. Then it is reasonable to seek a solution $y = y(x)$ in this neighborhood which is also **real-analytic** in that for some $\rho > 0$ and *real* constants a_k, $k = 0, 1, 2, \ldots$

$$y(x) = \sum_{k=0}^{\infty} a_k \frac{(x-a)^k}{k!}, \qquad |x - a| < \rho. \tag{5}$$

Such *functions* y are infinitely differentiable in the interval $(a - \rho, a + \rho)$, and the coefficients a_k are given by

$$a_k = y^{(k)}(a), \qquad k = 0, 1, 2, \ldots . \tag{5'}$$

On the other hand, for such *solutions* y the left side of equation (4), Y, say, can be differentiated successively as often as desired. Upon subsequent evaluation at $x = a$ we obtain the following results.

$$Y(a) = R(a)y''(a) + P(a)y'(a) + Q(a)y(a) = 0$$

$$Y'(a) = R(a)y'''(a) + R'(a)y''(a) + P(a)y''(a) + P'(a)y'(a)$$
$$+ Q(a)y'(a) + Q'(a)y(a) = 0$$

$$Y^{(k)}(a) = R(a)y^{(k+2)}(a) + \cdots + Q^{(k)}(a)y(a) = 0, \qquad k = 2, 3, \ldots .$$

(The intermediate terms for the higher derivatives are given by the Leibniz rule for differentiating products uv, namely

$$(uv)^{(k)} = uv^{(k)} + ku'v^{(k-1)} + \frac{k(k-1)}{2} u''v^{(k-2)} + \cdots + u^{(k)}v, \qquad k = 1, 2, \ldots, \tag{6}$$

which resembles the binomial formula and can be verified inductively.) The latter equations constitute relations that the derivatives $y^{(k)}(a)$ — or, equivalently, the coefficients a_k — of an *analytic solution* y must satisfy. Conversely, if numbers a_k can be found that satisfy these relations, then the series y of (5), *if convergent*, will be a solution to (4). [Indeed Y, the left side of (4), is then an analytic function near $x = a$, *all* of whose derivatives at a vanish by construction. Hence $Y(x) \equiv Y(a) = 0$.]

Whether nontrivial analytic solutions y exist depends on the behavior of P, Q, and R near $x = a$, as follows:

(B.2) Theorem:

(i) *When $R(a) \neq 0$, then equation (4) has a pair of linearly independent solutions that are analytic in a neighborhood of $x = a$.*

(ii) *Suppose $R(a) = 0$, but $(x - a)p(x)$ and $(x - a)^2 q(x)$ have finite limits p_0 and q_0, respectively, as $x \to a$. Then in some interval (a, b), for at least one root r of the* **indicial equation**

$$r(r-1) + p_0 r + q_0 = 0, \tag{7}$$

(4) *has a solution of* **Frobenius type** $(x - a)^r y(x)$ *where y is given by* (5), *with $y(a) \neq 0$. When r is a nonnegative integer, such solutions are analytic in a neighborhood of $x = a$.* ■

When $a = 0$, case (ii) of this theorem is exemplified by **Euler's equation**

$$x^2y'' + p_0xy' + q_0y = 0, \qquad x > 0, \tag{8}$$

which always has the solution $y_1(x) = x^r$, if r satisfies (7). This also follows from the observation that $v(t) \stackrel{\text{def}}{=} y(e^t)$ satisfies

$$\ddot{v} + (p_0 - 1)\dot{v} + q_0v = 0, \tag{8'}$$

an equation with constant coefficients.

*Frobenius Solutions**

When $a = 0$ in case (ii) of Theorem B.2 we can differentiate the product $x^ry(x)$ twice and substitute in (1′) to show that it is a Frobenius solution to (4) provided that its *analytic part* y satisfies the auxiliary equation

$$L(y) \stackrel{\text{def}}{=} x^2y'' + [2r + xp(x)]xy' + [r(r-1) + xp(x)r + x^2q(x)]y = 0. \tag{9}$$

In particular, upon evaluation at 0, we must have

$$[r(r-1) + p_0r + q_0]y(0) = 0,$$

and we see that r is chosen to satisfy (7) in order that a solution with $y(0) \neq 0$ is possible. A recursion formula for the remaining coefficients a_k in (5′) may be obtained through successive differentiation and subsequent evaluation as before. For $k \geq 1$, it takes the form

$$(2r + k + p_0 - 1)a_k = (\qquad)a_{k-1} + \cdots + (\qquad)a_0, \tag{9'}$$

where $r = r_1$ or r_2 is a root of (7), and we assume that $\mathscr{R}_e r_1 \geq \mathscr{R}_e r_2$. Observe that $p_0 - 1 = -r_1 - r_2$, so that

$$\left\{\begin{matrix}\text{the term}\\ \text{multiplying } a_k\end{matrix}\right\} \quad \text{is} \quad \left\{\begin{matrix} k + (r_1 - r_2), & \text{if } r = r_1 \\ k - (r_1 - r_2), & \text{if } r = r_2\end{matrix}\right. .$$

Thus unless $r_1 - r_2$ is an integer, the coefficients can be successively determined in either case, and they generate linearly independent Frobenius solutions $x^{r_1}y_1(x)$ and $x^{r_2}y_2(x)$ in some common interval $(0, \mathscr{b})$. These solutions are complex valued unless r_1 and r_2 are real numbers, but they can be replaced by equivalent real-valued solutions.

However, if $r_1 - r_2 = N$ is a *nonnegative integer*, then although we can always find nontrivial coefficients for the *first* solution, we may not be able to compensate for the fact that the left side of (9′) vanishes when $k = N$ and $r = r_2$. In such cases, we use reduction of order and seek a second linearly independent solution to (4) in the form

$$x^ry(x) + (\ln|x|)x^{r_1}y_1(x), \qquad (r = r_2 = r_1 - N).$$

Through substitution and simplification it can be shown that for such solutions of (4), y must satisfy the *inhomogeneous equation*

$$L(y) = -x^N[2xy_1' + (xp(x) + 2r_1 - 1)y_1],$$

where L is defined in (9). Since the coefficients of y_1 are assumed known recursively, those for a nontrivial analytic y are thereby determined.† Convergence for all of these series can be established a priori. (See [C-L].)

---

† Uniquely, if we take $y^{(N)}(0) = 0$ and $y(0) = 1$.

Each of these possibilities is illustrated by **Bessel's equation** of order $\nu \geq 0$,

$$x^2y'' + xy' + (x^2 - \nu^2)y = 0,$$

for *some* choice of ν. Here, the indicial equation is $r^2 - \nu^2 = 0$, with roots $r_1 = \nu$ and $r_2 = -\nu$. When ν is not an integer, Bessel's equation has for $x > 0$ linearly independent solutions of Frobenius type conventionally denoted J_ν and $J_{-\nu}$, even if $2\nu = N$ is an *odd* integer. However, if $\nu = m$ is an integer, then there is one solution analytic near $x = 0$, denoted J_m, and a linearly independent solution of the form $Y_m(x) = x^{-m}y(x) + \ln|x| J_m(x)$, for $x > 0$. Some details are given in Section 7.4.

Qualitative Behavior; Comparison

In Theorem B.6 we will prove that there exists a *unique* solution y to equation (1′)

$$y'' + py' + qy = 0, \qquad \text{on } [a, b],$$

with specified values of y and y' at any given point $x_0 \in [a, b]$. In particular

$$y(x_0) = y'(x_0) = 0 \Rightarrow y(x) \equiv 0 \qquad \text{on } [a, b].$$

It follows that a *nontrivial* solution y must change sign at any point $x_0 \in (a, b)$ where it vanishes. Moreover, a nontrivial y cannot vanish at an infinite sequence of points in the *compact interval* $[a, b]$, for by Theorem A.2, there would be a (sub)sequence x_n, $n = 1, 2, \ldots$ of these points with a limit point $x_0 \in [a, b]$. But at x_0, both

$$y(x_0) = \lim_{n\to\infty} y(x_n) = 0,$$

and

$$y'(x_0) = \lim_{n\to\infty} \frac{y(x_n) - y(x_0)}{x_n - x_0} = 0.$$

We can gain insight into the nature of solutions to equation (1′) above by comparing them with solutions to simpler related equations. In particular, when $p = 0$, we have the following.

(B.3) Comparison Theorem: (Sturm) *On $[a, b]$: suppose that*

$$(i)\ y \text{ is a solution of } y'' + qy = 0,$$

$$\text{and } (ii)\ y_0 \text{ is a solution of } y_0'' + q_0 y_0 = 0$$

$$\text{with } y_0(a) = y_0(b) = 0, \text{ and } y_0(x) > 0,\ x \in (a, b),$$

where q and q_0 are **different** *continuous functions with $q(x) \geq q_0(x)$, $x \in [a, b]$. Then y vanishes at some point $x_0 \in (a, b)$.*

Proof: Assume, say, that $y(x) > 0$ at all $x \in (a, b)$. Then $y(a) \geq 0$ and $y(b) \geq 0$, while also

$$y_0'(a) = \lim_{x \searrow a} \frac{y_0(x) - y_0(a)}{x - a} > 0 > y_0'(b),$$

since neither derivative can vanish. But this contradicts the fact that

$$(yy_0')(b) - (yy_0')(a) = \int_a^b \frac{d}{dx}(yy_0' - y_0y')\,dx$$
$$= \int_a^b (yy_0'' - y_0y'')\,dx = \int_a^b (q - q_0)yy_0\,dx$$

is *positive*, in view of our assumptions. ■

Remark: The general equation (1′) can be reduced to one in $\tilde{y} = \sqrt{\tau}y$ that is covered by this theorem, when $\tau(x) = \exp(\int p\,dx)$.

As an application, we note that if J_ν is Bessel's function of order $\nu \geq 0$, then $y(x) = \sqrt{x}J_\nu(x)$ satisfies the equation

$$y'' + \left(1 - \frac{(\nu^2 - \frac{1}{4})}{x^2}\right)y = 0, \qquad x > 0.$$

Now, if

$$0 \leq \nu \leq \tfrac{1}{2} \quad : \quad q(x) = 1 - \frac{(\nu^2 - \frac{1}{4})}{x^2} > 1,$$

and we can take $q_0(x) = 1$. For such ν, we compare y with $y_0(x) = \sin(x - a)$, which satisfies the equation, $y_0'' + 1y_0 = 0$, and see that y (and so J_ν) vanishes at some point in each positive interval $(a, a + \pi)$. If $\nu > 1/2$, we reverse the roles of q and q_0 and conclude that J_ν can vanish at most *once* in each such interval. In either case, similar comparisons show that the distance between successive zeros of J_ν on $[a, \infty)$ approaches π as $a \to +\infty$. (See Section 7.4.)

B.2

Two-Point Boundary Value Problems; The Green Function

We can use the integrating factor $\tau(x) = \exp(\int p\,dx)$ to rewrite equation (1) in the self-adjoint form

$$(\tau y')' + \tau q y = \tau f.$$

Conversely, when $\tau \in C^1[a, b]$ is positive, the simpler equation

$$L(y) = f, \qquad \text{where} \qquad L(y) \stackrel{\text{def}}{=} (\tau y')' + qy = \tau y'' + \tau' y' + qy, \tag{10}$$

may be expressed in the form of (1),

$$y'' + \frac{\tau'}{\tau}y' + \frac{q}{\tau}y = \frac{f}{\tau}.$$

If q and f are continuous on $[a, b]$, Theorem B.6 guarantees existence of a unique solution to equation (10) with specified values of y and y' at a (or at b), but it does not follow that we can find solutions to two-point problems in which one such value is prescribed at each endpoint. Let's restrict our attention to the case of interest in Chapter 7 and seek nontrivial solutions y to (10) such that at each endpoint e of $[a, b]$, y'/y assumes a prescribed value E, possibly infinite. The condition on y at e can be expressed in terms of a linear operator B_e as follows:

$$0 = B_e(y) \overset{\text{def}}{=} \begin{cases} y'(e) - Ey(e), & \text{if } E \text{ is finite} \\ y(e), & \text{if } E \text{ is infinite} \end{cases} \tag{10'}$$

As we show below, such solutions can be obtained with the help of a Green function that is constructed from a linearly independent pair of solutions α and β to the associated *homogeneous equation*

$$L(y) = (\tau y')' + qy = 0 \tag{10''}$$

where α meets the boundary condition (10′) at a, and β meets that at b.

The existence of such solutions is guaranteed, and each of these solutions is unique within a constant factor. (See Section B.3.) However, α and β need not be linearly independent even for a problem as simple as $y'' + y = 0$ on $[0, \pi]$ with $y(0) = y(\pi) = 0$, for which both α and β must be multiples of $\sin x$. Here, the homogeneous problem has a nontrivial (and hence nonunique) solution. Conversely, if the homogeneous problem has only the trivial solution, it follows that nontrivial α and β cannot be scalar multiples of one another and must be linearly independent.

Now, assuming that relevant linearly independent solutions α and β of (10″) have been obtained, we note that the solution of (10) satisfying the two-point conditions (10′) is given for some constant c_1 by

$$y(x) = c_1 \alpha(x) + y_f(x), \tag{11}$$

where y_f is obtained from (3′) with $y_1 = \alpha$, $y_2 = \beta$, and with W replaced by

$$W_0 = \tau(\alpha\beta' - \alpha'\beta). \tag{11'}$$

W_0 is a nonzero constant, since its derivative vanishes, and *we can assume that* $W_0 = 1$, since this normalization can be achieved by redefining α or β. As defined by (11), y, satisfies the same homogeneous conditions as α at a (because $y_f(a) = y_f'(a) = 0$), and c_1 can be selected to make $B_b(y) = 0$. Hence, we find that

$$y(x) = \beta(x) \int_a^x \alpha(\xi) f(\xi)\, d\xi + \alpha(x) B(x) \tag{12a}$$

for an appropriate function $B(x)$. Similarly, if we replace (11) by

$$y(x) = c_2 \beta(x) + \tilde{y}_f(x)$$

where $\tilde{y}_f$ is obtained by replacing $\int_a^x$ in (3′) *by* $-\int_x^b$ and keeping all other quantities as above, we conclude that

$$y(x) = \alpha(x) \int_x^b \beta(\xi) f(\xi)\, d\xi + \beta(x) A(x) \tag{12b}$$

for an appropriate function $A(x)$.

Upon comparing (12a) and (12b), we are led to attempt solution of (10) in the form

$$\left.\begin{aligned} y(x) &= \beta(x)\int_a^x \alpha f\,d\xi + \alpha(x)\int_x^b \beta f\,d\xi \\ \text{or}\quad y(x) &= \int_a^b g(x,\xi)f(\xi)\,d\xi, \end{aligned}\right\} \tag{13}$$

$$\text{where}\quad g(x,\xi) \stackrel{\text{def}}{=} \begin{cases} \alpha(\xi)\beta(x), & \xi \le x \\ \alpha(x)\beta(\xi), & x \le \xi \end{cases}, \qquad (W_0 = 1) \tag{13'}$$

is the **Green function** for (10) under boundary conditions (10′).[†] The proof that y given by (13) does solve equation (10) uniquely under these boundary conditions is straightforward.

First, when (13) is differentiated, then by the Fundamental Theorem of Calculus, we get

$$y'(x) = \beta'(x)\int_a^x \alpha f\,d\xi + \alpha'(x)\int_x^b \beta f\,d\xi$$

since the additional terms cancel. By repeating the process and using (11′) with $W_0 = 1$, we find that

$$y''(x) = \beta''(x)\int_a^x \alpha f\,d\xi + \alpha''(x)\int_x^b \beta f\,d\xi + \frac{f}{\tau}.$$

Now, multiply the last equation by τ, the previous equation by τ', equation (13) by q, and combine the results to form $L(y)$ in (10). After collecting like terms, we see that

$$L(y) = L(\beta)\int_a^x \alpha f\,d\xi + L(\alpha)\int_x^b \beta f\,d\xi + f,$$

but $L(\alpha) = L(\beta) = 0$ so that $L(y) = f$ as desired. Finally from (13),

$$y(a) = \alpha(a)\int_a^b \beta f\,d\xi$$

while as above

$$y'(a) = \alpha'(a)\int_a^b \beta f\,d\xi.$$

Since α satisfies the given homogeneous boundary condition (10′) at a, we see that y must satisfy this condition as well; a similar argument shows that y meets the given condition at b.

More generally, if α and β are *any* linearly independent solutions of (10″) for which the endpoint boundary operators in (10′) satisfy the condition

$$B_a(\alpha)B_b(\beta) \neq B_a(\beta)B_b(\alpha),$$

then we can find constants c_1 and c_2 so that with y_f as in (11),

$$y = c_1\alpha + c_2\beta + y_f$$

[†] This Green function is negative to that used by some authors.

will be the unique solution of (10) satisfying $B_a(y) = a_1$, and $B_\ell(y) = b_1$ for any given constants a_1 and b_1. It too can be expressed in a form related to (13). (See [Wein].)

B.3*

First-Order Systems

Consider the following first-order linear system for a vector-valued function $\mathbf{y} = \mathbf{y}(x)$:

$$\mathbf{y}' = \mathrm{p}\mathbf{y} + \mathbf{f} \tag{14}$$

where $\mathbf{f}$ is a vector-valued function with continuous components, and p is a square-matrix function with continuous elements p_{ij}, all given on some compact interval I. For convenience, we assume that $I = [0, \ell]$.

Upon integration, we see that a C^1 solution $\mathbf{y}$ of (14) with $\mathbf{y}(0) = \mathbf{A}$ must satisfy the vector-valued integral equation

$$\mathbf{y}(x) = \mathbf{A} + \int_0^x [\mathbf{f} + \mathrm{p}\mathbf{y}](t)\, dt, \qquad 0 \le x \le \ell. \tag{15}$$

Conversely, from the Fundamental Theorem of Calculus, each continuous solution of (15) is a C^1 solution of (14) with $\mathbf{y}(0) = \mathbf{A}$.

Now, if m is a bound for $|\mathbf{f}|$ and P denotes a bound for $(\sum_{i,j} p_{ij}^2)^{1/2}$ on $[0, \ell]$, then from Cauchy's inequality, we see that

$$|\mathrm{p}(t)\,\mathbf{y}(t)| \le P|\mathbf{y}(t)|, \qquad 0 \le t \le \ell,$$

and from (15), with $|\mathbf{A}| = a$, we get the estimates

$$|\mathbf{y}(x)| \le a + \int_0^x |\mathbf{f}|(t)\, dt + \int_0^x |\mathrm{p}(t)\,\mathbf{y}(t)|dt$$
$$\le a + m\ell + P\int_0^x |\mathbf{y}(t)|\, dt = \psi(x), \qquad \text{say.}$$

This integral inequality affords important a priori estimates for any possible continuous solution of (15). Following Gronwall (1919), we differentiate ψ to obtain the inequality

$$\psi'(x) = P|\mathbf{y}(x)| \le P\psi(x),$$
$$\text{or} \qquad (e^{-Px}\psi(x))' \le 0.$$

Hence, upon integration, we conclude that if $0 \le x \le \ell$:

$$|\mathbf{y}(x)| \le \psi(x) \le \psi(0)e^{Px} \le (a + m\ell)\, e^{P\ell} \stackrel{\text{def}}{=} Y. \tag{16}$$

(B.4) Proposition: *For given* $\mathbf{A}$ *and* $\mathbf{f}$, (15) *has at most one solution.*

Proof: The difference $\mathbf{y}$ of two solutions of (15) satisfies the homogeneous equation

$$\mathbf{y}(x) = \int_0^x (\mathbb{p}\mathbf{y})(t)\,dt,$$

and thus (16), with $a = m = 0$. Therefore $|\mathbf{y}(x)| \equiv 0$, and this implies that $\mathbf{y}(x) \equiv 0$, as desired. ■

Now, unlike (14), the integral equation (15) lends itself to attempted solution via iteration. For example, we can use the initial trial solution $y_0(\mathbf{x}) = \mathbf{A}$ to replace $\mathbf{y}$ in the integral and consider the resulting right side of (15) as *defining* the next approximation. This process can be repeated as often as desired, and in fact the sequence of successive approximations so defined converges uniformly on $[0, \ell]$ to a continuous limit function $\mathbf{y}$ that therefore satisfies (15). (See, for example, [B–R] or [Tr, App 5].)

In this manner, we obtain theoretical assurance that a solution $\mathbf{y}$ exists, but the underlying process cannot readily be used to approximate the solution in practice. However, for each $k = 1, 2, \ldots$, let's consider the kth **Tonelli approximant**,

$$\mathbf{y}_k(x) = \begin{cases} \mathbf{A}, & 0 \le x \le \ell/k \\ \mathbf{A} + \displaystyle\int_0^{x-(\ell/k)} [\mathbf{f} + \mathbb{p}\mathbf{y}_k](t)\,dt, & \ell/k \le x \le \ell \end{cases} \tag{17}$$

i.e., on each successive subinterval of length ℓ/k, $\mathbf{y}_k$ is determined by its values on the *previous* subintervals. We can compute these approximants rather easily for each k, and we will show that they *must* converge uniformly on $[0, \ell]$ to the theoretical solution, whose existence is assumed provisionally.

(B.5) Proposition: *Let* $\mathbf{y}$ *be a solution to* (14) *on* $[0, \ell]$ *with* $\mathbf{y}(0) = \mathbf{A}$ *given. Then for each* $k = 1, 2, \ldots,$

$$|\mathbf{y}(x) - \mathbf{y}_k(x)| \le M/k, \qquad 0 \le x \le \ell, \tag{18}$$

where $\quad M = (m + PY)\,\ell e^{P\ell}, \quad$ and $\quad Y = (|\mathbf{A}| + m\ell)e^{P\ell}.$

Proof: We know that $\mathbf{y}$ satisfies (15) and (16), and when we compare with (17), we get the following estimates:

When $0 \le x \le \ell/k$, then

$$|\mathbf{y}(x) - \mathbf{y}_k(x)| = |\mathbf{y}(x) - \mathbf{A}| = \left|\int_0^x [\mathbf{f} + \mathbb{p}\mathbf{y}](t)\,dt\right|$$

$$\le \int_0^x (m + PY)\,dt \le \frac{\ell}{k}(m + PY).$$

Similarly, when $\ell/k \le x \le \ell$, then

$$|\mathbf{y}(x) - \mathbf{y}_k(x)| \le \left|\int_{x-(\ell/k)}^x [\mathbf{f} + \mathbb{p}\mathbf{y}](t)\,dt\right| + \left|\int_0^{x-(\ell/k)} [\mathbb{p}(\mathbf{y} - \mathbf{y}_k)](t)\,dt\right|$$

or $\quad |\mathbf{y}(x) - \mathbf{y}_k(x)| \le \dfrac{\ell}{k}(m + PY) + P\displaystyle\int_0^x |\mathbf{y}(t) - \mathbf{y}_k(t)|\,dt = \psi(x), \quad$ say.

We see that the last integral inequality is also valid in the previous interval, and we can apply the Gronwall technique used in deriving (16) to obtain the desired estimate (18). ■

We can use similar Gronwall estimates to show that on $[0, \ell]$ each $|\mathbf{y}_k(x)| \le Y$, $k = 1, 2, \ldots$ and that $|\mathbf{y}_n(x) - \mathbf{y}_k(x)| \le M/k$, if $n > k$. It follows that these $\{\mathbf{y}_k\}$ form a *Cauchy sequence* that necessarily converges uniformly on $[0, \ell]$ to a continuous limit function $\mathbf{y}$. (See [Ru].) It is straightforward to see that if we let $k \to \infty$ in (17), then at each x, $\mathbf{y}$ satisfies (15). This outlines an existence proof that is independent of the iterative scheme mentioned previously and, in view of Proposition B.4, indicates a proof of the following.

(B.6) Theorem: *There exists a unique solution* $\mathbf{y}$ *of system* (14) *on* $[0, \ell]$ *with* $\mathbf{y}(0) = \mathbf{A}$ *prescribed.* ■

Remarks: When an ordinary linear differential equation of higher order (or a system of such equations) can be converted to a first-order system such as (14) on *some* interval, these results ensure the existence of unique solutions to the equation (or system) on this interval. In fact, corresponding Tonelli approximants can be considered for *nonlinear* systems in the normal form

$$\mathbf{y}'(x) = \mathbf{F}(x, \mathbf{y}(x)), \qquad 0 \le x \le \ell, \tag{19}$$

where $\mathbf{F}$ has continuous components, and $\mathbf{y}(0)$ is prescribed. The resulting $\mathbf{y}_k$ satisfy conditions on a *sufficiently small interval* $[0, \ell]$ guaranteeing that a *subsequence* converges uniformly to a local solution of (19) with the given initial condition. The solution need not be unique unless, for example, $\mathbf{F}$ has C^1 components; then $\mathbf{y}$ is the limit of the sequence $\{\mathbf{y}_k\}$.

However, on an interval in which such systems cannot be expressed in normal form (14) or (19), with continuous $\mathbf{F}$ and/or $\mathbb{p}$, there need not exist solutions with prescribed values at some point. These singular cases require individual analysis. (See, for example, [C–L] or [B–R].)

Appendix C
Pointwise Behavior of Bessel Series

In this Appendix we indicate how to adapt the proof of Theorem 9.24 to Bessel series of the form

$$\sum_{n=1}^{\infty} C_n J_m(\omega_{mn} x) = \sum_{n=1}^{\infty} c_n y_n(x), \qquad 0 < x < 1, \tag{20}$$

where

$$\int_0^1 x y_n^2(x)\, dx = 1, \qquad y_n(1) = 0,$$

and

$$c_n = \int_0^1 x f(x)\, y_n(x)\, dx, \qquad n = 1, 2, \ldots, \tag{21}$$

for an f that is sectionally smooth on $[0, 1]$.

For simplicity, we just look at the case $m = 0$ (where $\omega_{mn} = \omega_n$) but the same approach yields corresponding results when $m = 1, 2, \ldots$ and when other regular boundary conditions are imposed at $x = 1$. We will employ the asymptotic notation of Section 8.6.

From Example 7 of Section 9.4, recall that for $t = x \geq \delta > 0$ and large n:

$$v_n(t) \stackrel{\text{def}}{=} \sqrt{t}\, y_n(t) = \sqrt{2} \cos\left(\omega_n t - \frac{\pi}{4}\right) + \frac{\sqrt{2}}{8\omega_n t} \sin\left(\omega_n t - \frac{\pi}{4}\right) + O\left(\frac{1}{n^2}\right),$$

where the last term depends on δ. Moreover, in Problem 9.3K, it is shown that

$$\omega_n = \left(n - \frac{1}{4}\right)\pi + \frac{1}{8n\pi} + O\left(\frac{1}{n^2}\right).$$

Hence, when $0 \leq t \leq 1$ and

$$\tau_n \stackrel{\text{def}}{=} \left(n - \frac{1}{4}\right)\pi t - \frac{\pi}{4},$$

we see that

$$\cos\left(\omega_n t - \frac{\pi}{4}\right) = \cos\tau_n \cos\left[\frac{t}{8n\pi} + O\left(\frac{1}{n^2}\right)\right] - \sin\tau_n \sin\left[\frac{t}{8n\pi} + O\left(\frac{1}{n^2}\right)\right]$$

$$= \cos\tau_n - \frac{t}{8n\pi}\sin\tau_n + O\left(\frac{1}{n^2}\right);$$

and similarly

$$\sin\left(\omega_n t - \frac{\pi}{4}\right) = \sin\tau_n + O\left(\frac{1}{n}\right),$$

if we use standard approximations for sine and cosine with small argument.

Thus, instead of (76) in Section 9.5, we have the following second-order asymptotic approximation when $t \in [\delta, 1]$:

$$v_n(t) = \sqrt{2}\cos\tau_n + \frac{R(t)}{n}\sin\tau_n + O\left(\frac{1}{n^2}\right)$$

$$= \phi_n(t) + \frac{R(t)}{n}\sin\tau_n + O\left(\frac{1}{n^2}\right), \tag{22}$$

$$\text{where} \qquad R(t) = \frac{\sqrt{2}}{8\pi}\left(\frac{1}{t} - t\right), \qquad \text{and}$$

$$\phi_n(t) = \sqrt{2}\cos\tau_n = \sqrt{2}\cos\left[\left(n - \frac{1}{4}\right)\pi t - \frac{\pi}{4}\right], \qquad n = 1, 2, \ldots. \tag{23}$$

Unfortunately, the latter sequence is not that for a Fourier series nor are its terms orthogonal on $[0, 1]$. But if f is *sectionally-smooth* on $[0, 1]$, the Nth partial sum

$$V_N(t) \stackrel{\text{def}}{=} \sum_{n=1}^{N} c_n v_n(t) = \int_0^1 \sqrt{s} f(s) \left[\sum_{n=1}^{N} v_n(s) v_n(t)\right] ds \tag{24}$$

may be compared directly with

$$\Phi_N(t) \stackrel{\text{def}}{=} \int_0^1 \sqrt{s} f(s) \left[\sum_{n=1}^{N} \phi_n(s)\phi_n(t)\right] ds. \tag{25}$$

Now, by standard trigonometric identities,

$$\Phi_N(t) = \int_0^1 \sqrt{s} f(s)\, ds \sum_{n=1}^{N} \left\{\cos\left(n - \frac{1}{4}\right)\pi(s - t) + \cos\left[\left(n - \frac{1}{4}\right)(s + t) - \frac{\pi}{2}\right]\right\}$$

$$= \int_0^2 \sqrt{s} f(s)\, ds \sum_{n=1}^{N} \cos\left(n - \tfrac{1}{4}\right)\pi(s - t),$$

if we define

$$f(s) = -\frac{\sqrt{2 - s}}{\sqrt{s}} f(2 - s), \qquad 1 < s \le 2. \tag{26}$$

Also, as in the derivation of (32′) in Section 1.8:

$$\sum_{n=1}^{N} \cos\left(n-\frac{1}{4}\right)\pi(s-t) = \frac{\sin\left(N+\frac{1}{4}\right)\pi(s-t) - \sin\frac{\pi}{4}(s-t)}{2\sin\frac{\pi}{2}(s-t)}. \tag{27}$$

Therefore,

$$\Phi_N(t) = \frac{1}{2}\int_0^2 \sqrt{s}\, f(s)\left[\frac{\sin N\pi(s-t)}{\sin\frac{\pi}{2}(s-t)}\cos\frac{\pi}{4}(s-t) + \frac{\cos N\pi(s-t)}{2\cos\frac{\pi}{4}(s-t)}\right] ds - f_1(t) \tag{28}$$

where

$$f_1(t) \overset{\text{def}}{=} \frac{1}{4}\int_0^2 \frac{\sqrt{s}\, f(s)}{\cos\frac{\pi}{4}(s-t)}\, ds, \qquad t > 0. \tag{28′}$$

In (28), the first term is essentially the Dirichlet formula for $\sqrt{s}\, f(s)$ at t (on $[0, 2]$), while as $N \to \infty$, the second approaches zero. (Recall our analysis of (33) and (33′) in Section 1.8.) Hence, as $N \to \infty$:

$$\Phi_N(t) + f_1(t) \to \sqrt{t}\,\frac{f(t+) + f(t-)}{2}, \qquad 0 < t \le 1, \tag{29}$$

and the convergence is uniform on intervals of continuity of f *as extended* in (26), for which $t \ge \delta > 0$.

It remains only to associate this Fourier series-like behavior to V_N, and it suffices to do so for $t \ge \delta > 0$. If we let

$$\sigma_n = \left(n - \frac{1}{4}\right)\pi s - \frac{\pi}{4}, \qquad n = 1, 2, \ldots,$$

we can use (22) and (23) to see that if $s \in [\delta, 1]$ and $t \in [\delta, 1]$, then

$$\Psi_N(s, t) \overset{\text{def}}{=} \sum_{n=1}^{N} [v_n(s) v_n(t) - \phi_n(s)\phi_n(t)]$$

$$= R(t)\sum_{n=1}^{N} \frac{\sin\tau_n \cos\sigma_n}{n} + R(s)\sum_{n=1}^{N}\frac{\sin\sigma_n\cos\tau_n}{n} + O\left(\frac{1}{n^2}\right) \tag{30}$$

is uniformly bounded in absolute value by Ψ, say, independently of N. (See Problem 9.5K.)

For sufficiently small $\delta > 0$, a given sectionally smooth f can be approximated within $\epsilon > 0$ by an $\tilde{f} \in \hat{C}^1[0, 1]$ with $\tilde{f}(1) = 0$ and $\tilde{f} = f$ on $[0, \delta]$ in that

$$\int_0^1 \sqrt{s}\,|f - \tilde{f}|\, ds < \epsilon. \tag{31}$$

According to Theorem 9.5, on $[\delta/2, 1]$, $\tilde{f}$ can be represented by a uniformly convergent Bessel series $\sum_{n=1}^{\infty} \tilde{c}_n y_n$. Hence, by (29), with obvious notation, as $N \to \infty$:

$$\tilde{V}_N(t) - \tilde{\Phi}_N(t) \to \tilde{f}_1(t) = \frac{1}{4}\int_0^2 \frac{\sqrt{s}\,\tilde{f}(s)}{\cos\frac{\pi}{4}(s-t)}\, ds$$

uniformly, where, as in (26),

$$\tilde{f}(s) \stackrel{\text{def}}{=} -\frac{\sqrt{2-s}}{\sqrt{s}}\,\tilde{f}(2-s), \qquad 1 < s \le 2.$$

Therefore, by the triangle inequality, if $t > \delta$:

$$\begin{aligned}
|V_N(t) - [\Phi_N(t) + f_1(t)]| &\le \int_\delta^1 |\Psi_N(s,t)|\sqrt{s}\,|f - \tilde{f}|(s)\,ds \\
&\quad + \left|\int_\delta^1 \Psi_N(s,t)\sqrt{s}\,\tilde{f}(s)\,ds - \tilde{f}_1(t)\right| + |f_1(t) - \tilde{f}_1(t)| \\
&\le \Psi\epsilon + |\tilde{V}_N(t) - \tilde{\Phi}_N(t) - \tilde{f}_1(t)| + \frac{1}{4}\int_\delta^2 \frac{\sqrt{s}\,|f - \tilde{f}|(s)}{\cos\frac{\pi}{4}(s-t)}\,ds \\
&\le \Psi\epsilon + \Delta\epsilon + \epsilon, \qquad \text{if } N \ge N(\epsilon,\delta), \qquad \text{where } \Delta = \frac{1}{2}\sec\frac{\pi}{4}(2-\delta).
\end{aligned}$$

We have sketched a proof for the following.

(C.1) Theorem: *If f is sectionally smooth on $[0,1]$, then*

$$\sum_{n=1}^{\infty} c_n y_n(t) = \frac{f(t+) + f(t-)}{2}, \qquad t \in (0,1]$$

and the convergence is uniform on each subinterval $[a, b] \subseteq (0,1]$ where f (as extended in (26)) is continuous. ∎

Remark: The last inequality holds for more general f and shows that the convergence behavior for Bessel series is essentially that for Fourier series.

Appendix D
Boundary Value Problems and Maple

In this appendix, we illustrate some of the ways that the computer algebra system (CAS) Maple 2016 can be used in a course on boundary value problems. This software can perform symbolic manipulations such as differentiation of arbitrary expressions, integration of elementary functions using the Risch algorithm, computation of Taylor series, obtain solutions to ordinary differential equations, etc. It also has a nice collection of special functions like Legendre, Chebyshev and Gegenbauer polynomials, the error function, the hypergeometric function, and others. In addition, it has arbitrary precision numerical algorithms and high quality plotting routines for displaying 2 and 3 dimensional graphs in several coordinate systems. For the rest of this appendix, Maple is in worksheet mode so that input commands are left-justified, preceeded by the greater than symbol (>) prompt and uses italic fonts. Maple output is always centered and follows immediately from the input. The output can be suppressed by terminating the input line with a colon (:). If one prefers to work in document mode, then input commands can appear anywhere and displayed in italic fonts and need not be terminated by a semi-colon or a colon.

§1 Mathematical Functions in Maple

Mathematical functions of one or more variable are represented as mappings and use the more natural arrow (→) notation. Once defined, these functions can be manipulated in the usual way like evaluation at symbolic or numeric values, differentiation, integration, etc. They can also be plotted.

$> f := x \to x^2 \sin(x);$

$$x \mapsto x^2 \sin(x)$$

$> f(2.0);$

$$3.637189707$$

$> \dfrac{f(a+h)-f(a)}{h};$

$$\frac{(a+h)^2\sin(a+h)-a^2\sin(a)}{h}$$

$> limit(\%, h=0);$

$$\cos(a)\,a^2+2\sin(a)\,a$$

The percent symbol (%) refers to the last output from Maple. Two percent symbols (%%) refers to the second to the last or penultimate output from Maple, and so on.

$> plot(f(x), x=-4...4);$

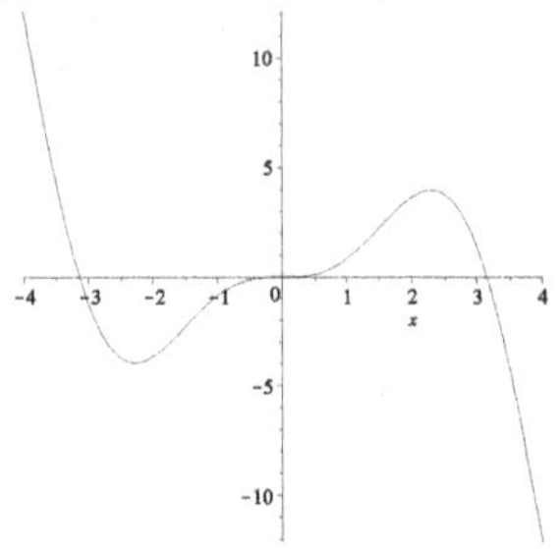

We make one important point here in order to spare the user from a lot of frustration. In Maple, there is a big difference between a function and an expression or a formula. Functions are defined either using the arrow $\rightarrow$ operator or a Maple procedure (**proc**). When a function f is applied to an argument x, we obtain a formula which is a Maple expression and is not a Maple function! Derivatives of functions are obtained using the **D** operator which produces another function. On the other hand, derivatives of expressions or formulas are obtained using the **diff** command which produces another expression or formula but not a function! An expression can be converted into a function using the **unapply** command. Here is an example showing the derivative function of f and an expression or formula for the second derivative of $f(x)$.

$> df := \mathrm{D}(f);$

$$x \mapsto 2x\sin(x)+x^2\cos(x)$$

$> df(\pi);$

$$-\pi^2$$

$> fprimeprime := diff(f(x), x\$2);$

$$2\sin(x)+4x\cos(x)-x^2\sin(x)$$

df is a function while *fprimeprime* is an expression and not a function and therefore behaves differently. If you tried to evaluate

$> fprimeprime\left(\dfrac{\pi}{2}\right);$

$$2\sin(x)\left(\frac{\pi}{2}\right)+4x\left(\frac{\pi}{2}\right)\cos(x)\left(\frac{\pi}{2}\right)-\left(x\left(\frac{\pi}{2}\right)\right)^2\sin(x)\left(\frac{\pi}{2}\right)$$

you get nonsense. The correct way to evaluate the expression *fprimeprime* is to use the **subs** command.

> *simplify* (*subs*($x=\pi/2$,*fprimeprime*));

$$2-\frac{\pi^2}{4}$$

We can convert the expression *fprimeprime* into a function named *ddf* using the **unapply** command.

> *ddf* := *unapply*(*fprimeprime*, x);

$$x\mapsto 2\sin(x)+4x\cos(x)-x^2\sin(x)$$

> *ddf*($\pi/2$);

$$2-\frac{1}{4}\pi^2$$

Functions of two or more variables are defined in a similar way.

> $g := (x,y)\to x\cdot\exp(-x^2-y^2)$;

$$(x,y)\mapsto xe^{-x^2-y^2}$$

We can easily plot the surface and even show contours by selecting the plot and using the appropriate menu item in the toolbar.

> *plot3d*($g(x,y), x=-2..2, y=-3..3$);

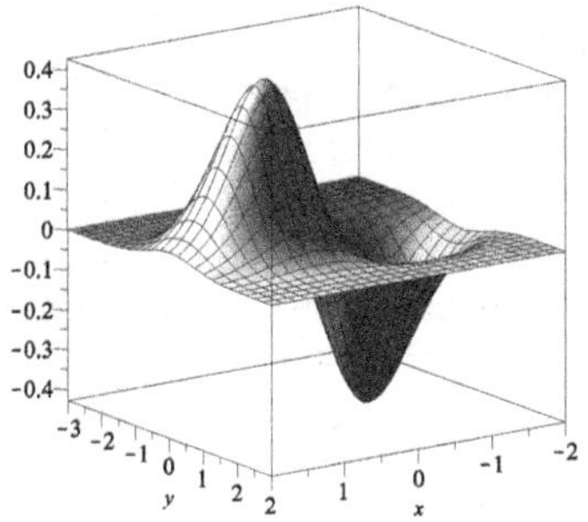

D[i](g) computes the partial derivative of g with respect to its i-th argument. **D[i,j](g)** computes the second order partial derivative of g with respect to the i-th argument first and then with respect to the j-th argument, and so on.

> $gy :=$ D[2](g);

$$(x,y)\mapsto -2yxe^{-x^2-y^2}$$

> $gyx := \mathrm{D}[2,1](g);$

$$(x,y) \mapsto -2ye^{-x^2-y^2} + 4x^2ye^{-x^2-y^2}$$

More complicated functions can be defined using a Maple procedure. For example, a piecewise-defined deadband function that commonly occurs in engineering problems can be defined using conditional statements.

```
> deadband := proc(x)
      if 1 < x then x − 1
          elif x < −1 then x + 1
          else 0
      fi
      end:
```

Maple can compute derivatives of piecewise defined functions.

```
> D(deadband);
```

proc(x) **if** $1 < x$ **then** 1 **elif** $x < -1$ **then** 1 **else** 0 **end if end proc**

Here is a plot of the deadband function and its derivative on the same axis.

```
> plot({deadband, D(deadband)}, −5..5);
```

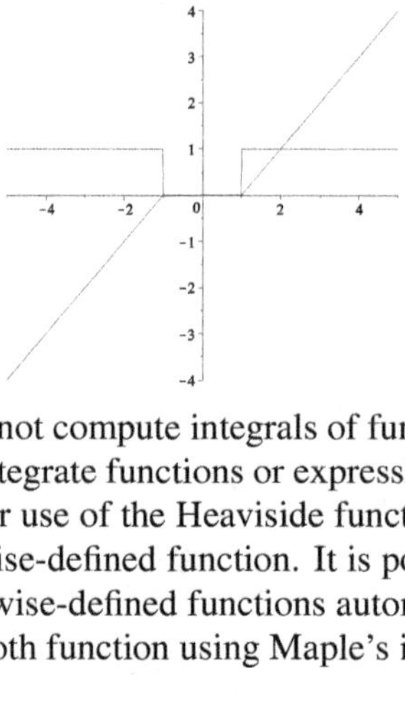

Unfortunately, Maple cannot compute integrals of functions defined using a Maple **proc**. However, Maple can integrate functions or expressions containing the Heaviside or unit step function. A clever use of the Heaviside function allows one to create even the most complicated piecewise-defined function. It is possible to write an integration routine that can handle piecewise-defined functions automatically. Here is an example of a more complicated sawtooth function using Maple's indexed function capability.

```
> sawtooth := proc(x)
      local a, b, c;
      if type(procname, 'indexed')
          then a := op(1, procname);
               b := op(2, procname);
               c := op(3, procname);
               c/π · arctan(a · sin(x), b · cos(x))
          else arctan(sin(x), cos(x))
      fi
      end:
```

$> plot(sawtooth[1,1,\pi], -4\pi..4\pi, scaling = constrained);$

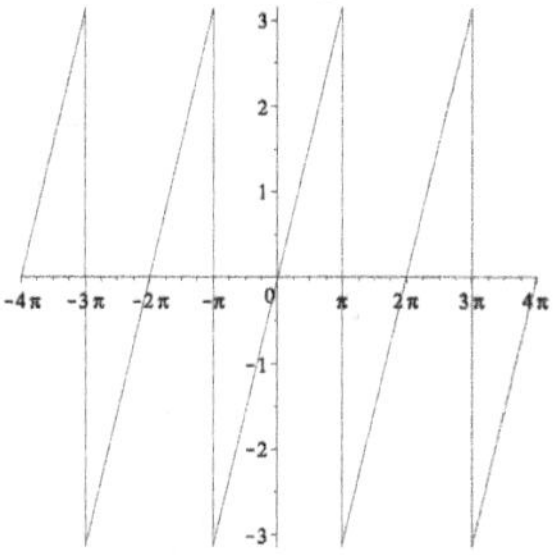

It would be wrong to enter $plot(sawtooth[1,1,\pi](x), x = -4\pi..4\pi)$ because of Maple's policy of full evaluation. A similar error would occur if we plot the deadband function using the command $plot(deadband(x), x = -5..5)$. However, we can prevent variables and functions from being evaluated prematurely by surrounding them with the unevaluation quote ($'$) symbol. Hence the following command produces the same plot result shown above.

$> plot('sawtooth[1,1,\pi](x)', 'x' = -4\pi..4\pi, scaling = constrained);$

For plotting purposes only, we will define a unit pulse function to control the domain of our functions.

$> unitpulse := (x,a,b) \rightarrow \text{Heaviside}(x-a) - \text{Heaviside}(x-b):$
$> plot(unitpulse(x,0,\pi), x = -2..5);$

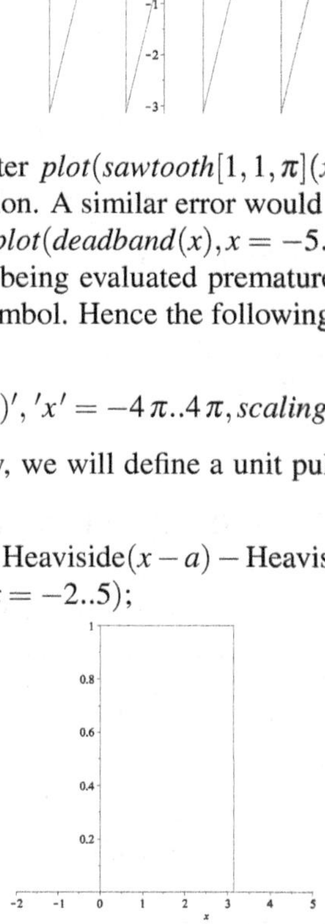

The code below produces a function (not an expression) that is the $2L$-periodic extension of $f(x)$ from the interval $(-L, L]$ to the whole real line. It must be assigned to a name. To use the unnamed function, you must use the Maple command **apply**.

$>$

```
pext := proc(f, L := π)
#f is a function of x, −L < x ≤ L and is entered as f(x)
#L is half the period with default value of π (if argument is omitted)
unapply(unapply(f,x)(x − 2L · floor((x+L)/(2L))),x)
end:
```

Example 1:

$> f := x \rightarrow x^2;$

$$x \mapsto x^2$$

$> fperiodic := pext(f(x));$

$$x \mapsto \left(x - 2\left\lfloor 1/2\,\frac{x+\pi}{\pi}\right\rfloor \pi\right)^2$$

$> plot(fperiodic, -10..10);$

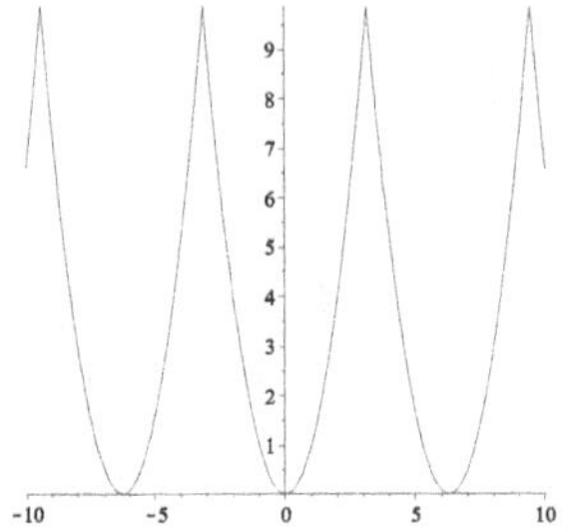

§2 Computing Fourier Coefficients and Partial Sums

Suppose we want to find the Fourier series of a periodic function $f(x)$ with period $2L$. It is not that difficult to write a Maple procedure (using indexed functions) that computes the Fourier coefficients c_n and s_n of the function. Given a $2L$-periodic function $f(x)$ with fundamental interval $-L < x \leq L$, the *fourierCosCoeff* code below computes the Fourier coefficients c_n if $n > 0$ and $\frac{c_0}{2}$ if $n = 0$ of the function. In all these codes, L is an optional parameter which, when omitted, assumes a default value of π. See Section 9 before continuing to read this section.

\>
```
fourierCosCoeff := proc(f, n :: nonnegint, L := π)
#f is a function of x, −L < x ≤ L and only the name of the function
        is entered
#n is the index of c_n
#L is half the period with default value of π (if argument is omitted)
local x, c;
if n > 0 then c := (1/L) ∫_{−L}^{L} f(x)·cos(n·π·x/L) dx; fi;
if n = 0 then c := (1/(2L)) ∫_{−L}^{L} f(x)dx; fi;
RETURN(c)
end:
```

The *fourierSinCoeff* code on the next page computes the Fourier coefficients s_n if $n > 0$.

```
>
fourierSinCoeff := proc(f, n :: posint, L := π)
#f is a function of x, −L < x ≤ L and only the name of the function
        is entered
#n is the index of s_n
#L is half the period with default value of π (if argument is omitted)
local x, s;
s := (1/L) ∫_{−L}^{L} f(x) · sin((n · π · x)/L) dx;
RETURN(s)
end:
```

The *fourierPoly* code below uses the two procedures above and computes $F_k(x)$, the k-th partial Fourier series expression for $f(x)$.

```
>
fourierPoly := proc(f, k :: nonnegint, L := π)
#f is a function of x, −L < x ≤ L and only the name of the function
        is entered
#k is the number of terms
#L is half the period with default value of π (if argument is omitted)
fourierCosCoeff(f, 0, L)
    + add(fourierCosCoeff(f, n, L) · cos((n · π · x)/L)
    + fourierSinCoeff(f, n, L) · sin((n · π · x)/L), n = 1..k);
end:
```

In the following examples, the function is assumed to be 2π-periodic even though we do not tell Maple that it is 2π-periodic. The code automatically assumes the correct fundamental interval! Here is an example:

Example 1: (cont)

This is the 2π-periodic function $f(x) = x^2, -\pi < x \leq \pi$ from the previous section.

$> f := x \rightarrow x^2 :$
$> F_5 := fourierPoly(f, 5);$

$$\frac{1}{3}\pi^2 - 4\cos(x) + \cos(2x) - \frac{4}{9}\cos(3x) + \frac{1}{4}\cos(4x) - \frac{4}{25}\cos(5x).$$

Example 2:

Consider the piecewise defined 2π-periodic function $g(x) = \begin{cases} 1, & -\pi < x < 0, \\ x, & 0 \leq x \leq \pi. \end{cases}$

$> g := x \rightarrow unitpulse(x, -\pi, \pi) \cdot piecewise(x < 0, 1, x \geq 0, x);$

$$x \mapsto unitpulse(x, -\pi, \pi) \cdot piecewise(x < 0, 1, x \geq 0, x)$$

Notice that the function as entered in Maple is not 2π-periodic and we have used the unit pulse function, even though it is required only for plotting.

> $> plot(pext(g(x)), -10..10);$

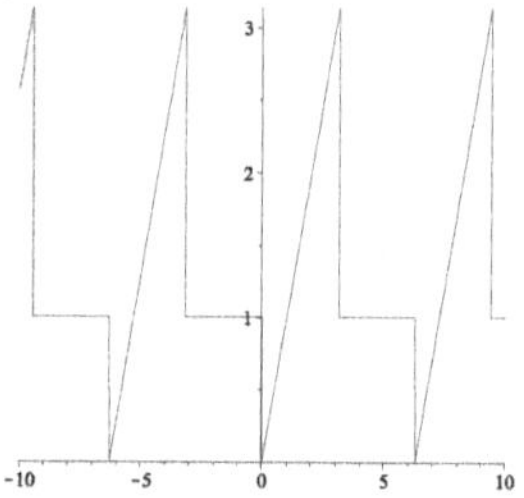

> $> fourierCosCoeff(g, 0);$

$$\frac{1}{2} \frac{\frac{1}{2}\pi^2 + \pi}{\pi}$$

> $> fourierCosCoeff(g, 1);$

$$-\frac{2}{\pi}$$

The two commands above returned $\frac{c_0}{2}$ and c_1 with $L = \pi$. The first 2 sine coefficients are computed below.

> $> fourierSinCoeff(g, 1);$

$$\frac{\pi - 2}{\pi}$$

> $> fourierSinCoeff(g, 2);$

$$-\frac{1}{2}$$

The first 4 cosine and sine terms of the Fourier series for $f(x)$ can now be generated using the *fourierPoly* code.

> $> G_4 := fourierPoly(g, 4);$

$$\frac{1}{2} \frac{1/2\pi^2 + \pi}{\pi} - 2\frac{\cos(x)}{\pi} + \frac{(\pi - 2)\sin(x)}{\pi} - \frac{1}{2}\sin(2x) - \frac{2}{9}\frac{\cos(3x)}{\pi} + \frac{(\pi/3 - 2/3)\sin(3x)}{\pi} - \frac{1}{4}\sin(4x)$$

If one wants the value of the n-th Fourier cosine coefficient of g, we could simply enter

> $> \frac{1}{\pi}\int_{-\pi}^{\pi} g(x) \cdot \cos\left(\frac{n \cdot \pi \cdot x}{\pi}\right) dx;$

$$\frac{n\sin(n\pi)\pi + \cos(n\pi) + n\sin(n\pi) - 1}{n^2\pi}$$

which is not the answer we were expecting even though it is correct. The difficulty here is that Maple has no idea that n is an integer. We have to help Maple simplify the resulting coefficients by providing the necessary information.

$> subs(\cos(n\pi) = (-1)^n, \sin(n\pi) = 0, \%);$

$$\frac{-1+(-1)^n}{n^2\pi}$$

§3 Graphing Partial Fourier Series and the Gibbs' Phenomenon

We can graph the partial sums of a Fourier series to see how well they approximate the original function. Here is a plot of $g(x)$ from Example 2 and the first 4 terms of its Fourier series G_4 (which in Maple is entered as $G[4]$).

$> plot([apply(pext(g(x)),x), G[4]], x = -10..10);$

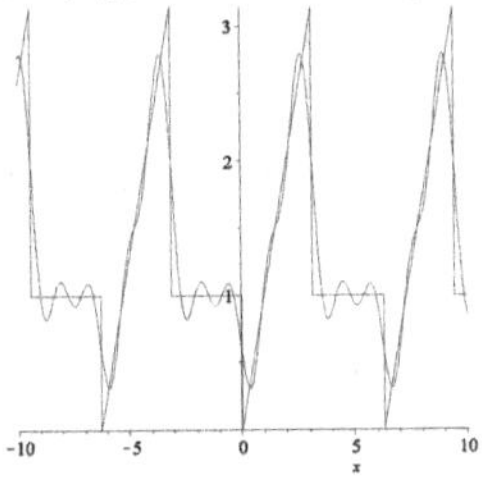

Example 3:

Consider the 2π-periodic function $h(x) = x, -\pi < x \leq \pi$. We can plot $h(x)$ and several partial sums of its Fourier series on the same axis to show the overshoot phenomenon.

$> h := x \to unitpulse(x, -\pi, \pi) \cdot x;$

$$x \mapsto (Heaviside\,(x+\pi) - Heaviside\,(x-\pi))\,x$$

$>plot([apply(pext(h(x)),x), fourierPoly(h,1), fourierPoly(h,5), fourierPoly(h,9)],$
$x = -2\pi..2\pi);$

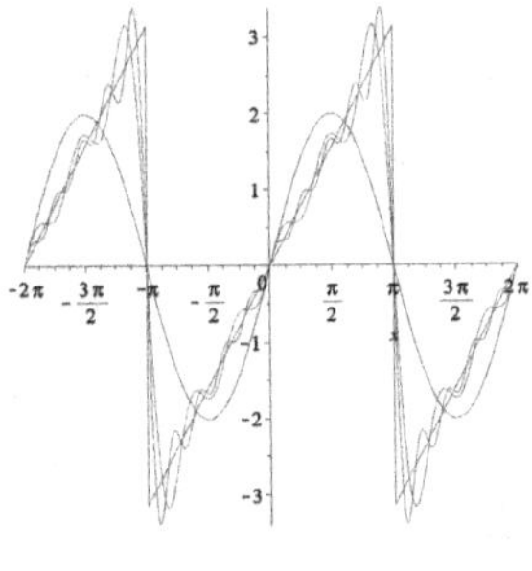

§4 Solving Ordinary Differential Equations

In the course of using separation of variables to solve a boundary value problem involving a partial differential equation, we will need to solve an ordinary differential equation for each of the unknown functions appearing in the product solution. Maple has fairly good differential equations solving capability and will even handle equations involving special functions like Bessel functions. Here is an example.

> $del1 := diff(y(t),t\$2)+6\cdot diff(y(t),t)+13\cdot y(t)=0;$

$$\frac{d^2}{dt^2}y(t)+6\frac{d}{dt}y(t)+13\,y(t)=0$$

> $dsolve(del1,y(t));$

$$y(t)=\_C1\,e^{-3t}\sin(2t)+\_C2\,e^{-3t}\cos(2t)$$

Maple uses the unusual symbols $\_C1$ and $\_C2$ to represent the arbitrary constants that appear in the solution. The use of the underscore (_) symbol is Maple's way of avoiding possible conflicts with user-defined variables. One should never use a variable preceeded by the underscore symbol in any input line or code. This solution of the ODE can be checked (and it is good advice to do this as much as possible).

> $subs(\%,del1);$

$$\frac{\partial^2}{\partial t^2}\left(\_C1\,e^{-3t}\sin(2t)+\_C2\,e^{-3t}\cos(2t)\right)$$
$$+6\frac{\partial}{\partial t}\left(\_C1\,e^{-3t}\sin(2t)+\_C2\,e^{-3t}\cos(2t)\right)$$
$$+13\_C1\,e^{-3t}\sin(2t)+13\_C2\,e^{-3t}\cos(2t)=0$$

> $expand(lhs(\%));$

$$0$$

Maple can easily handle initial or boundary conditions.

> $sol := dsolve(\{del1,y(0)=1,D(y)(0)=0\},y(t));$

$$y(t)=3/2\,e^{-3t}\sin(2t)+e^{-3t}\cos(2t)$$

At this point, we may be tempted to plot $y(t)$ using the usual **plot** command. That would result in an error because although Maple has informed us what the solution is, Maple's computational engine or kernel (its current memory) does not know what the solution is! Maple's computational engine has to be told what $y(t)$ is. It is best to use functions to indicate this relation and this can be accomplished using the **unapply** command.

> $y := unapply(rhs(sol),t);$

$$t\mapsto 3/2\,e^{-3t}\sin(2t)+e^{-3t}\cos(2t)$$

$> plot(\{y, \mathrm{D}(y)\}, -1/4..2);$

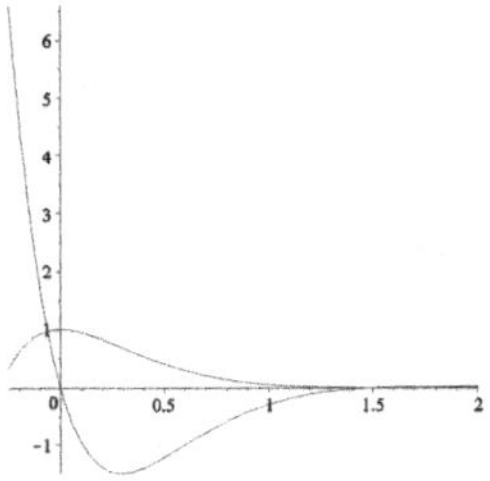

Sometimes, it is necessary to clear Maple's memory of any quantity previously defined. The **restart** command will do that.

Maple can even solve special equations like the Cauchy or the equi-dimensional equation.

$> cauchy := r^2 \cdot diff(R(r), r\$2) + r \cdot diff(R(r), r) - n^2 \cdot R(r) = 0;$
$> dsolve(cauchy, R(r));$

$$R(r) = \_C1\, r^n + \_C2\, r^{-n}$$

§5 Bessel Functions

Bessel functions are solutions to Bessel's differential equation of order n stated in equation (32) in Section 7.4.

$> besseleqn := x^2 \cdot diff(u(x), x\$2) + x \cdot diff(u(x), x) + (x^2 - m^2) \cdot u(x) = 0;$

$$x^2 \frac{\mathrm{d}^2}{\mathrm{d}x^2} u(x) + x \frac{\mathrm{d}}{\mathrm{d}x} u(x) + \left(-m^2 + x^2\right) u(x) = 0$$

The solution is given in terms of the built-in functions **BesselJ**(m,x) and **BesselY**(m,x) which represent the mathematical functions $J_m(x)$ (known as Bessel functions of the first kind stated in equation (32′) in Section 7.4) and $Y_m(x)$ (known as Bessel functions of the second kind), respectively. In addition, Maple also knows the modified Bessel functions of the first and second kinds, denoted by **BesselI**(m, x) and **BesselK**(m, x), respectively. These functions can be differentiated, simplified, evaluated, plotted, etc.

$> diff(\mathrm{BesselJ}(m,x), x);$

$$-\mathrm{BesselJ}(m+1,x) + \frac{m\ \mathrm{BesselJ}(m,x)}{x}$$

$> series(\mathrm{BesselJ}(0,x), x = 0, 12);$

$$1 - \frac{1}{4}x^2 + \frac{1}{64}x^4 - \frac{1}{2304}x^6 + \frac{1}{147456}x^8 - \frac{1}{14745600}x^{10} + O\left(x^{12}\right)$$

We can ask for the series solution directly from the differential equation using the 'series' option.

> $dsolve(besseleqn, u(x), series);$

$$u(x) = \_C1\, x^{-m}\left(1+\frac{x^2}{4m-4}+\frac{x^4}{(8m-16)(4m-4)}+O\left(x^6\right)\right)$$
$$+\_C2\, x^{m}\left(1+\frac{x^2}{-4m-4}+\frac{x^4}{(-8m-16)(-4m-4)}+O\left(x^6\right)\right)$$

Note that Maple returned the solution as series expansions of the built-in functions **BesselY**(m,x) and **BesselJ**(m,x) respectively. Here is a plot of several Bessel functions of different orders.

> $plot(\{\mathrm{BesselJ}(0,x), \mathrm{BesselJ}(1,x), \mathrm{BesselJ}(2,x)\}, x = -20..20);$

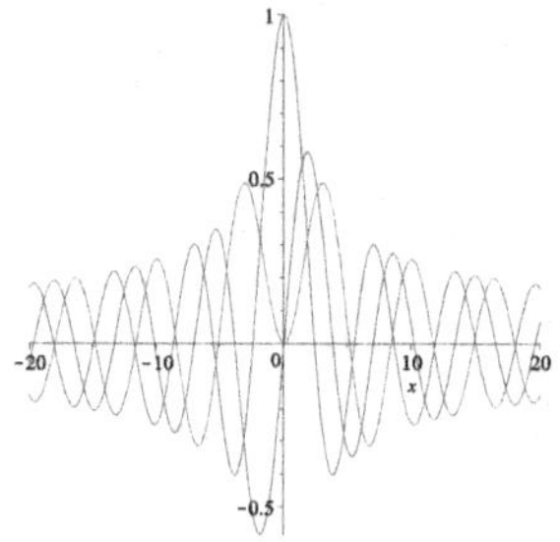

We can also see how well the function

> $j := x \to \sqrt{\frac{2}{\pi x}} \cdot \cos(x - \pi/4);$

$$x \mapsto \sqrt{\frac{2}{\pi x}} \cdot \cos\left(x - \frac{\pi}{4}\right)$$

given in equation (82) in Section 8.6 approximates $J_0(x)$ for large x.

> $plot(\{\mathrm{BesselJ}(0,x), j(x)\}, x = .1..30);$

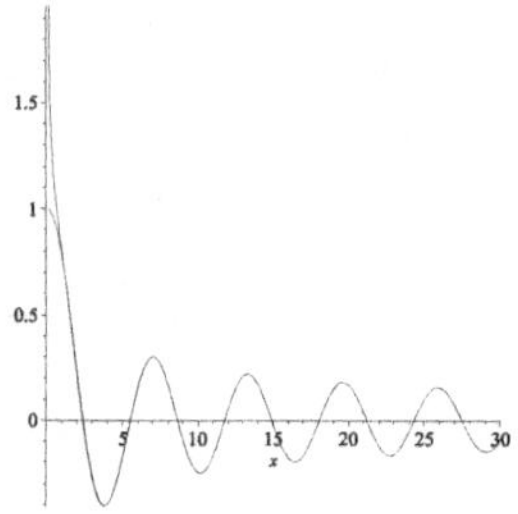

§6 The Vibrating Rectangular Membrane

Maple can be used to plot 3-dimensional graphs of the vibrating rectangular membrane discussed in Example 4 of Section 7.3. Consider the case of a rectangular membrane with dimensions $a = b = \pi$. The standing wave solutions are given by equation (22) in Section 7.3.

$$> u := (m,n,x,y,t) \rightarrow \sin(m \cdot x) \cdot \sin(n \cdot y) \cdot \cos\left(\sqrt{m^2+n^2} \cdot t\right);$$

$$(m,n,x,y,t) \mapsto \sin(m\,x)\sin(n\,y)\cos\left(\sqrt{m^2+n^2}\,t\right)$$

We can display the various mode shapes for any instant as surfaces.

$> plot3d(u(1,2,x,y,0), x = 0..\pi, y = 0..\pi);$

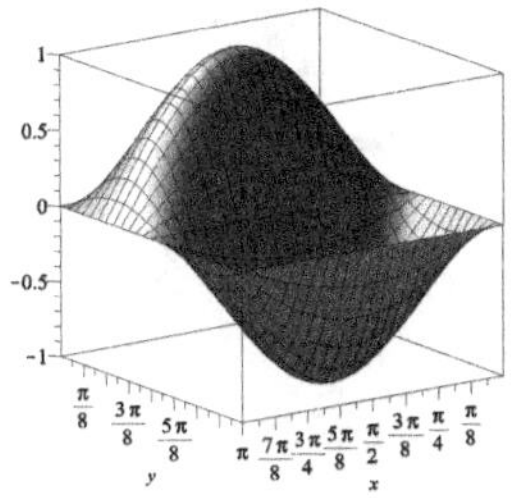

The menus in the graphics window allows us to manipulate the above plot and change the viewing orientation, the style, the color and various other parameters that are too numerous to mention here. The contour option shows the various zones of the membrane that are separated by nodal lines. Here is a more interesting mode

$> plot3d(u(2,2,x,y,0), x = 0..\pi, y = 0..\pi);$

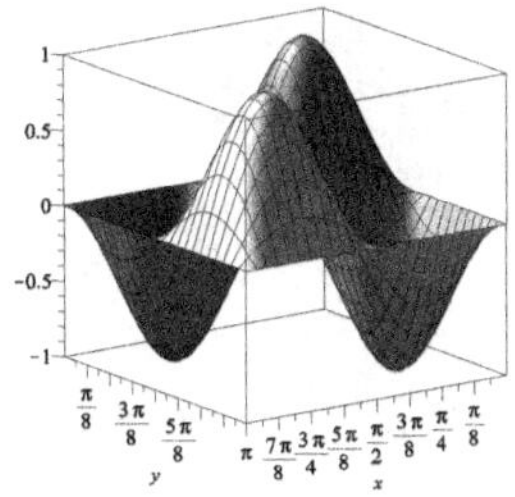

and its corresponding contour plot.

$> plot3d(u(2,2,x,y,0), x = 0..\pi, y = 0..\pi, style = contour);$

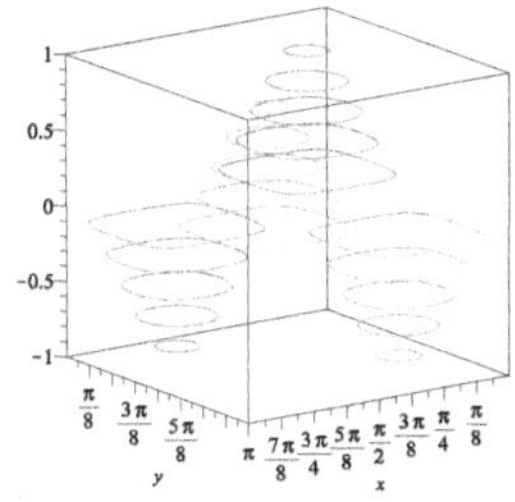

We can also look at the circular membrane using cylindrical coordinates as discussed in Section 7.4.

$> v := (m, w, r, \theta, t) \to \mathrm{BesselJ}\,(m, w \cdot r) \cdot \cos\,(m \cdot \theta) \cdot \cos\,(w \cdot t);$

$$(m, w, r, \theta, t) \mapsto \mathrm{BesselJ}\,(m, w\, r)\ \cos\,(m\ \theta)\ \cos\,(w\, t)$$

Here is a plot corresponding to the third mode $w = 8.654$ of J_0 (See Fig. 7.8).

$> plot3d([r, \theta, v(0, 8.654, r, \theta, 0)], r = 0..1, \theta = 0..2\,\pi, coords = cylindrical);$

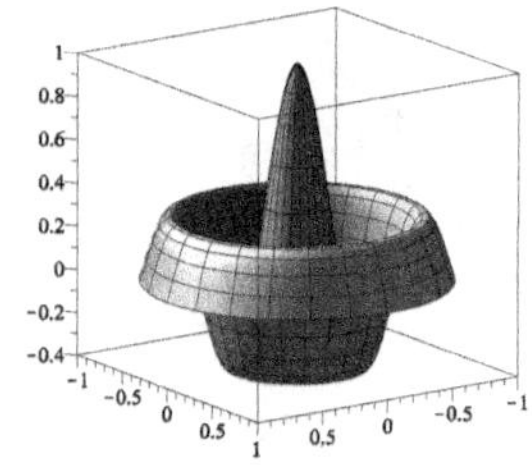

Here is a more interesting mode and its corresponding contour plot showing the nodal curves.

$> plot3d([r, \theta, v(1, 7.01559, r, \theta, 0)], r = 0..1, \theta = 0..2\,\pi, coords = cylindrical);$

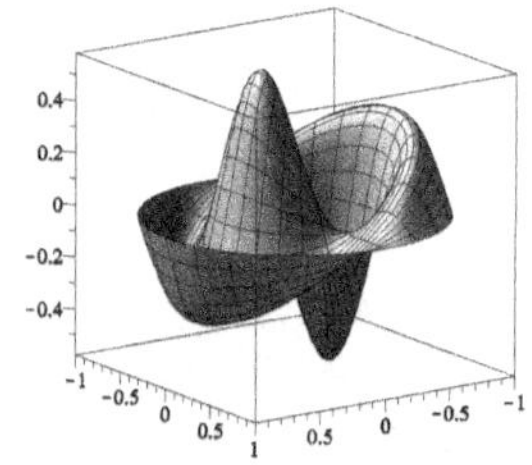

$> plot3d([r, \theta, v(1, 7.01559, r, \theta, 0)], r = 0..1, \theta = 0..2\,\pi, coords = cylindrical,$
$style = contour);$

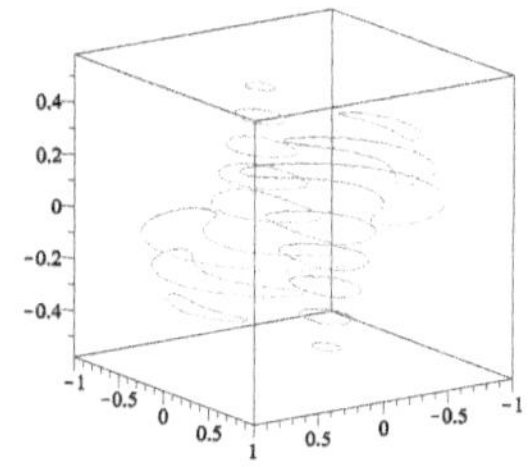

We can even see the motion in time of any of these modes by using the **animate** command in the **plots** package. Be aware that on older computers, this will require a good amount of memory and may take a while to generate. DVD-like controls allow you to replay these plots in sequence at any desired speed and direction in a window on the computer screen to create a standing wave effect.

$> with(plots);$

[*animate, animate3d, animatecurve, arrow, changecoords, complexplot, complexplot3d, conformal, conformal3d, contourplot, contourplot3d, coordplot, coordplot3d, densityplot, display, dualaxisplot, fieldplot, fieldplot3d, gradplot, gradplot3d, implicitplot, implicitplot3d, inequal, interactive, interactiveparams, intersectplot, listcontplot, listcontplot3d, listdensityplot, listplot, listplot3d, loglogplot, logplot, matrixplot, multiple, odeplot, pareto, plotcompare, pointplot, pointplot3d, polarplot, polygonplot, polygonplot3d, polyhedra_supported, polyhedraplot, rootlocus, semilogplot, setcolors, setoptions, setoptions3d, shadebetween, spacecurve, sparsematrixplot, surfdata, textplot, textplot3d, tubeplot*]

$> animate(plot3d, [[r, \theta, v(1, 7.01559, r, \theta, t)], r = 0..1, \theta = 0..2\pi, coords = cylindrical], t = 0..3);$

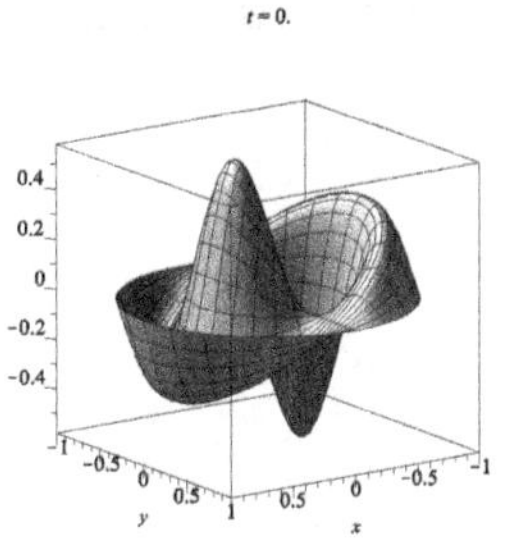

§7 Integral Transforms

Maple can also calculate Fourier, inverse Fourier, Laplace, inverse Laplace and Mellin integral transforms of certain functions. The Maple procedure *fourier* can calculate the Fourier transform of any expressions that only contains sums of rational functions of polynomials. It applies convolution methods, look-up tables and definite integration to handle special functions such as trigonometric functions, the Dirac and Heaviside functions, and Bessel functions. We need to load the **intrans** package first.

$> with(inttrans);$

[*addtable,fourier, fouriercos, fouriersin, hankel, hilbert, invfourier, invhilbert invlaplace, invmellin, laplace, mellin, savetable*]

Here are the Fourier transforms of a few familiar functions.

$> assume(0 < \sigma);$
$> fourier(\exp(-(x-\mu)^2/(2\cdot\sigma^2)), x, \omega);$

$$\sqrt{2}\sqrt{\pi\,\sigma\text{\textasciitilde}^2}\,e^{-\frac{1}{2}\frac{\mu^2}{\sigma\text{\textasciitilde}^2}-\frac{1}{2}\left(\omega+\frac{I\mu}{\sigma\text{\textasciitilde}^2}\right)^2\sigma\text{\textasciitilde}^2}$$

$> fourier(\sin(x^2), x, \omega);$

$$\frac{1}{2}\sqrt{\pi}\sqrt{2}\left(\cos\left(\frac{1}{4}\omega^2\right)-\sin\left(\frac{1}{4}\omega^2\right)\right)$$

$> fourier(\mathrm{BesselJ}(0,x),x,\omega);$

$$2\,\frac{Heaviside\,(\omega+1)-Heaviside\,(\omega-1)}{\sqrt{-\omega^2+1}}$$

The tilde (~) symbol next to a variable indicates that there are assumptions (like greater than 0) on that variable. Here is a more complicated calculation.

$> fourier\left(\frac{1}{x^3+1},x,\omega\right);$

$$\frac{1}{3}\mathrm{I}\,\mathrm{e}^{\mathrm{I}\omega}\pi\,(\mathrm{Heaviside}\,(-\omega)-\mathrm{Heaviside}\,(\omega)) + \frac{1}{3}\pi\left(-\mathrm{I}+\sqrt{3}\right)\mathrm{Heaviside}\,(-\omega)\,\mathrm{e}^{-\frac{1}{2}\mathrm{I}\omega+\frac{1}{2}\sqrt{3}\omega} + \frac{1}{3}\pi\left(\sqrt{3}+\mathrm{I}\right)\mathrm{Heaviside}\,(\omega)\,\mathrm{e}^{-\frac{1}{2}\mathrm{I}\omega-\frac{1}{2}\sqrt{3}\omega}$$

$> invfourier(\%,\omega,x);$

$$\frac{1}{(x+1)\,(x^2-x+1)}$$

$> normal(\%,expanded);$

$$(x^3+1)^{-1}$$

Maple knows certain properties of the Fourier transform. We can use these to solve the heat equation stated in equation (13) in Section 3.3.

$> heateqn := diff(u(x,t),t) = k\cdot(diff(u(x,t),x,x));$

$$\frac{\partial}{\partial t}u\,(x,t) = k\frac{\partial^2}{\partial x^2}u\,(x,t)$$

Let us denote the Fourier transform of $u(x,t)$ with respect to x by $U\,(\omega,t)$.

$> alias(U = fourier(u(x,t),x,omega)):$

$> fourier(lhs(heateqn),x,\omega) = fourier(rhs(heateqn),x,\omega);$

$$\frac{\partial}{\partial t}U = -k\omega^2 U$$

If we remove temporarily from the notation for U the explicit dependence on ω, this is just a first order ordinary differential equation for $U(\cdot,t)$. Maple can easily solve this equation.

$> dsolve(diff(U(t),t) = -k\cdot\omega^2\cdot U(t),U(t));$

$$U\,(t) = \_C1\,\mathrm{e}^{-k\omega^2 t}$$

Putting back the dependence on ω, we get the solution $U(\omega,t) = \_C1 \cdot e^{-k\cdot\omega^2\cdot t}$.

Here are other examples of transform computation.

$> mellin(\frac{1}{1+t}, t, s);$

$$\pi \csc(\pi s)$$

$> laplace(exp(3\cdot t)\cdot \cos(t), t, s);$

$$\frac{s-3}{(s-3)^2+1}$$

$> invlaplace(\%, s, t);$

$$e^{3t}\cos(t)$$

The next example shows how integral transforms can be used to enlarge the class of integration problems that Maple can solve.

$> v := t \to \int_0^t \mathrm{BesselJ}(1,x)\,dx;$

$$t \to \int_0^t \mathrm{BesselJ}(1,x)\,dx$$

When Maple returns an integral unevaluated as above, it means that it cannot perform the required integration. Nevertheless, Maple knows the properties of the Laplace transform.

$> laplace(v(t), t, s);$

$$\frac{1}{s} - \frac{1}{\sqrt{s^2+1}}$$

$> invlaplace(\%, s, t);$

$$1 - \mathrm{BesselJ}(0,t)$$

The main applications of Laplace transforms is in the area of ordinary differential equations and integral equations which are not the main subject of this book. You can define integral transforms of your own functions and add them to the look-up table.

Consider the function $myfunction := x \to \frac{\sinh(a\cdot x)}{\sinh(\pi\cdot x)}$ where $0 < a < \pi$.

$> fourier(\sinh(a\cdot x)/\sinh(x\cdot\pi), x, \omega);$

$$fourier\left(\frac{e^{ax}}{e^{\pi x}-e^{-\pi x}}, x, \omega\right) - fourier\left(\frac{e^{-ax}}{e^{\pi x}-e^{-\pi x}}, x, \omega\right)$$

This indicates that Maple does not know its Fourier transform. But we can calculate this using residue theory in complex function theory to get an infinite series expression in ω that simplifies to $\frac{\sin(a)}{\cosh(\omega)+\cos(a)}$. We can add this to the look-up table.

$> addtable(fourier, myfunction(x), \frac{\sin(a)}{\cosh(\omega)+\cos(a)}, x, \omega);$
$> fourier(myfunction(x), x, \omega);$

$$\frac{\sin(a)}{\cosh(\omega)+\cos(a)}$$

$> fourier(x^2 \cdot myfunction(x), x, \omega);$

$$\frac{\sin(a)\left(-2(\sinh(\omega))^2+(\cosh(\omega))^2+\cosh(\omega)\cos(a)\right)}{(\cosh(\omega)+\cos(a))^3}$$

§8 Orthogonal Polynomials

Fourier series are the simplest examples of eigenfunction expansions arising from the solution of a Sturm-Liouville problem. However, there are other eigenfunctions known as special functions and orthogonal polynomials. Maple has a package that allows for the manipulation of the following orthogonal polynomials: the n-th Gegenbauer ultraspherical polynomial $G(n,a,x)$, the n-th Hermite polynomial $H(n,x)$, the n-th Laguerre polynomial $L(n,x)$, the n-th generalized Laguerre polynomial $L(n,a,x)$, the n-th Legendre polynomial $P(n,x)$, the n-th Jacobi polynomial $P(n,a,b,x)$ with $a \neq 0, b \neq 0$, the n-th Chebyshev polynomial of the first kind $T(n,x)$, and the n-th Chebyshev polynomial of the second kind $U(n,x)$. These functions are loaded using the **orthopoly** package.

$> with(orthopoly);$

$$[G,H,L,P,T,U]$$

Here are the first 5 Legendre polynomials and their graphs on the interval $[-1,1]$.

$> seq(P(n,x), n=0..4);$

$$1,\, x,\, -\frac{1}{2}+\frac{3}{2}x^2,\, \frac{5}{2}x^3-\frac{3}{2}x,\, \frac{3}{8}+\frac{35}{8}x^4-\frac{15}{4}x^2$$

$> plot(\{\%\}, x=-1..1);$

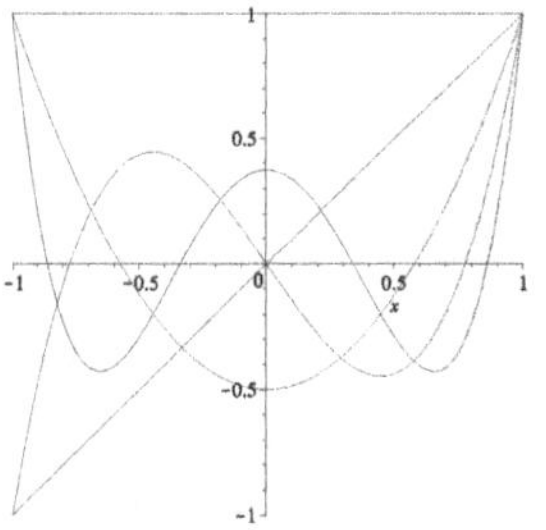

$> seq\left(\int_{-1}^{1} P(n,x)^2\,dx, n=0..5\right);$

$$2, \frac{2}{3}, \frac{2}{5}, \frac{2}{7}, \frac{2}{9}, \frac{2}{11}$$

These lead us to conjecture that for any integer n,

$$\int_{-1}^{1} P(n,x)^2 \, dx = \frac{2}{2n+1}$$

Here is a plot of the absolute value of the Legendre polynomial $P(2,\cos(\varphi))$ showing its azimuthal symmetry. As n increases, $P(n,x)$ acquires more lobes, which results from the increasing number of zeros of $P(n,x)$ as a function of x.

> $> plot3d(\text{abs}(P(2,\cos(\varphi))), \theta = 0..2\pi, \varphi = 0..\pi, coords = spherical);$

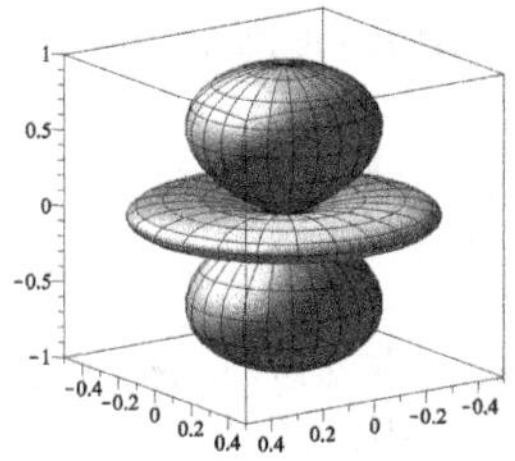

> $> plot3d(\text{abs}(P(3,\cos(\varphi))), \theta = 0..2\pi, \varphi = 0..\pi, coords = spherical);$

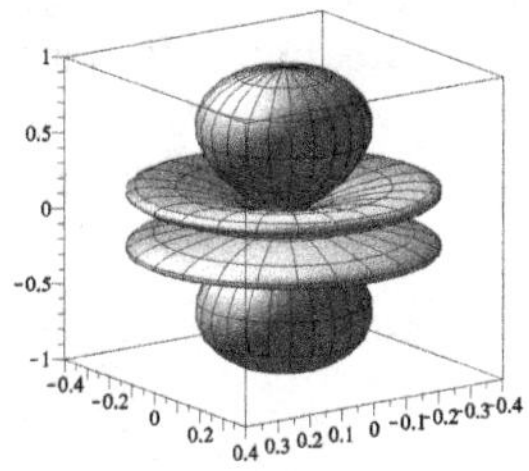

> $> plot3d(\text{abs}(P(4,\cos(\varphi))), \theta = 0..2\pi, \varphi = 0..\pi, coords = spherical);$

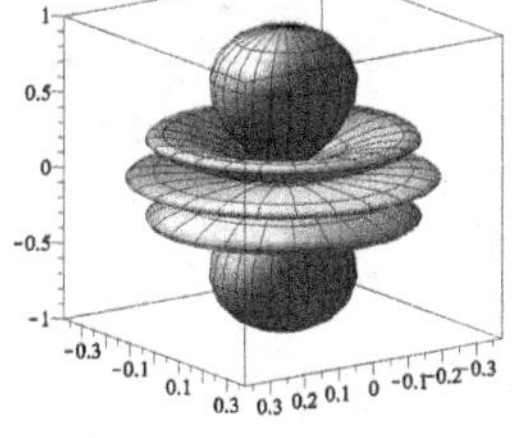

> $> plot3d(\text{abs}(P(5,\cos(\varphi))), \theta = 0..2\pi, \varphi = 0..\pi, coords = spherical);$

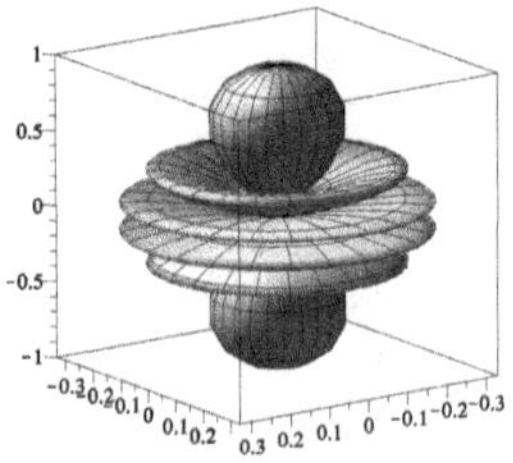

§9 Orthogonal Expansions

When this appendix was written as a supplement to the original edition of the book, the **OrthogonalExpansions** package was not available. With this package, Maple 2016 now includes the ability to calculate Fourier series and other orthogonal series expansions of functions. This makes the 3 procedures defined in Section 2 unnecessary and the calculation of Fourier series expansions in Maple more straightforward.

> *with(OrthogonalExpansions)*;

[*BesselSeries, ChebyshevTSeries, ChebyshevUSeries, FourierSeries, GegenbauerSeries, GramSchmidtL2, Haar, HaarSeries, HarmonicWavelet, HarmonicWaveletSeries, HermiteSeries, JacobiSeries, LaguerreSeries, LegendreSeries, MixedSeries, Rational, RationalSeries, RectSeries, SincSeries, SincWavelet, SincWaveletSeries, SphericalSeries, Walsh, WalshSeries, Zernike, ZernikeSeries*]

If this command produces an error message, that means that your Maple installation does not have this package. You will have to go to the MapleSoft Application Center at **http://www.maplesoft.com/applications/view.aspx?SID=7256** to download the package. Instructions on installing it are in the accompanying documentation. There are other user-contributed packages that you may want to add to your Maple installation. Be aware that the location of this website may change in the future.

Using this package, the Fourier series of the piecewise-defined function g in Example 2 is easily calculated.

> $FourierSeries(piecewise(x < 0, 1, x >= 0, x), x = -\pi..\pi, n, 'Coefficients');$

$$\frac{1}{2}\frac{\pi + \frac{1}{2}\pi^2}{\pi} + \sum_{i=1}^{n}\left(\frac{(-1)^i - 1}{i^2\pi}\cos(ix) + \frac{\frac{(-1)^i - 1}{i} - \frac{(-1)^i\pi}{i}}{\pi}\sin(ix)\right)$$

The parameter n can be replaced by '*infinity*' and the Fourier coefficients are assigned to the Maple variable *Coefficients*. You can change this variable to another name of your choice.

> *Coefficients*;

$$\left[\frac{1}{2}\frac{\pi + \frac{1}{2}\pi^2}{\pi}, \%seq\left(\left[\frac{(-1)^i - 1}{i^2\pi}, \frac{\frac{(-1)^i - 1}{i} - \frac{(-1)^i\pi}{i}}{\pi}\right], i = 1..n\right)\right]$$

Answers to Selected Problems

Chapter 1

1.1A (b.p. $\equiv$ base period) ***1.*** b.p. $= L$; ***4.*** b.p. $= 2\pi/m$; ***7.*** b.p. $= \pi$; ***8.*** b.p. $= 1$

B 2

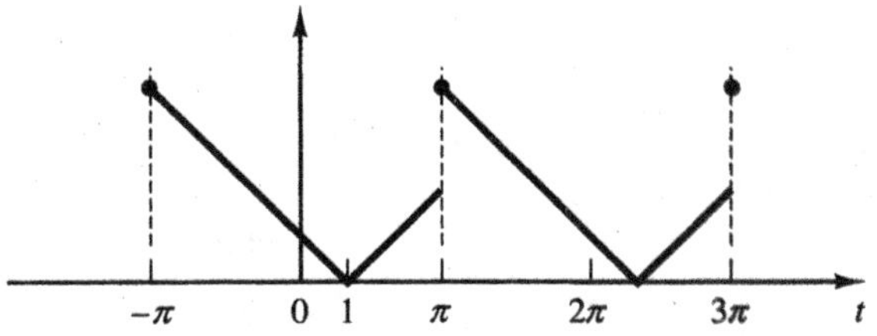

3

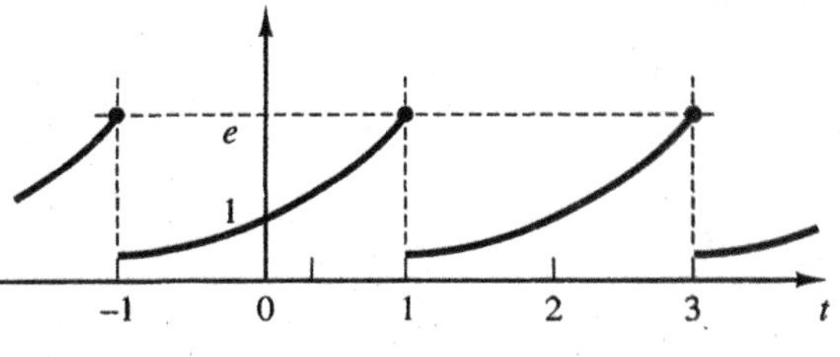

C 1. $f(t) = t - 2\pi; t + 2\pi$. ***2.*** $f(t) = 2\pi - |t|$.

D L/a

G 1. Graph has tangent lines with the same slope at t and $t + L$.

2. $f(t) = 1$; then $F(t) = t$.

H 1. No

1.2B

$$I = \int_{-\pi}^{\pi} \cos nt \cos mt \, dt = \frac{1}{m}\left[\cos nt \sin mt\Big|_{-\pi}^{\pi} + n\int_{-\pi}^{\pi} \sin nt \sin mt \, dt\right]$$

$$= \frac{n}{m^2}\left[-\sin nt \cos mt\Big|_{-\pi}^{\pi} + nI\right] = \frac{n^2}{m^2}I, \qquad (m \neq 0)$$

Hence, if $m \neq 0$ and $n \neq m$, then $I = 0$, and the other cases can be handled similarly.

E No, since the periodic extension of t^2 has a corner at $\pm\pi, \pm 3\pi, \ldots$.

F *1.* Let $t = 2\tau$. Then if $m \neq n$,

$$\int_0^\pi \sin\left(\frac{nt}{2}\right)\sin\left(\frac{mt}{2}\right)dt = 2\int_0^{\pi/2} \sin n\tau \sin m\tau \, d\tau$$

$$= \frac{\sin\left((n-m)\frac{\pi}{2}\right)}{n-m} - \frac{\sin\left((n+m)\frac{\pi}{2}\right)}{n+m}$$

and this is zero when $n - m$ and $n + m$ are even, which occurs if both n and m are even or if both are odd.

H *1.* $\sin^4 t = \frac{3}{8} - \frac{1}{2}\cos 2t + \frac{1}{8}\cos 4t$ **2.** True

1.3A *1.* $F(t) = 1$ *3.* $t(\pi - t) \sim -\frac{\pi^2}{3} + \sum_{n=1}^{\infty}(-1)^{n+1}\left[\frac{4}{n^2}\cos nt + \frac{2\pi}{n}\sin nt\right]$

5. $F(t) = \frac{\pi}{4} - \frac{2}{\pi}\sum_{n=1,3,\ldots}^{\infty}\frac{\cos nt}{n^2} + \sum_{n=1}^{\infty}\frac{(-1)^{n+1}}{n}\sin nt$

7. $\sin t \cos 2t = -\frac{1}{2}\sin t + \frac{1}{2}\sin 3t$ *8.* $|\sin t| \sim \frac{2}{\pi} - \frac{4}{\pi}\sum_{n=2,4,\ldots}^{\infty}\frac{\cos nt}{n^2 - 1}$

10. $e^{|t|} \sim \frac{e^\pi - 1}{\pi} + \frac{2}{\pi}\sum_{n=1}^{\infty}\frac{[(-1)^n e^\pi - 1]}{n^2 + 1}\cos nt$

C *3.* For $f(t) = t^3$, $c_n = 0$, $\quad n = 0, 1, 2, \ldots$.

1.4A *1.* and *6.* are even; *4.*, *8.*, and *10.* are neither.

B *2.*, *3.*, *5.*, *7.*, and *9.* generate sine series.

C *4.* $\phi_E(t) = t^2$; $\phi_O(t) = 3t$ *8.* $\Phi_E(t) = \cos t$; $\Phi_O(t) = \sin t$

10. $\Upsilon_E(t) = \frac{1}{1 - t^2}$; $\Upsilon_O(t) = \frac{-t}{1 - t^2}$, $\quad (t \neq \pm 1)$

E *1.* $t^3 \sim -2\sum_{n=1}^{\infty}\frac{(-1)^n}{n}\left[\pi^2 - \frac{6}{n^2}\right]\sin nt$

2. $t^4 \sim \frac{\pi^4}{5} + 8\sum_{n=1}^{\infty}\frac{(-1)^n}{n^2}\left[\pi^2 - \frac{6}{n^2}\right]\cos nt$

3. $F(t) = \frac{4}{\pi}\sum_{n=1,3,}^{\infty}\frac{\sin nt}{n}$

4. $F(t) = \frac{1}{2} + \frac{2}{\pi}\sum_{n=1}^{\infty}\frac{\sin\frac{n\pi}{2}}{n}\cos nt = \frac{1}{2} + \frac{2}{\pi}\left[\cos t - \frac{\cos 3t}{3} + \frac{\cos 5t}{5} - \ldots\right]$

5. $F(t) = \frac{\pi^2}{6} + \sum_{n=1}^{\infty}\left\{\frac{2(-1)^n}{n^2}\cos nt + \left[\frac{\pi(-1)^{n+1}}{n} + \frac{2((-1)^n - 1)}{\pi n^3}\right]\sin nt\right\}$

6. $F(t) = \frac{1}{2} + \frac{1}{2}\cos 2t$

F Extension is continuous at $\pm\pi$ only for *2.*, *4.*, and *6.*

G Odd extensions: for *1.*, *3.*, and *5.* see answers to Problem 1.4E.

2. $s_n = \frac{2}{\pi}\left[\frac{\pi^4}{n} - \frac{12\pi^2}{n^3} + \frac{24}{n^5}\right](-1)^{n+1} + \frac{48}{\pi n^5}$

4. $s_n = \frac{2}{\pi n}\left(1 - \cos\frac{n\pi}{2}\right)$

6. $s_n = \frac{2}{\pi}\left[\frac{1}{n} + \frac{n}{n^2-4}\right], \qquad n = 1, 3, 5, \ldots; \; s_2 = s_4 = s_6 = \cdots 0$

Even extensions: For *2.*, *4.*, and *6.*, see answers to Problem 1.4E.

1. $F(t) = \frac{\pi^3}{4} + \frac{6}{\pi}\sum_{n=1}^{\infty}\left[\frac{2}{n^4} + \left(\frac{\pi^2}{n^2} - \frac{2}{n^4}\right)(-1)^n\right]\cos nt$

3. $F(t) = 1$ *5.* $F(t) = \frac{\pi^2}{3} + 4\sum_{n=1}^{\infty}\frac{(-1)^n}{n^2}\cos nt$

H *1.* $F(t) = 2\sum_{n=1}^{\infty}\frac{1}{n}\sin nt$

2. $F(t) = \frac{\pi}{2} + \frac{4}{\pi}\sum_{n=1,3,}^{\infty}\frac{1}{n^2}\cos nt$

I *1.* $F(t) = \frac{8}{\pi}\sum_{n=1,3,}^{\infty}\frac{1}{n^3}\sin nt$

2. and *3.* $F(t) = \frac{\pi^2}{6} - 4\sum_{n=2,4,}^{\infty}\frac{1}{n^2}\cos nt$

J *1.* $F(t) = \frac{4}{\pi}\sum_{n=2,4,}^{\infty}\frac{n}{n^2-1}\sin nt$ *2.* $|\sin t| \sim \frac{2}{\pi} - \frac{4}{\pi}\sum_{n=2,4,}^{\infty}\frac{1}{n^2-1}\cos nt$

L *2.* $g(t) = -g(\pi - t)$

M $\cosh t \sim \frac{2}{\pi}\sinh \pi\left\{\frac{1}{2} + \sum_{n=1}^{\infty}\frac{(-1)^n}{1+n^2}\cos nt\right\}$

$\sinh t \sim \frac{-2}{\pi}\sinh \pi\sum_{n=1}^{\infty}\frac{(-1)^n n}{1+n^2}\sin nt$

O *2.* *No*; f odd $\Rightarrow f'$ even, but $f(t) = t + 1$, which is not odd, has the even derivative $f'(t) = 1$.

1.5A

(*b*)

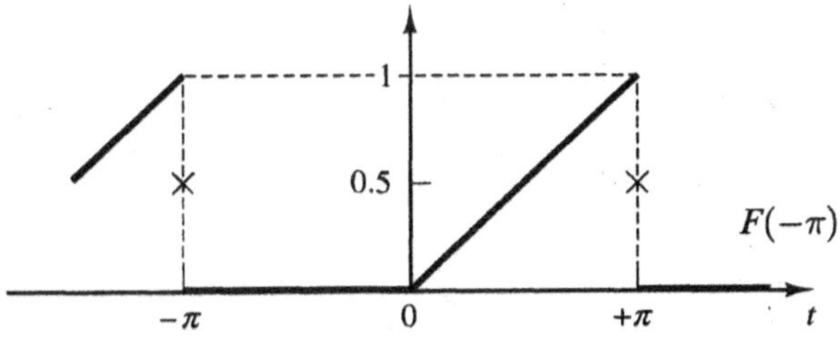

$F(-\pi) = F(+\pi) = \frac{1}{2}$

(*d*)

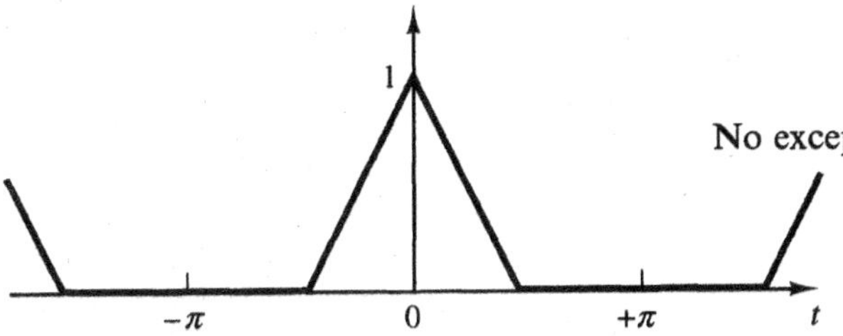

No exceptional value.

B *2.* $F(t) = t^4, \qquad -\pi \le t \le \pi$

5. $F(t) = \begin{cases} t^2, & 0 \le t < \pi \\ 0, & -\pi < t \le 0 \end{cases}$ $\qquad F(-\pi) = F(\pi) = \dfrac{\pi^2}{2}$

C *2.* $12\sum_{n=1}^{\infty}\frac{(-1)^n}{n^3}\sin\frac{n\pi}{2} = 12\left(-1+\frac{1}{3^3}-\frac{1}{5^3}+\dots\right) = \frac{-3\pi^3}{8}$

D *1.* (see ***B***): All except ***3.*** and ***4.***

2. All except ***2.***, ***3.***, ***5.***, and ***9.***

3. Example 1: $\frac{\pi^2}{2} + \frac{16}{\pi^2}\sum_{n=1,3,}^{\infty}\frac{1}{n^4} = \frac{2}{3}\pi^2.$

Example 2: $\frac{4}{\pi^2}\sinh^2\pi\left\{\frac{1}{2}+\sum_{n=1}^{\infty}\frac{1}{1+n^2}\right\} = \frac{1}{\pi}\sinh 2\pi$

E *1.* Since $\sinh\pi \ne \sinh(-\pi)$, *cannot* differentiate series for $\sinh t$ to get that for $\cosh t$.

2. $\frac{4}{\pi^2}\sinh^2\pi\left\{\frac{1}{2}+\sum_{n=1}^{\infty}\frac{1}{(1+n^2)^2}\right\} = 1 + \frac{\sinh 2\pi}{2\pi};$

$\frac{4}{\pi^2}\sinh^2\pi\sum_{n=1}^{\infty}\frac{n^2}{(1+n^2)^2} = -1 + \frac{\sinh 2\pi}{2\pi}$

H $f(t) = |t|^{3/4} \to 0$ as $|t| \to 0$, but $f'(t) = \frac{3}{4}t^{-1/4} \to +\infty$ as $t \searrow 0$

1.6A *2.* $|\sin x| = (1-c) - 2\sum_{n=1}^{\infty}\frac{(1-(-1)^n c)}{n^2\pi^2-1}\cos n\pi x, \quad |x| \le 1,$ where $c = \cos 1$

4. $x - x^2 = \frac{t}{\pi} - \frac{t^2}{\pi^2} = -\frac{1}{3} - \frac{2}{\pi}\sum_{n=1}^{\infty}\frac{(-1)^n}{n}\left[\frac{2}{n\pi}\cos n\pi x + \sin n\pi x\right], \quad (|x| < 1)$

6. $\cos x = 2\pi\sum_{n=1}^{\infty}\frac{[1-(-1)^n\cos 1]}{n^2\pi^2-1}n\sin n\pi x, \qquad (0 < x < 1)$

B

(3)

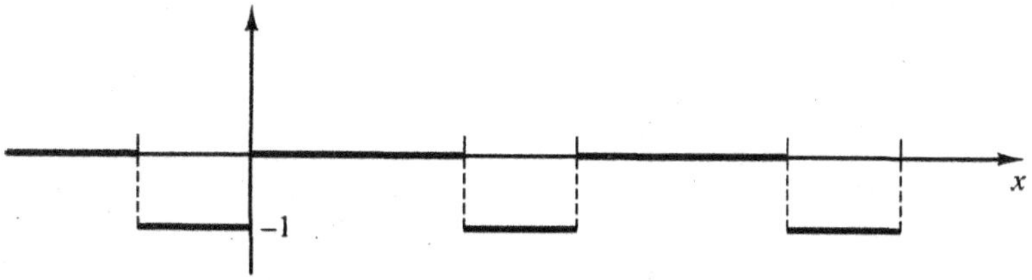

(5)

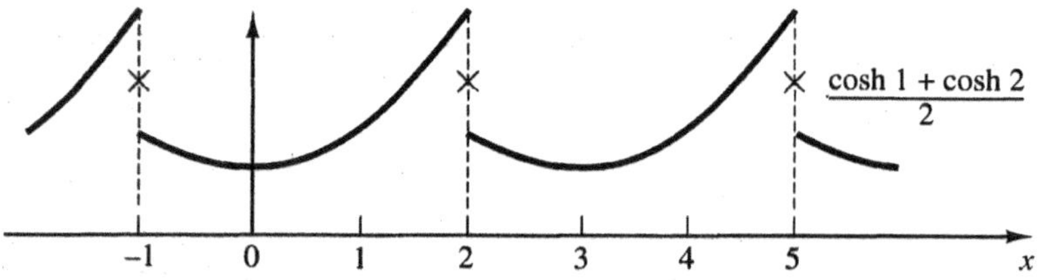

(6)

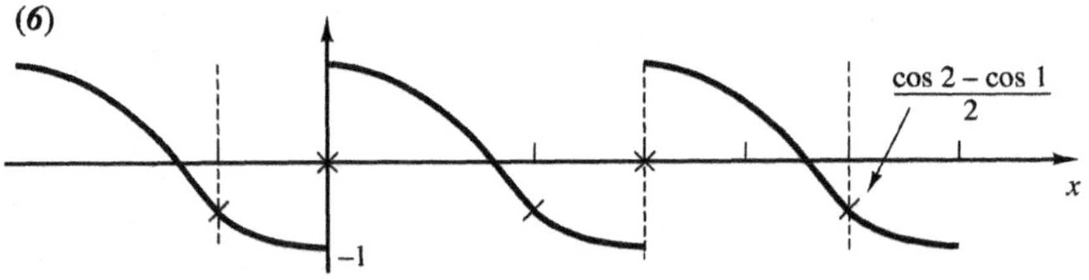

C *1.* $\pi x - x^2 = \frac{2}{\pi}\sum_{n=1}^{\infty}\left[(1-\pi)(-1)^n + 2\frac{(1-(-1)^n)}{n^2\pi^2}\right]\frac{\sin n\pi x}{n}, \qquad (0 \le x < 1)$

3. $\pi x - x^2 = \left(\frac{\pi}{2} - \frac{1}{3}\right) - \frac{4}{\pi^2}\sum_{n=2,4,}^{\infty}\frac{\cos n\pi x}{n^2} + \frac{2(1-\pi)}{\pi}\sum_{n=2,4}^{\infty}\frac{\sin n\pi x}{n}, \quad 0 < x < 1$

D 2. $\pi - x = \pi - 1 + \frac{8}{\pi^2}\sum_{n=1,3,}^{\infty}\frac{1}{n^2}\cos\left(\frac{n\pi x}{2}\right), \quad 0 \le x \le 2$

3. $\pi - x = \pi - 1 + \frac{4}{\pi}\sum_{n=2,4,}^{\infty}\frac{1}{n}\sin\left(\frac{n\pi x}{2}\right), \quad 0 < x < 2$

E $g(t) = \begin{cases} 2t, & 0 \le t \le \pi/2 \\ 2(\pi - t), & \pi/2 \le t \le \pi \end{cases}$ has a *continuous* odd extension of period 2π, so

$x = \frac{8}{\pi}\sum_{n=1,3,}^{\infty}\frac{1}{n^2}\sin\frac{n\pi}{2}\sin\frac{nx}{2}, \quad -\pi \le x \le \pi,$

and the series converges uniformly.

2. $g(t) = \sin 2t$ on $[0, \pi]$ has cosine series with $c_n = 0$, $n = 0, 2, 4, \ldots$ and for n odd, $c_n = \frac{-8}{\pi}\frac{1}{n^2 - 4}$.

1.7C $F_2(t) = .206 - .026\cos t - .084\sin t - .055\cos 2t - .024\sin 2t$.

D *1.* $F_2(t) = .616 - .470\cos t - .145\cos 2t$

2. $F_2(t) = 1.38 - .726\cos t - .510\sin t - .652\cos 2t - .128\sin 2t$.

3. $F_2(t) = 9.475 - 3.325\cos t - 10.87\sin t - 6.15\cos 2t - 2.565\sin 2t$.

1.9A Must have $-x'(t) = y(t) = c_1\cos t + s_1\sin t$.

B *1.* $p_3(x) = 4x^3 - 3x$; $p_5(x) = 16x^5 - 20x^3 + 5x$

3. $\cos^5 t = \frac{1}{16}(10\cos t + 5\cos 3t + \cos 5t)$

D 2. $\sin^3 t = \frac{1}{4}(3\sin t - \sin 3t)$

$\sin^5 t = \frac{1}{16}(10\sin t - 5\sin 3t + \sin 5t)$

Chapter 2

2.2B *3.* $(x + iy)^2 = (x^2 - y^2) + i(2xy)$

$(x + iy)^3 = (x^3 - 3xy^2) + i(3x^2y - y^3)$

$u_{n+1} + iv_{n+1} = (x + iy)(u_n + iv_n) = (xu_n - yv_n) + i(xv_n + yu_n)$

C *1.* $u(r,\theta) = \frac{1}{2} + \frac{1}{2}r^2\cos 2\theta$

3. $u(r,\theta) = \frac{2}{\pi} - \frac{4}{\pi}\sum_{n=1}^{\infty}\frac{(-1)^n r^{2n}}{4n^2 - 1}\cos 2n\theta$

E *1.*, 2. $u(r,\theta) = \frac{8}{\pi}\sum_{n=1,3,}^{\infty}\frac{1}{n^3}\left(\frac{r}{R}\right)^n\sin n\theta$

F Define $f(\theta) = -f(\pi - \theta), \quad \pi/2 \le \theta \le \pi$. Then

$u(r,\theta) = \frac{2}{\pi}\sum_{n=1,3,}^{\infty}\frac{1}{n^3}\left(\frac{r}{R}\right)^{2n}\sin 2n\theta.$

2.3B Only 2. and *3.*

C 2. (*c*) *4.* (*b*) 6. (*a*)

D *1.* (*b*) *3.* (*a*) 5. (*c*) 7. (*a*)

E 2. $u(x,t) = x^2 + t^2$; $u(x,t) = x^3 + 3xt^2$

4. Only polynomial solution is a constant.

2.4A 2. $u_n(x,t) = e^{-(n/2)^2 t} \sin\frac{nx}{2}, \qquad n = 1, 3, 5, \ldots$

B 2. $u_n(x,t) = e^{-(n\pi/\ell)^2 t} \sin\frac{n\pi x}{\ell}, \qquad n = 1, 2, \ldots$

3. $u_n(x,t) = \cosh ny \cos nx, \qquad n = 0, 1, 2, \ldots$

4. $u_n(x,t) = \sin nt \cos nx, \qquad n = 1, 2, 3, \ldots, u_0(x,t) = t$

5. $u_n(x,t) = (e^{-n^2 t} - 1) \sin nx, \qquad n = 1, 2, 3, \ldots$

C Only product solution is $u(x,t) = 0$.

D 4. $\omega_0 = 0$ is a solution.

I 2. $u_{mn}(x,y,t) = e^{-[(m^2/4)+n^2]t} \sin\left(\frac{mx}{2}\right) \cos ny, \qquad m = 1, 3, 5, \ldots \quad n = 0, 1, 2, \ldots$

J $u_{mn}(x,y,z) = [a_{mn} \cosh \omega_{mn} z + b_{mn} \sinh \omega_{mn} z] \cos\left(\frac{m\pi x}{\ell}\right) \sin\left(\frac{n\pi y}{\ell}\right)$
where

$$\omega_{mn} = \sqrt{\left(\frac{m\pi}{\ell}\right)^2 + \left(\frac{n\pi}{\ell}\right)^2}, \qquad m = 0, 1, 2, \ldots \quad n = 1, 2, \ldots$$

2.5B 1. $t = \ell\bar{t}/k;\ \bar{u}_{\bar{t}\bar{t}} = \bar{u}_{\bar{x}\bar{x}}$

3. $t = \ell\bar{t}\sqrt{\ell/k};\ \bar{u}_{\bar{t}\bar{t}} = \bar{u}_{\bar{x}\bar{x}\bar{x}}$

C $u = B\bar{u},\ t = \bar{t}/AB$

D 3. $\gamma = \sqrt{RC/GL} + \sqrt{GL/RC}$

Chapter 3

3.1A 1. $(\nu u)_t = (Ku_x)_x + (Ku_y)_y + Q$

3. $\tilde{u}_t = k\left(\tilde{u}_{rr} + \frac{1}{r}\tilde{u}_r + \frac{1}{r^2}\tilde{u}_{\theta\theta}\right) + q$

B 4. $\nu u_t = (Ku_x)_x + Q$, where Q is the *linear* density of heat sources per unit time.

C 3. $\nu u_t = (Ku_x)_x - \frac{2}{r}Kh(u - U) + Q$

D $u(x,t) = 28 + 7e^{-ax}\cos\left(\frac{\pi}{12}t - ax\right)$, where $a = \sqrt{\frac{\pi}{24}\frac{10}{1.4}\frac{1}{36}}$

G 4(*a*) $u_t = (\kappa u_x)_x + Q$, where Q is the linear density of the mass of dye generated per unit time.

3.2D 3. $\frac{d}{dt}(\nu u^2) - 2\ u^2 = \nu e^{at}\frac{d}{dt}(e^{-at}u^2)$, where $a = 2/\nu$, and the arguments go through when ν is constant.

3.3C 1. $U(x,t) = 2\sum_{n=1}^{\infty}\frac{(-1)^{n+1}}{n}e^{-kn^2 t}\sin nx$; converges uniformly for x near 0, but *not* for x near π.

3. $U(x,t) = \frac{\pi}{2} - \frac{4}{\pi}\sum_{n=1,3,}^{\infty}\frac{e^{-kn^2 t}}{n^2}\cos nx$; converges uniformly for $t \geq 0$.

5. $U(x,t) = \frac{8}{\pi}\sum_{n=1,3,}^{\infty}\frac{1}{n^2}\sin\frac{n\pi}{2}e^{-k(n/2)^2 t}\sin\frac{nx}{2}$; converges uniformly for $t \geq 0$.

D 2. $U(x,t) = \frac{4\ell}{\pi^2}\sum_{n=1,3,}^{\infty}\frac{\sin\frac{n\pi}{2}}{n^2}e^{-k(n\pi/\ell)^2 t}\sin\frac{n\pi x}{\ell}$

4. $U(x,t) = \frac{3\ell}{8} + \frac{2\ell}{\pi^2}\sum_{n=1}^{\infty} \frac{\left(\cos\frac{n\pi}{2} - 1\right)}{n^2} e^{-k(n\pi/\ell)^2 t} \cos\frac{n\pi x}{\ell}$

6. $U(x,t) = +\frac{4}{\pi}\sum_{n=1,3,}^{\infty} \left[\frac{\sin\frac{n\pi}{2} - \sin\frac{n\pi}{4}}{n}\right] e^{-k(n\pi/2\ell)^2 t} \cos\frac{n\pi x}{2\ell}$

[convergence corresponds to that for problems in part C.]

E 2. For $t_0 > 0$: $n^2 e^{-kn^2 t_0} \leq n^2 e^{-knt_0}$ and $\sum_{n=1}^{\infty} n^2 e^{-knt_0}$ converges by the ratio test.

3.4A 1. $V(x,t) = 100x + \frac{198}{\pi}\sum_{n=1}^{\infty} \frac{(-1)^n}{n} e^{-2(n\pi)^2 t} \sin n\pi x$

3. $V(x,t) = e^{-t}\sin x + \sum_{n=1,3,}^{\infty} s_n e^{-(n/2)^2 t} \sin\frac{nx}{2}$, where

$$s_n = \frac{2}{\pi}\int_0^{\pi/2} \sin\frac{nx}{2}\,dx - \frac{2}{\pi}\int_0^{\pi} \sin x \sin\frac{nx}{2}\,dx$$

$$= \frac{4}{\pi n}\left(1 - \cos\frac{n\pi}{4}\right) + \frac{8}{\pi}\frac{\sin\frac{n\pi}{2}}{(n^2-4)}, \qquad n = 1, 3, \ldots$$

5. $V(x,t) = \frac{4}{\pi}\sum_{n=1}^{\infty}\left[\frac{(-1)^n - 1}{n^3} - \frac{\pi^2(-1)^n}{2n}\right]\left(\frac{e^t - e^{-n^2 t}}{n^2+1}\right) \sin nx$

B $\mathscr{s}(x,t) = -\frac{1}{6k}x + \frac{1}{6k}(x^3 + 6kxt)$

C 2. $u(x,t) = 5 + v(x,t)e^{-4t}$, where $v(x,t) = 4\sum_{n=1}^{\infty} \frac{(-1)^{n+1}}{n} e^{-n^2 t} \sin nx$

D Find v: $v_t = kv_{xx}$, $0 < x \leq 100$, $t > 0$ with $v(0,t) = 100$; $v(100,t) = 0$ and $v(x,0) = x$.

E 3. Whenever $\ell \neq n\pi\sqrt{k/c}$, $n = 1, 2, \ldots$.

3.5E 2. Yes 3. If dye could not be removed internally.

Chapter 4

4.1A 1. $\phi(x,y) = -30x - 40y$

C $\phi(\mathbf{x},\mathbf{y}) = -\gamma q Q/|\mathbf{x} - \mathbf{y}|$

4.3A For $u(x,y) = xy$; $u_x = y$ and $u_y = x$.

$\frac{\partial u}{\partial n}(0,y) = -u_x(0,y) = -y$; $\frac{\partial u}{\partial n}(a,y) = u_x(a,y) = y$.

$\frac{\partial u}{\partial n}(x,0) = -u_y(x,0) = -x$; $\frac{\partial u}{\partial n}(x,b) = u_y(a,b) = x$.

B $Y(y) = \frac{\sinh \pi y}{\sinh \pi}$

C 1. $p(x,y) = bxy$ 2. $a + c = 0$, so $b^2 - 4ac > 0$.

D 1. $U(x,y) = \sum_{n=1}^{\infty} s_n \frac{\sinh ny}{\sinh nb} \sin nx$, where in (i)

$$s_n = \frac{2}{\pi n}\left\{\frac{2[(-1)^n - 1]}{n^2} - \pi^2(-1)^n\right\}, \qquad n = 1, 2, \ldots$$

and series converges nonuniformly near $(\pi, 0)$. In (ii)

$$s_n = \frac{4}{\pi}\frac{(1-(-1)^n)}{n^3}, \qquad n = 1, 2, \ldots;$$

series converges uniformly for $0 \le y \le \ell$, and U is *the* solution to the problem. In (iii)

$$s_n = \frac{2}{\pi n}\left(1 - \cos\frac{n\pi}{2}\right), \qquad n = 1, 2, \ldots;$$

series converges nonuniformly near $(0,0)$ and $(\pi/2, 0)$.

3. $U(x,y) = \frac{c_0}{2}\frac{y}{\ell} + \sum_{n=1}^{\infty} c_n \frac{\sinh ny}{\sinh n\ell}\cos nx$, where in (i)

$$c_0 = \frac{2\pi^2}{3} \text{ and } c_n = \frac{4}{n^2}(-1)^n, \qquad n = 1, 2, \ldots$$

and U_x series converges nonuniformly near $(\pi, 0)$.

5. $U(x,y) = \sum_{n=1,3,}^{\infty} s_n \frac{\sinh \frac{ny}{2}}{\sinh \frac{n\ell}{2}} \sin\frac{nx}{2}$, where

$$s_n = \frac{8}{\pi n^3}\left[4 - n\pi \sin\frac{n\pi}{2}\right], \qquad n = 1, 3, 5, \ldots$$

and U_x series converges nonuniformly near $(0,0)$ and $(\pi, 0)$.

E *1.* $0 = \nabla \cdot f = -K\nabla \cdot (u_x, 4u_y) = -K(u_{xx} + 4u_{yy})$.

2. $\bar{u}(x,\bar{y}) \overset{\text{def}}{=} u(x, 2\bar{y})$ satisfies Laplace's equation in the rectangle $0 < x < \pi$, $0 < \bar{y} < \ell/2$.

F $\phi(r) = (A - B)\frac{\log r}{\log 4} + A$

G *1.* $u(r,\theta) = c_0 + \frac{8}{\pi}\sum_{n=1,3,}^{\infty}\frac{r^n}{n^4}\sin n\theta, \qquad 0 \le r \le 1$ *2.* No.

J $u(r,\theta) = \frac{\log r}{\log R} + \frac{8}{\pi}\sum_{n=1,3,}^{\infty}\left[\frac{r^n}{1-R^{2n}} + \frac{r^{-n}}{1-R^{-2n}}\right]\frac{\sin n\theta}{n^3} \qquad 1 \le r \le R$

L *2.* $u(r,\theta) = \frac{R}{8\pi}\sum_{n=1,3,}^{\infty}\frac{1}{n^4}\frac{r^{4n}+r^{-4n}}{R^{4n}-R^{-4n}}\sin 4n\theta, \qquad 1 \le r \le R, \quad 0 \le \theta \le \pi/4$

(since $\tilde{u}_x = ru_r$).

4.4B *5.* $\phi(x,y) = x^3 - 3xy^2$; $\psi(x,y) = 3x^2y - y^3$

D *1.* $\cosh z = \frac{e^z + e^{-z}}{2}$ is analytic everywhere since $e^z \neq 0$.

2. A ratio of polynomials in z; it is analytic where the denominator is nonvanishing.

E *3.* $f(z) = -iz^4 + z + 1$ *4.* $i\cos z$

G ψ is not single-valued.

I *1.* Stagnation points occur at $z = \pm 1$.

L *1.* $\frac{1}{\pi}\mathcal{R}_e \sin^{-1}(t^2) + \frac{1}{2}$ *2.* $1 - \frac{1}{\pi}\mathcal{I}_m(\log z) = 1 - \frac{\operatorname{Arg} z}{\pi}$

3. $\frac{1}{\pi}\mathcal{R}_e \sin^{-1}\left\{i\frac{1+\zeta}{1-\zeta}\right\} + \frac{1}{2}$ *4.* $1 + \mathcal{I}_m(\sqrt{z}) = 1 + \sqrt{r}\sin\frac{\theta}{2}$

M *3.* $\frac{1}{\pi}\mathcal{I}_m \log\left(\frac{z-1}{z+1}\right)\left(\text{since } \frac{i-1}{i+1} = i\right)$

4.5A *1.* For given f, the formal solution is

$$V(x,y) = 1 + \frac{2x}{\pi} + \frac{2}{\pi}\sum_{n=1}^{\infty}\left[\frac{2}{n^2}\sin\frac{n\pi}{2} + \frac{3(-1)^n - 1}{n}\right]\frac{\sinh ny}{\sinh n\ell}\sin nx$$

B *1.* $V(x,y) = \left(\frac{\ell\sinh y}{\sinh \ell} - y\right)\sin x (= v(x,y))$

3. $V(x,y) = \frac{\pi^2}{6}y(y-\ell) + 8\sum_{n=1}^{\infty}\frac{(-1)^n}{n^4}\left[\sinh^2\frac{ny}{2} - \left(\sinh^2\frac{n\ell}{2}\right)\frac{\sinh ny}{\sinh n\ell}\right]\cos nx$

5. $V(x,y) = \left\{\frac{32}{81}\left[\cosh\frac{3y}{2} - 1 + \left(\frac{3\ell}{2} - \sinh\frac{3\ell}{2}\right)\frac{\sinh\frac{3y}{2}}{\cosh\frac{3\ell}{2}}\right] - \frac{4}{9}y^2\right\}\sin\frac{3x}{2}.$

C *1.* $v(r,\theta) = \frac{r}{3}(r-1)\sin\theta$

2. $V(r,\theta) = \frac{\pi^2}{24}(1-r^4) + \frac{r^4}{32}(\log r)\cos 4\theta + 4\sum_{\substack{n=1\\ n\neq 4}}^{\infty}\frac{(-1)^n}{n^2}\frac{(r^n - r^4)}{n^2 - 4^2}\cos n\theta$

3. $v(r,\theta) = \frac{1}{15}r(r^3 - 4)\cos\theta$

E *1.* Take $u(0,y) = 0$; $u(x,\pi) = -x$. Then,

$$U(x,y) = 2\sum_{n=1}^{\infty}\frac{(-1)^n}{n\sinh n\pi}[\sinh ny\sin nx - \sinh nx\sin ny].$$

2. Take $u(0,y) = 0$, $u(x,\pi) = x$, and replace $\sin nx$ by $-\sin nx$ in solution to 1.

Chapter 5

5.3A *1.* $U(x,t) = -2\sum_{n=1}^{\infty}\frac{(-1)^n}{n}\sin nx\cos nct, \qquad 0 < x < \pi$

4. $U(x,t) = \frac{\ell^2}{3} + \frac{4\ell^2}{\pi^2}\sum_{n=1}^{\infty}\frac{(-1)^n}{n^2}\cos\frac{n\pi x}{\ell}\cos\frac{n\pi ct}{\ell}, \qquad 0 \le x \le \ell.$

5. $U(x,t) = \frac{8}{\pi}\sum_{n=1,3,}^{\infty}\frac{\sin\frac{n\pi}{2}}{n^2}\sin\frac{nx}{2}\cos\frac{nct}{2}, \qquad 0 \le x \le \pi.$

B Multiply $c_0 = \frac{\pi^3}{6}$ by t, replace $\cos nct$ by $\sin nct$, and divide $c_n = -\frac{8}{\pi n^4}$ by nc.

C *2.* $f(x) = x^3(\pi - x)^3$ since f and f'' vanish at both 0 and π.

E Frequency $\omega_n = n\omega_1$ does not contribute when np/ℓ is an integer.

5.4A *3.* Once with respect to x or t.

C *2.* $v(x,t) = -A\left[\frac{\cos\omega t - \cos\omega_1 t}{\omega^2 - \omega_1^2}\right]\sin\frac{\pi x}{\ell}$ *3.* Yes, but for $n = 1$ only.

D *2.* $V(x,t) = \frac{8\ell^4}{\pi^5}A\sum_{n=1,3,}^{\infty}\left(\frac{\pi^2}{n^3} - \frac{12}{n^5}\right)\left[\frac{\cos\omega t - \cos\omega_n t}{\omega^2 - \omega_n^2}\right]\sin\frac{n\pi x}{\ell}$

F $U(x,t) = \frac{8}{\pi}\sum_{n=1,3,}^{\infty}\frac{1}{n^3}T_n(t)\sin nx$, where T_n are those in (28).

G *1.* Replace $\sin nx$ by $\cos nx$ and add the term $\frac{c_0}{4\delta}(1 - e^{-2\delta t})$ to the series in (29).

2. Change initial conditions on T_n to $T_n(0) = 1$, $T_n'(0) = 0$.

H *1.* $v(x,t) = -AH'(c)\sin x$, where for $\alpha \in \mathbb{C}$, and fixed positive t and ω, let

$$H(\alpha) \stackrel{\text{def}}{=} \int_0^t e^{-\alpha s}\cos\omega(t-s)\,ds = \frac{\omega\sin\omega t + \alpha\cos\omega t - \alpha e^{-\alpha t}}{\omega^2+\alpha^2}, \quad \alpha^2 \neq -\omega^2.$$

2. $V(x,t) = -\frac{8}{\pi}A\sum_{n=1,3,}^{\infty}\left[\frac{\pi^2}{n^3} - \frac{12}{n^5}\right]T_n(t)\sin nx$, where, as in part 1,

$$T_n(t) = \frac{1}{2\beta_n}\begin{cases} H(\delta-\beta_1) - H(\delta+\beta_1) & \text{for } n = 1 \\ -iH(\delta - i\beta_n) + iH(\delta + i\beta_n), & \text{for } n = 3, 5, \ldots \end{cases}$$

and $\beta_n = |4 - n^2|^{1/2}c$

5.6B *1.* $\ell/2$ **2.** *(a)* 3

C δ is even with period 2π. (See Problem 1.4O).

D δ is odd with period 4π.

F If $V(x) = \int_0^x v(s)\,ds$, then $u(0,t) = 0$ gives $c\delta_E(x) + V_O(x) = 0$, so that both δ_E and V_O are zero. (See Section 1.4.)

G *1.* Both δ and v are even with period 2ℓ.

H *3.* $u(x,t) = \frac{1}{96c^2}(x^2 - c^2t^2)(7x^2 - c^2t^2 + 12t) + \phi(x+ct) + \psi(x-ct)$.

K *4.* $\phi(t) = h(t)/1 + e^{-2}$, so

$$u(x,t) = \frac{e^{-t}}{2\cosh 1}[e^{1-x}\sin\pi(x+t) - e^{x-1}\sin\pi(x-t)]$$

6. $\phi(x + 2\ell) = -\phi(x)$, so $u(x,t) = \phi(x+ct) - \phi(ct - x)$.

5.7F *1.* $u(x,t) = 3xe^{-t}$ *2.* $u(x,t) = \psi(xe^{-t})$

G *1.* $u(x,t) = 3xe^{-t}e^{-t^2/2}$ *2.* $u(x,t) = x^2e^{-2t}e^{-t^2/2}$

M $u(x,t) = 3xt$

Chapter 6

6.1B *1.* $y_\pm = (3 \pm \sqrt{6})x + c_\pm$

3. $y_+ = c_1e^{2x} - \frac{x}{2} - \frac{1}{4}$; $y_- = \frac{x^2}{2} + c_2 \quad (y \neq 0)$

5. $y_\pm = e^{\pm x} + c_\pm$, $(x \neq 0)$

6.2A *1.* elliptic *3.* hyperbolic

5. For $x > 0$ or $x < 0$: hyperbolic when $|y| < 1$; elliptic when $|y| > 1$. Only one characteristic direction when $|y| = 1$ or $x = 0$.

D *3.* $\tilde{u}_{\xi\xi} + \tilde{u}_{\eta\eta} = 1$

6.3F *2.* $\rho\theta_{tt} = (\lambda + 2\mu)\nabla^2\theta + \boldsymbol{\nabla}\cdot\mathbf{F}$

$\rho\xi_{tt} = \mu\nabla^2\xi + \boldsymbol{\nabla}\times\mathbf{F}$, where $\xi = \boldsymbol{\nabla}\times\mathbf{u}$.

6.4G To obtain (33), differentiate $c^2 = \gamma p/\rho$ with respect to ρ.

6.5C *2.* $u_{tt} = abu_{xx}$ has discriminant $\delta = ab$.

D *2.* Characteristics are straight lines with unit slope and equilateral hyperbolas asymptotic to t-axis.

3. $w = u$ since system is already in normal form.

I *4.* $uv \pm c\sqrt{u^2 + v^2 - c^2}$

Chapter 7

7.1A *1.* Regular: $\tau = x$, $q = 1$, $\rho = e^x$ and $\| y \|^2 = \int_1^2 e^x y^2(x)\,dx$.

2. Singular since $\tau(0) = 0$: $\| y \|^2 = \int_0^1 e^x y^2(x)\,dx$.

4. Regular: $\tau = \sin x$, $q = x^2$, $\rho = 1$ and $\| y \|^2 = \int_{\pi/4}^{3\pi/4} y^2(x)\,dx$.

B *1.* The eigenfunctions y satisfy $y'' + (\lambda - q)y = 0$ with $y'(0) = y'(\ell) = 0$. Consequently, $y(x) = A\cos\sqrt{\lambda - q}\,x$ for appropriate λ.

2. π^2/ℓ^2 *3.* $y_n(x) = \sqrt{\dfrac{2}{\ell}}\cos\dfrac{n\pi}{\ell}x$. *4.* No; Yes

C Take $P(x) = e^{\int p\,dx}$ so that $P' = pP$.

D $x^2y'' + 2xy' + \lambda y = 0$ $(0 < x \le 1)$ becomes $\ddot{v} + \dot{v} + \lambda v = 0$ $(-\infty < t < 0)$ when $v(t) = y(e^t)$ so that if $v(0) = 0$, then $v(t) = Ae^{-t/2}\sin\sqrt{\lambda - \frac{1}{4}}\,t$ for $\lambda > \frac{1}{4}$, and as $t \to -\infty$: $|v(t)| \to +\infty$.

E *2.* If $A' = 0$, make $v(a) = 1$; if $A = 0$, make $v'(a) = 1$; and if $AA' \ne 0$ make $v(a) = 1$, $v'(a) = A/A'$. Then $B_a(v) \ne 0$.

H *2.* $h/g \le 0$ wherever $g \ne 0$.

7.2A *1.* Separation in the form $u_n(x,t) = e^{-k\omega_n^2 t}\sin\omega_n x$ is feasible when $\tan\omega_n = \omega_n/B$. $U(x,t) = \sum_{n=1}^{\infty} C_n u_n(x,t)$ is a formal solution when

$$C_n = \beta_n^2 \int_0^1 x\sin\omega_n x\,dx = \frac{(2\cos\omega_n)(1 - B)}{\omega_n(B - \cos^2\omega_n)}, \qquad n = 1, 2, \ldots$$

and the series does converge absolutely and uniformly for $t \ge 0$.

2. If $B > 0$, heat is flowing into the bar at $x = 1$ at a rate proportional to the existing temperature, and this requires an artificial feedback system.

B *1.* The base shear-force is proportional to the base deflection. Hence, $\mu u_x(0,t) = ku(0,t)$, and as $k \to \infty$, $u(0,t) \to 0$, which is reasonable.

2. Separation in the form $u_n(x,t) = \cos\omega_n(\ell - x)\sin\omega_n t$ is feasible when $\cot\omega_n\ell = \omega_n\ell/2$, and this equation has solutions $\omega_n > 0$, $n = 1, 2, \ldots$. Hence, $U(x,t) = \sum_{n=1}^{\infty} b_n u_n(x,t)$ will give a formal solution when

$$b_n = 2\frac{\sin\omega_n\ell}{\omega_n^2\ell}\Big/\left(1 + \frac{\sin^2\omega_n\ell}{2}\right), \qquad n = 1, 2, \ldots .$$

7. When $A = -1$, then negative eigenvalues $\lambda = -\omega^2$ must satisfy the equation $(1 - \omega^2)\sinh\omega = 2\omega$, and this has no solution except $\omega = 0$; (For $0 < \omega < 1$, compare the power series of algebraic and hyperbolic functions.)

7.3A *1.* $y\sin x = \dfrac{\pi}{2}\sin x - \dfrac{4}{\pi}\sum_{n=1,3,}^{\infty}\dfrac{1}{n^2}\sin x\cos ny$

2. $xy = \pi\sum_{m=1}^{\infty}\dfrac{(-1)^{m+1}}{m}\sin mx + \dfrac{8}{\pi}\sum_{\substack{m=1\\n=1,3,}}^{\infty}\dfrac{(-1)^m}{mn^2}\sin mx\cos ny$

3. Let $C(m) = \dfrac{4}{\pi^2}[1 - (-1)^m]$. Then $c_{m0} = \dfrac{C(m)}{2m^3}$, $c_{mm} = \dfrac{C(m)}{m^2}$, $m = 1, 2, \ldots$

and for $m \ne n$, $c_{mn} = \dfrac{1}{mn^2}\left[\dfrac{m^2\,C(m+n)}{m^2 - n^2} - C(m)\right]$, $m, n = 1, 2, \ldots$

B **1.** $\sin\frac{m\pi x}{2}\sin\frac{n\pi y}{2}$, $\lambda_{mn} = (m^2+n^2)\frac{\pi^2}{4}$, $m = 1, 3, 5, \ldots$ $n = 1, 2, 3, \ldots$

2. $\sin m\pi x\cos\frac{n\pi y}{2}$, $\lambda_{mn} = \left(m^2+\frac{n^2}{4}\right)\pi^2$, $m = 1, 2, 3, \ldots$ $n = 0, 1, 2, \ldots$

3. $\cos\frac{m\pi x}{2}\sin\frac{n\pi y}{4}$, $\lambda_{mn} = \left(m^2+\frac{n^2}{4}\right)\frac{\pi^2}{4}$, $m = 1, 3, 5, \ldots$ $n = 1, 3, 5, \ldots$

4. $\sin m\pi x\cos\mu_n y$, $\lambda_{mn} = (m^2\pi^2+\mu_n^2)$, $m = 1, 2, 3, \ldots$ $n = 1, 2, 3, \ldots$

where $\tan(2\mu_n) = \mu_n^{-1} > 0$.

C **1.** $\cos\frac{m\pi x}{a}\cos\frac{n\pi y}{b}$, $\nu_{mn} = \left(\frac{m^2}{a^2}+\frac{n^2}{b^2}\right)\pi^2$, $m, n = 0, 1, 2, \ldots$

2. $\cos\frac{m\pi x}{a}\sin\frac{n\pi y}{2b}$, $\mu_{mn} = \left(\frac{m^2}{a^2}+\frac{n^2}{4b^2}\right)\pi^2$, $m = 0, 1, 2, \ldots$ $n = 1, 3, 5, \ldots$

If $\lambda_{mn} = \left(\frac{m^2}{a^2}+\frac{n^2}{b^2}\right)\pi^2$, $m, n = 1, 2, 3, \ldots$

then for $a < b$: $0 = \nu_{00} < \mu_{01} < \nu_{01} < \nu_{10} < \mu_{11} < \nu_{11} = \lambda_{11}$, etc.

$\lambda_{mn} = \lambda_{MN}$ iff $M^2 - m^2 = \frac{a^2}{b^2}(n^2 - N^2)$, which is not possible for distinct pairs (m, n) and (M, N) when a^2/b^2 is irrational.

7.3D **2.** $c_{mn} = \frac{32}{\pi^6}\frac{(1+2(-1)^m)}{m^3}\frac{(-1)^n-1}{n^3}$, $m, n, = 1, 2, \ldots$; series converges absolutely and uniformly and sum is equal to f since this is true for series in each variable.

3. $c_{mn} = \frac{8}{\pi^2}\frac{(-1)^{m+n}}{mn}$. Series does not converge uniformly.

E $\lambda_{mn} = m^2+\frac{n^2}{4}$, $c_{mn} = \frac{2}{\pi^2}\int_0^{\pi}dx\int_0^{2\pi}dy\,f(x, y)\cos mx\sin\frac{ny}{2}$, $m = 0, 1, 2, \ldots$, $n = 1, 2, \ldots$
but each c_{0n} must be halved to recover U.

G $U(x, y, t) = \sum_{m,n=1}^{\infty} c_{mn}\sin mx\sin ny\, e^{-(m^2+n^2)t}$

H **4.** $W(x, y, t) = 4\sum_{m,n=1}^{\infty}\frac{(-1)^{m+n}}{mn\beta_{mn}}\sin mx\sin ny\, e^{-t}\sin\beta_{mn}t$,

where $\beta_{mn} = \sqrt{m^2+n^2-1}$, $m, n = 1, 2, \ldots$.

7.4C **2.** $U(r, t) = 20\sum_{n=1}^{\infty}C_nJ_0\left(\frac{\omega_n}{2}r\right)\cos\left(\frac{\omega_n}{2}t\right)$ where

$C_n = \int_0^{\omega_n}xJ_0(x)\,dx\Big/\int_0^{\omega_n}xJ_0^2(x)\,dx = \frac{2\omega_nJ_1(\omega_n)}{[J_0'(\omega_n)]^2} = \frac{2\omega_n}{J_1(\omega_n)}$ if we use results of Problems 7.4E, F.

E **3.** $j_n = [J_0'(\omega_n)]^2/2$

J **2.** Need that $J_0(\omega)Y_0(2\omega) = J_0(2\omega)Y_0(\omega)$.

7.5A **2.** $\Psi(r, s) = \sum_{n=0}^{\infty}C_n\left(\frac{r}{2}\right)^nP_n(s)$, where $C_n = \frac{2n+1}{2}\int_{-1}^{1}e^sP_n(s)\,ds$, $n = 0, 1, 2, \ldots$ can be calculated exactly (and recursively: see Problem 7.5D).

F *1., 2.* $2C_n = \int_{-1}^{1} f'(P_{n-1} - P_{n+1})\,ds = \int_{-1}^{1} f''\left[\frac{P_{n+2}-P_n}{2n+3} - \frac{P_n - P_{n-2}}{2n-1}\right] ds$

Uniform convergence of series implies that its sum F is continuous and that $F - f$ is orthogonal to every P_n. But recall Theorem 1.17.

7.6C *2.* $u(r,\phi,\theta) = \frac{5}{3}\left(\frac{r}{2}\right)^2 \sin^2\phi P_2''(\cos\phi)\cos 2\theta$

E *1.* $u(r,\phi,\theta) = \frac{1}{3}\left(\frac{r}{2}\right)^{-3} \sin^2\phi P_2''(\cos\phi)\cos 2\theta$

F *2.* $U(r,\phi,\theta) = \sum_{0\le m\le n}^{\infty} \left(\frac{r}{3}\right)^{-n-1} P_n^m(\cos\phi)(c_{mn}\cos m\theta + s_{mn}\sin m\theta)$

I *1.* $U(r,\phi,\theta) = \sum_{0\le m\le n}^{\infty} \left(\frac{r}{2}\right)^n P_n^m(\cos\phi) s_{mn}\sin m\theta$, where $s_{mn} = 0$ for m odd; i.e., extend f as odd function in θ with $f(\pi-\theta) = -f(\theta)$ as in Problem 1.4L.

7.7A $U(x,t) = 2\sum_{n=1}^{\infty} \frac{(-1)^{n+1}}{n}(t + e^{-n^2 t})\sin nx - 2\sum_{n=1}^{\infty} \frac{(-1)^n}{n^3}(e^{-n^2 t} - 1)\sin nx$, and the series multiplying t is just that for x on $[0,\pi)$. The remaining series and their derivatives converge uniformly for $t \ge t_0 > 0$.

B $U(x,t) = \frac{2e^{-t}}{\pi}\sum_{n=1}^{\infty}\frac{n}{n^2+1}\sin nx + \frac{2}{\pi}\sum_{n=1}^{\infty}\frac{(\sin nt - n\cos nt)}{n^2+1}\sin nx$

$$+\sum_{n=1}^{\infty}\frac{2(-1)^{n+1}}{n}\cos nt\sin nx$$

C *2.* $U(x,t) = e^{-t}\sin x + \frac{1}{\pi}\left(1 - e^{-t} - \frac{t^2}{2}\right) - \frac{2}{\pi}(t - 1 + (t+1)e^{-t})\cos x$

$$+\frac{2}{\pi}\sum_{n=2}^{\infty}\left[\frac{(-1)^n}{n^2-1}(e^{-t} - e^{-n^2 t}) + \frac{1 - e^{-n^2 t}}{n^4} - \frac{t}{n^2}\right]\cos nx$$

D *2.* $U(r,\theta) = \frac{2}{\pi}\sum_{n=1,3,}^{\infty}\left[\frac{2n(r^2 - r^{n/2})}{n^2-16} + \frac{4\sin\left(\frac{n\pi}{2}\right)}{n^2-4}(r - r^{n/2}) + \frac{2r^{n/2}}{n}\right]\sin\frac{n\theta}{2}$

Chapter 8

8.1A $s(\omega) = \frac{\omega}{\pi(1+\omega^2)} = \omega\cdot c(\omega)$. (Integrate by parts.)

B *1.* $s(\omega) = 0;\ c(\omega) = \frac{2}{\pi}\frac{(1-\cos\omega)}{\omega^2}, \qquad \omega > 0;\ c(0) = \frac{1}{\pi}$

$$F(x) = \frac{2}{\pi}\int_0^{\infty}\left[\frac{1-\cos\omega}{\omega^2}\right]\cos\omega x\,d\omega = \begin{cases} 1 - |x|, & |x| \le 1 \\ 0, & |x| \ge 1 \end{cases}$$

2. $s(\omega) = \frac{2}{\pi\omega^2}(\sin 2\omega - 2\omega\cos 2\omega), \qquad \omega > 0,\ s(0) = 0;\ c(\omega) = 0$

$$F(x) = \frac{2}{\pi}\int_0^{\infty}\frac{(\sin 2\omega - 2\omega\cos 2\omega)}{\omega^2}\sin\omega x\,d\omega = \begin{cases} x, & |x| < 2 \\ \pm 1, & x = \pm 2 \\ 0, & |x| > 2 \end{cases}$$

3. $F(x) = \dfrac{1}{\pi}\displaystyle\int_0^\infty \left\{ \left[\frac{\sin \omega b - \sin \omega a}{\omega}\right] \cos \omega x + \left[\frac{\cos \omega a - \cos \omega b}{\omega}\right] \sin \omega x \right\} d\omega$

$$= \begin{cases} 1, & a < x < b \\ \frac{1}{2}, & x = a \text{ or } x = b \\ 0, & \text{otherwise} \end{cases}$$

4. $F(x) = \dfrac{1}{\pi}\displaystyle\int_0^\infty \frac{\omega}{\omega^2 - 1}[(1 + \cos \omega\pi) \sin \omega x - \sin \omega\pi \cos \omega x]\, d\omega$

$$= \begin{cases} \cos x, & 0 < x < \pi \\ \frac{1}{2}, & x = 0 \\ -\frac{1}{2}, & x = \pi \\ 0, & \text{otherwise} \end{cases}$$

5. $s(\omega) = \dfrac{2}{\pi}\dfrac{\omega}{1 + \omega^2}$, so both integrals $= \dfrac{\pi e^{-x}}{2}$ for $x > 0$.

6. $F(x) = \dfrac{1}{\pi}\displaystyle\int_0^\infty \left[\frac{1}{1+(\omega+1)^2} + \frac{1}{1+(\omega-1)^2}\right] \cos \omega x\, d\omega = e^{-|x|} \cos x,\quad x > 0.$

7. $F(x) = \dfrac{1}{\pi}\displaystyle\int_0^\infty \left[\frac{\omega + 1}{1 + (\omega + 1)^2} - \frac{\omega - 1}{1 + (\omega - 1)^2}\right] \sin \omega x\, d\omega$

$$= \begin{cases} e^{-x} \cos x, & x > 0 \\ -e^{x} \cos x, & x < 0 \\ 0, & x = 0 \end{cases}$$

8.2A **1.** $u_1(x, y) = \dfrac{1}{\pi}\left[\tan^{-1}\left(\dfrac{b - x}{y}\right) + \tan^{-1}\left(\dfrac{x - a}{y}\right)\right] \to \dfrac{1}{2}$ as $y \searrow 0$ at $x = a$ or $x = b$.

2. $u(x, y) = u_1(x, y) + cy$ for any real $c \neq 0$.

3. $\displaystyle\lim_{b\to\infty} u_1(x, y) = \frac{1}{2} + \frac{1}{\pi}\tan^{-1}\left(\frac{x - a}{y}\right) \to 1$ as $a \to -\infty$.

B $|U(x, y)| \leq \dfrac{My}{\pi}\displaystyle\int_{-\infty}^\infty \frac{d\xi}{(\xi - x)^2 + y^2} = \frac{M}{\pi}\tan^{-1}\left(\frac{\xi - x}{y}\right)\Bigg|_{-\infty}^\infty = M$

C $U(x, y) = \dfrac{1}{2\pi}\displaystyle\int_0^\infty \log\left[\frac{y^2 + (\xi + x)^2}{y^2 + (\xi - x)^2}\right] f(\xi)\, d\xi$

D **2.** $U(x, y) = \dfrac{2}{\pi}\displaystyle\int_0^\infty \frac{1}{1 + \omega^2}\frac{\cosh \omega y}{\cosh \omega}\cos \omega x\, d\omega, \qquad (0 < y < 1)$

E $U(x, y) = \dfrac{2}{\pi}\displaystyle\int_0^\infty \frac{1 - \cos \omega}{\omega}\frac{\sinh \omega y}{\sinh \omega}\sin \omega x\, d\omega, \qquad (0 < y < 1)$

F **2.** $U(x, y) = \dfrac{2}{\pi}\displaystyle\int_0^\infty \frac{\cosh \omega y \cos \omega x}{\omega \sinh \omega + \cosh \omega}\, d\omega \int_0^\infty f(\xi) \cos \omega\xi\, d\xi$

8.3A $\hat{f}(\omega) = -(i/\omega)(e^{i\omega b} - e^{i\omega a}), \omega \neq 0; \hat{f}(0) = b - a$. Inversion recovers f except at $x = a$ and b.

8.4A $s(\lambda) = \hat{h}(\lambda)/\omega$, so

$$U(x, t) = \frac{2}{\pi}\int_0^\infty \sin \omega x \hat{f}(\omega^2) \cos \omega t\, d\omega + \frac{2}{\pi}\int_0^\infty \sin \omega x\, \hat{h}(\omega^2)\omega^{-1} \sin \omega t\, d\omega.$$

B Replace $\sin \omega x$ by $\cos \omega x$.

E *2.* $\int_0^\infty J_{m+1}(a\omega)J_m(r\omega)\,d\omega = \left\{\begin{array}{ll} r^m/a^{m+1}, & 0 \le r < a \\ 1/2a, & r = a \\ 0, & r > a \end{array}\right\}$

3. $U(r,t) = a^{m+1}\int_0^\infty J_0(r\omega)J_{m+1}(a\omega)\cos\omega t\,d\omega$

8.6F†

9.1A *3.*, *4.*, *9.*, *10.*, and *11.* are not regular. *1.*, *5.*, *6.*, and *13.* are not fully positive.

B Zero is an eigenvalue.

C *2.* $U(x,t) = \sum_{n=1}^\infty c_n e^{-\lambda_n t} y_n(x)$, where $c_n = \int_0^\ell e^{-3x} y_n f\,dx$ when $\int_0^\ell e^{-3x} y_n^2\,dx = 1$, $n = 1, 2, \ldots$

3. If $v(x) = -3$, problem is regular but not fully positive.

F *1.* Problem is fully positive but $y = 1$ is not an eigenfunction.

2. See (28) of section 5.4.

G *3.*, *4.*, *5.*: $\lambda = I(y,y) \Big/ \int_0^\ell \rho y^2\,dx \ge 0$, and λ cannot be zero for nontrivial y since then $y''(x) \equiv 0$.

9.3A *1.* $\lambda_1 \le 10$ *3.* $\lambda_1 \le \frac{1}{4}$

B *2.* $[(n-1)\pi]^2 2e^{-2} \le \lambda_n \le [(n-1)\pi]^2 2 + 1, \qquad n = 1, 2, \ldots$

I *4.* $8 \le \lambda_1 \le \pi^2$

9.4A *1.* $|\omega_n \ell - n\pi| \le M/n$, where $2\ell = \sqrt{2} + \log(1 + \sqrt{2})$

5. $|\omega_n \ell - n\pi| \le M/n$, where $\ell = \int_{\pi/4}^{3\pi/4} \sqrt{\csc x}\,dx$

7. $|\omega_n - \ell(n - \frac{1}{2})\pi| \le M/n$, where $\ell = 1/\log 2$

13. Either $|\omega_n - (n-1)\pi| \le M/n$ or $|\omega_n - n\pi| \le M/n$.

B *1.* $|y_n(x) - \sqrt{2/\ell}(1 + x^2)^{-1/4}\sin\left(\frac{n\pi}{\ell}t(x)\right)| \le M/n$, where

$$t(x) = \int_0^x \sqrt{1 + \xi^2}\,d\xi = \tfrac{1}{2}[x\sqrt{1 + x^2} + \log(x + \sqrt{1 + x^2})]$$

5. $|y_n(x) - \sqrt{2/\ell}(\sin x)^{-1/4}\sin\left(\frac{n\pi}{\ell}t(x)\right)| \le M/n$, where

$$t(x) = \int_{\pi/4}^x \sqrt{\csc \xi}\,d\xi$$

7. $|y_n(x) - \sqrt{2}\,(x \log 2)^{-1/2}\cos((n - {}^1/_2)\pi \log x/\log 2)| \leqq M/n$

13. $|y_n(x) - \sqrt{2}\cos\omega_n x| \le M/n$, where either $\omega_n \approx (n-1)\pi$ for all n or $\omega_n \approx n\pi$ for all n.

I $\cos\tilde{\omega}_{mn} = O\left(\frac{1}{\omega_{mn}}\right)$ where $\tilde{\omega}_{mn} = \omega_{mn} - \frac{\pi}{4} - \frac{m\pi}{2}$ so that

$$\sin\tilde{\omega}_{mn} = \sqrt{1 - \cos^2\tilde{\omega}_{mn}} = 1 + O\left(\frac{1}{\omega_{mn}^2}\right).$$

†**8.6F** *1.* $(\pi/x)^{1/2} e^{i\pi/4} + O(1/x)$

3. $(2\pi/x)^{1/2} e^{i\pi/4}(e^{-ix} - ie^{ix}) + O(1/x)$

Bibliography

[Ap] T. M. Apostol, *Mathematical Analysis*, Addison-Wesley, Reading, 1957.

[B–C] B. B. Baker and E. T. Copson, *The Mathematical Theory of Huygens' Principle*, 2nd ed., Oxford University Press, Fair Lawn, New Jersey, 1950.

[Ba] H. Bateman, *Partial Differential Equations*, Cambridge University Press, London, 1964.

[Bir] G. Birkhoff, *Hydrodynamics: A Study in Logic, Fact and Similitude*, Princeton University Press, 1960.

[B–R] G. Birkhoff and G. C. Rota, *Ordinary Differential Equations*, 3rd ed., Wiley, New York, 1978.

[B–F] R. Burden and J. Faires, *Numerical Analysis*, 4th ed., PWS Publishing Co., Boston, 1989.

[Ca] L. Carleson, *On convergence and growth of partial sums of Fourier series*, Acta Math, v 116, 1966.

[C–J] H. S. Carslaw and J. C. Jaeger, *Conduction of Heat in Solids*, 2nd ed., Oxford University Press, New York, 1959.

[C–L] E. Coddington and N. Levinson, *Theory of Ordinary Differential Equations*, McGraw-Hill, New York, 1955.

[Co] D. Colton, *Partial Differential Equations: An Introduction*, Random House, 1988.

[C–F] R. Courant and K. Friedrichs, *Supersonic Flow and Shock Waves*, Wiley-Interscience, New York, 1948.

[C–H.1] R. Courant and D. Hilbert, *Methods of Mathematical Physics*, Vol 1, Interscience, New York, 1953.

[Do] G. Doetsch, *Handbuch der Laplace Transformationen*, Birkhauser, Basel, 1950–56.

[D–S] N. Dunford and J. T. Schwartz, *Linear Operators, Part I: General Theory*, Interscience, New York, 1964.

[Du] P. Duren, *Univalent Functions*, Springer, New York, 1983.

[D–M] H. Dym and H. P. McKean, *Fourier Series and Integrals*, Academic Press, New York, 1972.

[Ed] C. H. Edwards, *Advanced Calculus of Several Variables*, Academic Press, New York, 1973.

[Edw] R. E. Edwards, *Fourier Series: A Modern Introduction*, Holt, New York, 1967.

[E–H] W. C. Elmore and M. A. Heald, *Physics of Waves*, McGraw-Hill, New York, 1969.

[Er] A. C. Eringen, *Nonlinear Theory of Continuous Media*, McGraw-Hill, New York, 1962.

[Fef] C. Fefferman, *The uncertainty principle*, Bull. AMS 9, 1983, 2, 129–206.

[Fl] J. A. Fleming, *Propagation of Electric Currents*, Van Nostrand, 1911.

[Fr] A. Friedman, *Partial Differential Equations of Parabolic Type*, Prentice-Hall, Englewood-Cliffs, NJ, 1964.

[Ga] P. Garabedian, *Partial Differential Equations*, 2nd ed., Chelsea, New York, 1986.

[G–T] D. Gilbarg and N. Trudinger, *Elliptic Partial Differential Equations of Second Order*, Springer-Verlag, Berlin, 1977.

[G–L] R. Guenther and J. Lee, *Partial Differential Equations of Mathematical Physics and Integral Equations*, Prentice-Hall, New Jersey, 1988.

[Gu] M. E. Gurtin, *Introduction to Continuum Mechanics*, Academic Press, New York, 1981.

[Hi] J. R. Higgins, *Completeness and Basic Properties of Sets of Special Functions*, Cambridge University Press, Cambridge, 1977.

[Ho] L. Hörmander, *Linear Partial Differential Operators*, Springer, New York, 1963.

[In] E. Ince, *Ordinary Differential Equations*, Dover Reprint, New York, 1956.

[Ja] D. Jackson, *Fourier Series and Orthogonal Polynomials*, Mathematical Association of America, 1943.

[J–E] E. Jahnke and F. Emde, *Tables of Functions*, Dover, New York, 1945.

[Jo] F. John, *Plane Waves and Spherical Means Applied to Partial Differential Equations*, Wiley-Interscience, New York, 1955.

[Ke] O. D. Kellogg, *Foundations of Potential Theory*, Dover, New York, 1953.

[K–P] R. J. Knops and L. Payne, *Uniqueness Theorems in Linear Elasticity*, Springer, New York, 1971.

[Kl] M. Kline, *Mathematical Thought from Ancient to Modern Times*, Oxford University Press, New York, 1972.

[Ko] H. Kober, *Dictionary of Conformal Representations*, Dover Publications, New York, 1957.

[Kor] T. W. Körner, *Fourier Analysis*, Cambridge University Press, Cambridge, 1988.

[La] P. Lax, *Hyperbolic Systems of Conservation Laws and the Mathematical Theory of Shock Waves*, Conf. Board Math. Sci. 11, SIAM, 1973.

[L–R] H. Liepmann and A. Roshko, *Elements of Gas Dynamics*, Wiley, New York, 1957.

[L–S] C. C. Lin and L. A. Segel, *Mathematics Applied to Deterministic Problems in the Applied Sciences*, Macmillan, New York, 1974.

[L–W] J. Lindmayer and C. Y. Wrigley, *Fundamentals of Semiconductor Devices*, Van Nostrand Co., Princeton, NJ, 1965.

[L–M] J. L. Lions and E. Magenes, *Non-homogeneous Boundary Value Problems and Applications*, 3 vols., Springer-Verlag, New York, 1972.

[Lo] L. Loeb, *Fundamentals of Electricity and Magnetism*, Dover, New York, 1947.

[Log] J. D. Logan, *Applied Mathematics: A Contemporary Approach*, Wiley-Interscience, New York, 1987.

[Mack] A. G. Mackie, *Boundary Value Problems*, Oliver & Boyd, London, 1965.

[M–W] J. Mathews and R. L. Walker, *Mathematical Methods of Physics*, W. A. Benjamin, New York, 1970.

[Mu] C. Müller, *Foundations of the Mathematical Theory of Electromagnetic Waves*, Springer-Verlag, New York, 1969.

[Ne] Z. Nehari, *Introduction to Complex Analysis*, Allyn and Bacon, New York, 1961.

[Ner] L. Nering, *Linear Algebra and Matrix Theory*, 2nd ed., Wiley, New York, 1970.

[Os] K. Oswatitsch, *Gas Dynamics*, Academic Press, New York, 1956.

[Pa] L. Payne, *Improperly Posed Problems in Partial Differential Equations*, SIAM, Philadelphia, 1975.

[Pe] T. Pedley, *The Fluid Mechanics of Large Blood Vessels*, Cambridge University Press, Cambridge, 1980.

[Pi] M. Pinsky, *Introduction to Partial Differential Equations with Applications*, 2nd ed., McGraw-Hill, New York, 1991.

[P–W] M. H. Protter and H. F. Weinberger, *Maximum Principles in Differential Equations*, Springer, New York, 1984.

[Rei] W. Reid, *Sturmian Theory for Ordinary Differential Equations*, Springer, New York, 1980.

[Ru] W. Rudin, *Principles of Mathematical Analysis*, 3rd ed., McGraw-Hill, New York, 1976.

[S] G. Sansone, *Orthogonal Functions*, R. E. Krieger, Melbourne, Fla., 1977.

[Sch] P. W. Schaefer, *Maximum Principles and Eigenvalue Problems*, Wiley, New York, 1988.

[Sn] I. N. Sneddon, *The Use of Integral Transforms*, McGraw-Hill, New York, 1972.

[Sog] C. Sogge, *Fourier Integrals in Classical Analysis*, Cambridge University Press, New York, 1993.

[So] I. S. Sokolnikoff, *Mathematical Theory of Elasticity*, 2nd ed., McGraw-Hill, New York, 1956.

[Sp] R. P. Sperb, *Maximum Principles and Their Applications*, Academic Press, New York, 1981.

[Sta] I. Stakgold, *Green's Functions and Boundary Value Problems*, Wiley, New York, 1980.

[St] J. J. Stoker, *Water Waves*, Wiley-Interscience, New York, 1957.

[S–F] G. Strang and G. Fix, *An Analysis of the Finite Element Method*, Prentice-Hall, Englewood Cliffs, NJ, 1973.

[Str] R. Street, *Analysis and Solution of Partial Differential Equations*, Wadsworth, Belmont, 1973.

[Ti] E. C. Titchmarsh, *Eigenfunction Expansions Associated with Second-Order Differential Equations*, Part I (1946) and Part II (1958), Oxford University Press, Oxford.

[Tre] F. Treves, *Basic Linear Partial Differential Equations*, Academic Press, New York, 1975.

[Tr] J. L. Troutman, *Variational Calculus and Optimal Control*, 2nd ed., Springer-Verlag, New York, 1996.

[Wa] P. R. Wallace, *Mathematical Analysis of Physics Problems*, Dover, New York, 1984.

[Wal] J. S. Walker, *Fourier Analysis*, Oxford University Press, New York, 1988.

[Wat] G. N. Watson, *A Treatise on the Theory of Bessel Functions*, 2nd ed., Cambridge University Press, Cambridge, 1966.

[Wein] H. F. Weinberger, *A First Course in Partial Differential Equations*, Blaisdell, New York, 1965.

[W–S] A. Weinstein and W. Stenger, *Methods for Intermediate Problems for Eigenvalues*, Academic Press, New York, 1972.

[Wh] G. B. Whitham, *Linear and Nonlinear Waves*, Wiley-Interscience, New York, 1974.

[Wi] D. Widder, *The Heat Equation*, Academic Press, New York, 1975.

[Z] A. Zygmund, *Trigonometric Series*, Cambridge University Press, Cambridge, 1968.

Index

A CATALOG OF SELECTED

DOVER BOOKS

IN SCIENCE AND MATHEMATICS

Mathematics-Bestsellers

HANDBOOK OF MATHEMATICAL FUNCTIONS: with Formulas, Graphs, and Mathematical Tables, Edited by Milton Abramowitz and Irene A. Stegun. A classic resource for working with special functions, standard trig, and exponential logarithmic definitions and extensions, it features 29 sets of tables, some to as high as 20 places. 1046pp. 8 x 10 1/2. 0-486-61272-4

ABSTRACT AND CONCRETE CATEGORIES: The Joy of Cats, Jiri Adamek, Horst Herrlich, and George E. Strecker. This up-to-date introductory treatment employs category theory to explore the theory of structures. Its unique approach stresses concrete categories and presents a systematic view of factorization structures. Numerous examples. 1990 edition, updated 2004. 528pp. 6 1/8 x 9 1/4. 0-486-46934-4

MATHEMATICS: Its Content, Methods and Meaning, A. D. Aleksandrov, A. N. Kolmogorov, and M. A. Lavrent'ev. Major survey offers comprehensive, coherent discussions of analytic geometry, algebra, differential equations, calculus of variations, functions of a complex variable, prime numbers, linear and non-Euclidean geometry, topology, functional analysis, more. 1963 edition. 1120pp. 5 3/8 x 8 1/2. 0-486-40916-3

INTRODUCTION TO VECTORS AND TENSORS: Second Edition--Two Volumes Bound as One, Ray M. Bowen and C.-C. Wang. Convenient single-volume compilation of two texts offers both introduction and in-depth survey. Geared toward engineering and science students rather than mathematicians, it focuses on physics and engineering applications. 1976 edition. 560pp. 6 1/2 x 9 1/4. 0-486-46914-X

AN INTRODUCTION TO ORTHOGONAL POLYNOMIALS, Theodore S. Chihara. Concise introduction covers general elementary theory, including the representation theorem and distribution functions, continued fractions and chain sequences, the recurrence formula, special functions, and some specific systems. 1978 edition. 272pp. 5 3/8 x 8 1/2. 0-486-47929-3

ADVANCED MATHEMATICS FOR ENGINEERS AND SCIENTISTS, Paul DuChateau. This primary text and supplemental reference focuses on linear algebra, calculus, and ordinary differential equations. Additional topics include partial differential equations and approximation methods. Includes solved problems. 1992 edition. 400pp. 7 1/2 x 9 1/4. 0-486-47930-7

PARTIAL DIFFERENTIAL EQUATIONS FOR SCIENTISTS AND ENGINEERS, Stanley J. Farlow. Practical text shows how to formulate and solve partial differential equations. Coverage of diffusion-type problems, hyperbolic-type problems, elliptic-type problems, numerical and approximate methods. Solution guide available upon request. 1982 edition. 414pp. 6 1/8 x 9 1/4. 0-486-67620-X

VARIATIONAL PRINCIPLES AND FREE-BOUNDARY PROBLEMS, Avner Friedman. Advanced graduate-level text examines variational methods in partial differential equations and illustrates their applications to free-boundary problems. Features detailed statements of standard theory of elliptic and parabolic operators. 1982 edition. 720pp. 6 1/8 x 9 1/4. 0-486-47853-X

LINEAR ANALYSIS AND REPRESENTATION THEORY, Steven A. Gaal. Unified treatment covers topics from the theory of operators and operator algebras on Hilbert spaces; integration and representation theory for topological groups; and the theory of Lie algebras, Lie groups, and transform groups. 1973 edition. 704pp. 6 1/8 x 9 1/4. 0-486-47851-3

Browse over 9,000 books at www.doverpublications.com

Mathematics-Probability and Statistics

BASIC PROBABILITY THEORY, Robert B. Ash. This text emphasizes the probabilistic way of thinking, rather than measure-theoretic concepts. Geared toward advanced undergraduates and graduate students, it features solutions to some of the problems. 1970 edition. 352pp. 5 3/8 x 8 1/2. 0-486-46628-0

PRINCIPLES OF STATISTICS, M. G. Bulmer. Concise description of classical statistics, from basic dice probabilities to modern regression analysis. Equal stress on theory and applications. Moderate difficulty; only basic calculus required. Includes problems with answers. 252pp. 5 5/8 x 8 1/4. 0-486-63760-3

OUTLINE OF BASIC STATISTICS: Dictionary and Formulas, John E. Freund and Frank J. Williams. Handy guide includes a 70-page outline of essential statistical formulas covering grouped and ungrouped data, finite populations, probability, and more, plus over 1,000 clear, concise definitions of statistical terms. 1966 edition. 208pp. 5 3/8 x 8 1/2. 0-486-47769-X

GOOD THINKING: The Foundations of Probability and Its Applications, Irving J. Good. This in-depth treatment of probability theory by a famous British statistician explores Keynesian principles and surveys such topics as Bayesian rationality, corroboration, hypothesis testing, and mathematical tools for induction and simplicity. 1983 edition. 352pp. 5 3/8 x 8 1/2. 0-486-47438-0

INTRODUCTION TO PROBABILITY THEORY WITH CONTEMPORARY APPLICATIONS, Lester L. Helms. Extensive discussions and clear examples, written in plain language, expose students to the rules and methods of probability. Exercises foster problem-solving skills, and all problems feature step-by-step solutions. 1997 edition. 368pp. 6 1/2 x 9 1/4. 0-486-47418-6

CHANCE, LUCK, AND STATISTICS, Horace C. Levinson. In simple, non-technical language, this volume explores the fundamentals governing chance and applies them to sports, government, and business. "Clear and lively ... remarkably accurate." – *Scientific Monthly*. 384pp. 5 3/8 x 8 1/2. 0-486-41997-5

FIFTY CHALLENGING PROBLEMS IN PROBABILITY WITH SOLUTIONS, Frederick Mosteller. Remarkable puzzlers, graded in difficulty, illustrate elementary and advanced aspects of probability. These problems were selected for originality, general interest, or because they demonstrate valuable techniques. Also includes detailed solutions. 88pp. 5 3/8 x 8 1/2. 0-486-65355-2

EXPERIMENTAL STATISTICS, Mary Gibbons Natrella. A handbook for those seeking engineering information and quantitative data for designing, developing, constructing, and testing equipment. Covers the planning of experiments, the analyzing of extreme-value data; and more. 1966 edition. Index. Includes 52 figures and 76 tables. 560pp. 8 3/8 x 11. 0-486-43937-2

STOCHASTIC MODELING: Analysis and Simulation, Barry L. Nelson. Coherent introduction to techniques also offers a guide to the mathematical, numerical, and simulation tools of systems analysis. Includes formulation of models, analysis, and interpretation of results. 1995 edition. 336pp. 6 1/8 x 9 1/4. 0-486-47770-3

INTRODUCTION TO BIOSTATISTICS: Second Edition, Robert R. Sokal and F. James Rohlf. Suitable for undergraduates with a minimal background in mathematics, this introduction ranges from descriptive statistics to fundamental distributions and the testing of hypotheses. Includes numerous worked-out problems and examples. 1987 edition. 384pp. 6 1/8 x 9 1/4. 0-486-46961-1

Mathematics-History

THE WORKS OF ARCHIMEDES, Archimedes. Translated by Sir Thomas Heath. Complete works of ancient geometer feature such topics as the famous problems of the ratio of the areas of a cylinder and an inscribed sphere; the properties of conoids, spheroids, and spirals; more. 326pp. 5 3/8 x 8 1/2. 0-486-42084-1

THE HISTORICAL ROOTS OF ELEMENTARY MATHEMATICS, Lucas N. H. Bunt, Phillip S. Jones, and Jack D. Bedient. Exciting, hands-on approach to understanding fundamental underpinnings of modern arithmetic, algebra, geometry and number systems examines their origins in early Egyptian, Babylonian, and Greek sources. 336pp. 5 3/8 x 8 1/2. 0-486-25563-8

THE THIRTEEN BOOKS OF EUCLID'S ELEMENTS, Euclid. Contains complete English text of all 13 books of the Elements plus critical apparatus analyzing each definition, postulate, and proposition in great detail. Covers textual and linguistic matters; mathematical analyses of Euclid's ideas; classical, medieval, Renaissance and modern commentators; refutations, supports, extrapolations, reinterpretations and historical notes. 995 figures. Total of 1,425pp. All books 5 3/8 x 8 1/2.
Vol. I: 443pp. 0-486-60088-2
Vol. II: 464pp. 0-486-60089-0
Vol. III: 546pp. 0-486-60090-4

A HISTORY OF GREEK MATHEMATICS, Sir Thomas Heath. This authoritative two-volume set that covers the essentials of mathematics and features every landmark innovation and every important figure, including Euclid, Apollonius, and others. 5 3/8 x 8 1/2. Vol. I: 461pp. 0-486-24073-8
Vol. II: 597pp. 0-486-24074-6

A MANUAL OF GREEK MATHEMATICS, Sir Thomas L. Heath. This concise but thorough history encompasses the enduring contributions of the ancient Greek mathematicians whose works form the basis of most modern mathematics. Discusses Pythagorean arithmetic, Plato, Euclid, more. 1931 edition. 576pp. 5 3/8 x 8 1/2.
0-486-43231-9

CHINESE MATHEMATICS IN THE THIRTEENTH CENTURY, Ulrich Libbrecht. An exploration of the 13th-century mathematician Ch'in, this fascinating book combines what is known of the mathematician's life with a history of his only extant work, the Shu-shu chiu-chang. 1973 edition. 592pp. 5 3/8 x 8 1/2.
0-486-44619-0

PHILOSOPHY OF MATHEMATICS AND DEDUCTIVE STRUCTURE IN EUCLID'S ELEMENTS, Ian Mueller. This text provides an understanding of the classical Greek conception of mathematics as expressed in Euclid's Elements. It focuses on philosophical, foundational, and logical questions and features helpful appendixes. 400pp. 6 1/2 x 9 1/4. 0-486-45300-6

BEYOND GEOMETRY: Classic Papers from Riemann to Einstein, Edited with an Introduction and Notes by Peter Pesic. This is the only English-language collection of these 8 accessible essays. They trace seminal ideas about the foundations of geometry that led to Einstein's general theory of relativity. 224pp. 6 1/8 x 9 1/4. 0-486-45350-2

HISTORY OF MATHEMATICS, David E. Smith. Two-volume history – from Egyptian papyri and medieval maps to modern graphs and diagrams. Non-technical chronological survey with thousands of biographical notes, critical evaluations, and contemporary opinions on over 1,100 mathematicians. 5 3/8 x 8 1/2.
Vol. I: 618pp. 0-486-20429-4
Vol. II: 736pp. 0-486-20430-8